Remote Sensing, Artificial Intelligence and Deep Learning in Hydraulic Structure Safety Monitoring

Remote Sensing, Artificial Intelligence and Deep Learning in Hydraulic Structure Safety Monitoring

Guest Editors

Chenfei Shao
Hao Gu
Yanxin Xu
Huixiang Chen
Xiangnan Qin
Guang Yang

MDPI

Basel • Beijing • Wuhan • Barcelona • Belgrade • Novi Sad • Cluj • Manchester

Guest Editors

Chenfei Shao
College of Water Conservancy and Hydropower Engineering
Hohai University
Nanjing
China

Hao Gu
College of Water Conservancy and Hydropower Engineering
Hohai University
Nanjing
China

Yanxin Xu
College of Water Conservancy and Hydropower Engineering
Hohai University
Nanjing
China

Huixiang Chen
College of Agricultural Science and Engineering
Hohai University
Nanjing
China

Xiangnan Qin
Division of Water Conservation and Hydropower Engineering
Zhengzhou University
Zhengzhou
China

Guang Yang
School of Water Conservancy
North China University of Water Resources and Electric Power
Zhengzhou
China

Editorial Office
MDPI AG
Grosspeteranlage 5
4052 Basel, Switzerland

This is a reprint of the Special Issue, published open access by the journal *Water* (ISSN 2073-4441), freely accessible at: https://www.mdpi.com/journal/water/special_issues/AH23FHBT1Q.

For citation purposes, cite each article independently as indicated on the article page online and as indicated below:

Lastname, A.A.; Lastname, B.B. Article Title. *Journal Name* **Year**, *Volume Number*, Page Range.

ISBN 978-3-7258-5445-5 (Hbk)
ISBN 978-3-7258-5446-2 (PDF)
https://doi.org/10.3390/books978-3-7258-5446-2

Cover image courtesy of Chenfei Shao

Contents

water

'ditorial

Research Progress in Remote Sensing, Artificial Intelligence and Deep Learning in Hydraulic Structure Safety Monitoring

'henfei Shao [1], Hao Gu [1], Yanxin Xu [1,*], Huixiang Chen [2], Xiangnan Qin [3] and Guang Yang [4]

[1] College of Water Conservancy& Hydropower Engineering, Hohai University, Nanjing 210098, China; shao_chen_fei@hhu.edu.cn (C.S.); ghao@hhu.edu.cn (H.G.)
[2] College of Agricultural Science and Engineering, Hohai University, Nanjing 211100, China; chenhuixiang@hhu.edu.cn
[3] Division of Water Conservation and Hydropower Engineering, Zhengzhou University, Zhengzhou 450001, China; qin_xn2105@zzu.edu.cn
[4] School of Water Conservancy, North China University of Water Resources and Electric Power, Zhengzhou 450045, China; yangguanghhu@outlook.com
* Correspondence: xu_yanxin@hhu.edu.cn

1. Introduction to the Special Issue

Ensuring the safety of hydraulic structure engineering is of paramount importance, as these infrastructures play a critical role in water management, flood control, and the provision of clean water for various human and ecological needs [1,2]. It is becoming increasingly difficult to ignore the impact of dam structural safety and operational reliability on public safety, environmental sustainability, and economic stability [3,4]. Failure to adequately safeguard these structures can lead to catastrophic consequences, including loss of life, environmental degradation, and significant financial losses [5]. Therefore, the continuous safety assessment, maintenance, and improvement of hydraulic structure engineering are essential to mitigate risks, adapt to changing climatic conditions, and sustain the socio-economic benefits they provide [6]. Robust safety monitoring methods and innovative health diagnostic technologies are crucial to ensuring that hydraulic structure engineering remains resilient and effective in the face of evolving challenges [7].

Consequently, it is critical to propose safety monitoring methods for hydraulic engineering, as well as guarantee their secure, stable, and effective operation. With the gradual transformation of hydraulic engineering from digitization and intelligence to WISDOM, remote sensing technology, artificial intelligence and deep learning methods have been widely used for automatic perception, processing, storage and analysis of hydraulic structure engineering monitoring data [8]. These novel methods and technologies, to some extent, compensate for the limitations of traditional safety monitoring approaches. The advent of remote sensing technologies such as three-dimensional tilt photography offers the opportunity to build an integrated hydraulic engineering monitoring and acquisition system capable of capturing all the details of hydraulic engineering. With the introduction of artificial intelligence and deep learning methods, the hydraulic engineering information was analyzed and exploited efficiently. Combined with the traditional hydraulic structure behavior analysis methods, such as geotechnical testing and numerical simulation, artificial intelligence and deep learning methods can help solve more complex hydraulic engineering problems by providing more accurate and professional intelligent analysis and ubiquitous hydraulic engineering services of great theoretical importance and application value in order to achieve the general improvement of safety monitoring of hydraulic structures [9].

The aim of this Special Issue is to focus on artificial intelligence, deep learning methods and remote sensing technologies in the safety monitoring of hydraulic structure engineering. Since the announcement of the call for papers in April 2023, a total of 15 original papers have been accepted for publication following a thorough peer review procedure

(Manuscripts 1–15); the contents cover the analysis methods of hydraulic engineering structural behavior, technologies of dam crack detection, gross error identification, artificial forecast of streamflow, application of UAVs in morphology measurement of rivers, temperature control index, and stability study of engineering structure. To enhance comprehension of this particular edition, we have succinctly outlined the key points of the published articles below.

2. Summary of the Contributions in This Special Issue

In the field of hydraulic structure engineering research, scholars have endeavored to ensure the long-term safety and health of projects through a variety of research methods and technological approaches. These studies pay attention to the applications of artificial intelligence, deep learning methods and remote sensing technologies in the safety monitoring of the entire system.

Gross error detection plays a vital role in ensuring accuracy of structural characteristic analysis [10]. In order to identify gross errors in dam monitoring data, Contribution 1 introduced a novel method that integrates the Fuzzy C-Means clustering algorithm (FCM), Ordering Points To Identify the Clustering Structure (OPTICS) and Local Outlier Factor (LOF) algorithm. The proposed approach operates in several stages. Initially, the $FCMalgorithm$ is employed to achieve the division of measurement point areas. Subsequently, the OPTICS and LOF algorithms are jointly utilized to identify the gross errors. The final step involves the confirmation of actual gross errors by analyzing the temporal coincidence of detected anomalies across measurement points within the same area. A case study demonstrated the efficacy of this method in accurately identifying spurious gross errors attributable to environmental anomalies. The findings highlight a significant enhancement in gross error detection accuracy and a notable reduction in the misjudgment rate of such errors.

In the realm of structure stability research, Contribution 2 systematically provided a comprehensive review of the advancements and achievements related to the foundational theories concerning the stability of coal pillar dams and artificial dams in CMURs. This review encompasses the mechanical properties under hydraulic coupling, the evolution mechanisms and impact relationships of the "three fields" (stress field, fracture field, and seepage field), and the optimized design of dam dimensions. Clarifying the instability and failure patterns of CMUR dam structures is crucial to ensuring dam structure stability. Contribution 3 delved into the stability assessment of a double-row steel sheet pile cofferdam structure under soft ground conditions. The stability of the cofferdam design solution was evaluated by using a model that incorporates factors such as the coordination of independent pile top displacement, as well as the m-value for backfilled sand and the thrown rock body. The internal force and displacement results of the cofferdam under different working conditions are obtained. And the entire construction process was analyzed using the finite element method. The findings assist in determining whether the overall and overturning stability of the cofferdam meet the pertinent safety requirements.

In the field of dam crack detection, Contribution 4 proposed an improved residual neural network (ResNet)-based algorithm for concrete dam crack detection through dynamic knowledge distillation, which was aimed at achieving higher accuracy for small models. Preliminary experiments on the mini-ImageNet dataset demonstrated the effectiveness of this approach. ResNet18 was trained by incorporating additional tasks to match the soft targets generated by ResNet50 under dynamically high temperatures. These pre-trained teacher and student models were then applied to concrete crack detection, achieving an accuracy of 99.85%, which is an improvement of 4.92%. Contribution 5 presented a pure Vision Transformer (ViT)-based dam crack segmentation network (DCST-net). The DCST-net employs an improved Swin Transformer (SwinT) block to enhance long-range dependencies within a SegNet-like encoder–decoder architecture. The Swin Transformer (SwinT) is a type of deep learning model based on the Transformer architecture, specifically designed for computer vision tasks. Developed by Microsoft Research Asia. Additionally, a weighted attention block to facilitate side fusion between the symmetric pair of encoders and decoders in each stage to sharpen the edge of crack is utilized. The DCST-net was tested on a self-built dam crack dataset and two publicly available datasets, outperforming mainstream methods in both visualization and most evaluation metrics, demonstrating its potential for practical application in dam safety inspection and maintenance.

In the construction of a structural behavior prediction model, Contribution 6 introduced an improved long short-term memory (LSTM) model and weighted Markov model to predict sluice deformation. In the method, the seagull optimization algorithm (SOA) is utilized to optimize the hyper-parameters of the LSTM, enhancing the model's accuracy. Subsequently, the relevant error sequences of the fitting results of SOA-LSTM model are classified and the Markovity of the state sequence is examined. Then, the autocorrelation coefficients and weights of each order are calculated to predict future random states of sluice deformation, resulting in a proposed SOA-LSTM-weighted Markov model. This model demonstrated superior prediction ability and accuracy compared to the SOA-LSTM and stepwise regression models, as evidenced by analysis of an actual sluice project in China. Contribution 7 established the MHHO-BiLSTM statistical prediction model of sluice seepage. This model incorporates water pressure, rainfall, and aging effects as input data. The parameters of the BiLSTM neural network are optimized using the improved Harris Hawks optimization algorithm. Empirical analysis using sluice seepage data indicated that the proposed optimization algorithm effectively searches for optimal parameters, enhancing the prediction accuracy and robustness of the BiLSTM model. Contribution 8 applied the random coefficient model of panel data into the analysis of sluice deformation, addressing the unobservable overall and individual effects. Moreover, the maximum entropy principle is used to approximate the probability distribution function of individual effect extreme values and to determine the early warning indicators, facilitating the assessment and analysis of non-uniform deformation states. A case study demonstrated the method's effectiveness in identifying overall deformation trends and deviations at individual measurement points. Contribution 9 aimed to propose a settlement prediction model for CFRDs combining the variational mode decomposition (VMD) algorithm, long short-term memory (LSTM) network, and support vector regression algorithm (SVR). Primarily, VMD is applied in the decomposition of dam settlement monitoring data to reduce its complexity. Furthermore, feature information on settlement time series is extracted. Secondly, the LSTM and SVR are optimized by the Harris Hawks optimization (HHO) algorithm and modified least square (PLS) method to mine the major influencing factors and establish the prediction model with higher precision. Applied to the Yixing CFRD, the model demonstrated superior predictive performance compared to other models, highlighting its potential for evaluating settlement trends and safety states of CFRDs. Contribution 10 proposed an artificial intelligence-based process for predicting concrete dam deformation. Initially, using the principles of feature engineering, the preprocessing of deformation safety monitoring data is conducted. Subsequently, employing a stacking model fusion method, a novel prediction process embedded with multiple artificial intelligence algorithms is developed. Moreover, three new performance indicators are introduced to provide a comprehensive assessment of the model's effectiveness. An engineering case study confirmed that the ensemble AI method outperformed traditional statistical models and single machine learning models in both fitting and predictive accuracy, providing a robust foundation for concrete dam deformation prediction and safety monitoring.

On certain occasions, the finite element method proves invaluable for analyzing dam structural characteristics [11]. Contribution 11 utilized a finite element model based on seepage–stress coupling theory to investigate the variations in the phreatic line, earth pressure, and deformation of a core wall rockfill dam due to rapid fluctuations in the reservoir's water level. Additionally, the results of the finite element simulation were compared with and analyzed alongside safety monitoring data, providing a scientific basis for assessing the health and ensuring the long-term safety of core wall rockfill dams in pumped storage power facilities. Furthermore, artificial intelligence algorithms have been applied to integrated time-dependent analyses of hydraulic structures. Contribution 12 developed a comprehensive superstructure–foundation–backfill model to investigate the time-dependent displacement and stress of a lock head project on a soft foundation during construction. Finite element analyses incorporated a transient thermal creep model for concrete and an elasto-plastic consolidation model for the soil. The modified Cam-clay model was used to describe the soil's elasto-plastic behavior. Subsequently, global sensitivity analyses were conducted to determine the relative importance of model parameters on the system's response, utilizing Garson's and partial derivative algorithms based on the backpropagation (BP) neural network.

Machine learning and statistical methods are increasingly utilized in streamflow forecasting. Contribution 13 conducted a rigorous investigation into the effectiveness of three machine learning techniques and two statistical approaches using streamflow data from the Göksu Stream in the Marmara Region of Turkey, covering the period from 1984 to 2022. This comparative analysis aimed to advance the methodologies used in streamflow prediction. The results revealed that the model generated using the XGBoost algorithm outperformed other machine learning and statistical techniques. Consequently, the models developed in this study demonstrate a high level of accuracy in predicting streamflow within the river basin system, offering significant contributions to the field of hydrological forecasting.

It is important to adopt scientific safety control criteria for dams in the construction period according to practical experience and theoretical calculation [12]. Contribution 14 synthetically introduced a novel methodology combining information entropy and a cloud model to develop in situ observation data-based temperature control indexes from a spatial field perspective. The order degree and the disorder degree of observation values are defined according to the probability principle. Information entropy and weight parameters are combined to describe the distribution characteristics of the temperature field, with weight parameters optimized via projection pursuit analysis (PPA), leading to the construction of temperature field entropy (TFE). Based on this framework, multi-level temperature control indexes were established using a cloud model. Finally, a case study was conducted to verify the performance of the proposed method. The calculation results indicated that the change law of TFEs aligned well with actual conditions, demonstrating the reasonableness of the established TFE. The cloud model showed broader applicability compared to typical small probability methods, and the derived temperature control indexes significantly enhanced the safety management level of high concrete dams. The research outcomes provide a scientific reference and technical support for temperature control standards in similar projects.

In order to analyze the morphometry of drifting ice on the water surface, Contribution 15 employed UAV technology as a cost-effective and efficient method for describing ice phenomena in lowland rivers. This methodology focused on measuring the area, average size, perimeter, and circularity of frazil ice floes. Individual frames captured by a UAV were analyzed using statistical techniques to obtain these measurements. In prior research, the team successfully assessed ice velocity on a similar test sample. By deriving the average velocity, surface area, and fundamental morphological characteristics of frazil ice, the study facilitated the automated segmentation, classification, and prediction of potential risks related to ice blockages in water routes. These risks include waterway obstructions, infrastructure damage, and potential threats to human safety.

3. Conclusions

This Special Issue not only proposes novel methods and technologies in the field of hydraulic structure safety monitoring, but also provides a wealth of practical experience and insights. The guest editors believe that the following issues in the field of hydraulic structure safety monitoring need continued attention.

(1) Combined with the UAV inclined photography technology, a more refined numerical calculation model must be established so as to improve the structural behavior analysis accuracy of hydraulic structure engineering.
(2) With the development of artificial intelligence technology, the intelligence of the hydraulic structure safety assessment method is also expected to be further improved.
(3) Further emphasis should be given to improving the prediction precision of the structural characteristic prediction model and reducing time complexity of the algorithm at the same time.

We believe that in future research, hydraulic structure safety monitoring technology will continue to make new breakthroughs, providing a strong guarantee for security and sustainable development of hydraulic engineering construction and operation.

Author Contributions: Conceptualization, C.S.; validation, Y.X.; resources, X.Q. and G.Y.; writing—original draft preparation, C.S.; writing—review and editing, H.C.; supervision, H.G. All authors have read and agreed to the published version of the manuscript.

Funding: This research was partially supported by the National Natural Science Foundation of China (Grant Nos. 52209159, U2243223, 52209170, 52109155), the Fundamental Research Funds for the Central Universities (Grant No. B230201011).

Data Availability Statement: Not applicable.

Conflicts of Interest: The authors declare no conflicts of interest.

List of Contributions:

1. Chen, L.; Gu, C.; Zheng, S.; Wang, Y. A Method for Identifying Gross Errors in Dam Monitoring Data. *Water* **2024**, *16*, 978.
2. Wang, Y.; Liu, F.; Kou, M.; Li, M. A Review of Stability of Dam Structures in Coal Mine Underground Reservoirs. *Water* **2024**, *16*, 1856.
3. Jiang, Y.; Guo, F.; Wang, W.; Yang, G.; Yue, J.; Huang, Y. Stability Study of a Double-Row Steel Sheet Pile Cofferdam Structure on Soft Ground. *Water* **2023**, *15*, 2643.
4. Zhang, J.; Bao, T. An Improved ResNet-Based Algorithm for Crack Detection of Concrete Dams Using Dynamic Knowledge Distillation. *Water* **2023**, *15*, 2839.
5. Zhou, J.; Zhao, G.; Li, Y. Vison Transformer-Based Automatic Crack Detection on Dam Surface. *Water* **2024**, *16*, 1348.
6. Peng, J.; Xie, W.; Wu, Y.; Sun, X.; Zhang, C.; Gu, H.; Zhu, M.; Zheng, S. Prediction for the Sluice Deformation Based on SOA-LSTM-Weighted Markov Model. *Water* **2023**, *15*, 3724.
7. Huang, Z.; Gu, C.; Peng, J.; Wu, Y.; Gu, H.; Shao, C.; Zheng, S.; Zhu, M. A Statistical Prediction Model for Sluice Seepage Based on MHHO-BiLSTM. *Water* **2024**, *16*, 191.
8. Ma, Z.; Lou, B.; Shen, Z.; Ma, F.; Luo, X.; Ye, W.; Li, X.; Li, D. A Deformation Analysis Method for Sluice Structure Based on Panel Data. *Water* **2024**, *16*, 1287.
9. Zheng, X.; Ren, T.; Lv, F.; Wang, Y.; Zheng, S. Settlement Prediction for Concrete Face Rockfill Dams Considering Major Factor Mining Based on the HHO-VMD-LSTM-SVR Model. *Water* **2024**, *16*, 1643.
10. Wu, W.; Su, H.; Feng, Y.; Zhang, S.; Zheng, S.; Cao, W.; Liu, H. A Novel Artificial Intelligence Prediction Process of Concrete Dam Deformation Based on a Stacking Model Fusion Method. *Water* **2024**, *16*, 1868.
11. Zheng, X.; Yan, B.; Wang, W.; Du, K.; Fang, Y. Seepage–Deformation Coupling Analysis of a Core Wall Rockfill Dam Subject to Rapid Fluctuations in the Reservoir Water Level. *Water* **2024**, *16*, 1621.
12. Xu, C.; Ye, L.; Pan, S.; Luo, W. Integrated Time-Dependent Analysis of a Hydraulic Structure on Soft Foundations during Construction. *Water* **2024**, *16*, 1375.
13. Ciner, M.N.; Güler, M.; Namlı, E.; Samastı, M.; Ulu, M.; Peker, İ.B.; Gülbaz, S. The Forecast of Streamflow through Göksu Stream Using Machine Learning and Statistical Methods. *Water* **2024**, *16*, 1125.
14. Yang, G.; Sun, J.; Zhang, J.; Niu, J.; Luan, B.; Huang, Z.; Zhao, A. Research on Temperature Control Index for High Concrete Dams Based on Information Entropy and Cloud Model from the View of Spatial Field. *Water* **2023**, *15*, 4023.

15. Błotnicki, J.; Jarzembowski, P.; Gruszczyński, M.; Popczyk, M. The Use of UAV for Measuring the Morphology of Ice Cover on the Surface of a River: A Case Study of the Low Head Dam and Fishway Inlet Area in the Odra River. *Water* **2023**, *15*, 3972.

References

1. Shao, C.; Zheng, S.; Gu, C.; Hu, Y.; Qin, X. A novel outlier detection method for monitoring data in dam engineering. *Expert Sys Appl.* **2022**, *193*, 116476. [CrossRef]
2. Zhu, X.; Bao, T.; Yeoh, J.; Jia, N.; Li, H. Enhancing dam safety evaluation using dam digital twins. *Struct. Infrastruct. Eng.* **2023**, 19 904–920. [CrossRef]
3. Su, H.; Chen, Z.; Wen, Z. Performance improvement method of support vector machine-based model monitoring dam safety *Struct. Control Health Monit.* **2016**, *23*, 252–266. [CrossRef]
4. Adamo, N.; Al-Ansari, N.; Sissakian, V.; Laue, J. Geophysical methods and their applications in dam safety monitoring. *J. Earth Sci. Geotech. Eng.* **2020**, *11*, 291–345. [CrossRef]
5. Wei, B.; Chen, L.; Li, H.; Yuan, D.; Wang, G. Optimized prediction model for concrete dam displacement based on signal residual amendment. *Appl. Math. Model.* **2020**, *78*, 20–36. [CrossRef]
6. Sang, L.; Wang, J.; Sui, J.; Dziedzic, M. A new approach for dam safety assessment using the extended cloud model. *Water Resour. Manag.* **2022**, *36*, 5785–5798. [CrossRef]
7. Ribas, J.R.; Pérez-Díaz, J.I. A multicriteria fuzzy approximate reasoning approach for risk assessment of dam safety. *Environ Earth Sci.* **2019**, *78*, 514. [CrossRef]
8. Cheng, J.; Xiong, Y. Application of extreme learning machine combination model for dam displacement prediction. *Procedia Comput. Sci.* **2017**, *107*, 373–378. [CrossRef]
9. Wang, L.; Mao, Y.; Cheng, Y.; Liu, Y. Deep learning-based diagnosing structural behavior in dam safety monitoring system *Sensors* **2021**, *21*, 1171. [CrossRef] [PubMed]
10. Meng, Z.; Wang, Y.; Zheng, S.; Wang, X.; Liu, D.; Zhang, J.; Shao, Y. Abnormal monitoring data detection based on matrix manipulation and the cuckoo search algorithm. *Mathematics* **2024**, *12*, 1345. [CrossRef]
11. Athani, S.S.; Solanki, C.H.; Dodagoudar, G.R. Seepage and stability analyses of earth dam using finite element method. *Aquat Procedia* **2015**, *4*, 876–883. [CrossRef]
12. Wang, S.; Gu, C.; Bao, T. Safety monitoring index of high concrete gravity dam based on failure mechanism of instability. *Math Probl. Eng.* **2013**, *2013*, 732325. [CrossRef]

water

MDPI

Review

A Review of Stability of Dam Structures in Coal Mine Underground Reservoirs

Yan Wang [1,2,*], Fei Liu [1,2], Miaomiao Kou [1,2] and Mingfei Li [2]

1 State Key Laboratory of Water Resource Protection and Utilization in Coal Mining, Beijing 102209, China; liufei@qut.edu.cn (F.L.); koumiaomiao@qut.edu.cn (M.K.)

2 School of Civil Engineering, Qingdao University of Technology, Qingdao 266520, China; limingfei@qut.edu.cn

* Correspondence: wangyantumu@qut.edu.cn

Abstract: Coal has remained the primary component of China's energy structure, and high-intensity extraction has continued in the central and western coal-producing regions of China. In contrast to the abundant coal resources, water resources have become extremely scarce in these regions, creating a conflict between coal resource extraction and water resource conservation. The coal mine underground reservoir (CMUR), as a typical technology for combined coal and water extraction and water-preserving coal mining, has been applied in numerous mines in central and western China. This effectively alleviates water resource shortages and achieves the goal of water resource conservation. The CMURs utilizes the goaf created by longwall mining as the water storage space. The reservoir dam structure comprises coal pillars, which serve as protective coal pillars in the mining area, and artificial dam structures that filled the gaps between these coal pillars. The stability of the dam structure under the complex stress effects of hydraulic coupling has been identified as the key to maintaining the safe operation of the CMUR. The mechanical properties, stress field, fracture field, and seepage field ("three fields") change mechanisms, as well as the research results on size optimization of coal pillar dams and artificial dams in CMURs, were systematically reviewed. The core content included the instability and failure mechanisms of dam structures under the comprehensive coupling effects of factors such as dry–wet cycles of mine water, long-term immersion, chemical effects of high-salinity water, dynamic and static loads, and cyclic loads. This paper is considered to have certain reference value for the study of the stability of dam structures in CMURs and to provide some guidance for the safe operation of CMURs.

Keywords: coal mine; underground reservoir; coal pillar dam; artificial dam; stability

1. Introduction

China has been recognized as the world's largest producer and consumer of coal. With the development of China's social economy, the demand for coal remained high. The overall energy resource situation in China has been characterized by a scarcity of oil, a shortage of natural gas, and a relative abundance of coal. In recent years, however, the proportion of coal as a primary energy source in China's energy structure has gradually decreased [1]. In Figure 1, the structure of China's energy consumption from 2013 to 2022 is depicted [2,3]. In 2023, the proportion of coal in China's total energy consumption remained as high as 55.3% [4]. For a considerable period in the future, coal is expected to remain the primary component of China's energy structure [5]. High-intensity coal mining has been found to damage groundwater and cause significant water resource waste. Approximately two tons of mine water is generated per ton of coal, resulting in about 8 billion m^3 of mine water annually in China, with a utilization rate of around 40%. Each year, about 5 billion m^3 of mine water is not effectively utilized, equivalent to 50% of the annual industrial and domestic water shortage in China (10 billion m^3) [6,7]. China's primary coal-producing regions, located in the central and western provinces (Shanxi, Shaanxi, Inner Mongolia,

Ningxia, and Xinjiang), hold over 80% of the nation's coal reserves and contribute more than 70% to its coal production. However, water resources in these regions only account for 6.7% of the national total and 0.3% of the global total [8,9]. A reverse distribution trend between coal resources and water resources has been observed, highlighting the prominent conflict between coal resource development and water scarcity in western China. Figure 2 illustrates the distribution of coal production and water resources across provinces in 2023. The coal-rich regions in central and western China are predominantly located in arid and semi-arid ecologically fragile zones, characterized by water scarcity and vulnerable surface ecosystems. Major coal bases experience severe water shortages, with mine area water supply primarily relying on groundwater extraction and the comprehensive utilization of mine water. Challenges in these mining areas included water sourcing, integrated water resource utilization, and water rights allocation. Therefore, the question of how to achieve both safe and efficient coal mining and the protection of groundwater resources has been identified as a major challenge faced by coal mining operations in central and western mining areas.

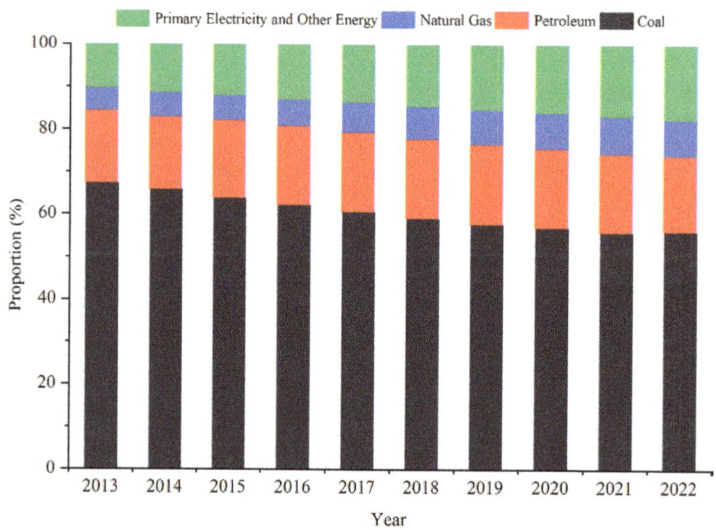

Figure 1. China's energy consumption structure (2013–2022).

Due to the shallow burial of coal seams, large seam thickness, and thin overlying bedrock in the central and western regions, the full-height mining of coal seams inevitably caused the overlying strata to fracture and connect with aquifers, resulting in significant underground water leakage and loss [10–12]. In response to this situation, traditional passive water-preserving mining techniques, characterized by "blocking" underground water leakage, were previously employed in coal mines. Techniques such as limited height mining, backfill mining, and partitioned mining were utilized to ensure that the overlying strata fractures caused by coal extraction did not connect with aquifers, thereby preventing damage to the aquifer structure [13–15]. Within the coal mine field, groundwater was found to converge towards the mining area (goaf). It is estimated that every 10,000 t of coal produces 1000 to 1600 m^3 of void space. These spaces provided conditions for storing mine water and were considered ideal locations for mine water storage [16]. Based on the characteristics of the collapsed space in the goaf and drawing on traditional underground reservoir construction concepts [17], a new concept was proposed by Gu to address the phenomenon of water resource evaporation and loss during coal mining in western China's mining areas. This concept involved constructing CMURs in the goaf caved zones to achieve underground storage and the utilization of mine water, representing water resource protection technology characterized by "conducting" [6,7,18,19]. This technology enabled the safe underground storage and utilization of mine water, preventing evaporation losses associated with surface discharge. The theoretical framework and technical system of CMURs were widely promoted and applied in mining areas such as Shendong. With a maximum storage capacity of 35 Mm^3, these reservoirs supplied over 95% of the production, domestic, and ecological water needs of the mining areas, and also provided water for nearby power plants and coal-to-oil projects, achieving a transition from major water consumers to major water suppliers [20].

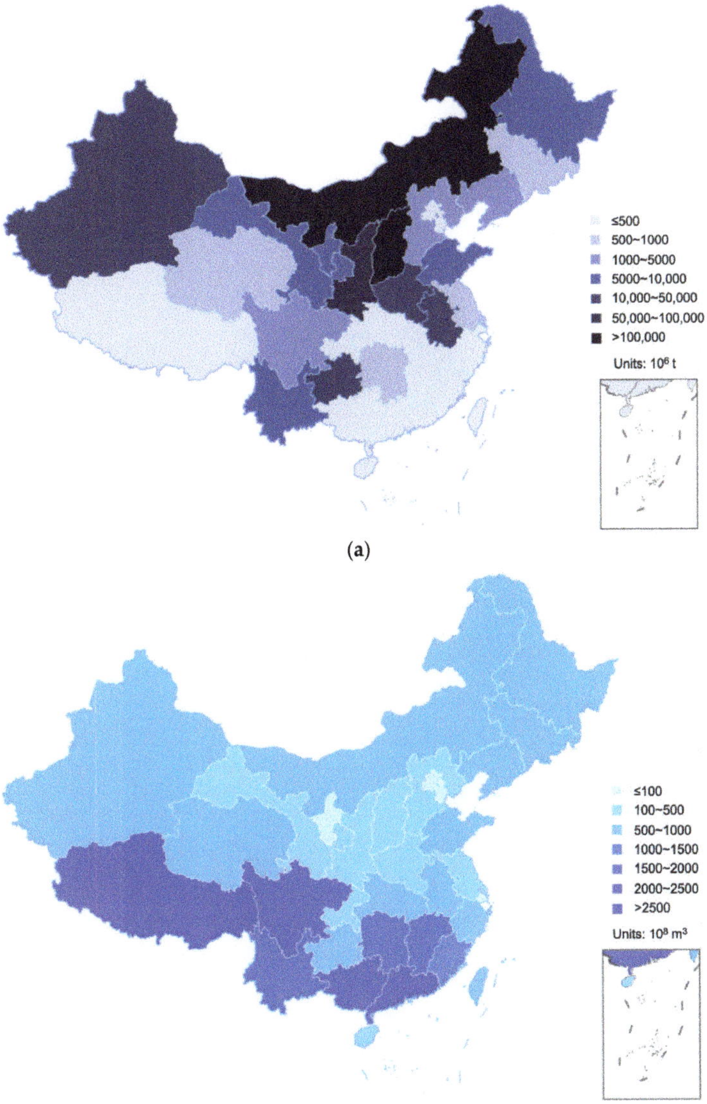

Figure 2. Coal production and water resource distribution by province in China. (**a**) Coal production. (**b**) Water resource distribution.

The concept of CMURs and the strategy of storing and using mine water underground adopted in China differed from the Western approach, which only used abandoned mine tunnels as water storage spaces. Underground reservoirs in Western countries primarily utilized abandoned mines. After these mines were abandoned, groundwater accumulated in the abandoned spaces through mining-induced fractures and seepage, ultimately forming flooded mines. Examples include the Martelange mine in Belgium [21], the Freiberg mine and Staßfurt mine in Germany [22,23], the Gaspé mine in Canada [24], the Barredo-Figaredo underground reservoir in Spain [25], and the abandoned mines in the Appalachian coal region of the United States [26]. Such abandoned flooded mines did not require consideration of mining impacts, and the storage of mine water did not necessitate pre-planned reservoir construction based on the mining sequence. Therefore, the artificial construction of dam structures for underground reservoirs was not necessary. In China, water was stored in the rock voids of the goaf formed after longwall coal mining. The reservoir dam structure consists of safety coal pillars supplemented by artificial dam structures. Water intake and extraction facilities were constructed simultaneously, making full use of the rock mass's natural ability to purify mine water in the goaf. The construction of CMUR projects, as shown in Figure 3, was undertaken [27]. Coal pillar dams and artificial dams were identified as the most critical components of CMURs, with the stability of these dam structures being crucial for the safety of the underground reservoirs.

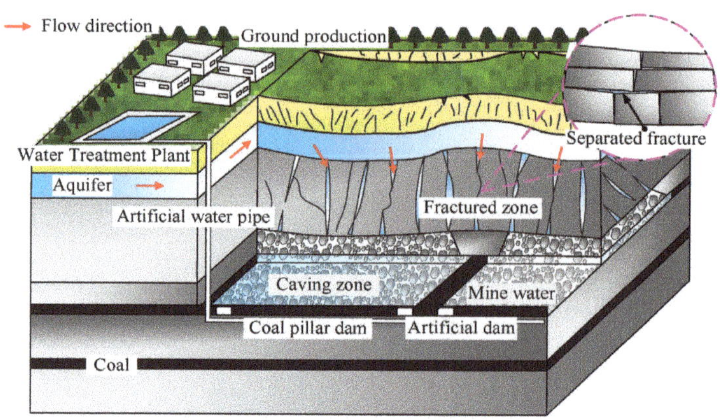

Figure 3. Schematic diagram of the CMUR (modified from [27]).

At the same time, the continued extraction of coal resources inevitably resulted in the creation of a substantial amount of abandoned coal mine underground spaces, primarily consisting of shafts, abandoned tunnels, and goafs. The effective utilization of these abandoned coal mine underground spaces was identified as an urgent issue that needed to be addressed. Currently, the primary approach for dealing with abandoned underground spaces in mines has been their conversion into pumped hydro storage (PHS) systems [28–31]. Due to the typical occurrence of coal seams in groups in China and the extremely high surface evaporation rates in central and western mining areas, effective water source replenishment could not be achieved. Therefore, PHS systems in China have primarily been implemented in the form of underground reservoirs. Figure 4 illustrates a PHS system utilizing the underground goaf in a coal mine [32]. The CMUR was identified as the most suitable technological approach for utilizing abandoned underground spaces in coal mines in central and western China. While addressing water resource conservation issues, this approach also provided an effective engineering reference for the efficient use of underground spaces.

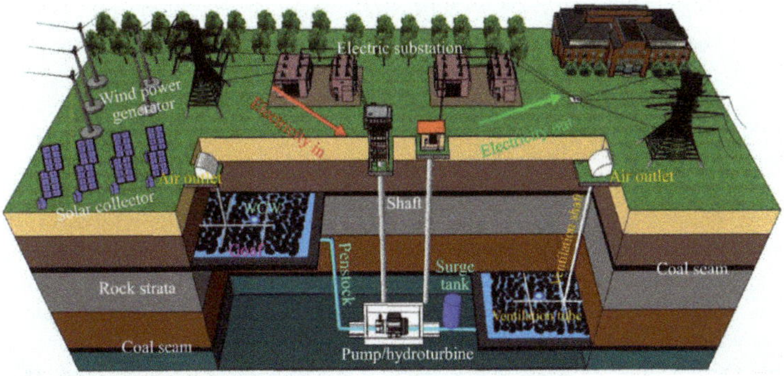

Figure 4. A pumped hydro-storage system (modified from [32]).

The above research and analysis indicates that, to address the issue of water scarcity in central and western mining areas and resolve the conflict between coal mining and water conservation, the construction of CMURs within the abandoned underground spaces of coal mines was a highly scientific and rational choice. CMURs had already become an important hydraulic infrastructure for the utilization of mine water. Since the CMUR primarily utilized the goaf created by longwall mining as the water storage space, it had a larger storage capacity and longer dam structure compared to underground reservoirs primarily utilizing tunnels. However, it was inevitably affected by mining activities on adjacent working faces and mine-induced seismic events during its service life. Therefore, several key issues were encountered during the construction of CMURs, such as the stability, deformation patterns, and safety of the dam structure under conditions of immersion, dry–wet cycles, high-concentration saltwater erosion, and complex stress environments. This paper, in conjunction with the development history of CMURs in China, provides a comprehensive review of the stability of the key structures, coal pillar dams, and artificial dam structures. Section 2 first introduces the main technologies and key structures involved in the construction of CMURs. Sections 3 and 4 primarily discuss the mechanical properties of coal pillar dam structures and artificial dam structures under complex hydraulic coupling and stress conditions (laboratory scale), as well as the changes in stress fields, fracture fields, and seepage fields ("three fields") and size optimization (engineering scale). Section 5 provides an outlook on the construction and future development prospects of CMURs. This paper aimed to serve as a reference for the construction of CMURs.

2. Main Technologies and Key Structures of CMURs

The construction of CMURs has been recognized as a complex system engineering project, involving the organic integration of multiple disciplines, including mining engineering, hydraulic engineering, engineering geology, hydrogeology, and environmental engineering. The design, construction, and operation of CMURs involves key technologies such as reservoir site selection, storage capacity calculation, water quality purification, and environmental impact assessment. The main structures include coal pillar dams, artificial dams, reservoir roofs, reservoir floors, water storage spaces, and drainage facilities. From a macro perspective, reservoir site selection and storage capacity calculation have been identified as the primary technologies for CMURs. From a structural perspective, coal pillar dams and artificial dams are considered the most critical structures. The reservoir site selection, storage capacity calculation, coal pillar dams, and artificial dams are introduced in this section.

2.1. Reservoir Site Selection

Reservoir site selection and storage capacity calculation have been identified as important tasks during the planning and design stages of CMURs. These tasks are typically carried out in a coordinated and intersecting manner. Reservoir site selection requires comprehensive consideration of factors such as mining height, surrounding rock properties, water storage range and level. The main factors influencing the performance of underground reservoirs are identified as storage coefficient, mine self-flow recharge, floor impermeability, and the safe distance from working faces. Identifying a suitable site is acknowledged as essential for building CMURs. However, limited research exists on the location selection of CMURs.

The aforementioned factors affecting the site selection of CMURs are numerous, and the ways and angles in which these factors influence the reservoirs are varied. Building CMURs by injecting water into abandoned goafs to form storage spaces inevitably disrupts the original dynamic balance of the groundwater system. Chen et al. [33] developed a detailed hydrogeological model for both the mining area and the underground reservoir. The impact of underground reservoir construction on the flow field was analyzed using factors like mining height at the reservoir site, storage range, overlying rock lithology, and water level. Sensitivity coefficients for each factor were determined. Numerical simulation was employed as a common method for establishing and analyzing groundwater systems. Figure 5 shows the groundwater system model of the Daliuta coal mine in the Shendong mining area [33]. Another method for considering the site selection of CMURs is based on water storage capacity. Distributed CMURs are suitable for coal mines with simple geological conditions. Faults, karst, and other geological structures have a significant impact on stress changes, hydraulic fracture propagation, and permeability distribution, which may lead to water seepage and even dam collapse. Therefore, the construction of CMURs was recommended to avoid areas with well-developed geological structures of this type. The storage capacity of CMURs was found to be positively correlated with mining height, roof permeability, and storage coefficient. Conversely, it was negatively correlated with the level of the coal seam floor and the floor permeability. A method for selecting sites for CMURs based on entropy weight and a calculation model for the storage capacity were developed by Zhang et al. [34]. Furthermore, the significance of hydrochemistry for the selection of CMUR sites was discussed by Guo et al. [35]. The durability of artificial concrete dams could be reduced under the dry–wet cycling effects of high-sulfate water. CMURs were not suitable for locations with poor water quality and a high tendency for water-rock interactions. The concentrations of Ca^{2+}, sodium adsorption ratio (SAR), and F^- in the water were found to be further increased under the influence of water-rock interactions, adversely affecting water quality improvement.

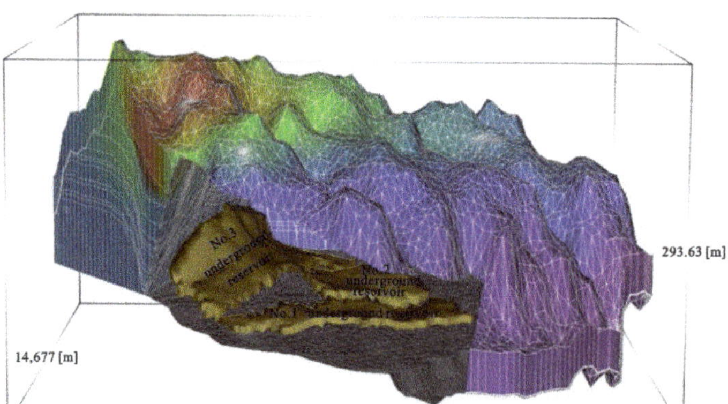

Figure 5. Simulation model of the groundwater system of Daliuta coal mine (modified from [33]).

2.2. Storage Capacity Calculation

CMURs primarily utilize the free voids between the fractured and collapsed rock masses in the goaf for water storage. The reservoir capacity is defined as the total amount of free voids within the storage range of the goaf that can be used for water storage [36]. The free voids mainly involve the development height of the "three zones" (caved zone, fractured zone, and bending zone) of the goaf, which are central to the design and estimation of reservoir capacity. The formation of the three zones and hydraulic migration resulting from mining is illustrated in Figure 6 [37]. The evaluation metric for the water storage capacity of CMURs is the water storage coefficient, which was the unit volume water storage capacity of the goaf. It is defined as the total voidage of the broken rock mass in the caved zone [38]. A close relationship has been found between the water storage coefficient and the mining height, mining methods, overburden pressure, and surrounding rock properties. It primarily depends on the spatial structural characteristics of the storage space [27]. Regarding spatial structural characteristics, the formation patterns of the spatial structure after working face mining have been studied through various approaches, including theoretical analysis [39,40], numerical simulation [41,42], and similar simulation experiments [27,36,43]. Additionally, the compaction characteristics of the gangue accumulation process are influenced by mining activities on adjacent working faces and the coal seam inclination, causing variations. Detailed calculations also consider the shape and accumulation characteristics of the gangue blocks. However, due to the significant randomness of gangue breakage and caving, research in this area remains in the exploratory stage [44].

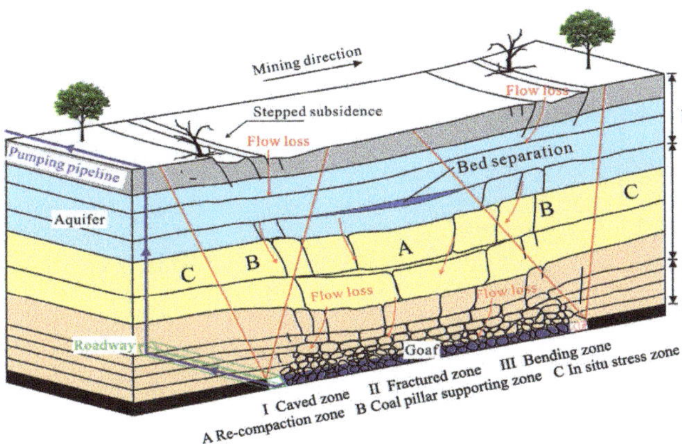

Figure 6. The "three zones" of the goaf and hydraulic migration (modified from [37]).

2.3. Coal Pillar Dam

The longwall mining method has typically been used in China's underground coal mining. This method established protective coal pillars along the boundary of the entire mining area to maintain overall stability and reduce mutual interference between adjacent mining areas [45]. In CMURs, boundary protective coal pillars are designed and modified to convert them into coal pillar dams. These are connected with artificial dam structures to form an integrated dam structure, serving as a permanent hydraulic construction for the reservoir [46,47]. Figure 7 illustrates the relationship between working face mining and the formation of coal pillar dams [48]. Coal pillar dams are major components of the dam structure in CMURs and play a crucial role in ensuring the safe operation of CMURs and coal mining activities [49,50]. Additionally, within the goafs converted into storage spaces, residual coal pillars have been identified as remnants left behind from the mining of adjacent working faces. Although residual coal pillars are typically not considered part of the dam structure due to their location within the reservoir, they play an important role in maintaining the stability of the roof of the water storage space [51].

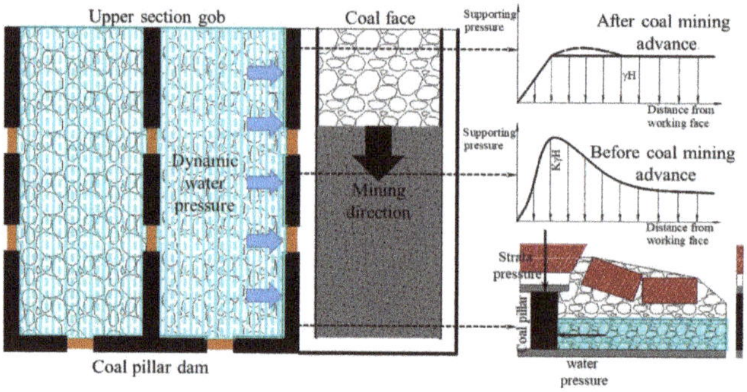

Figure 7. Relationship between working face mining and the formation of coal pillar dams (modified from [48]).

2.4. *Artificial Dam*

In addition to the reserved coal pillar dams, artificially constructed non-coal dam structures, known as artificial dams, have also been constructed as part of the dam structure of CMURs. Artificial dams are primarily used to connect the gaps between adjacent coal pillar dams, forming a complete and sealed water storage space maintenance structure together with the coal pillar dams. Together, they constitute the dam structure of the CMUR [52,53]. The artificial dam structure has played an important role in protecting the stability and durability of the underground reservoir. Unlike coal pillar dams, which can reach thicknesses of tens of meters, most artificial dams in CMURs have been designed with a thickness of approximately 1–3 m [54]. These structures are typically constructed as reinforced concrete monolithic beams [55], predominantly in flat or "T" shapes. Due to the relatively thin thickness of artificial dams, their waterproof performance is crucial. Therefore, high-grade concrete, such as C30, C40, C50, and engineered cementitious composites (ECC), have often been used when constructing artificial dams [45,56]. Figure 8 illustrates a typical artificial dam structure [56]. There are also cases where the waterproof performance has been improved by altering the structural form. For example, at the Daliuta coal mine, the artificial dam was designed as a combination of an impermeable wall and a bearing wall. The impermeable wall was constructed from brick and filled with impermeable materials such as loess, while the bearing wall used I-beams as the framework and concrete as the bonding filler [57].

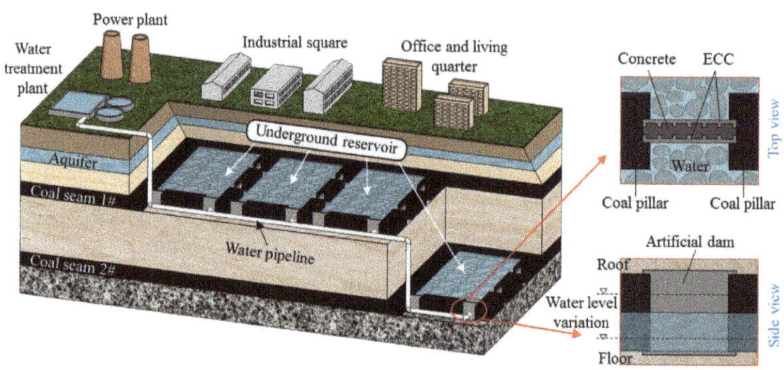

Figure 8. Artificial dam structure (adapted from [56], with permission from Elsevier, 2024).

3. Key Theories and Technologies for the Stability of Coal Pillar Dams

As the core component of the dam structure in CMURs, coal pillar dams are not only influenced by dynamic and static loads such as roof breakage, fracturing, and slipping during the mining process but also subject to long-term water erosion due to contact with mine water on the water-facing side. This is particularly significant during the repeated rising and falling of the storage water level caused by the water extraction process in CMURs [58,59]. After the CMUR was filled with water, the coal pillar dam is subjected to a complex environment of combined dynamic and static stress fields and water immersion. Due to water immersion, significant changes occurr in the mechanical parameters of the coal, such as strength, elastic modulus, and cohesion. Under hydraulic coupling, the coal pillar dam is more susceptible to damage and even instability [60].

Following the mining of a working face, the overlying rock strata will experience roof collapse and stress redistribution. The stress environment of the coal pillar dam changes from a stable triaxial stress environment to a complex stress environment, which includes stress from the overlying rock pressure, stope bearing pressure, lateral pressure from the fractured roof, and superimposed microseismic activity from the mine. Figure 9 illustrates the stress environment of the coal pillar dam [61]. Based on the types of loads acting on the coal pillar dam, the stress can be categorized into static load, mining-induced load, and mine seismic load, with the latter two summarized as vibrational loads. The impacts of water immersion on coal pillar dams were categorized into two main types: dynamic water level fluctuations causing dry–wet cycles and variations in immersion time (water content); and long-term creep effects under stable water level conditions, along with erosion induced by saline mine water. Therefore, clarifying the mechanical deformation characteristics, as well as the changes in stress field, fracture field, and seepage field of coal pillar dams under complex hydraulic coupling and stress conditions, is an important prerequisite for maintaining the long-term stability of coal pillar dams in CMURs and for determining their reasonable width and dimensions.

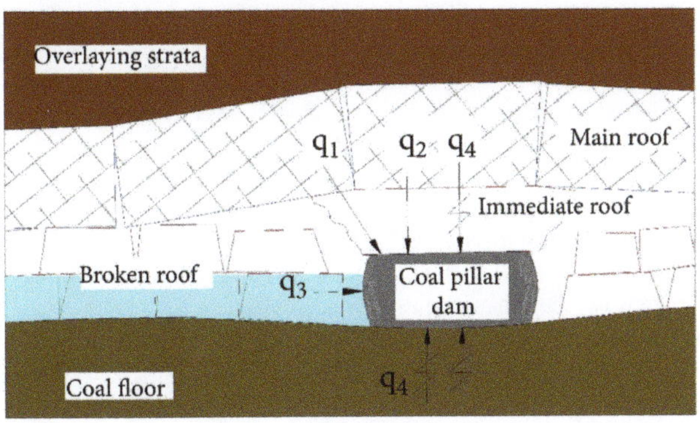

Figure 9. Stress environment of coal pillar dams (modified from [61]). (q1, side abutment pressure; q2, overlying strata gravity; q3, hydrostatic pressure; and q4, dynamic stress).

3.1. Mechanical Properties of Coal Pillar Dams

As shown in Figure 7, the coal pillar dams in CMURs are composed of multiple segments of boundary protective coal pillars. After the goaf has filled with water, one side of the coal pillar dam is exposed while the other side is subjected to water pressure, resulting in a very complex stress state. Laboratory tests are an important means of studying the mechanical properties of coal pillar dam samples. Through uniaxial compression, triaxial compression, and cyclic loading tests on coal pillar dam samples, researchers have replicated the mechanical response and failure patterns of coal pillar dams under load-bearing conditions on a laboratory scale. Since coal pillar dams in CMURs are significantly different from ordinary protective coal pillars due to the substantial impact of water, most studies have considered the effects of water immersion.

(1) Mechanical properties of coal pillar dams affected by water immersion

During the use of underground reservoirs, the cyclic stages of water storage and drainage inevitably cause changes in the water level. Coal pillar dams periodically experience alternating states of drying and water loss, and water saturation. Research conducted on coal pillar dam samples under repeated water immersion more closely approximate the actual environmental changes of underground reservoirs. Currently, laboratory-scale experiments consider the mechanical properties of coal pillar dam samples under the influence of cyclic water storage and drainage primarily use the dry–wet cycle (repeated water immersion) mode. In these experiments, coal samples are repeatedly subjected to drying and water saturation treatments. An oven has typically been used as the dehydration equipment, with a non-destructive water immersion device used as the water saturation equipment, as shown in Figure 10 [62]. The number of dry–wet cycles has been shown to significantly affect the saturation water content of coal pillar dam samples. Overall, the water content of the coal samples initially increased rapidly and then gradually stabilized as the number of dry–wet cycles increased. The saturation water content of the coal samples showed a logarithmic increase with the dry–wet cycle numbers. Figure 11 presents the curve of coal sample water content w versus the number of dry–wet cycles, as summarized from various studies [59,62–64]. During the dry–wet cycle process, the complex physical and chemical interactions between water and coal promoted the further expansion of the primary fractures in the coal samples, as well as the formation of additional pores and voids, thereby increasing their saturation water content. The mechanical properties of coal pillar dam samples under dry–wet cycles have typically been tested using uniaxial compression tests and triaxial compression tests. Additionally, acoustic emission (AE) non-destructive testing has been used to analyze the mechanical damage mechanisms and energy evolution characteristics of the coal samples during failure. The conclusions indicated that under uniaxial compression, the uniaxial compressive strength gradually decreased with the increase in the number of dry–wet cycles. The peak stress of the coal samples exhibited an exponential decline with the increase in the number of cycles, while the peak strain gradually increased, and the elastic modulus also gradually decreased [63,65,66]. Under triaxial compression, the triaxial compressive strength of the coal samples decreased with the increase in the number of water immersion cycles at different confining pressures. The most significant decrease in triaxial compressive strength occurred after a single immersion, with the rate of decline gradually slowing thereafter. The AE cumulative ringing counts of the samples also decreased with the increase in the number of dry–wet cycles, indicating a reduction in the energy required for the coal samples to fracture [64].

During the water storage stage, as the water level rises, the immersion time of the coal pillar dam continually increases, and the water content will also change, gradually transitioning from a natural water content state to a saturated water content state. The uniaxial compression stress–strain curve of water-saturated coal samples can be segmented into five phases: (a) the fracture compaction phase, (b) the elastic deformation phase, (c) the stable fracture propagation phase, (d) the unstable fracture propagation phase, and (e) the unloading failure phase. Under uniaxial compression, as the water content increased, the compaction stage of the coal sample curve becomes more gradual. This is due to the softening of the coal sample and clay minerals upon water immersion, which prolongs the compaction process. The peak stress of the coal sample gradually decreases, while the peak strain gradually increases [58,67]. The elastic modulus also decreases with the increase in water content. The primary reason is that water has penetrated the coal sample particles and interacted with the soluble substances in the particle cement, leading to hydrolysis reactions that reduced the bonding force between particles. Additionally, the lubricating effect of water decreases the friction between particles, which macroscopically manifests as a reduction in the friction of the fracture surface [68–72]. Under the combined stress of compression and shear, the microcrack propagation in the coal sample is controlled by shear cracks. At a low water content, the macroscopic fracture mode of the coal sample gradually shifts from tensile failure to shear failure as the loading angle increases. At a moderate water content, the coal sample mostly exhibits a combination of tensile and shear failure. At a high water content, the coal sample predominantly exhibits shear failure [60,73,74]. Many coal pillar dams experienced multiple disturbances from the mining of adjacent working faces before the underground reservoir stored water. In experiments, coal samples have been subjected to multiple loading and unloading cycles, which accumulates stress damage, exhibiting characteristics of decreased peak stress, increased elastic modulus, and increased residual deformation with prolonged immersion time [75].

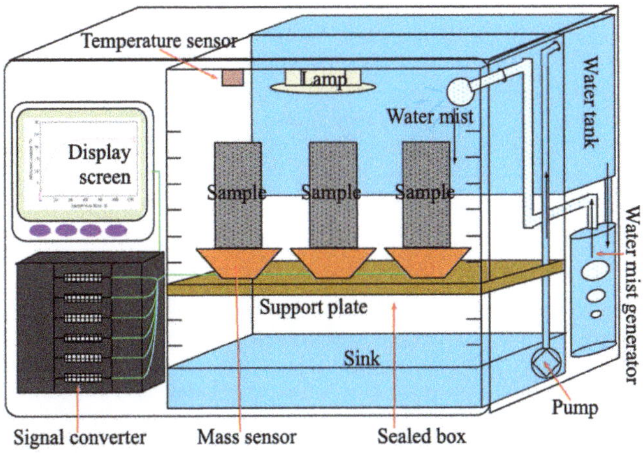

Figure 10. Nondestructive water immersion device (adapted from [62], with permission from Springer Nature, 2024).

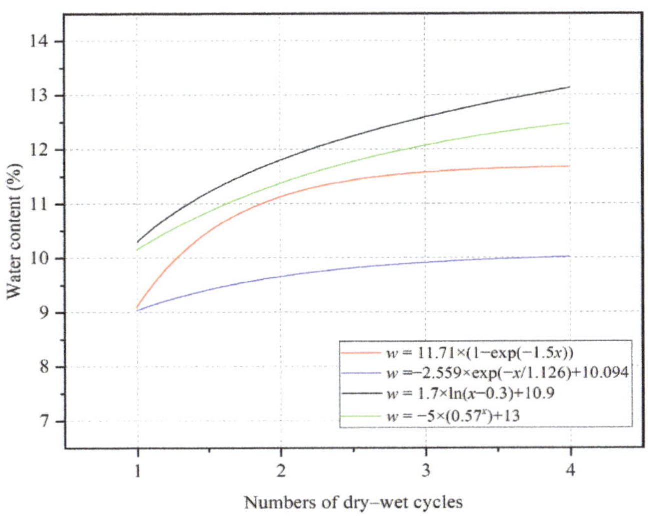

Figure 11. Relationship between coal sample water content w and dry–wet cycle numbers.

During the stable service stage, the coal pillars located at the lower part of the dam remain in a long-term water-saturated state. Coal pillar dams are susceptible to creep failure under the long-term influence of overburden load. Studying the creep fracture behavior of water-saturated coal samples can more accurately reveal the instability mechanism of coal pillars. The dissolution of coal's microstructure by water has been identified as the primary cause of coal softening, leading to the propagation of macroscopic cracks during immersion [76,77]. Under axial load, the coal pillar samples have been shown to experience instantaneous elastic strain, followed by decelerating creep deformation, with the creep rate gradually decreasing as strain increases. Once the creep rate reaches a certain level, the coal pillar samples enter the steady creep stage. Finally, the samples transition into an accelerated creep phase, characterized by a rapid increase in creep strain that follows an approximate exponential relationship with time [78,79].

In some mines where CMURs were constructed, the mine water was of the Cl·SO_4-K·Na type, with high salinity, resulting in the coal pillar dams being immersed in a high-concentration saline water environment for extended periods [47]. In these studies, the process by which high salinity ions move within the coal pillar dam, and ions' effect on the degradation of its structural integrity, was shown to be crucially important. During uniaxial compression tests, it has been shown that the peak strength and elastic modulus of the coal pillar samples exponentially decreases with increasing saline water concentration and immersion time, while the peak strain before failure increases significantly. Similar patterns have been observed during dry–wet cycling tests [47,80,81]. At the microstructural level, the quantity of mineral particles on the coal surface significantly decreases, along with an increase in the internal structural connectivity and porosity [82].

(2) Mechanical properties of coal pillar dams affected by vibrational loads

Researchers have addressed the issue of dynamic load disturbance to coal pillar dams caused by mining activities and mine-induced seismic events. It has been shown that the impact loads generated by mining disturbances and mine seismic events propagate as stress waves through the rock layers, causing damage to the coal pillar dams. Additionally, the weakening effect of water erosion has made the study of the stability of coal pillar dams under dynamic loads more complex. The mechanical properties of coal pillar dams under the influence of vibrational loads have mainly been studied using split Hopkinson pressure bar (SHPB) tests to examine the dynamic mechanical characteristics at high strain rates. Here, the dynamic compressive strength and elastic modulus followed exponential and logarithmic functions of the strain rate, respectively, showing a significant positive correlation. As the water content increased, the dynamic elastic modulus increased almost linearly, while the compressive strength gradually decreased. Under the same impact load, samples with higher water content failed more quickly, and the expansion of parallel cracks into intersecting cracks exacerbated the damage [68,83]. However, some studies indicated that the presence of water pressure had a positive effect on the samples, increasing the dynamic strength of the coal pillar samples and reducing the failure strain [84]. Figure 12 shows an innovative SHPB experimental apparatus that considers water pressure [84].

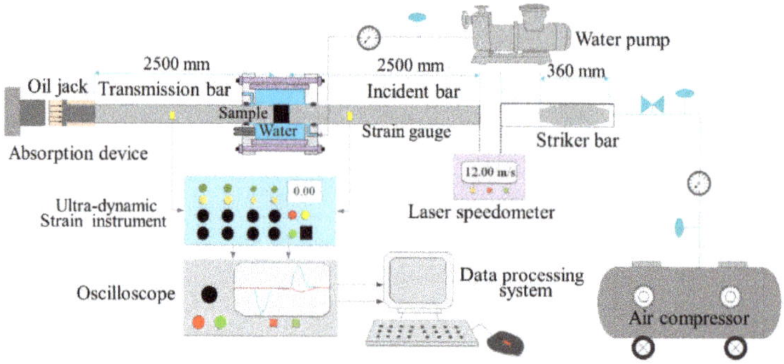

Figure 12. A coupled water static–dynamic test system (adapted from [84], with permission from Elsevier, 2024).

3.2. Three-Field Patterns of Coal Pillar Dams

Based on the findings of laboratory mechanical testing, the mechanical properties of coal pillar dam samples subjected to mine water have been observed to undergo a certain degree of weakening. Therefore, their response on an engineering scale has warranted significant attention. The coal pillar dam is situated in a typical hydraulic coupling stress environment, and the stress distribution patterns, fracture development patterns, and seepage patterns of coal pillar dams under complex stress conditions have consistently been key research issues. The study of the "three-field" patterns (stress field, fracture field, and seepage field) of coal pillar dams in CMURs has primarily employed similar simulation experiments and numerical simulation methods, although some researchers have also conducted analyses using theoretical methods. Due to the highly complex stress environment of coal pillar dams, theoretical derivations have struggled to fully reflect hydraulic coupling conditions. Typically, the elastic mechanics equation and hydraulics equations (Darcy's law, Biot equation) were coupled, often from the perspectives of permeability and hydraulic conductivity [85,86], and validated through numerical models.

Similar simulation experiments, which can directly reflect deformation and stress phenomena on an engineering scale, have been widely used to study the movement patterns of overlying rock in coal mining fields. In studies on the stability of coal pillar dams considering the water environment, some scholars have also adopted this approach. Similar simulation experiments conveniently simulate the disturbances to coal pillar dams caused by different coal seam mining activities. However, the impact of mine water is simplified to the horizontal water pressure applied to the coal pillar dam through the horizontal load of the experimental apparatus, and the hydrochemical effects of water on the coal pillar cannot be considered. Some scholars have adopted the direct water injection method to study the seepage patterns of the rock layers in the dam floor [87]. Figure 13 illustrates a similar simulation experiment system for a CMUR [86]. During the advancement of the lower working face towards the upper underground reservoir, the interlayer rock undergoes tensile and shear failure, with fractures gradually extending towards the coal pillar dam. The interlayer fracture network connectivity during multi-seam mining of underground reservoirs is progressive, differing from the sudden connectivity observed in laboratory rock mechanics experiments. During the water storage process in CMURs, as the water storage height increases, the horizontal stress borne by the coal pillar dam gradually increases. It has been shown that displacement changes do not occur before reaching the ultimate strength. However, after reaching the ultimate strength, displacement gradually increases. In these studies, repeated processes of water injection and extraction caused damage to the coal pillar dam. Even after the water level dropped (reducing horizontal stress), the water-blocking capacity of the coal pillar dam was diminished. To ensure the safety of the underground reservoir, it was considered not advisable for them to store water again [86,88].

An important cause of damage to coal pillar dams after water immersion is the softening effect of water on coal. To investigate the damage and degradation characteristics of coal pillar dams under water softening, it has been necessary to obtain real-time feedback the relationship between the mechanical parameters of the coal pillar and the water content and pore water pressure into the research model [89,90]. Therefore, many scholars have adopted numerical simulation methods [49,63,91,92]. Numerical simulation allows for the convenient, simultaneous output of the stress patterns, fracture patterns, and seepage patterns of coal pillar dams. Detailed comparative analysis of the stress distribution, pore water pressure distribution, and plastic zone expansion patterns of coal pillar dams, with and without the influence of water immersion, can intuitively reflect the impact of water immersion on coal pillar dams. In these works, the strength of the coal pillar dam decreased after water immersion, leading to a gradual increase in the plastic zone of the coal pillar. The development of the plastic zone caused stress redistribution and transfer to the deeper parts of the coal pillar. Damage and degradation allowed water to infiltrate the coal pillar dam, further reducing the coal pillar's strength. During the immersion process following mining disturbances, the coal pillar dam exhibited progressive failure, with the depth of the plastic zone development corresponding to the location of the vertical peak stress. The coal pillar exhibited a "damage–seepage–accumulated damage" effect in a water immersion-stress coupling environment. As water immersion deepened, the permeability in the elastic zone of the coal pillar decreased, the weakening effect of water immersion gradually diminished, and the damage to the coal pillar stabilized [61,75]. Figure 14 provides a detailed illustration of the stress patterns and plastic zone expansion of coal pillar dams with and without the influence of water immersion [75]. Additionally, under the influence of dynamic loads, it has been shown that the stress concentration within the coal pillar dam increases and the range of the high-stress zone expands and gradually extends towards the elastic zone and the floor. The disturbance effect of near-field mine seismic loads on the coal pillar dam is significant. The stress concentration level increases after the dynamic load ended, leading to more severe damage to the dam [68].

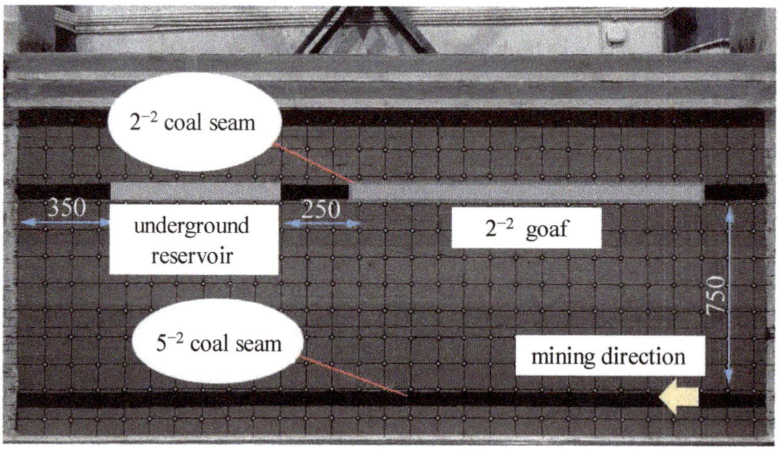

Figure 13. A similar simulation experiment system (modified from [86]).

3.3. Size Optimization of Coal Pillar Dams

Coal pillar dams in CMURs were created from sectional coal pillars, which often had an excessive width. In early coal mine designs, the conversion of goafs into underground reservoirs was not considered, resulting in sectional coal pillars with widths that were overly large for use as coal pillar dams in underground reservoirs. In current designs for new mines in central and western regions of China, the construction of CMURs has been preemptively planned. Therefore, reasonably determining the width of early sectional coal pillars, and thus the width of coal pillar dams in CMURs, is crucial for the efficient recovery of coal resources. After plastic deformation occurs on both sides of the coal pillar, there must be a sufficiently wide elastic zone in its internal area. From the perspective of the stress environment and structure, the coal pillar dam is divided into the reservoir side and the roadway side, as shown in Figure 15. Therefore, the width of the coal pillar dam is the sum of the widths of the plastic zone on the reservoir side L_{Cp}, the elastic zone L_{Ce}, and the plastic zone on the roadway side L_{Hp} [93]. Yao et al. [65] and Chen et al. [93] derived the calculation formula for the width L of coal pillar dams from the perspective of elastoplastic theory, considering the stress environment and structural characteristics of coal pillar dams. As shown in Table 1, the latter specifically provided a formula for determining the recommended width of the elastic zone. However, this derivation treated coal as a continuous and homogeneous medium, whereas in reality, coal is a type of typical sedimentary rock with distinct layered characteristics. Therefore, the calculation results have a certain degree of error. Zhang et al. [94] suggested that the ideal width of coal pillar dams could be defined by their deformation and stress state. Smaller coal pillar sizes were more prone to stress concentration, which was more likely to "erode" the coal pillar dam. A correlation was observed between the dimensions of the coal pillar dam and the extent of crack development. As the coal pillar dimensions increased, the stress concentration effect became less pronounced, and fractures were less likely to penetrate both the interior and exterior of the dam. Zhi et al. [95] identified the maximum extent of mine water infiltration into coal pillars at various water storage levels using seepage experiments.

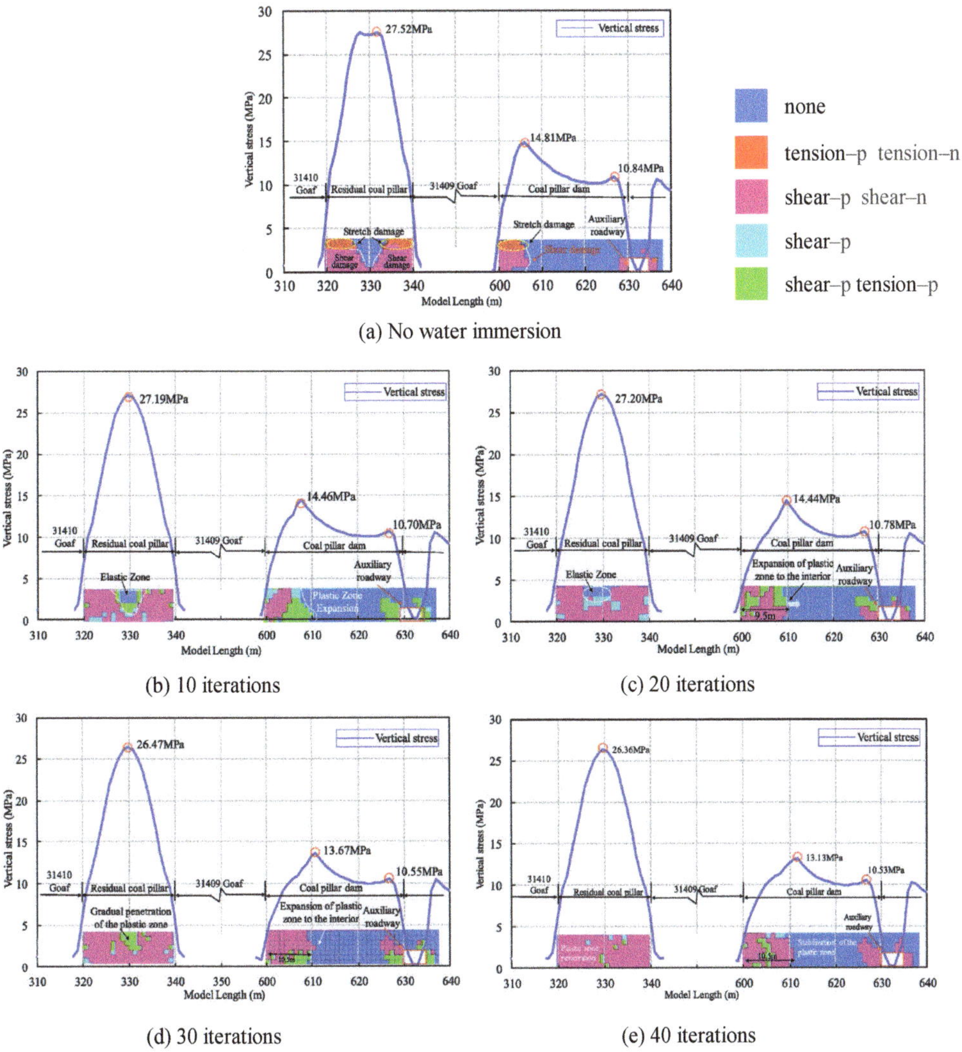

Figure 14. Distribution of vertical stress and plastic zone of coal pillar dam after water immersion weakening (adapted from [75], with permission from Springer Nature, 2024).

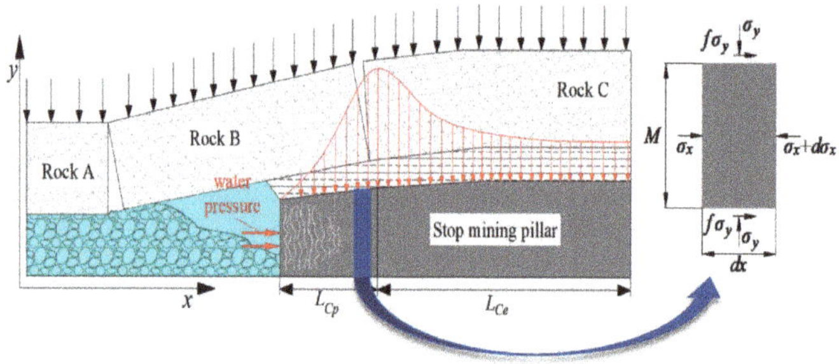

Figure 15. Stress region division of coal pillar dams (modified from [93]).

Table 1. Theoretical calculation formula for the width of coal pillar dam.

	Yao [65]	Chen [93]
L_{Cp}	$L_{Cp} = \frac{M}{2\lambda f} \ln\left[\frac{K_1\gamma H}{\lambda p+\sigma_r} - \frac{S_m S_g(1-\eta)}{2\lambda f(\lambda p+\sigma_r)}\right];\ \eta = \exp\left[\frac{2\lambda f}{S_m S_g}(\sigma_c - \sigma_r)\right]$	
L_{Hp}	$L_{Hp} = \frac{M}{2\lambda f} \ln\left\{\frac{K_2\gamma H}{\sigma r} - S_m S_g\left[1 - \exp\left(\frac{2\lambda f}{S_m S_g}(\sigma_c - \sigma_r)\right)\right]/2\lambda f \sigma_r\right\}$	
L_{Ce}	$L_{Ce} = kM$	$L_{Ce} = 0.5kM\sqrt{\frac{3p}{K_p}}$
L	$L = L_{Cp} + L_{Hp} + L_{Ce}$	

Note(s): where M is the thickness of the coal seam, m; f is the friction factor of the interface between the coal seam and the roof and floor; K_1 is the stress concentration coefficient of the reservoir side; K_2 is the stress concentration factor of the roadway side; σ_c is the uniaxial compressive strength of coal, MPa; σ_r is the residual strength of coal, MPa; S_m is the softening modulus of the coal mass, MPa; S_g is the strain gradient of the coal mass in the plastic zone; γ is the volume force; p is the water pressure, MPa; k is the safety factor, usually 2–5; K_p is the tensile strength of coal, MPa; and $\lambda = (1 + \sin\phi)/(1 - \sin\phi)$, ϕ is the effective internal friction angle of coal.

4. Key Theories and Technologies for the Stability of Artificial Dams

Similar to coal pillar dams, artificial dams in CMURs are also subjected to the coupled effects of stress and seepage. In addition to the pressure from the overlying rock layer, mining-induced stress, and water pressure, artificial dams, used to connect the spaces between adjacent sectional coal pillars, are also subjected to friction and compression forces from the coal pillars on both sides (referred to as coal pillar forces) [96,97], as shown in Figure 16 [98]. Additionally, unlike the material properties of coal, the primary construction material for artificial dams is concrete. As a typical porous material, concrete is significantly affected by water seepage. This effect is particularly notable when mine water is highly saline, as the high concentration of salt ions in the water can react with the components of the concrete, further damaging the strength of the artificial dam.

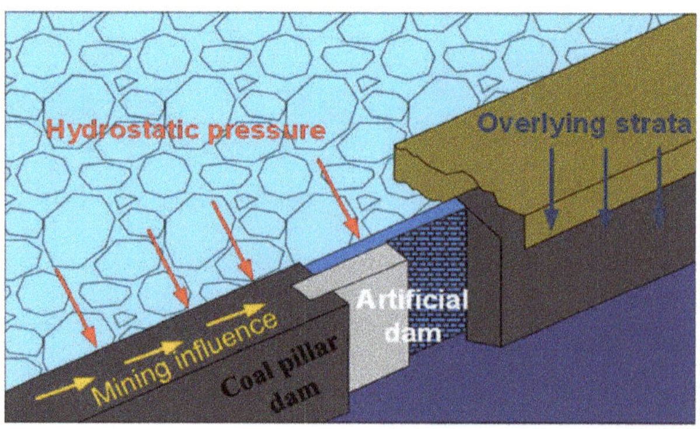

Figure 16. Schematic diagram of an artificial dam under the influence of different loads (modified from [98]).

4.1. Mechanical Properties of Artificial Dams

At the laboratory scale, the research methods for studying the mechanical properties of artificial dams have been similar to those for coal pillar dams. In addition to conventional dynamic and static mechanical tests, due to the particularity of the artificial dam material, many studies have borrowed methods from concrete durability research, focusing on the microscopic level of concrete samples. Nuclear magnetic resonance imaging (NMRI), scanning electron microscopy (SEM), CT scanning, and three-dimensional digital reconstruction have commonly been used to characterize the mechanical properties and structural changes of artificial dam samples after saline water erosion from a microscopic perspective [56,99].

(1) Mechanical properties of artificial dams affected by water immersion

During the operation stage of CMURs, the stability of artificial dams is influenced by the sustained, repeated infiltration of mine water. Similar to the coal pillar dams, the water content under repeated dry–wet cycles has also been found to exhibit a logarithmic relationship with the cycle numbers, but the increase was significantly lower. Under the action of dry–wet cycles with saline water, the anisotropic concrete artificial dam samples exhibited complex mechanical properties and damage mechanisms. This altered the macro-micro crack evolution patterns and the geometric distribution characteristics of macro cracks in the samples. The percentage of micro shear cracks in the samples changed from initially decreasing and then increasing to a continuous increasing trend, and the complexity of the crack geometric distribution significantly increased. The physicochemical changes in the artificial dam samples led to alterations in the pore structure, a reduction in density, and the connection of weak structural surfaces. These were the reasons for the decrease in uniaxial compressive strength and the changes in the macro- and micro-crack evolution patterns in the concrete. The chemical damage caused by ion exchange was the primary reason for the changes in the mechanical properties and the macro- and micro-crack evolution patterns of the concrete under dry–wet cycles [98,99]. Figure 17 clearly illustrates the cracking changes in the samples through fractal dimension analysis [99].

Wang et al. [56] and Yang et al. [100] further determined the degradation mechanism of the mechanical properties of concrete samples from artificial dams from a micro-ion chemical reaction perspective. As the concentration of salt ions increased, the concentrated sulfate solution continuously transformed portlandite into gypsum and ettringite phases. The adhesion of concrete to $Ca(OH)_2$ gradually weakened and even detached from its original stable structure, which was the main reason for the degradation of concrete's mechanical properties. Adding nanomaterials could change the failure mode, crack propagation mode, and load-bearing capacity of saturated concrete. Adding 3 wt% nano-Al_2O_3 to the concrete of artificial dams significantly increased its uniaxial compressive strength and effectively reduced crack formation [101].

(2) Mechanical properties of artificial dams affected by vibrational loads

Since the size of the artificial dam is much smaller than that of the coal pillar dam and is constructed through post-mining artificial pouring, vibrations and mine seismic activities caused by mining operations frequently occur. These uncertain loads, along with geostress and water pressure, have become common load forms [102]. The study of the seismic performance of artificial dams is identified as a crucial aspect in ensuring safety throughout its lifecycle and forms the basis for monitoring and early warning in field engineering practice [103]. Unlike the study of the mechanical properties of coal pillar dams under vibrational loads, which commonly uses SHPB tests, the smaller size of artificial dams allows for the use of similar simulation combined with shaking table tests to study the mechanical properties of artificial dams under different vibrational loads. These findings can then be validated through numerical simulation tests. Figure 18 illustrates a reasonable test system combining similar simulation and shaking table tests [53].

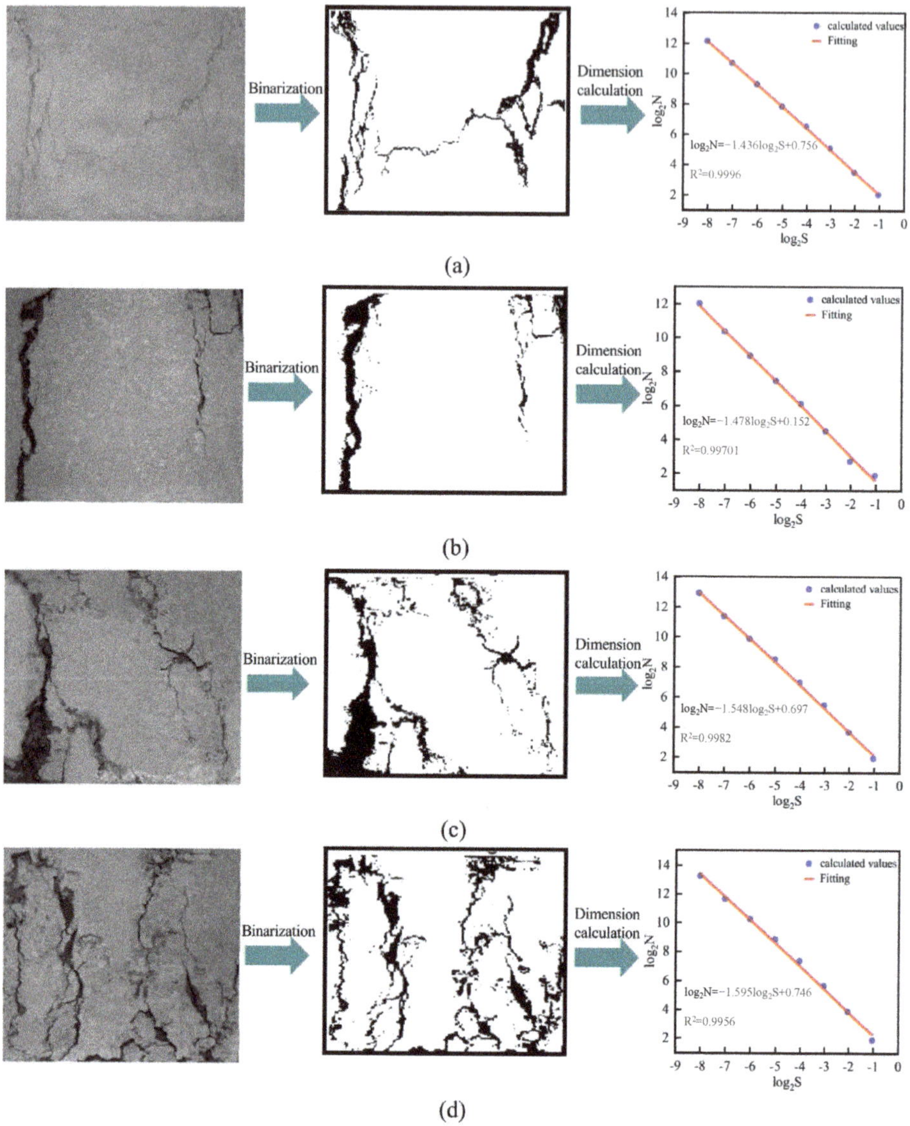

Figure 17. Macro failure fractal dimension of concrete samples: (**a**) dry, (**b**) 1 dry–wet cycle, (**c**) 3 dry–wet cycles, and (**d**) 5 dry–wet cycles (adapted from [99], with permission from Elsevier, 2024).

Chi et al. [53] conducted vibration similar simulation tests on the seismic performance of flat-type artificial dams. The results indicated that, under the combined action of vibration and external loads on the top, the bottom of the artificial dam was the stress concentration area. Under the combined effects of impact and compression from the crushed stones in the goaf, the reservoir-side stress was approximately 10% higher than the roadway-side stress. Under increased upper loads, the artificial dam exerted a "supporting" effect, leading to uneven compressive deformation of the coal pillar dam. Particularly, due to the differences in materials between the artificial dam and the coal pillar dam, the acceleration generated by vibrational loads differed. Under the same seismic wave, the displacement peak of the artificial dam was greater than that of the coal pillar dam, with the displacement of the artificial dam being approximately 1.14 times that of the coal pillar dam. The difference in velocity caused stress concentration at the connection between the coal pillar dam and the artificial dam, making it a weak point in the reservoir dam structure during an earthquake. Comparing the seismic performance of flat-type artificial dams and arch-type artificial dams under the same geological and seismic intensity conditions, the seismic safety performance of arch-type artificial dams was significantly superior to that of flat-type artificial dams. The primary reason is that the arch effect generated partial compressive stress in the reaction force of the coal pillar on the artificial dam. Part of the vibrational load on the arch-type artificial dam was transferred to the coal pillar dam, reducing the shear stress on the artificial dam caused by mine seismic events and improving the seismic performance of the arched artificial dam [104]. However, from a practical engineering perspective, since the construction of an arch-type artificial dam requires slotting on the coal pillar dam structure before pouring, the existing seepage prevention methods are less applicable to the arch-type artificial dam structure compared to the flat-type artificial dam structure.

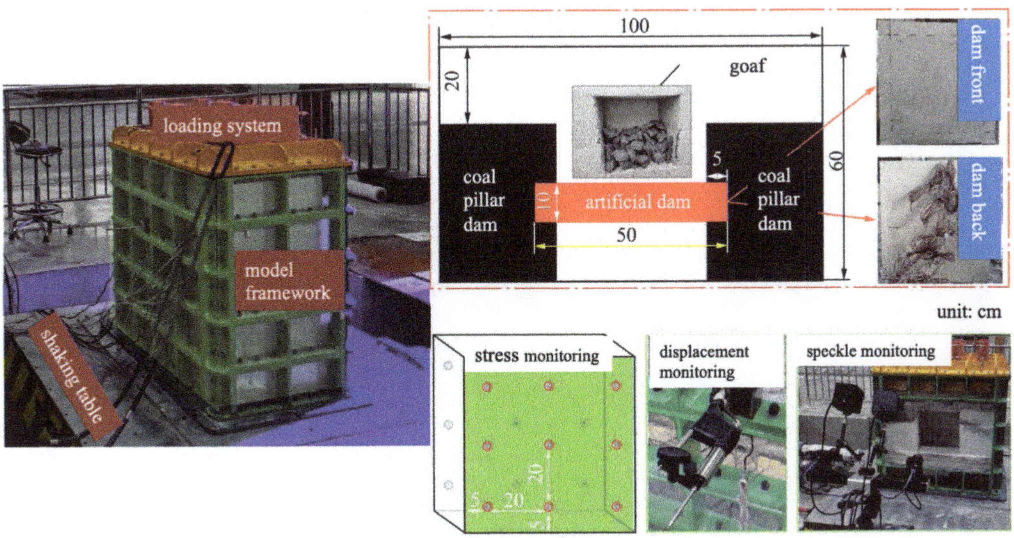

Figure 18. A test system combining a similitude simulation platform and a shaking table (modified from [53]).

4.2. Three-Field Patterns of Artificial Dams

As previously mentioned, the thickness of the coal pillar dam reaches tens of meters, whereas the thickness of the artificial dam is only a few meters. Additionally, because the artificial dam is poured at a later stage, the connections between the concrete structure and the roof, floor, and adjacent coal pillar dams are highly susceptible to leakage under water pressure and require special attention. For example, a slight leakage occurred in the MB-1 artificial dam at the Wulanmulun mine, as shown in Figure 19 [54]. The coal pillar dam, with a thickness of approximately 80 m, exhibited good impermeability, and therefore, no leakage was observed. Preliminary analysis suggested that the leakage in the artificial dam stemmed from the incomplete collapse of the goaf, which led to weak connections between the dam and the surrounding rock in the roof. Additionally, the deformation and crack propagation patterns of the artificial dam under different water pressures directly affected the seepage patterns. Therefore, for artificial dams, much of the research has focused on their seepage patterns.

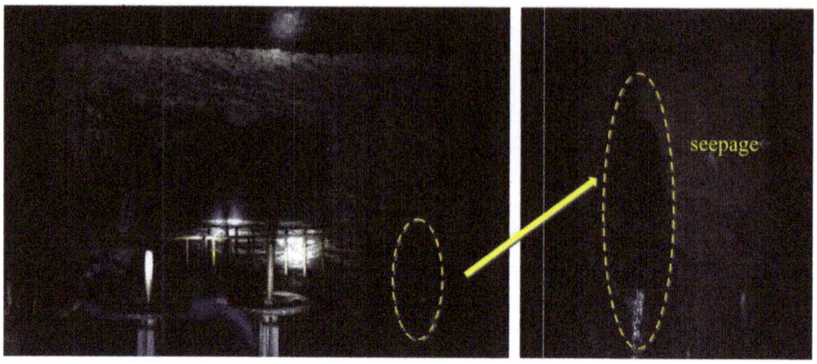

Figure 19. Seepage in the artificial dam (modified from [54]).

Kong et al. [54] and Fang [55] conducted similar experimental studies on the stress, fracture, and seepage patterns of artificial dams under the influence of water pressure and mining pressure using a self-developed nested multi-coupled test system. The test system is shown in Figure 20 [54]. The results indicated that an increase in water pressure and model top load both led to an increase in the deformation of the artificial dam surface. Under the influence of water pressure, seepage occurred sequentially at the following locations: the bottom slot of the artificial dam, the top slot of the artificial dam, the coal pillar dam, the reservoir floor, the reservoir roof, the roadway roof, the roadway floor, and the surface of the artificial dam. As the water pressure increased, the permeation boundary range on the model surface gradually expanded. The failure of the artificial dam was characterized by compression-shear failure, with cracks extending from the lower outer side of the artificial dam model to near the upper inner vertex, penetrating the entire artificial dam at an approximate 45° angle to the cross-section, as shown in Figure 21.

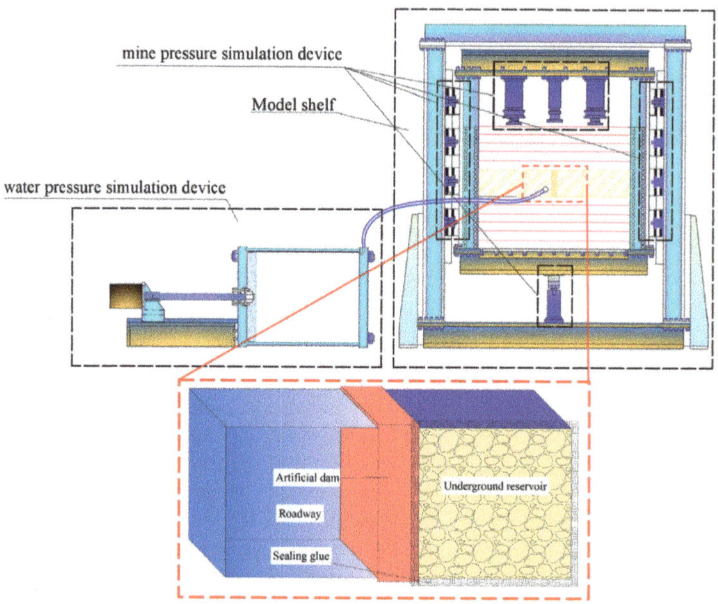

Figure 20. A nested multi-coupled test system for CMUR (modified from [54]).

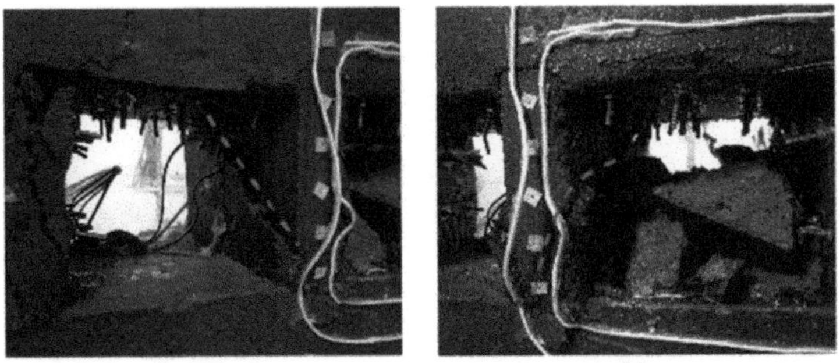

(a) (b)

Figure 21. Failure characteristics of the artificial dam (modified from [54]): (**a**) near the roadway and (**b**) near the reservoir.

After the CMUR is filled with water, the original failure zone and plastic zone of the coal pillar dam expand under long-term water storage and repeated immersion. At this time, the original bearing pressure environment changes, and both the magnitude and location of the bearing pressure peak points change. The artificial dam located in the middle of the coal pillar dam acts as a local support body and cannot deform in coordination with the surrounding rock. The maximum displacement on the outer surface of the artificial dam has been shown to have a clear linear correlation with the water storage height under different roof and floor pressures [105]. Ma et al. [106] further studied the displacement and stress patterns of artificial dams at different water storage heights. The stress distribution caused by different water storage heights was basically the same, but the stress increased with the increase in water storage height. The strength of the artificial dam played a critical role.

4.3. Size Optimization of Artificial Dams

Unlike the primary goal of optimizing coal pillar dam size, which is to maximize the recovery of coal resources, the main purpose of optimizing the size of artificial dams is to enhance their impermeability and improve the stability of the slot structures between the artificial dam and the adjacent coal pillar dams. However, some studies, through theoretical derivation of mechanical models for specific CMUR cases, suggested that the vertical support wall of "T"-shaped artificial dams could be removed, and the dam wall thickness could be reduced to within the range of 0.6–0.7 m. This would save materials and improve the economic efficiency of dam construction [57]. The connection between the artificial dam and the coal pillar dam is a vulnerable point in the dam structure of a CMUR. Currently, slotting is used as the connection method between the coal pillar dam and the artificial dam. When lateral loads, such as water pressure, are applied to the artificial dam, they are transmitted from the artificial dam to the coal pillar dam at the slot position. The slot depth dictates the stress conveyed to the coal pillar, influencing the stability of the entire dam structure. Because of mining disturbances, a certain plastic zone had already formed in the adjacent sections of the coal pillar dam prior to the artificial dam's construction. These plastic zones undoubtedly provided pathways for water seepage, so slotting operations should avoid the plastic zone as much as possible [52]. Kong et al. [107,108] comprehensively analyzed the stress, displacement, plastic zone distribution, and seepage patterns of the artificial dam and coal pillar dam at the slot connection through similar simulation and numerical simulation. They concluded that increasing the slot depth within a certain range could improve the stability of the underground reservoir dam. However, when the slot depth exceeded a critical value, further increases in slot depth were detrimental to the reservoir's stability. They recommended a slot depth of 0.6 m. In addition to flat, "T"-shaped, and arch-shaped forms, artificial dams also come in wedge and inverted cone shapes, among other forms. However, there has been little research on specially shaped artificial dams. Zhao [109] used numerical simulation to demonstrate that inverted truncated cone-shaped dams had better stress conditions and smaller displacements when subjected to longitudinal and lateral stresses. These dams could achieve uniform pressure distribution on the coal pillar under high water pressure, thereby reducing the deformation of the coal pillar. Compared to flat-type artificial dams, wedge-shaped artificial dams also have certain advantages in terms of stress distribution and displacement deformation. Flat-type artificial dams are more suitable for use under smaller water pressures.

5. Conclusions and Prospects

With the depletion of coal resources in the eastern mining areas of China, large-scale coal mining has shifted to the central and western regions. The intensity of coal mining in these regions is substantial (>10 million t/a), which starkly contrasts with their characteristics of water scarcity and high surface evaporation. Government departments and coal companies have gradually recognized the importance of water-preserving coal mining. During the coal mining process, it is necessary not only to protect mine water from being damaged and wasted but also to make full use of mine water [7]. Since the completion of the first goaf water storage facility at the Daliuta Coal Mine in 1998, nearly 20 years of engineering practice have led to the development of CMUR water storage technology, which utilizes goafs from longwall mining as water storage spaces. This established the concept of storing mine water underground in coal mines [16]. According to statistics, more than 35 CMURs have been constructed in the central and western regions of China [1]. The total number of constructed, under-construction, and planned CMURs will exceed 50 [11]. A CMUR technology demonstration zone will be established in the Inner Mongolia–Shaanxi border area. Based on the development history of CMUR construction in China, this paper systematically reviewed the research progress and achievements in the foundational theories of the stability of coal pillar dams and artificial dams in CMURs. This includes the mechanical properties under hydraulic coupling, the evolution mechanisms and impact relationships of the "three fields" (stress field, fracture field, and seepage field), and the optimized design of dam dimensions. Clarifying the instability and failure patterns of CMUR dam structures is essential to maintaining dam structure stability. However, engineering practice in CMUR technology has outpaced technical research, and technology has advanced beyond theoretical exploration. To further apply this technology for the protection and utilization of mine water resources under different geological and working conditions, increased technical exploration and theoretical research are necessary. Based on a systematic review of the stability research of CMUR dam structures in China, the main conclusions and outlook are as follows:

(1) Reservoir site selection and storage capacity calculation are prerequisites for the construction of CMURs. The water storage space in CMURs consists of the voids in the caved rock masses of the goaf, with the core element being the determination of the water storage coefficient. Further theoretical and predictive research on the development patterns of overburden fracture zones and groundwater migration mechanisms during coal seam mining should be conducted. A systematic understanding of the spatial and temporal evolution patterns of void spaces in caved rock masses within the goaf, as well as the seepage paths and migration patterns of mine water in the goaf and overburden, is necessary.

(2) As core components of the dam structure in CMURs, coal pillar dams and artificial dams should address key issues during the design, construction, and operation stages. These include understanding the mechanical and hydrochemical mechanisms of coal pillar dams and artificial dams under complex hydraulic coupling stress conditions, such as overlying rock load, mining-induced load, and mine water pressure. It is essential to investigate the micro- to macro-scale failure, seepage, and instability mechanisms. Special attention should be given to the creep effects of coal pillar dams and the durability of artificial dams under long-term exposure to corrosive mine water.

Additionally, the issue of steel reinforcement corrosion in artificial dams due to the influence of corrosive ions such as chloride and sulfate ions needs to be addressed.

(3) Currently, the constructed CMURs are all near-horizontal reservoirs, with coal seams deposited at near-horizontal angles (0°–8°). There has been limited application and research under gently inclined coal seam conditions (8°–25°), with only the Lingxin Coal Mine having applications in gently inclined coal seams (10°–15°). At present, there are no engineering cases of CMURs constructed under complex steeply inclined coal seam conditions. The applicability and systematic technical standards for steeply inclined coal seams remain to be studied.

(4) In eastern mining areas, due to resource depletion and capacity withdrawal, abandoned mines are increasing year by year. However, the water resources in these abandoned mines have not been effectively utilized. For abandoned mines, the impact of future coal mining on the mine reservoir does not need to be considered. Therefore, it is of great significance to fully utilize the underground space of coal mines for the construction of CMURs. The underground spaces of abandoned mines should be fully utilized to explore the application of CMUR technology, thereby achieving full utilization of mine water resources in abandoned mines.

Author Contributions: Conceptualization, Y.W. and F.L.; Methodology, Y.W., F.L. and M.K.; Investigation, M.K. and M.L.; Writing—original draft, Y.W.; Writing—review & editing, F.L., M.K. and M.L.; Supervision, F.L.; Funding acquisition, Y.W. All authors have read and agreed to the published version of the manuscript.

Funding: This work is supported by Open Fund of the State Key Laboratory of Water Resource Protection and Utilization in Coal Mining (Grant No. GJNY-21-41-04), Shandong Provincial Natural Science Foundation (Grant No. ZR2023QE136 and ZR2022QD102) and Chunhui Program of Ministry of Education (Grant No. 202201798).

Data Availability Statement: Dataset available on request from the authors.

Conflicts of Interest: The authors declare no conflicts of interest.

References

1. Zhang, C.; Wang, F.; Bai, Q. Underground space utilization of coalmines in China: A review of underground water reservoir construction. *Tunn. Undergr. Space Technol.* **2021**, *107*, 103657. [CrossRef]
2. The State Council Information Office of the People's Republic of China. Energy in China's New Era. Available online: https://english.mee.gov.cn/Resources/publications/Whitep/202012/t20201222_814160.shtml (accessed on 3 May 2024).
3. National Bureau of Statistics of China. National Data of Total Energy Consumption. Available online: https://data.stats.gov.cn/english/easyquery.htm?cn=C01 (accessed on 3 May 2024).
4. National Bureau of Statistics of China. Statistical Communiqué of The People's Republic of China on the 2023 National Economic and Social Development. Available online: https://www.stats.gov.cn/english/PressRelease/202402/t20240228_1947918.html (accessed on 3 May 2024).
5. Yuan, L. Chalenges and countermeasures for high quality development of China's coal industry. *China Coal* **2020**, *46*, 6–12. (In Chinese)
6. Gu, D.Z.; Cao, Z.G.; Li, J.F.; Wu, B.Y.; Zhang, Y.; Jiang, B.B.; Guo, Q.; Wang, H.P.; Wu, Y.; Shi, X.M.; et al. Original Experimental Platform System and Application of Underground Coal Mine Reservoirs. *J. China Coal Soc.* **2024**, *49*, 100–113. (In Chinese)
7. Gu, D.Z.; Li, J.F.; Cao, Z.G.; Wu, B.Y.; Jiang, B.B.; Yang, Y.; Yang, J.; Chen, Y.P. Technology and engineering development strategy of water protection and utilization of coal mine in China. *J. China Coal Soc.* **2021**, *46*, 3079–3089. (In Chinese)
8. Gu, D.Z.; Li, T.; Li, J.F.; Guo, Q.; Jiang, B.B.; Bian, W.; Bao, Y.X. Current status and prospects of coal mine water treatment technology in China. *Coal Sci. Technol.* **2021**, *49*, 11–18. (In Chinese)
9. Gu, D.Z. Water resource and surface ecology protection technology of modern coal mining in China's energy "Golden Triangle". *Strateg. Study CAE* **2013**, *15*, 102–107. (In Chinese)
10. Peng, S.P.; Bi, Y.L. Strategic consideration and core technology about environmental ecological restoration in coal mine areas in the Yellow River basin of China. *J. China Coal Soc.* **2020**, *45*, 1211–1221. (In Chinese)
11. Shi, X.C. Research progress and prospect of underground mines in coal mines. *Coal Sci. Technol.* **2022**, *50*, 216–225. (In Chinese)
12. Meng, Z.P.; Shi, X.C.; Li, G.Q. Deformation, failure and permeability of coal-bearing strata during longwall mining. *Eng. Geol.* **2016**, *208*, 69–80. [CrossRef]
13. Qian, M.G.; Miu, X.X.; Xu, J.L. Green mining of coal resources harmonizing with environment. *J. China Coal Soc.* **2007**, *32*, 1–7. (In Chinese)

14. Qian, M.G.; Miu, X.X.; Xu, J.L. Resources and Environment Harmonics (Green) Mining and Its Technological System. *J. Min. Sa Eng.* **2006**, *23*, 1–5. (In Chinese)
15. Miu, X.X.; Wang, A.; Sun, Y.J.; Wang, L.G.; Pu, H. Research on Basic Theory of Mining with Water Resources Protection and Its Application to Arid and Semi-Arid Mining Areas. *Chin. J. Rock. Mech. Eng.* **2009**, *28*, 217–227. (In Chinese)
16. Chen, S.S.; Huang, Q.X.; Xue, G.; Li, R.Q. Technology of underground reservoir construction and water resource utilization in Daliuta Coal Mine. *Coal Sci. Technol.* **2016**, *44*, 21–28. (In Chinese)
17. Cairney, T. Utilisation of disused coal mines as water storage reservoirs. *J. Hydrol.* **1973**, *19*, 251–258. [CrossRef]
18. Gu, D.Z.; Zhang, Y.; Cao, Z.G. Technical progress of water resource protection and utilization by coal mining in China. *Coal Sci. Technol.* **2016**, *44*, 1–7. (In Chinese)
19. Gu, D.Z. Theory framework and technological system of coal mine underground reservoir. *J. China Coal Soc.* **2015**, *40*, 239–246 (In Chinese)
20. Gu, D.Z.; Yan, Y.G.; Zhang, Y.; Wang, E.Z.; Cao, Z.G. Experimental study and numerical simulation for dynamic response of coal pillars in coal mine underground reservoir. *J. China Coal Soc.* **2016**, *41*, 1589–1597. (In Chinese)
21. Kitsikoudis, V.; Archambeau, P.; Dewals, B.; Pujades, E.; Orban, P.; Dassargues, A.; Pirotton, M.; Erpicum, S. Underground Pumped-Storage Hydropower (UPSH) at the Martelange Mine (Belgium): Underground Reservoir Hydraulics. *Energies* **2020**, *13* 16. [CrossRef]
22. Kranz, K.; Dillenardt, J. Mine Water Utilization for Geothermal Purposes in Freiberg, Germany: Determination of Hydrogeological and Thermophysical Rock Parameters. *Mine Water Environ.* **2010**, *29*, 68–76. [CrossRef]
23. Renz, A.; Rühaak, W.; Schätzl, P.; Diersch, H.J.G. Numerical Modeling of Geothermal Use of Mine Water: Challenges and Examples. *Mine Water Environ.* **2009**, *28*, 2–14. [CrossRef]
24. Raymond, J.; Therrien, R. Low-temperature geothermal potential of the flooded Gaspe Mines, Quebec, Canada. *Geothermics* **2008** *37*, 189–210. [CrossRef]
25. Ordóñez, A.; Jardón, S.; Alvarez, R.; Andrés, C.; Pendás, F. Hydrogeological definition and applicability of abandoned coal mines as water reservoirs. *J. Environ. Monit.* **2012**, *14*, 2127–2136. [CrossRef] [PubMed]
26. Watzlaf, G.R.; Ackman, T.E. Underground Mine Water for Heating and Cooling using Geothermal Heat Pump Systems. *Mine Water Environ.* **2006**, *25*, 1–14. [CrossRef]
27. Wang, W.A.; Yao, Q.L.; Xu, Q.; Chen, X.Y.; Liu, H.Y.; Li, X.H. Experimental Study on the Evolution Law of Coal Mine Underground Reservoir Water Storage Space under the Disturbance and Water-Rock Interaction Effect. *Minerals* **2022**, *12*, 21. [CrossRef]
28. Menéndez, J.; Loredo, J.; Galdo, M.; Fernández-Oro, J.M. Energy storage in underground coal mines in NW Spain: Assessment of an underground lower water reservoir and preliminary energy balance. *Renew. Energy* **2019**, *134*, 1381–1391. [CrossRef]
29. Xie, H.; Zhao, J.W.; Zhou, H.W.; Ren, S.H.; Zhang, R.X. Secondary utilizations and perspectives of mined underground space. *Tunn. Undergr. Space Technol.* **2020**, *96*, 14. [CrossRef]
30. Pujades, E.; Orban, P.; Bodeux, S.; Archambeau, P.; Erpicum, S.; Dassargues, A. Underground pumped storage hydropower plants using open pit mines: How do groundwater exchanges influence the efficiency? *Appl. Energy* **2017**, *190*, 135–146. [CrossRef]
31. Winde, F.; Kaiser, F.; Erasmus, E. Exploring the use of deep level gold mines in South Africa for underground pumped hydroelectric energy storage schemes. *Renew. Sust. Energ. Rev.* **2017**, *78*, 668–682. [CrossRef]
32. Jiang, D.Y.; Chen, S.; Liu, W.H.; Ren, Y.W.; Guo, P.Y.; Li, Z.Z. Underground Hydro-Pumped Energy Storage Using Coal Mine Goafs: System Performance Analysis and a Case Study for China. *Front. Earth Sci.* **2021**, *9*, 15. [CrossRef]
33. Chen, Y.M.; Zhang, Y.; Xia, F.; Xing, Z.; Wang, L.C. Impacts of Underground Reservoir Site Selection and Water Storage on the Groundwater Flow System in a Mining Area-A Case Study of Daliuta Mine. *Water* **2022**, *14*, 19. [CrossRef]
34. Zhang, H.; Li, G.Q.; Xing, Z.G.; Wang, L.; Zou, W.K. A Geological Site Selection Method of a Coal Mine Underground Reservoir and Its Application. *Water* **2023**, *15*, 21. [CrossRef]
35. Guo, Y.N.; Li, G.Q.; Wang, L.; Zhang, Z. Hydrochemical Characteristics of Mine Water and Their Significance for the Site Selection of an Underground Reservoir in the Shendong Coal Mining Area. *Water* **2023**, *15*, 20. [CrossRef]
36. Ju, J.F.; Xu, J.L.; Zhu, W.B. Storage capacity of underground reservoir in the Chinese western water-short coalfield. *J. China Coal Soc.* **2017**, *42*, 381–387. (In Chinese)
37. Fan, K.F.; Xie, P.; Liu, Z.C. Analytical method of reservoir capacity of underground reservoir in coal mine: A case study. *Water Supply* **2023**, *23*, 581–598. [CrossRef]
38. Song, H.Q.; Xu, J.J.; Fang, J.; Cao, Z.G.; Yang, L.Z.; Li, T.X. Potential for mine water disposal in coal seam goaf: Investigation of storage coefficients in the Shendong mining area. *J. Clean. Prod.* **2020**, *244*, 15. [CrossRef]
39. Fang, J.; Song, H.Q.; Xu, J.J.; Yang, L.Z.; Li, Z.Y. Storage coefficient calculation model of coal mine underground reservoir considering effect of effective stress. *J. China Coal Soc.* **2019**, *44*, 3750–3759. (In Chinese)
40. Pang, Y.H.; Li, Q.S.; Cao, G.M.; Zhou, B.J. Analysis and calculation method of underground reservoir water storage space composition. *J. China Coal Soc.* **2019**, *44*, 557–566. (In Chinese)
41. Wen, Z.J.; Jiang, P.F.; Song, Z.Q.; Jiang, Y.J.; Wen, J.H.; Jing, S.L. Structural model and capacity determination of underground reservoir in goaf: A case study of Shendong mining area in China. *Geomech. Geophys. Geo-Energy Geo-Resour.* **2023**, *9*, 18 [CrossRef]
42. Zhang, C.H.; Luo, B.; Xu, Z.M.; Sun, Y.J.; Feng, L. Research on the Capacity of Underground Reservoirs in Coal Mines to Protect the Groundwater Resources: A Case of Zhangshuanglou Coal Mine in Xuzhou, China. *Water* **2023**, *15*, 14. [CrossRef]

43. Wang, B.F.; Zhou, D.; Jiang, J.Q.; Liang, B. Storage coefficient calculation model of coal mine underground reservoir considering the effects of coal mine water velocity and time. *J. Min. Saf. Eng.* **2019**, 1–12. (In Chinese)
44. Qin, X.; Cao, Z.G.; Wei, L.C.; Li, P.; Sun, H. Spatial Structure Characteristics of Underground Reservoir Water Storage Space in Coal Mines Considering Shape Characteristics of Crushed Rock. *Processes* **2023**, *11*, 17. [CrossRef]
45. Li, Y.M. Research on Reservoir Dam Structure Stability and Anti-seepage Technology in Mined-out Area. Master's Thesis, Inner Mongolia University of Science & Technology, Baotou, China, 2020. (In Chinese)
46. Wu, B.Y. Reasonable Arrangement Mode About Coal Pillar Dam of Distributed Groundwater Reservoir in Coal Mine. *Saf. Coal Mines* **2018**, *49*, 68–72+78. (In Chinese)
47. Wang, M. Study on Damage Mechanism and Time-Dependent Stability of Coal Pillar in Coal Mine Underground Reservoir. Ph.D. Thesis, University of Science and Technology Beijing, Beijing, China, 2022. (In Chinese)
48. Xie, H.X.; Yao, Q.L.; Yu, L.Q.; Shan, C.H. Study on Damage Characteristics of Water-Bearing Coal Samples under Cyclic Loading-Unloading. *Sustainability* **2022**, *14*, 16. [CrossRef]
49. Liu, X.S.; Song, S.L.; Wu, B.Y.; Li, X.B.; Yang, K. Study on Deformation and Fracture Evolution of Underground Reservoir Coal Pillar Dam under Different Mining Conditions. *Geofluids* **2022**, *2022*, 11.
50. He, Q.Y.; Li, Y.C.; Li, D.Q.; Zhang, C.G. Microcrack fracturing of coal specimens under quasi-static combined compression-shear loading. *J. Rock. Mech. Geotech. Eng.* **2020**, *12*, 1014–1026. [CrossRef]
51. Zhang, C.; Wang, X.J.; Han, P.H.; Bai, Q.S.; Chen, J.H. Failure analysis of residual coal pillar under the coupling of mining stress and water immersion in the goaf underground water reservoir. *Environ. Earth Sci.* **2023**, *82*, 19. [CrossRef]
52. Zhi, G.J.; Liu, R.; Yang, R.G.; Cao, Z.G.; Ning, S.; Zhang, C.P.; Ju, J.F. Research on Optimization of Cutting Parameters for Structuring Artificial Dam around Underground Reservoir in Coal Mine. *Coal Technol.* **2022**, *41*, 18–22. (In Chinese)
53. Chi, M.B.; Li, P.; Cao, Z.G.; Wu, Y.; Zhang, Y.; Li, H.X.; Wu, B.Y. Seismic performance analysis of flat artificial dam of underground reservoir in coal mine. *J. China Coal Soc.* **2023**, *48*, 1179–1191. (In Chinese)
54. Kong, X.S.; Huang, J.W.; Shan, R.L.; Yang, W.; Fang, J. Elaboration and application of the nested multi-coupling model test system for an artificial dam seepage simulation under combined action of dynamic and static loads. *Geomech. Geophys. Geo-Energy Geo-Resour.* **2024**, *10*, 16. [CrossRef]
55. Fang, J. Study on Deformation Failure Mechanism and Stability Evaluation of Artificial Dam of Underground Reservoir in Coal Mine. Ph.D. Thesis, China University of Mining & Technology-Beijing, Beijing, China, 2020. (In Chinese)
56. Wang, Z.B.; Sun, P.; Zuo, J.P.; Liu, C.; Han, Y.D.; Zhang, Z.S. Long-term properties and microstructure change of engineered cementitious composites subjected to high sulfate coal mine water in drying-wetting cycles. *Mater. Des.* **2021**, *203*, 13. [CrossRef]
57. Fang, Z.Y.; Ju, J.F.; Cao, Z.G.; Wang, X.Z.; Xu, J.L. Structural optimization of artificial dam body in coal mine underground reservoir based on orthotropic plate model. *J. China Coal Soc.* **2020**, *45*, 1375–1384. (In Chinese)
58. Xu, Q.; Yao, Q.L.; Yu, L.Q.; Zhu, L.; Chong, Z.H.; Li, Y.H.; Li, X.H. Mechanical properties and damage constitutive model of coal specimen with different moisture under uniaxial loading. *Int. J. Damage Mech.* **2023**, *32*, 462–481. [CrossRef]
59. Yu, L.Q.; Yao, Q.L.; Chong, Z.H.; Li, Y.H.; Xu, Q.; Xie, H.X.; Ye, P.Y. Mechanical and micro-structural damage mechanisms of coal samples treated with dry-wet cycles. *Eng. Geol.* **2022**, *304*, 14. [CrossRef]
60. Chen, L.; Zhang, D.S.; Yao, N.; Wang, L.; Fan, G.W.; Wang, X.F.; Zhang, W. Coupling Influence of Inclination Angle and Moisture Content on Mechanical Properties and Microcrack Fracture of Coal Specimens. *Lithosphere* **2021**, *2021*, 23. [CrossRef]
61. Wang, F.T.; Liang, N.N.; Li, G. Damage and Failure Evolution Mechanism for Coal Pillar Dams Affected by Water Immersion in Underground Reservoirs. *Geofluids* **2019**, *2019*, 2985691. [CrossRef]
62. Tang, C.J.; Yao, Q.L.; Xu, Q.; Shan, C.H.; Xu, J.M.; Han, H.; Guo, H.T. Mechanical Failure Modes and Fractal Characteristics of Coal Samples under Repeated Drying-Saturation Conditions. *Nat. Resour. Res.* **2021**, *30*, 4439–4456. [CrossRef]
63. Yang, P. Study on Strength Damage Mechanism of Coal Pillar Dam in Coal Mine Groundwater Reservoir. Master's Thesis, China University of Mining and Technology, Xuzhou, China, 2019. (In Chinese)
64. Wang, B.F.; Zhou, D.; Zhang, J.; Liang, B. Study on the strength deterioration characteristic and damage model of coal pillar dams with repeated water immersion in underground reservoirs. *Sci. Rep.* **2024**, *14*, 19. [CrossRef] [PubMed]
65. Yao, Q.L.; Hao, Q.; Chen, X.Y.; Zhou, B.J.; Fang, J. Design on the width of coal pillar dam in coal mine groundwater reservoir. *J. China Coal Soc.* **2019**, *44*, 891–899. (In Chinese)
66. Wu, Y.; Yao, Q.L.; Wu, B.Y.; Xie, H.X.; Yu, L.Q.; Li, Y.H.; Wang, L.J. Strength Damage and Acoustic Emission Characteristics of Water-Bearing Coal Pillar Dam Samples from Shangwan Mine, China. *Energies* **2023**, *16*, 20. [CrossRef]
67. Tang, C.J.; Yao, Q.L.; Chen, T.; Shan, C.H.; Li, J. Effects of water content on mechanical failure behaviors of coal samples. *Geomech. Geophys. Geo-Energy Geo-Resour.* **2022**, *8*, 17. [CrossRef]
68. Chen, N. Study on Damage Characteristics of Coal Pillar Dam Body of Groundwater Reservoir under Dynamic and Static Load. Master's Thesis, China University of Mining and Technology, Xuzhou, China, 2020. (In Chinese)
69. Li, J.H. Experimental Study of Water Storage Soaking of Coal Mine Underground Reservoir to Coal Pillar Dam Body Strength. *Coal Min. Technol.* **2018**, *23*, 15–17+19. (In Chinese)
70. Yao, Q.L.; Chen, T.; Tang, C.J.; Sedighi, M.; Wang, S.W.; Huang, Q.X. Influence of moisture on crack propagation in coal and its failure modes. *Eng. Geol.* **2019**, *258*, 12. [CrossRef]
71. Yao, Q.L.; Zheng, C.K.; Tang, C.J.; Xu, Q.; Chong, Z.H.; Li, X.H. Experimental Investigation of the Mechanical Failure Behavior of Coal Specimens with Water Intrusion. *Front. Earth Sci.* **2020**, *7*, 13. [CrossRef]

72. Yao, Q.L.; Zheng, C.K.; Shang, X.B.; Yan, L.; Shan, C.H.; Wang, W. Evolution of Mechanical and Acoustic Emission Characteristics of Coal Samples under Different Immersion Heights. *Nat. Resour. Res.* **2023**, *32*, 2273–2288. [CrossRef]
73. Wang, L.; Wu, P.; Li, M.; Mao, X.B.; Chen, L. Mechanical Properties and Acoustic Emission Characteristics of Water-Bearing Coal Specimens under a Coupled Compression-Shear Load. *Minerals* **2022**, *12*, 22. [CrossRef]
74. Yao, Q.L.; Tang, C.J.; Xia, Z.; Liu, X.L.; Zhu, L.; Chong, Z.H.; Hui, X.D. Mechanisms of failure in coal samples from underground water reservoir. *Eng. Geol.* **2020**, *267*, 10. [CrossRef]
75. Wang, F.T.; Wei, X.Q.; Shao, D.L.; Zhang, C. The progressive failure mechanism for coal pillars under the coupling of mining stress and water immersion in underground reservoirs. *Bull. Eng. Geol. Environ.* **2023**, *82*, 13. [CrossRef]
76. Guo, J.; Zhang, T.; Pan, H.; Wu, J. Experimental investigation of the creep damage evolution of coal rock around gas extraction boreholes at different water contents. *PLoS ONE* **2023**, *18*, 18. [CrossRef] [PubMed]
77. Yu, L.Q.; Yao, Q.L.; Chong, Z.H.; Li, Y.H.; Xu, Q.; Liu, Z.C. Experimental study on the moisture migration and triaxial mechanical damage mechanisms of water-bearing coal samples. *Int. J. Rock Mech. Min. Sci.* **2022**, *160*, 13. [CrossRef]
78. Li, G.; Wanyan, Q.; Li, Z.S.; Yi, H.Y.; Ren, F.F.; Chen, Z.; Liu, Y. A Fractional-Order Creep Model of Water-Immersed Coal. *Appl. Sci.* **2023**, *13*, 16. [CrossRef]
79. Zhao, Z.; Wu, P.; Wang, L.; Li, M.; Mao, X.B.; Chen, L. Influence of Moisture Content on Creep Mechanical Characteristic and Mic-Fracture Behavior of Water-Bearing Coal Specimen. *Geofluids* **2022**, *2022*, 17. [CrossRef]
80. Wang, M.; Guo, Q.F.; Tian, Y.K.; Dai, B. Physical and Mechanical Properties Evolution of Coal Subjected to Salty Solution and a Damage Constitutive Model under Uniaxial Compression. *Mathematics* **2021**, *9*, 19. [CrossRef]
81. Wang, M.; Xi, X.; Guo, Q.F. Deterioration Mechanism and Constitutive Model of Coal Subjected to Drying-Wetting Cycles of Salty Mine Water. *Lithosphere* **2022**, *2022*, 12. [CrossRef]
82. Shan, C.H.; Yao, Q.L.; Cao, S.G.; Xie, H.X.; Xu, Q.; Zheng, C.K.; Chen, X.Y. Measurement of fracture development evolution of coal samples under acid-alkaline by three-dimensional reconstruction and AE time-frequency characteristic analysis. *Measurement* **2023**, *217*, 23. [CrossRef]
83. Yao, Q.L.; Yu, L.Q.; Chen, N.; Wang, W.N.; Xu, Q. Experimental Study on Damage and Failure of Coal-Pillar Dams in Coal Mine Underground Reservoir under Dynamic Load. *Geofluids* **2021**, *2021*, 18. [CrossRef]
84. Wang, K.; Feng, G.R.; Bai, J.W.; Guo, J.; Shi, X.D.; Cui, B.Q.; Song, C. Dynamic behaviour and failure mechanism of coal subjected to coupled water-static-dynamic loads. *Soil Dyn. Earthq. Eng.* **2022**, *153*, 13. [CrossRef]
85. Bai, L.W.; Wu, D.; Tang, Z.X.; Wang, Y.; ZHao, R.K. Study on the Stability of Coal Mine Underground Reservoir Dam under the Coupling Action of Stress and Seepage. *China Mine Eng.* **2021**, *50*, 1–6. (In Chinese)
86. Wang, L.J.; Cao, Z.G.; Cheng, J.C.; Liu, S.G.; Zhou, H.W.; Wu, Y.; Wang, J.G.; Ouyang, D.; Xue, D.J. Failure analysis of rock strata between upper and lower coals under underground reservoir in coal mine and its critical percolation model of jumping permeability. *J. China Coal Soc.* **2023**, *48*, 1192–1208. (In Chinese)
87. Wen, Z.J.; Jiang, P.F.; Jing, S.L.; Cao, Z.G.; Guan, Y.T. Development and verification of simulation testing system for floor seepage in coal mine underground reservoir. *J. China Coal Soc.* **2021**, *46*, 1487–1497. (In Chinese)
88. Chi, M.B.; Wu, B.Y.; Cao, Z.G.; Li, P.; Liu, X.Q.; Li, H.X.; Zhang, Y.; Yang, Y. Research on Instability Mechanism and Precursory Information of Coal Pillar Dam of Underground Reservoir in Coal Mine. *Coal Sci. Technol.* **2023**, *51*, 36–49. (In Chinese)
89. Zhang, C.; Bai, Q.S.; Han, P.H.; Wang, L.; Wang, X.J.; Wang, F.T. Strength weakening and its micromechanism in water-rock interaction, a short review in laboratory tests. *Int. J. Coal Sci. Technol.* **2023**, *10*, 15. [CrossRef]
90. Han, P.H.; Zhang, C.; Wang, W. Failure analysis of coal pillars and gateroads in longwall faces under the mining-water invasion coupling effect. *Eng. Fail. Anal.* **2022**, *131*, 23. [CrossRef]
91. Fan, J.Y.; Li, Z.; Feng, G.R.; Zhang, H.D.; Qi, C.E.; Zhang, J.Y. Failure analysis of coal pillars and overburden from underground water reservoir under the mining-water invasion coupling effect. *Eng. Fail. Anal.* **2023**, *151*, 19. [CrossRef]
92. Wang, P.T.; Qi, Z.W.; Ma, C.; Cai, M.F. Anisotropic behavior of the seepage-stress coupling mechanical model of coal pillars of underground reservoirs. *Geomech. Geophys. Geo-Energy Geo-Resour.* **2023**, *9*, 20. [CrossRef]
93. Chen, Z.H.; Zhao, X.D.; Han, Z.; Ji, Y.; Qiao, Z.J. Reasonable Size Design and Influencing Factors Analysis of the Coal Pillar Dam of an Underground Reservoir in Daliuta Mine. *Processes* **2023**, *11*, 13. [CrossRef]
94. Zhang, B.; Ni, W.; Hao, X.Q.; Li, H.Q.; Shen, Y.P. A Study on the Development and Evolution of Fractures in the Coal Pillar Dams of Underground Reservoirs in Coal Mines and Their Optimum Size. *Processes* **2023**, *11*, 21. [CrossRef]
95. Zhi, G.J.; Liu, R.; Yang, R.G.; Qin, W.; Ju, J.F. Study on hydraulic connection and seepage law of adjacent goaf of underground reservoir in coal mine. *Min. Saf. Environ. Prot.* **2022**, *49*, 9–15. (In Chinese)
96. Bai, D.Y. Stability Analysis of Underground Reservoir Dam in Li Jiahao Mine. Master's Thesis, China University of Mining and Technology, Xuzhou, China, 2018. (In Chinese)
97. Bai, D.Y.; Ju, J.F.; Xu, J.L.; Li, J.H. Stability analysis of mine underground reservoir artificial dam in Lijiahao Mine. *J. China Coal Soc.* **2017**, *42*, 1839–1845. (In Chinese)
98. Shan, C.H.; Yao, Q.L.; Zheng, C.K.; Xie, H.X.; Xu, Q.; Xia, Z.; Zou, S. Strength Degradation and Fracture Propagation of Repeatedly Immersed Artificial Dam Samples under Uniaxial Cyclic Loading-Unloading. *ACS Omega* **2023**, *8*, 2538–2555. [CrossRef]
99. Yang, H.Z.; Wang, X.R.; Wang, E.Y.; Song, Y.; Chen, D.; Zhang, Y.D.; Liu, W.Z. Macro-micro crack and damage evolution characteristics of concrete: After the action of acidic drying-saturation cycle. *Constr. Build. Mater.* **2024**, *411*, 15. [CrossRef]

100. Yang, H.Z.; Wang, E.Y.; Cheng, D.Q.; Wang, X.R.; Li, B.L.; Song, Y.; Wang, D.M.; Liu, W.Z. Characteristics of energy evolution and acoustic emission response of concrete under the action of acidic drying-saturation processes cycle. *J. Build. Eng.* **2023**, *74*, 21. [CrossRef]
101. Hao, X.J.; Wei, Y.N.; Zhang, T.; Chen, Z.Y.; Zhang, C.; Teng, T. Comparative study of the effects of different nanomaterials on the failure process of saturated concrete in an underground reservoir of a coal mine. *Energy Explor. Exploit.* **2022**, *40*, 1788–1805. [CrossRef]
102. Yan, Y.G. Experimental Study on Seismic Resistance of Underground Reservoir in Level Coal Seam. Master's Thesis, Tsinghua University, Beijing, China, 2016. (In Chinese)
103. Zhang, Y.; Cao, Z.G.; Wang, L.J.; Zha, E.R.; Li, S.B.; Chu, Z.F. Seismic Safety Analysis of Interlaminar Rock Mass in the Distributed Underground Reservoir of a Coal Mine. *Water* **2024**, *16*, 19. [CrossRef]
104. Cao, Z.G. Study on aseismic performance of different artificial dam structures of coal mine underground reservoir. *Coal Sci. Technol.* **2020**, *48*, 237–243. (In Chinese)
105. Fan, F.; Liu, C.Y.; Xu, J.S. Stability analysis of artificial dam body in coal mine groundwater reservoir. *Coal Eng.* **2020**, *52*, 120–125. (In Chinese)
106. Ma, J.P.; Zhang, G.C.; Zhou, G.L.; Zhang, Y.; Meng, X.J.; Zhao, Y.Q.; Chen, M. Stability analysis of artificial dam in coal mine underground water reservoir based on the hydro-mechanical damage model. *Geomat. Nat. Hazards Risk* **2023**, *14*, 16. [CrossRef]
107. Kong, X.S.; Xu, Z.Z.; Shan, R.L.; Liu, S.; Xiao, S.C. Investigation on groove depth of artificial dam of underground reservoir in coal mines. *Environ. Earth Sci.* **2021**, *80*, 17. [CrossRef]
108. Kong, X.S.; Fang, J.; Xu, Z.Z.; Wang, S.; Liu, S. Design of Cutting Depth of Artificial Dam of Underground Reservoir in Wulanmulun Mine. In Proceedings of the International Conference on Advances in Civil Engineering, Energy Resources and Environment Engineering (ACCESE), Jilin Jianzhu Univiversity, Coll Civil Engineering, Changchun, China, 28–30 June 2019.
109. Zhao, R.J. Research on Seepage Flow and Design of Artificial Dam in Lingxin Coal Mine. Master's Thesis, North China University of Technology, Beijing, China, 2020. (In Chinese)

water

MDPI

Article

Settlement Prediction for Concrete Face Rockfill Dams Considering Major Factor Mining Based on the HHO-VMD-LSTM-SVR Model

Xueqin Zheng [1], Taozhe Ren [1], Fengying Lv [1], Yu Wang [1] and Sen Zheng [2,3,4,5,*]

1 Pump-Storage Technological & Economic Research Institute, State Grid Xinyuan Group Co., Ltd., Beijing 100761, China
2 The National Key Laboratory of Water Disaster Prevention, Hohai University, Nanjing 210098, China
3 College of Water Conservancy and Hydropower Engineering, Hohai University, Nanjing 210098, China
4 National Engineering Research Center of Water Resources Efficient Utilization and Engineering Safety, Hohai University, Nanjing 210098, China
5 ENAC/IIC/LHE, Ecole Polytechnique Fédérale de Lausanne, 1015 Lausanne, Switzerland
* Correspondence: zheng_sen@hhu.edu.cn; Tel.: +86-15150688990

Abstract: Some important discoveries have been revealed in some studies, including that the settlement of concrete face rockfill dams (CFRDs) may cause cracks in the concrete face slabs, which may lead to dam collapse. Therefore, deformation behavior prediction of CFRDs is a longstanding and emerging aspect of dam safety monitoring. This paper aims to propose a settlement prediction model for CFRDs combining the variational mode decomposition (VMD) algorithm, long short-term memory (LSTM) network, and support vector regression algorithm (SVR). Firstly, VMD is applied in the decomposition of dam settlement monitoring data to reduce its complexity. Furthermore, feature information on settlement time series is extracted. Secondly, the LSTM and SVR are optimized by the Harris hawks optimization (HHO) algorithm and modified least square (PLS) method to mine the major influencing factors and establish the prediction model with higher precision. Finally, the proposed model and other models are applied to predict the deformation behavior of the Yixing CFRD. Prediction results indicate that the proposed method possesses particular advantages over other models. The proposed VMD-LSTM-SVR model might help to evaluate the settlement trends and safety states of CFRDs.

Keywords: settlement prediction model; concrete face rockfill dam; variational mode decomposition; Harris hawks optimization; support vector regression; long short-term memory; factor mining

1. Introduction

For decades, concrete face rockfill dams (CFRDs) have been widely adopted in the field of dam construction. Recent engineering applications have revealed that CFRDs become the most competitive reservoir dam type due to their safety, economy, and construction convenience [1–3]. As shown in Figure 1, CFRDs adopt a rockfill body as the support structure. The concrete slab is an important part of a concrete face rockfill dam, its major role is to prevent seepage. The findings seem to demonstrate the concrete panel is much stiffer than the rockfill body, which leads to myriad problems in compatibility and deformation of the two. The foremost problems are the facts that a large or uneven settlement under load appears in the dam [4,5]. A challenging problem that arises in this domain is that large settlements can cause cracks in the concrete panel and damage the water stop, which then affects the dam safety [6,7]. On 27 August 1993, a catastrophic dam failure occurred at the Gouhou CFRD in Qinghai Province, China [8]. This incident is the first failure of a high CFRD worldwide, resulting in widespread destruction of farmland and homes and claiming over 300 lives. Numerous dams have experienced structural cracks and extrusion damage in their face slabs. Notable examples include the Aguamilpa dam in Mexico [9], the

Tianshengqiao-I dam in China [10], and the Xingo dam in Brazil [11], all of which exhibited various degrees of structural cracking. Similar extrusion damages have been observed in several international CFRD projects. For instance, during the initial impoundment stage of the Mohale dam, the face slab along the riverbed developed extrusion damage extending to the bottom of the wave wall. Similarly, the Barra Grande and Campos Novos dams exhibited significant extrusion damage at the central vertical joints of the face slabs under initial water pressure [12]. These structural cracks and extrusion damage are primarily caused by excessive and uneven deformation of the dam body. Dams in the early stage, such as the early Salt Spring dam, the Barra Grande dam, and the Campos Novos dam during their development stages, all experienced leakage rates exceeding 1000 L/s. The leakage rates of Alto Anchacaya and Shiroro dams in the modern stage both exceeded 1800 L/s [13,14]. The occurrence of these issues is closely linked to excessive deformation of the dam body. The findings indicate that excessive deformation does not significantly impact the dam body itself, it directly damages the face slabs and joint waterstop structures, leading to substantial leakage and potentially catastrophic seepage failure. Despite better control of deformations in modern CFRDs compared to earlier counterparts, the increasing height of dams and the interplay of various complex geological conditions continue to face challenges, including excessive deformation and face slab damage.

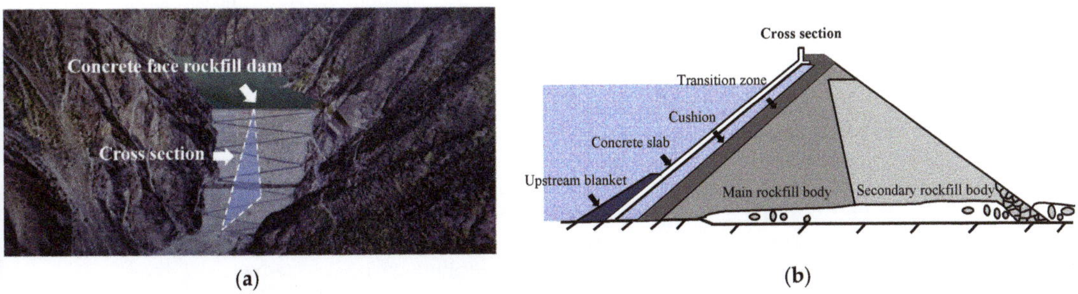

Figure 1. (**a**) Digital three-dimensional model of a CFRD; (**b**) the structure of the CFRD.

Consequently, a large number of alternative research have been developed over the past few years to analyze the deformation behaviors of CFRDs [15–17]. Among them, the dam prediction models enjoy great advantages in reflecting the internal relationship between settlement and influencing factors compared to other means [18,19]. As a result, various settlement prediction models for CFRDs have been developed.

Generally, one of the most popular prediction models in the dam safety monitoring area is the monitoring data-driven model. This type of prediction model can be divided into two primary categories: statistical models and artificial intelligence models [20,21]. As a common statistical model, regression models such as linear regression and polynomial regression models are utilized universally due to their quick running speed. However, it should be noted that they have low interpretability for nonlinear series, which affects their prediction accuracy. In order to rectify the problem, artificial intelligence technology is introduced to combine with the statistical models. As a result, many artificial intelligence prediction models demonstrate particular advantages in nonlinear time series prediction in the CFRD safety monitoring field over other methods [22,23].

For instance, Su et al. [24] used the wavelet support vector machine to reveal the dam settlement characteristics. The results indicated that the prediction model owed high prediction accuracy. Owing to the definite ability to reflect nonlinear relationships between variables, the artificial neural network (ANN) algorithms are utilized to establish the prediction models. The prediction results point to the likelihood that the ANN model possesses better learning ability in extracting deformation trends from dam monitoring data [25–27]. Recently, characterized by the specialized architecture and gating mechanisms, the long short-term memory (LSTM) algorithm offers distinct advantages over conventional neural network algorithms [28]. Consequently, many researchers introduced LSTM to establish the prediction model for dam deformation. Hu et al. [29] adopted an optimized LSTM model to construct the settlement prediction model for CFRD by integrating multiple monitoring data. The findings would seem to imply that the model has a better prediction performance. Xu et al. [30] considered the impact of random error on dam deformation and utilized the LSTM model to propose a prediction model for CFRDs. The results showed that the model showed perfect long-term prediction performance.

Furthermore, according to [14], the ANN model is easy to fall into the local minima, and its convergence speed of the learning process is slow due to the CFRD monitoring data with complex features. However, the support vector regression (SVR) model can be introduced to resolve the regression problem for this kind of monitoring data under the influence of various influencing factors. Therefore, some scholars have used the SVR algorithm to predict the deformation performance of dams. Li et al. [31] presented a settlement prediction model for CFRDs by introducing an improved hybrid SVR algorithm. To improve the efficiency of the prediction model, Su et al. [32] presented an optimization method for the SVM model. Despite its advantages, these models suffer from several major drawbacks: slow convergence and weak generalizability. Due to the highly nonlinear and complexity of the monitoring deformation for CFRDs, machine learning models exhibit limitations in terms of prediction accuracy, generalization capability, and robustness [33–35]. Therefore, in our work, the deep learning models and SVR model are combined to predict the deformation of CFRDs. In particular, few studies, to our knowledge, have focused on this research topic.

Moreover, scholars have delved into mining the primary influencing factors of dam deformation by analyzing the correlation between influencing factors and deformation monitoring sequences [36]. This aims to avoid mitigating the adverse effects on prediction performance caused by excessive redundant information. For instance, Ma et al. [37] adopted the fuzzy neural network to mine the main factor affecting the dam performance. Gu et al. [38] presented a factor mining model based on a random forest algorithm and evidence theory. In recent years, some artificial algorithms such as clustering algorithms, neural networks, and genetic algorithms have been utilized in factor mining. Among these, one of the most popular ideas in factor mining is the partial least squares (PLS) method, as a multivariate statistical analysis technique, it retains the correlations among variables, reduces the dimensionality of influencing factors, and eliminates the effects of multicollinearity among factors.

Furthermore, studies have focused on the application of signal decomposition algorithms in the establishment of settlement prediction models. It has played an important role in reducing the nonlinearity of time series and refining features of monitoring data, which contributes to the interpretability of models [39,40]. Considering that analyzing the correlation between influencing factors and decomposed sub-sequences of settlement monitoring data can enable easier factor mining, in this study, the variational mode decomposition (VMD) algorithm is applied in reducing the nonlinearity and complexity of settlement monitoring time series at first. Subsequently, the improved PLS method is adopted to assess the correlation between influencing factors and decomposed sub-sequences, which could contribute a lot to enhancing the prediction accuracy of the model.

In this paper, a prediction model based on the HHO, VMD, LSTM, and SVR algorithms is proposed. Primarily, the VMD method optimized by HHO is applied to decompose the settlement sequence, thereby reducing data complexity. Secondly, PLS is employed to assess the main influencing factors as input variables for the prediction model. Subsequently, an integrated model combining the optimized LSTM and SVR models is established for dam settlement prediction. The proposed model sufficiently takes the correlation between influencing factors and dam settlement into account. In this way, by leveraging the advantages of two artificial intelligence algorithms, the settlement safety of CFRDs can be analyzed and evaluated accurately. Finally, the developed model is able to be applied to a practical CFRD health monitoring system.

The overall structure of the study takes the form of six sections, including this introductory section. In Section 2, the Hydrostatic-season-time (HST) model for the settlement of CFRDs is introduced. In Section 3, the framework of the presented model in this paper is illustrated. In Section 4, the monitoring data of a practical engineering project is utilized to test and verify the prediction performance of the proposed model. And three common machine learning models are applied in the case study to be taken as the comparisons. In Section 5, comparison results and prediction results are discussed. Ultimately, conclusions are discussed in Section 6.

2. The HST Model for the Settlement of CFRDs

The Hydrostatic-season-time (HST) model stands out as a frequently utilized statistical framework for the interpretation of dam settlement based on structural analysis and mechanical theory. The specific formula of the HST model is expressed as follows [41].

$$\delta(h,t,s) = \delta_H(h,t) + \delta_S(s) + \delta_T(t) + \varepsilon \tag{1}$$

For rockfill materials, settlement primarily occurs due to soil compression, and dam settlement typically depends on the load and time. The settlement of CFRDs during operation is similarly influenced by water pressure (δ_H), aging effect (δ_S), and temperature effect (δ_T). Therefore, the settlement at a certain point of the dam can be expressed using Equation (1), with the expressions for each component as follows:

(1) Settlement caused by water pressure (δ_H)

Water pressure is the predominant factor contributing to dam settlement. As water is impounded, the dam deformation due to water pressure can be categorized into three primary types: upward buoyancy deformation, water pressure deformation, and deformation due to water saturation. The upstream water pressure can be decomposed into horizontal and vertical forces, and the vertical force induces the dam settlement. Meanwhile, the dam body below the phreatic line experiences the action of buoyancy, which reduces its downward settlement. Additionally, the pore water within the rockfill decreases friction, allowing the particles to rearrange and compress under their own weight, resulting in further settlement. In summary, the settlement caused by water pressure can be represented by a Taylor series expansion, considering the first three terms and incorporating the impact of creep under water pressure over time. The influence factor is defined as the average value of upstream water levels in the early stage, the expression δ_H can be demonstrated as follows:

$$\delta_H = f(H, \overline{H}) = \sum_{i=0}^{3} a_{1i} H^i + \sum_{i=0}^{m_1} a_{2i} \overline{H}^i \tag{2}$$

where H represents the upstream water level, $\overline{H}$ denotes the average value of upstream water levels in the early stage, and m_1 indicates the dates considered as the previous stage.

(2) Settlement caused by temperature (δ_T)

The settlement induced by the linear expansion of the dam is related to temperature variations, especially in cold regions where soil tends to undergo frost heaving, leading to significant settlement. Since temperature changes generally follow an annual cycle, the relationship between settlement caused by frost heaving and temperature can be represented by a periodic function of time. Therefore, sinusoidal and cosine functions are employed to represent the temperature components. The expression for the temperature component can be demonstrated as follows:

$$\delta_T = \sum_{i=1}^{m_2} \left(b_{1i} \sin \frac{2\pi i t}{365} + b_{2i} \cos \frac{2\pi i t}{365} \right) \tag{3}$$

where m_2 can be taken as 1 for CFRD, b_{1i}, b_{2i} are pending coefficients, t means the time.

(3) Settlement caused by time effect (δ_S)

Analysis of monitoring data from constructed panel rockfill dams indicates that after impoundment, the dam settlement appears due to the water load, the settlement usually persists for 2–3 years. In the later stages of dam settlement, the rate of settlement increase gradually decreases over time, eventually stabilizing. The functional form should conform to the actual settlement characteristics of CFRDs, thereby the expression for the aging component can be exhibited as follows:

$$\delta_S = c_1\theta + c_2 \ln \theta \tag{4}$$

where c_1 and c_2 are pending coefficients; t is the time, $\theta = t/100$.

Above all, the formula of the HST model for the settlement of CFRDs can be described as follows:

$$\delta = \delta_H + \delta_S + \delta_T + cons \tag{5}$$

where *cons* denotes the constant term.

3. VMD-LSTM-SVR Model for the Settlement Prediction of CFRDs

In this section, the HHO algorithm, the VMD algorithm, and the LSTM and SVR models are introduced to establish the settlement prediction of CFRDs.

3.1. Improved VMD Algorithm Based on HHO Algorithm

3.1.1. HHO Algorithm

Research conducted by Seyedali Mirjalili in 2017 found that the hunting behavior of Harris' hawks exhibits a unique social hunting strategy involving cooperation and competition. Inspired by the hunting behavior, the Harris' hawk optimization (HHO) algorithm, which is a kind of metaheuristic algorithm [42,43], has been utilized in different fields. The HHO algorithm consists of three stages: exploration, transition, and exploitation.

(1) Exploration stage

During the exploration stage, Harris' hawks, randomly perched in a certain location, search for prey through two strategies. When iterating, they update their positions with a probability q.

$$X(t+1) = \begin{cases} X_{rand}(t) - r_1|X_{rand}(t) - 2r_2X(t)| & q \geq 0.5 \\ \left[X_{R(t)} - X_m(t)\right] - r_3[lb + r_4(ub - lb)] & q < 0.5 \end{cases} \tag{6}$$

where $X(t+1)$ and $X(t)$ represent the positions of individuals at the next iteration and the current iteration, respectively, t denotes the iteration count, $X_{rand}(t)$ represents the position of a randomly selected individual of Harris' hawks, $X_R(t)$ represents the position of the prey, r_1, r_2, r_3, r_4, and q are random numbers, ranging between 0 and 1, $X_m(t)$ denotes the average position of individuals, which can be expressed as follows:

$$X_m(t) = \sum_{k=1}^{M} \frac{X_k(t)}{M} \tag{7}$$

where $X_k(t)$ represents the position of the k-th individual in the population, and M denotes the population size.

(2) Transition stage

Since the energy expended by the prey during escape can be converted between exploration and different exploitation behaviors, the HHO algorithm defines its escape energy E as follows:

$$E = 2E_0\left(1 - \frac{t}{T}\right) \tag{8}$$

where E_0 represents the initial escape energy of the prey, the value of E_0 can be selected randomly within the range [−1, 1]. When $-1 < |E_0| < 0$, the prey is in a state of weak energy; when $0 \leq |E_0| < 1$, the prey is in a state of energy recovery. The escape energy E

exhibits a decreasing trend during the iteration process. When $|E| < 1$, Harris' hawks search for the position of prey in different areas, the HHO algorithm executes the exploration stage. When $|E| < 1$, the HHO algorithm performs local search on adjacent solutions, and executes the exploitation stage. At this point, the update of individual positions varies according to the value of E.

(3) Exploitation stage

Define a random number z within the range of [0, 1]. Different exploitation strategies can be selected based on the random number z. When $0.5 < |E| < 1$ and $z \geq 0.5$, the algorithm adopts soft besieging technique for location updating:

$$X(t+1) = \Delta X(t) - E|JX_R(t) - X(t)| \tag{9}$$

where $\Delta X(t) = X_R(t) - X(t)$ represents the difference between the prey's position and the current position of the individual, J is a random number between 0 and 2, indicating the jump distance of the prey during its escape process.

When $|E| < 0.5$ and $z \geq 0.5$, the algorithm adopts the hard besieging method for location updating:

$$X(t+1) = X_R(t) - E|\Delta X(t)| \tag{10}$$

When $0.5 \leq |E| < 1$ and $z < 0.5$, the algorithm adopts the soft besieging strategy with asymptotic rapid descent for updates. The first update strategy is as follows:

$$Y = X_R(t) - E|JX_R(t) - X(t)| \tag{11}$$

If the first strategy is ineffective, the second strategy is executed:

$$Z = Y + S \times LF(D) \tag{12}$$

Therefore, the final update strategy for this stage is as follows:

$$X(t+1) = \begin{cases} Y & f(Y) < f[X(t)] \\ Z & f(Z) < f[X(t)] \end{cases} \tag{13}$$

where $f(\ )$ represents the fitness function, Y and Z denote the current fitness positions of individuals, D denotes the dimensionality of the problem, S represents a $1 \times D$ random vector, and $LF(\ )$ is the mathematical expression for Levy flight.

When $|E| < 0.5$ and $z < 0.5$, the algorithm adopts the hard besieging strategy with asymptotic rapid descent for location updating:

$$X(t+1) = \begin{cases} Y & f(Y) < f[X(t)] \\ Z & f(Z) < f[X(t)] \end{cases} \tag{14}$$

$$Y = X_R(t) - E|JX_R(t) - X_m(t)| \tag{15}$$

$$Z = Y + S \times LF(D) \tag{16}$$

3.1.2. Improved VMD Algorithm

The VMD algorithm involves numerous parameters, and the selection of the number of modal components K and the penalty factor α play an important role in the decomposition effectiveness. If the number of modal components K is too small, it may fail to capture all the details and features of the sequence, leading to information loss. Conversely, if K is too large, it may result in excessive information redundancy, increasing computational and analytical complexity. For the penalty factor α, an excessively large value can overly smooth the decomposed modal components, leading to the loss of crucial information. On the contrary, if α is configured to be too small, it may introduce significant instability and roughness to the sub-sequences, exacerbating the impact of data noise on the decomposition results. Hence, determining appropriate values for K and α is essential before performing data decomposition. Typically, these parameters are determined empirically, which is subjective and can affect the generality of the model. In this paper, the HHO algorithm is applied to the parameter selection of the VMD algorithm by selecting minimum average envelope entropy.

In the signal decomposition, the smaller the envelope entropy of each sub-sequence, the lower the complexity of the decomposed sequence. Consequently, the average envelope entropy is utilized as the fitness function. Specifically, for a signal $X(t)$, the average envelope entropy of its decomposed sequences can be calculated through the following steps:

(1) Calculate the envelope $E(t)$ of a signal using the Hilbert transform.

(2) Segment the signal into M data blocks according to a fixed length.

(3) For each data block, compute its probability density function $p_i(n)$, where n represents the envelope falling within the n-th interval.

(4) Calculate the envelope entropy using the following formula:

$$H_E = -\sum_{i=1}^{M}\sum_{n=1}^{X} p_i(n)\ln_2(p_i(n)) \quad (17)$$

where X denotes the number of internals.

In this study, the variance percentage (VP) indicator is applied to calculate the proportion of each sub-sequence in the original sequence. Firstly, the variance of each component and the total variance of the original sequence are computed to obtain the proportion of each sub-sequence in the original sequence. Secondly, the proportion of each sub-sequence is multiplied by the envelope entropy of the sub-sequence and summed to obtain the average envelope entropy of the decomposed sub-sequences. The main steps of the optimization process of the VMD method based on HHO are as follows:

(1) Initialize the parameters of the HHO algorithm, including the number of population quantities, maximum iteration times, and the population location.

(2) Set the fitness function as the average envelope entropy and calculate the fitness value of each hawk to determine the optimal location of an individual.

(3) Estimate $|E|$ and z. Update the location according to Equations (9)–(16).

(4) Calculate the updated fitness values of the Harris hawks' individual. If the fitness of the new individual surpasses that of its predecessor, the original location is replaced by the new location. Otherwise, the original location is retained.

(5) Adjust the search range and direction of the particle swarm adaptively through the algorithm's fitness. Continuously search for the minimum fitness value according to the fitness function.

(6) Evaluate if the maximum iteration times has been reached. If affirmative, output the current optimal individual's location. If negative, repeat operations from step (2) to step (5).

(7) Obtain the parameters corresponding to the minimum fitness as the optimal parameters for the VMD algorithm.

Subsequently, the optimized VMD algorithm can be applied to decompose the settlement monitoring data series of CFRD.

3.2. HHO-LSTM-SVR Prediction Model for the Settlement of CFRD

3.2.1. LSTM Network

To resolve the problem of vanishing or exploding gradients during model training, scholars have introduced "gate" control structures on the hidden layer nodes of Recurrent Neural Networks (RNNs). This innovation aims to regulate the flow of information, proposing the LSTM network. The LSTM network exhibits strong temporal dependencies that are capable of storing relevant information over arbitrary time intervals, thereby effectively extracting patterns from historical data. Moreover, it has demonstrated promising results in practical applications [44,45].

As depicted in Figure 2, the architecture of the LSTM network operates with a unidirectional hidden layer. In Figure 2, x_t represents the input value at time step t, y_t denotes the output value at time step t, S_t signifies the cell state at time step t, h_t signifies the output value of the hidden layer at time step t.

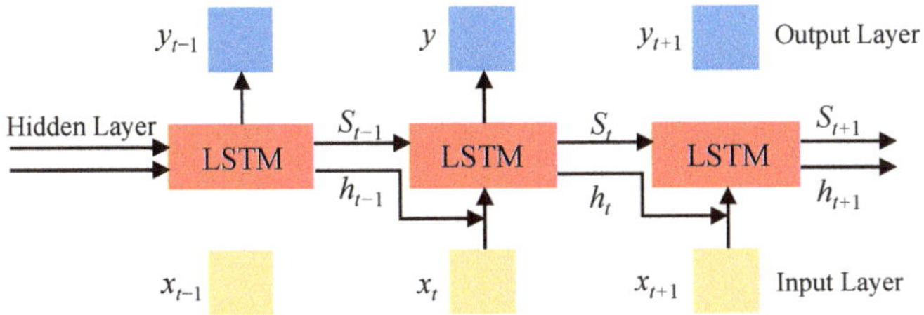

Figure 2. The structure of the LSTM network.

3.2.2. SVR Algorithm

The support vector regression (SVR) algorithm employs a nonlinear mapping technique to project training samples into a high-dimensional space, facilitating linear regression in this space without introducing additional computational complexity [46]. The SVR algorithm does well in addressing nonlinear regression prediction tasks, particularly demonstrating strong performance with moderate- to small-scale datasets. The SVR algorithm exhibits robustness and is effective in mitigating the issue of local extremum. Moreover, the SVR algorithm shows promising results in practical applications [47].

The computational expression of the SVR model can be represented as follows:

$$y(x) = \sum_{j=1}^{N}\left[\left(\alpha_j^* - \alpha_j\right) \times K\left(x_j, x\right)\right] + d \tag{18}$$

where x is the input vector, y is the corresponding output value, α_j and α_j^* are Lagrange multipliers, $K(x_j, x)$ is the kernel function, d is the bias term, N is the total number of samples in the dataset.

3.2.3. HHO-LSTM-SVR Model

With the aim of determining the parameters of the LSTM and SVR models effectively, the HHO algorithm is utilized. In this work, the HHO-LSTM and HHO-SVR models are presented, respectively. By integrating multiple homogeneous or heterogeneous learners, the prediction performance of individual learners can be enhanced effectively [48]. Consequently, the merits of HHO-LSTM and HHO-SVR models can be further integrated by using partial least squares (PLS) regression. The HHO-LSTM-SVR model is proposed to predict the dam settlement with high accuracy.

The computational workflow of the HHO-LSTM-SVR model begins with the parallel computation of the HHO-LSTM and HHO-SVR models. Subsequently, the sum of squared prediction errors of the two single models and a penalty term of a weighting coefficient is selected as the objective function. Then, the optimal weights for each single model are computed to minimize the objective function, as shown in Equation (19). Finally, on the basis of the weights, the HHO-LSTM-SVR model is tested. The PLS regression can effectively address the issue of excessive bias weighting due to differences in the predictive accuracy of single models in ensemble combination models. The computational process of the HHO-LSTM-SVR model is illustrated in Figure 3.

$$\min Q = \sum_{k=1}^{m}[\xi y_{1k}^* + (1-\xi)y_{2k}^* - y_k]^2 + \lambda\|\xi\|^2 \tag{19}$$

where y_k is the actual value, y_{1k}^* and y_{2k}^* are the prediction values of the HHO-LSTM model and the HHO-SVR model, respectively; ξ presents the weight of the model, λ is the coefficients of the penalty term, and m is the number of the sample.

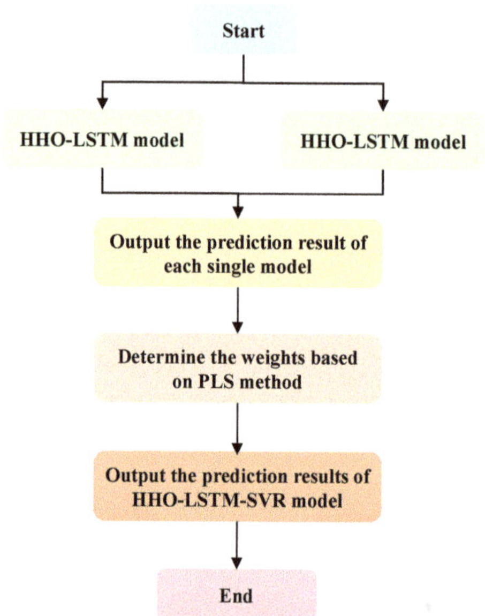

Figure 3. The calculation flow of the HHO-LSTM-SVR model.

3.3. HHO-VMD-LSTM-SVR Model Considering Factor Mining for Settlement Prediction of CFRD

Based on the methods and theories discussed above, the proposed workflow framework for the settlement prediction model of CFRDs is illustrated in Figure 4. The specific steps of the process are outlined as follows:

(1) Select the target monitoring points and preprocess the monitoring data of water level, temperature, dam settlement, and time series data. The settlement sequences are decomposed using the VMD method optimized by the HHO algorithm, yielding several IMF sub-sequences (refer to Modules 1 and 2 in Figure 4).

(2) Determine the optimal number of input influencing factors for each IMF sub-sequence through cross-validation. Subsequently, the PLS algorithm is utilized to extract the optimal input influencing factor set for each IMF sub-sequence, achieving the mining of influencing factors (refer to Module 3 in Figure 4).

(3) Analyze the local characteristics of each IMF sub-sequence and establish the HHO-LSTM-SVR prediction model. The iterations of these steps are executed to construct and optimize the HHO-LSTM-SVR prediction model for each IMF component (refer to Module 4 in Figure 4).

(4) Predict each IMF sub-sequence component using the HHO-LSTM-SVR model sequentially to obtain output sequences. These sequences are then superimposed to form the final prediction results (refer to Module 4 in Figure 4).

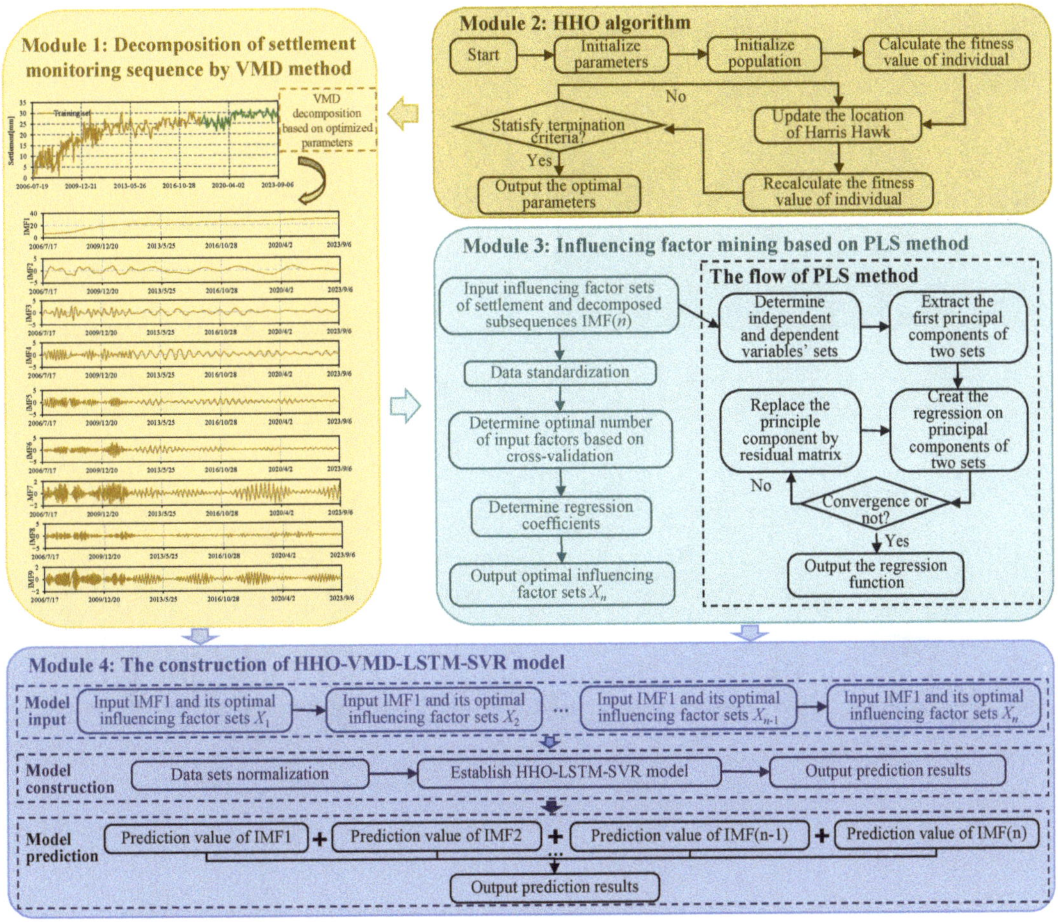

Figure 4. The flowchart of the HHO-VMD-LSTM-SVR model construction of settlement prediction for CFRDs (in Module 1, yellow line represents the fitting data sets, and green line represents the prediction data sets).

4. Project Overview

The Yixing pumped-storage power station (PSPS) is situated on the southwest side of Tongguan Mountain in Yixing city, Jiangsu Province. The station comprises upper and lower reservoirs, water transmission systems, underground powerhouses, and switchyards. With a total installed capacity of 1000 MW and an annual electricity generation of 1.49 billion kWh, the station primarily serves peak shaving, valley filling, frequency regulation, phase adjustment, and emergency backup functions. The upper reservoir is located in the valley on the north side of Tongguan Mountain, surrounded by the main dam, auxiliary dam, and surrounding mountains. The main dam of the upper reservoir is a concrete face rockfill dam, with a crest elevation of 474.20 m and a crest length of 494.90 m. The schematic of the dam structure and distribution of monitoring points of deformation for Yixing PSPS are exhibited in Figure 5b,c, separately.

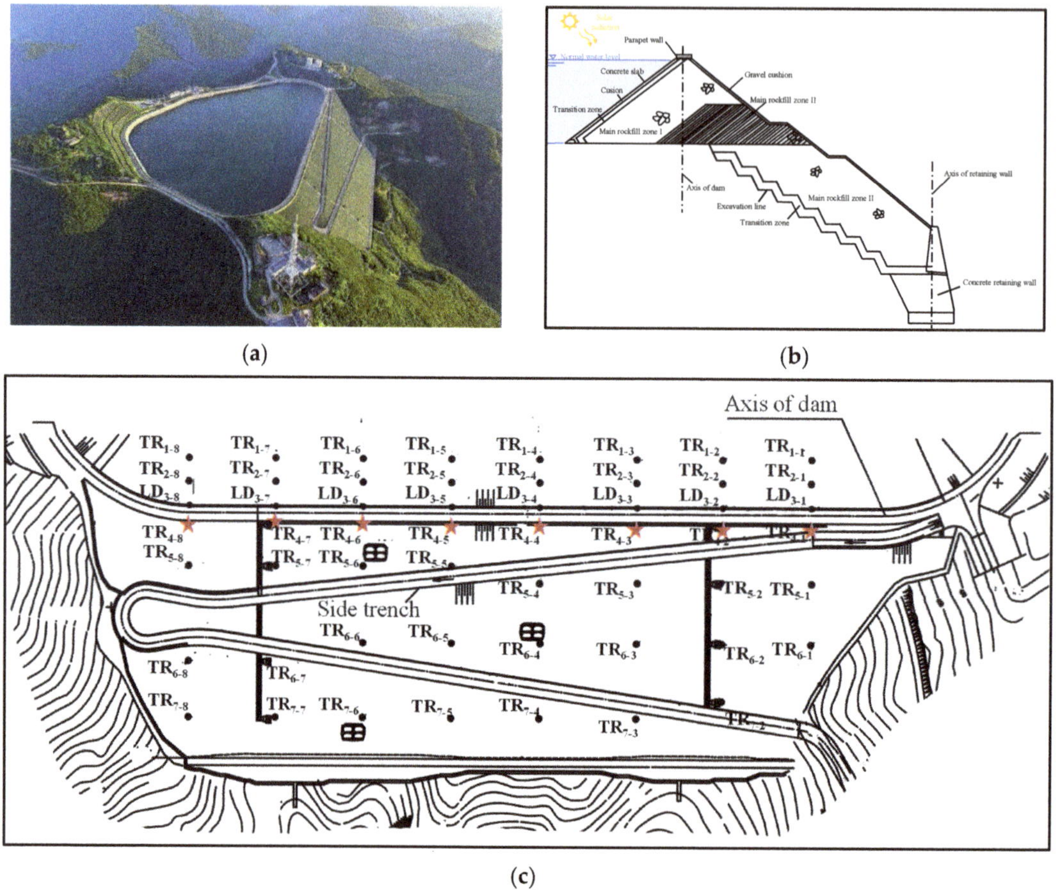

Figure 5. (**a**) Photo of the upstream reservoir and CFRD, (**b**) schematic of structure of CFRD, (**c**) schematic of monitoring points of deformation of CFRD for Yixing PSPS (the locations of the selected monitoring points ($TR_{4\text{-}1}$ to $TR_{4\text{-}8}$) are marked with red star-shaped labels).

The settlement data of $TR_{4\text{-}1}$ to $TR_{4\text{-}8}$ from 4 December 2006 to 6 September 2023 are selected for analysis in this case study. The locations of the monitoring points ($TR_{4\text{-}1}$ to $TR_{4\text{-}8}$) are marked in Figure 5c with red star-shaped labels. After data preprocessing, the valid settlement data of the selected monitoring points are shown in Figure 6. The upstream water level and the environmental temperature in Yixing PSPS are illustrated in Figure 7. It can be seen that all the settlement monitoring data increase slowly, and tend to converge with time.

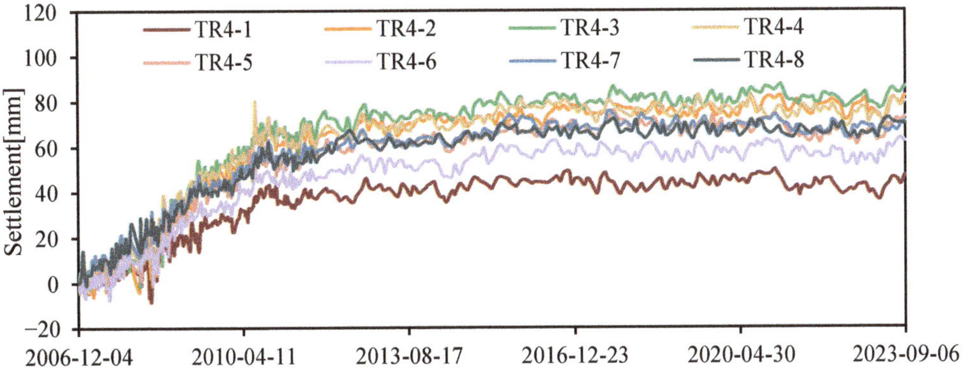

Figure 6. Settlement monitoring data sequences of the selected monitoring points.

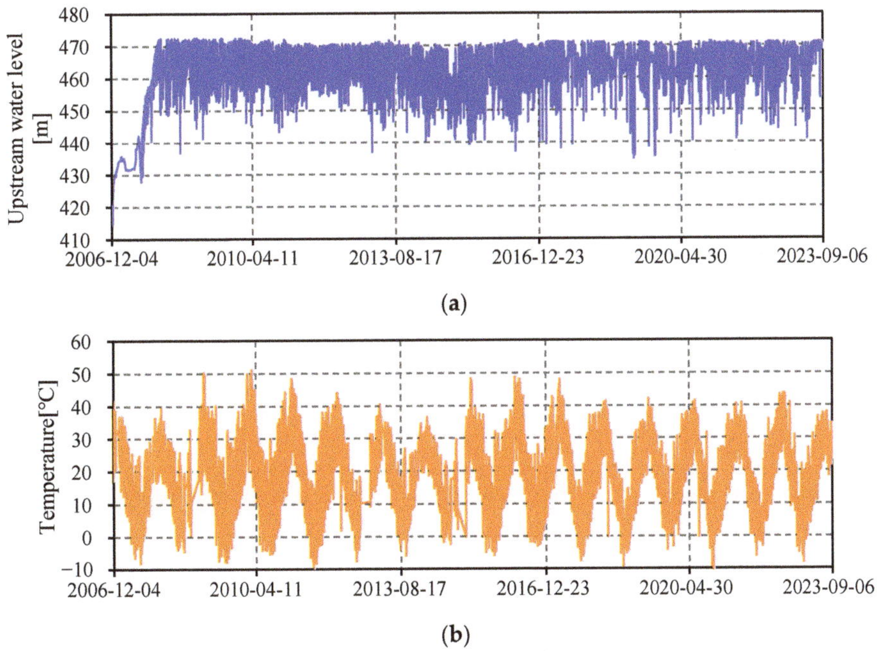

Figure 7. Monitoring data series: (**a**) upstream water level, (**b**) environmental temperature.

5. Results and Discussion

5.1. Decomposition of Settlement Sequences Based on VMD Method

The parameters of the VMD method are determined by using the HHO algorithm. The optimization process aims to synthesize the most suitable parameters based on the complexity of the data sequences, employing the average envelope entropy as the fitness function. Ultimately, the optimal number of modal components k is set at 9, and the penalty factor α is set at 664. The optimization process of the HHO algorithm is exhibited in Figure 8.

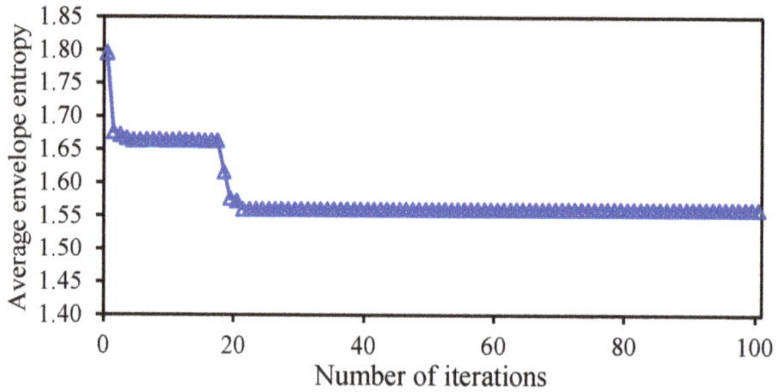

(a)

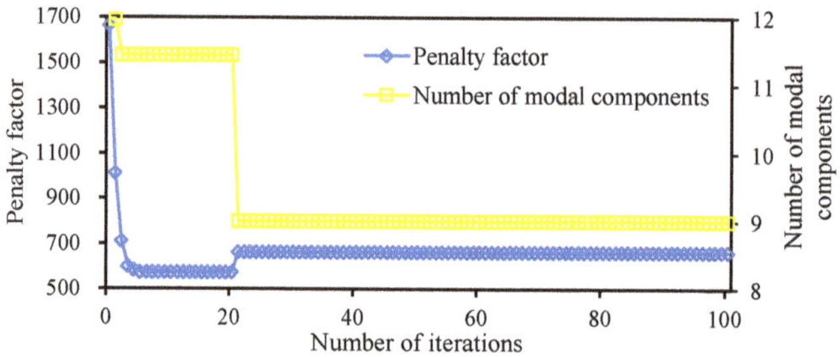

(b)

Figure 8. The optimization process of the HHO algorithm. (**a**) The variation process of average envelope entropy, (**b**) the optimization process of penalty factor and number of modal components.

For example, Figure 9 depicts the decomposition results by the VMD method of the settlement monitoring sequence at monitoring point $TR_{4\text{-}7}$. Based on the characteristic features of the decomposition sub-sequences, IMF1 is defined as the trend component of the original sequence, portraying the sustained deformation of the rockfill body of the Yixing CFRD underwater load, particularly under prolonged duration and high water levels. The deformation characteristic exhibits a rheological characteristic with a trending tendency. IMF2 and IMF4 represent periodic sequences gradually, influenced significantly by cyclic factors such as upstream water level and temperature changes. IMF5 to IMF9 demonstrate characteristics of short-period and low-amplitude oscillations, possibly attributed to rapid and transient deformation components induced by factors like seismic events, self-weight, or monitoring errors in dam structures. Due to the differing primary influencing factors of each component, it is essential to construct and optimize the factor sets for the decomposed sub-sequences, subsequently establishing a corresponding prediction model.

5.2. Major Factor Mining Based on PLS Method

To simultaneously consider the correlation between multiple independent variables and dependent variables, partial least squares (PLS) regression is introduced for morphological sequence influence factor mining. To prevent overfitting, the first 80% of the data is selected for training to obtain the input primary influence factors for each decomposed sub-sequence. Furthermore, to accurately determine the number of input impact factors for different decomposed sub-sequences and to improve the fitting effects of sub-sequences, cross-validation is employed to obtain the mean absolute error, serving as the criterion for determining the optimal number of input factors for each sub-sequence.

Finally, the optimal number of input impact factors and the corresponding regression coefficients are determined. By computing the mean absolute error under different numbers of input impact factors, the primary influencing factors are identified according to the minimum mean absolute error. When the mean absolute errors are close, fewer input impact factors should be selected to enhance the modeling efficiency. Taking the determination process of the primary influence factor mining for IMF1 and IMF2 as examples, Figure 10 illustrates the mean absolute error of IMF1 and IMF2 under different numbers of impact factors. It is evident that IMF1 achieves the minimum mean absolute error of 0.748 when the number of input factors is 3. As for IMF2, the minimum mean absolute error is attained when the number of input influence factors is 13 or 15. However, it is noted that the mean absolute error is close to the minimum value when the number of influence factors is 3. Considering training and prediction efficiency, the number of influence factors should be taken as 3. Similarly, the optimal number of input influence factors and mean absolute errors for each sequence are obtained, as shown in Table 1.

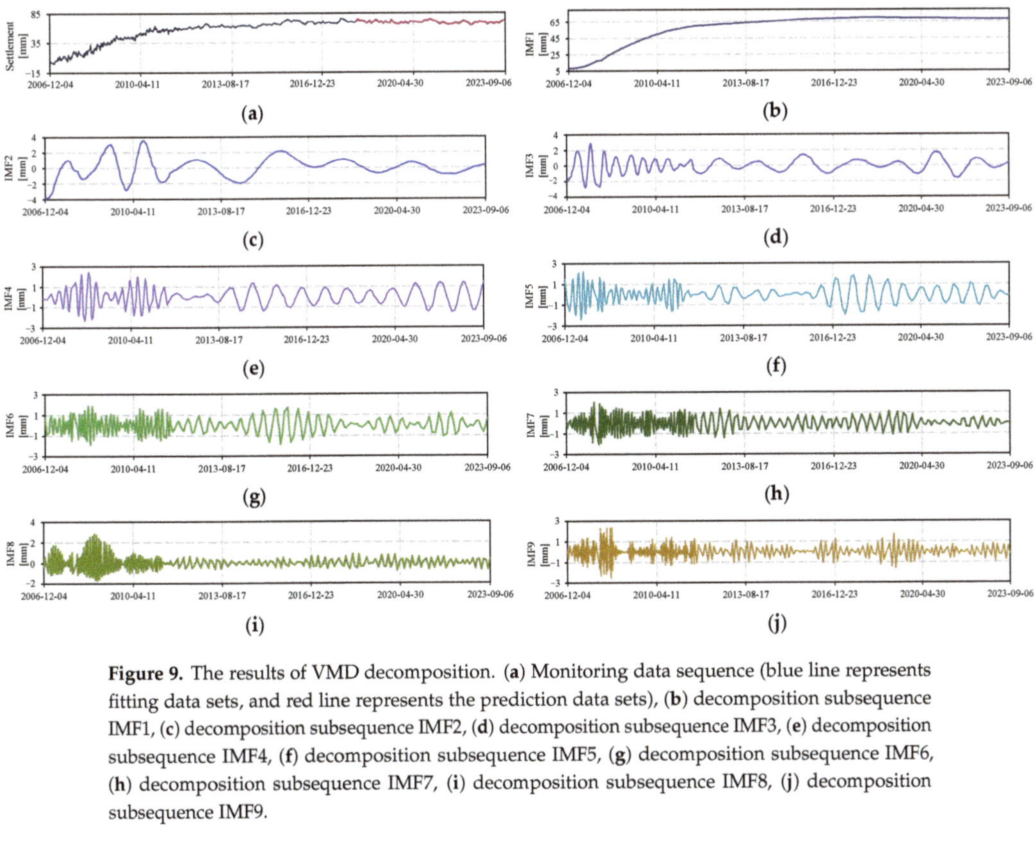

Figure 9. The results of VMD decomposition. (**a**) Monitoring data sequence (blue line represents fitting data sets, and red line represents the prediction data sets), (**b**) decomposition subsequence IMF1, (**c**) decomposition subsequence IMF2, (**d**) decomposition subsequence IMF3, (**e**) decomposition subsequence IMF4, (**f**) decomposition subsequence IMF5, (**g**) decomposition subsequence IMF6, (**h**) decomposition subsequence IMF7, (**i**) decomposition subsequence IMF8, (**j**) decomposition subsequence IMF9.

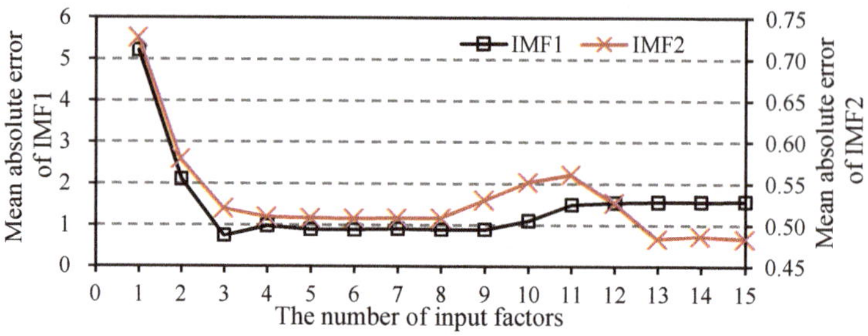

Figure 10. Mean absolute errors of IMF1 and IMF2 corresponding to various numbers of input factors.

Table 1. Number of the optimal input influence factors of the decomposition sequence and the corresponding average absolute errors.

Decomposition Sequence	Mean Absolute Error	Number of the Optimal Input Influence Factors	Decomposition Sequence	Average Absolute Error	Number of the Optimal Input Influence Factors
IMF1	0.748	3	IMF6	0.051	1
IMF2	0.520	4	IMF7	0.059	4
IMF3	0.419	3	IMF8	0.063	5
IMF4	0.157	5	IMF9	0.061	6
IMF5	0.062	1	-	-	-

5.3. Fitting and Prediction Results

On the basis of parameter optimization by the HHO algorithm, the selected hyperparameters of the LSTM model are set as follows: hidden layers can be taken as 2 (each hidden layer has 8 neurons), iteration period is 1000, initial learning rate is 0.05, and descending factor is 0.15. In the same way, the hyperparameters of the SVR model are set as follows: penalty parameter *C* can be selected as 0.2, kernel function parameter γ can be taken as 0.8. 80% of the monitoring data are applied to train the prediction model.

Take the monitoring settlement data of monitoring point $TR_{4\text{-}7}$ again, all the prediction results of the HHO-LSTM-SVR model for each sub-sequence are illustrated in Figure 11. Comparisons between the true monitoring values and prediction values of settlement reveal that the model exhibits great fitting and prediction accuracy.

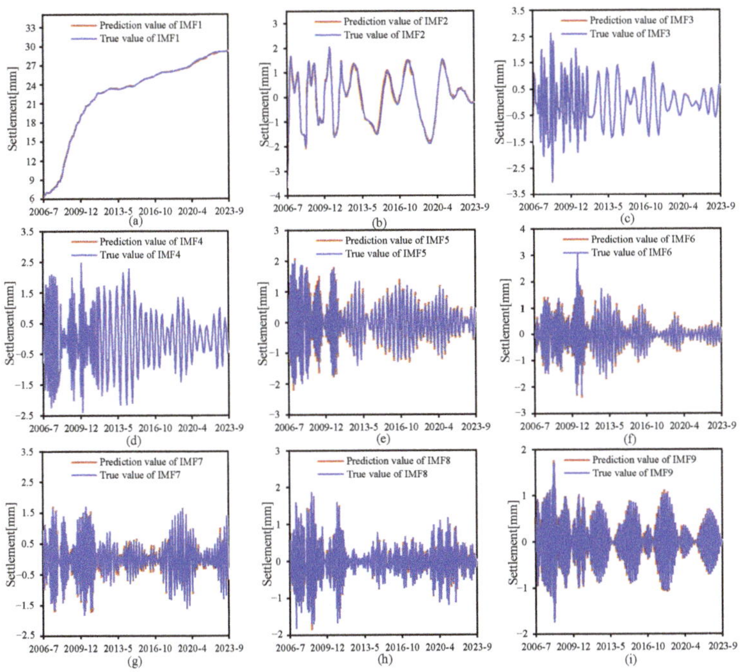

Figure 11. Comparisons between true monitoring values and prediction values of (**a**) decomposed subsequence IMF1, (**b**) decomposed subsequence IMF2, (**c**) decomposed subsequence IMF3, (**d**) decomposed subsequence IMF4, (**e**) decomposed subsequence IMF5, (**f**) decomposed subsequence IMF6, (**g**) decomposed subsequence IMF7, (**h**) decomposed subsequence IMF8, (**i**) decomposed subsequence IMF9.

In order to evaluate the fitting and prediction ability of the VMD-LSTM-SVR model, two evaluation indicators, the coefficient of determination R^2 and the root mean square error *RMSE*, are utilized. The calculation formula for the above indicators is as follows:

$$R^2 = \frac{\sum_{i=1}^{N} \left(\hat{\delta}_i - \overline{\delta}\right)}{\sum_{i=1}^{N} \left(\delta_i - \overline{\delta}\right)} \tag{20}$$

$$RMSE = \sqrt{\frac{1}{M}\sum_{j=1}^{M} \left(\delta_j - \hat{\delta}_j\right)^2} \tag{21}$$

where δ_i represents the training sets of the settlement monitoring data, $\overline{\delta}$ is the average value of the training datasets, $\hat{\delta}_i$ denotes the fitting results of the training datasets, δ_j indicates the test sets of the settlement monitoring data, $\hat{\delta}_j$ is the prediction results of the test datasets, M and N represent the numbers of data of training and test datasets, respectively.

At the same time, we introduced three other models: the stepwise regression (SR) model, LSTM optimized by the HHO (HHO-LSTM) model, and SVR optimized by the HHO (HHO-SVR) model to establish a prediction model for dam monitoring settlement. The comparison results of R^2 and *RMSE* of all the selected monitoring points of different prediction models are exhibited in Table 2 and Figure 12, respectively. It should be noted that the results of R^2 of the SR model, the HHO-LSTM model, the HHO-SVR model, and the proposed model vary from 0.801 to 0.862, 0.886 to 944, 0.873 to 0.943, and 0.980 to 0.995, respectively. Therefore, the proposed model possesses high fitting accuracy. As exhibited in Figure 12, the values of *RMSE* of the SR model, the HHO-LSTM model, the HHO-SVR model, and the proposed model vary from 3.5355 to 5.3826, 2.1386 to 3.6316, 2.4835 to 3.8471 and 1.5543 to 2.1537, respectively. It is obvious that the values of *RMSE* of the proposed model for all the selected monitoring points are the smallest.

Table 2. R^2 of different prediction models for all the selected monitoring points.

Monitoring Point	SR Model	HHO-LSTM Model	HHO-SVR Model	The Proposed Model
TR_{4-1}	0.854	0.931	0.936	0.995
TR_{4-2}	0.862	0.927	0.899	0.982
TR_{4-3}	0.801	0.894	0.887	0.981
TR_{4-4}	0.847	0.886	0.873	0.980
TR_{4-5}	0.835	0.921	0.915	0.981
TR_{4-6}	0.829	0.907	0.924	0.982
TR_{4-7}	0.813	0.939	0.943	0.985
TR_{4-8}	0.817	0.944	0.938	0.986

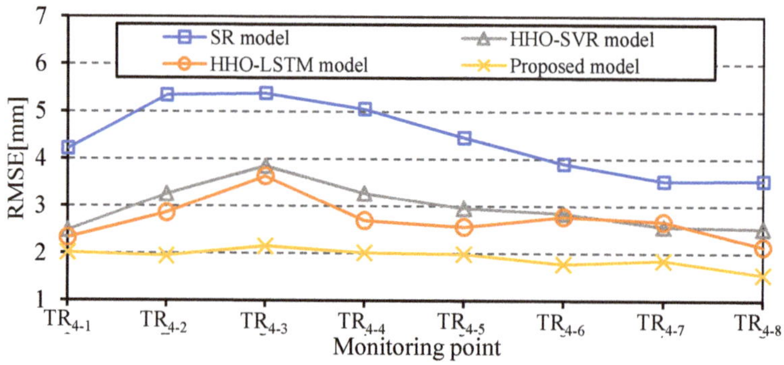

Figure 12. *RMSE* of different prediction models for all the selected monitoring points.

The settlement monitoring data from 4 December 2006 to 30 April 2020 was utilized to establish the prediction model. Then, the settlement data from 1 May 2020 to 6 September 2023 is applied in model testing. As shown in Figure 13, for all the settlement monitoring points, the fitting and prediction performance of the SR model is the poorest, while the proposed model demonstrates significantly better fitting and prediction performance, with results closer to the actual monitoring values. Despite significant fluctuations in the settlement monitoring data of all monitoring points, the fitting and prediction results of the proposed model closely match the actual monitoring data, demonstrating a high level of accuracy in both fitting and prediction outcomes for all the monitoring points.

Furthermore, the prediction residual was introduced as an evaluation index to assess the effectiveness of the models. Figure 14 implies the boxplots of prediction residuals for all the monitoring points of different prediction models. For all the monitoring points, the residuals of the SR model, the HHO-LSTM model, the HHO-SVR model, and the proposed model vary from −22.43 to 22.05, −7.10 to 6.07, −18.46 to 20.55, and −3.4 to 4.10, respectively.

5.4. Discussion

In statistics and data analysis, the coefficient of determination R^2 is a measure that indicates the strength and direction of the linear relationship between a set of predictor variables and a response variable. The value of R^2 ranges from 0 to 1, and a larger value suggests the higher prediction accuracy of the models. Therefore, as given in Table 2, the prediction model possesses the largest values of R^2 for all the monitoring points, which indicates a stronger ability in prediction performance. Moreover, the root mean square error *RMSE* as another statistical indicator was introduced, and the smaller value of *RMSE* means higher prediction precision of the models. As depicted in Figure 12, *RMSE* is adopted to quantitatively assess the fitting and prediction ability of the models. The smallest value range of *RMSE* implies the proposed model is reliable.

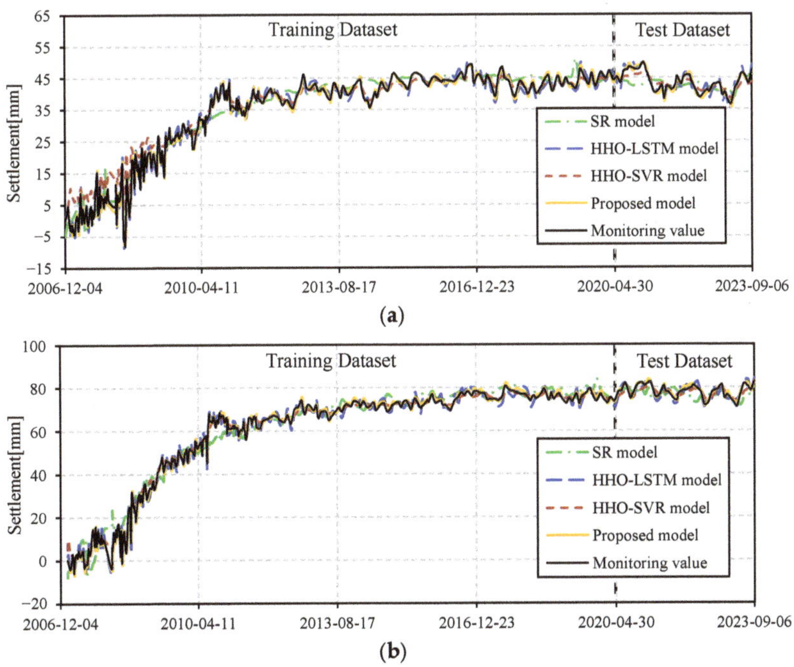

Figure 13. *Cont.*

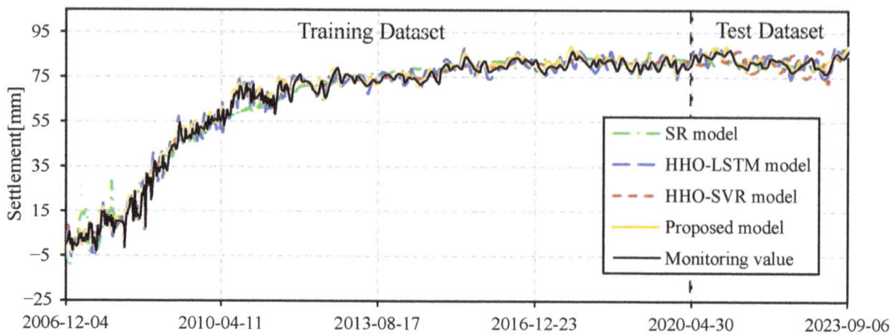

(**c**)

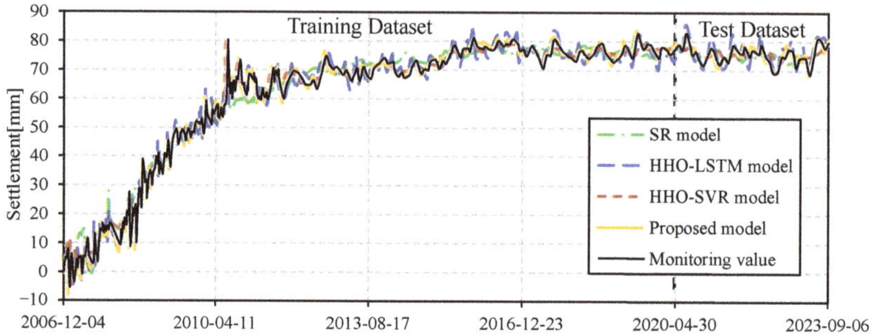

(**d**)

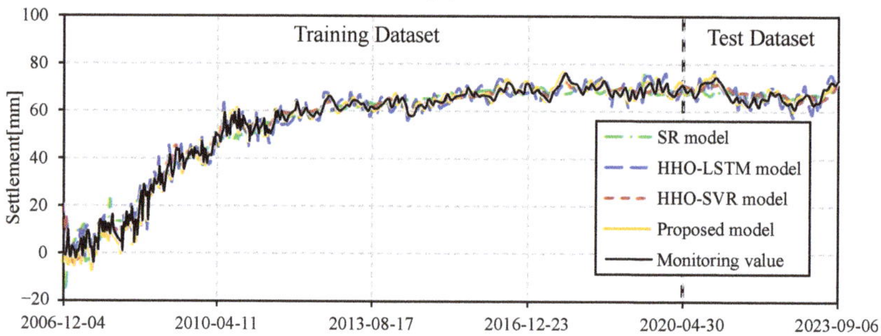

(**e**)

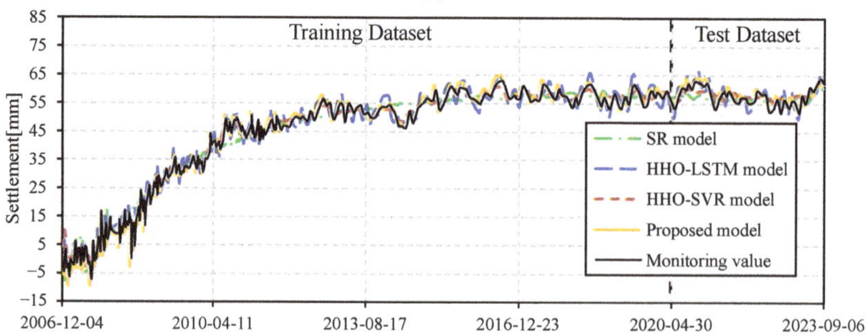

(**f**)

Figure 13. *Cont.*

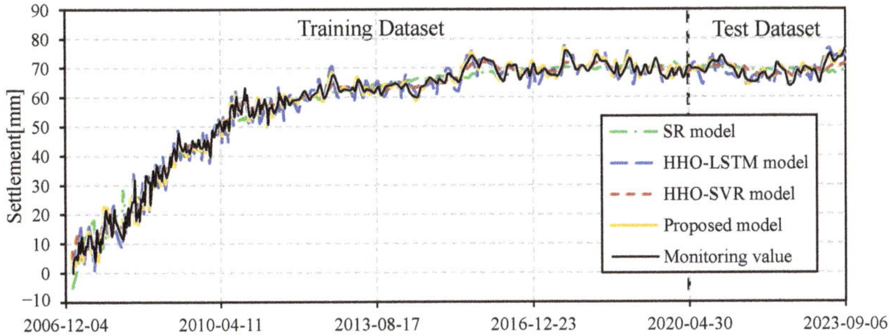

(g)

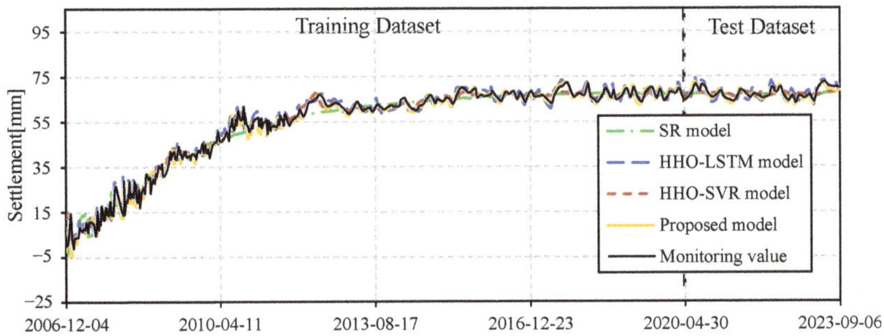

(h)

Figure 13. Fitting and prediction results of different prediction models for all the selected monitoring points: (**a**) $TR_{4\text{-}1}$, (**b**) $TR_{4\text{-}2}$, (**c**) $TR_{4\text{-}3}$, (**d**) $TR_{4\text{-}4}$, (**e**) $TR_{4\text{-}5}$, (**f**) $TR_{4\text{-}6}$, (**g**) $TR_{4\text{-}7}$, (**h**) $TR_{4\text{-}8}$.

The same conclusion is further confirmed in Figure 13, the prediction model exhibits excellent fitting and prediction ability for all the monitoring points. Figure 14 shows that the absolute values of the prediction residuals for the SR model are relatively large, the SR model has the poorest prediction performance. In Figure 14, it is evident that the prediction residual distributions of the proposed models for all the selected monitoring points are tighter, with overall smaller residuals, and notably fewer extreme outliers compared to other models. Consequently, for different monitoring sequences at various points, the proposed model exhibits excellent prediction performance, indicating its strong generalizability.

With the combination of utilizing the HHO algorithm, the LSTM model, and the SVR model, for the total selected 8 monitoring points, the proposed shows the strong nonlinear data mining ability. In this case study, the fitting and prediction results indicate the higher prediction accuracy and generalization ability of the proposed model, which is expected to be applied to the health diagnosis of CFRDs.

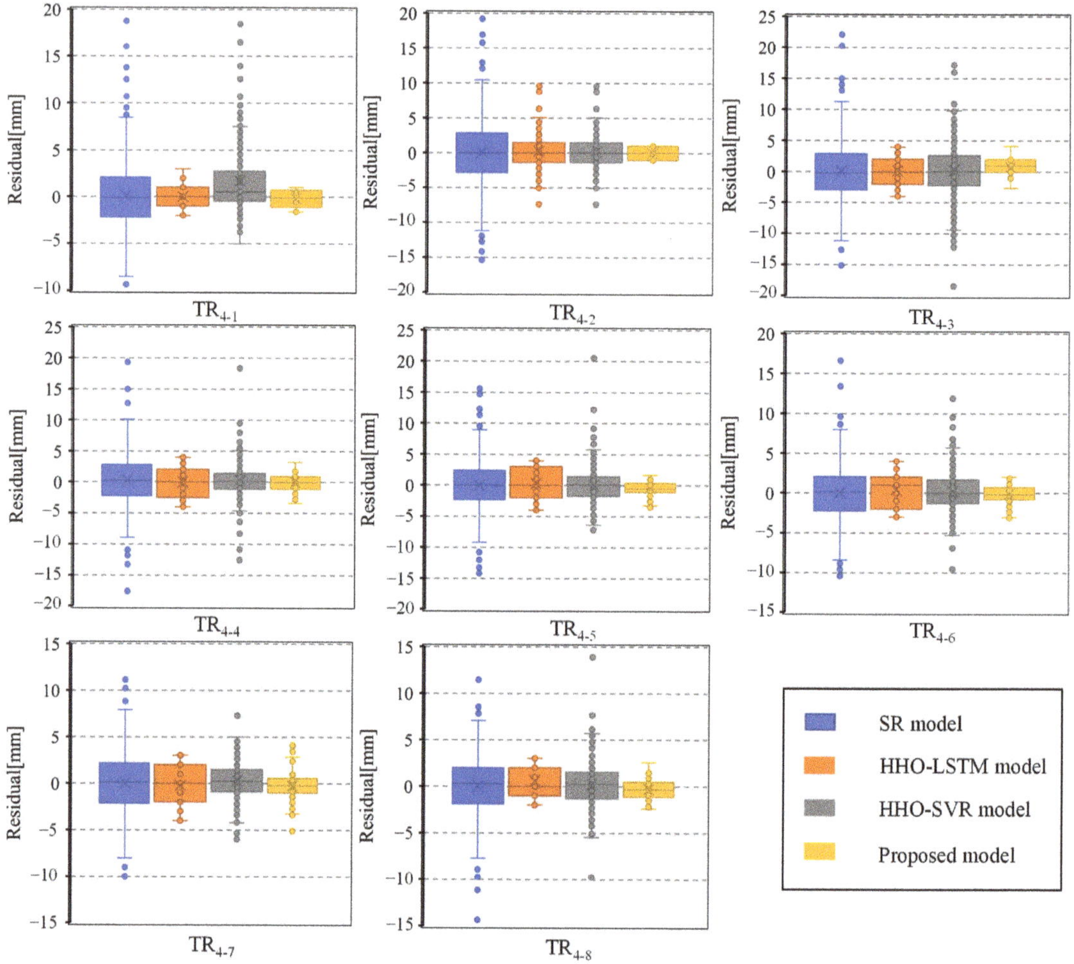

Figure 14. Boxplots of fitting and prediction residuals of different prediction models for all the selected monitoring points.

6. Conclusions

In summary, this paper has presented a useful study of a settlement prediction model for CFRDs. In this study, the VMD signal processing combined with LSTM and SVR models was introduced to establish the settlement prediction model of CFRDs. Through optimizing parameters of the above machine learning algorithms, the dam settlement monitoring data were processed via multiscale decomposition, enabling the accurate major influencing factors mining and subsequent time series prediction. In constructing the SR model, the HHO-LSTM model, and the HHO-SVR model, the fitting and prediction performance of the proposed model was evaluated. Some conclusions can be drawn as follows.

(1) On the basis of introducing the HHO algorithm, the training efficiency of the LSTM model and SVR model has been significantly enhanced. The integration of the VMD algorithm effectively decomposes the dam deformation time series, transforming the complex nonlinear sequence into multiple relatively stable sub-sequences with different frequency scales, thereby substantially improving the prediction ability of the model. Utilizing the PLS method to analyze the decomposed deformation components allows for

the major influencing factors mining, the correlation between the influencing factors and the deformation can be improved. It not only improves prediction accuracy but also reduces the input dimensionality of the prediction model. The better prediction performance is a consequence of accurate major influencing factor mining. In addition, the model achieves improved prediction accuracy by significantly reducing the input dimensions of the prediction model.

(2) The combined application of the LSTM and SVR models results in the proposed model demonstrating advantages over comparative models in terms of prediction accuracy, residual distribution, and operational efficiency. The proposed model can play an important role in decomposing the deformation monitoring data, extracting data features, mining major influencing factors, and high prediction efficiency. Consequently, the proposed model exhibits certain advantages in contrast to other comparative models in practical engineering applications for the safety monitoring of CFRDs.

(3) It should be noted that the proposed model only focuses on data from individual monitoring points, without considering the interrelationships between different monitoring points. Therefore, it seems that considering the spatiotemporal relationships among various monitoring points when establishing a prediction model may be capable of improving the prediction accuracy in future work.

Author Contributions: Conceptualization, X.Z. and S.Z.; methodology, X.Z.; software, T.R.; validation, F.L., Y.W. and S.Z.; formal analysis, X.Z.; investigation, T.R.; data curation, F.L.; writing—original draft, X.Z. and S.Z.; writing—review and editing, T.R., Y.W. and F.L.; supervision, X.Z.; funding acquisition, X.Z. All authors have read and agreed to the published version of the manuscript.

Funding: This research was funded by the National Natural Science Foundation of China (Grant Nos. U2243223, 52209159), the Fundamental Research Funds for the Central Universities (Grant No. B230201011), the Water Conservancy Science and Technology Project of Jiangsu (Grant No. 2022024), the Anhui Provincial Natural Science Foundation "Water Sciences" Joint Fund (Grant No. 2208085US17), and the State Grid Xinyuan (Holding) Company Ltd. Science and Technology Project (Grant No. SGXYKJ-2023-059).

Data Availability Statement: The data presented in this study are available on request from the corresponding author.

Conflicts of Interest: X.Z., T.R., F.L., and Y.W. were employed by the Pumped-storage Technological & Economic Research Institute State Grid Xinyuan Group Co., Ltd. The remaining authors declare that the research was conducted in the absence of any commercial or financial relationships that could be construed as potential conflicts of interest.

References

1. Shao, C.; Gu, C.; Meng, Z.; Hu, Y. Integrating the finite element method with a data-driven approach for dam displacement prediction. *Adv. Civ. Eng.* **2020**, *2020*, 4961963. [CrossRef]
2. Shao, C.; Gu, C.; Yang, M.; Xu, Y.; Su, H. A novel model of dam displacement based on panel data. *Struct. Control Health Monit.* **2018**, *25*, e2037. [CrossRef]
3. Qin, X.; Gu, C.; Shao, C.; Chen, Y.; Luis, V.; Zhao, E. Safety evaluation with observational data and numerical analysis of Langyashan reinforced concrete face rockfill dam. *Bull. Eng. Geol. Environ.* **2020**, *79*, 3497–3515. [CrossRef]
4. Zhang, G.; Zhang, J. Large-scale Apparatus for Monotonic and Cyclic Soil-Structure Interface Test. *Geotech. Test. J.* **2006**, *29*, 401–408. [CrossRef]
5. Meng, Z.; Wang, Y.; Zheng, S.; Wang, X.; Liu, D.; Zhang, J.; Shao, Y. Abnormal Monitoring Data Detection Based on Matrix Manipulation and the Cuckoo Search Algorithm. *Mathematics* **2024**, *12*, 1345. [CrossRef]
6. Liu, Y.; Zheng, D.; Cao, E.; Wu, X.; Chen, Z. Cracking risk analysis of face slabs in concrete face rockfill dams during the operation period. *Structures.* **2022**, *40*, 621–632. [CrossRef]
7. Jia, J.; Xu, Y.; Hao, J.; Zhang, L. Localizing and quantifying leakage through CFRDs. *J. Geotech. Geoenviron. Eng.* **2016**, *142*, 06016007. [CrossRef]
8. Chen, Q.; Zhang, L. Three-dimensional analysis of water infiltration into the Gouhou rockfill dam using saturated unsaturated seepage theory. *Can. Geotech. J.* **2006**, *43*, 449–461. [CrossRef]
9. Ma, H.; Chi, F. Technical progress on researches for the safety of high concrete-faced rockfill dams. *Engineering* **2016**, *2*, 332–339. [CrossRef]

10. Zhang, B.; Wang, J.; Shi, R. Time-dependent deformation in high concrete-faced rockfill dam and separation between concrete face slab and cushion layer. *Comput. Geotech.* **2004**, *31*, 559–573. [CrossRef]
11. Seo, M.; Ha, I.; Kim, Y.; Olson, S. Behavior of concrete-faced rockfill dams during initial impoundment. *J. Geotech. Geoenviron. Eng.* **2009**, *135*, 1070–1081. [CrossRef]
12. Won, M.; Kim, Y. A case study on the post-construction deformation of concrete face rockfill dams. *Can. Geotech. J.* **2008**, *45*, 845–852. [CrossRef]
13. Bodtman, W.; Wyatt, J. Closure to "Design and Performance of Shiroro Rockfill Dam" by William L. Bodtman and John D. Wyatt. *J. Geotech. Eng.* **1987**, *113*, 1164–1171. [CrossRef]
14. Wen, L.; Li, Y.; Zhao, W.; Cao, W.; Zhang, H. Predicting the deformation behaviour of concrete face rockfill dams by combining support vector machine and AdaBoost ensemble algorithm. *Comput. Geotech.* **2023**, *161*, 105611. [CrossRef]
15. Wen, L.; Li, Y.; Chai, J. Multiple nonlinear regression models for predicting deformation behavior of concrete-face rockfill dams. *Int. J. Geomech.* **2021**, *21*, 04020253. [CrossRef]
16. Wang, P.; Song, W. Analysis on the measured settlement and prediction modeling of statistic settlement of Shuibuya CFRD. *J. Hydroelectr. Eng.* **2009**, *28*, 81–86.
17. Chen, Y.; Gu, C.; Shao, C.; Qin, X. Parameter sensitivity and inversion analysis for a concrete face rockfill dam based on CS-BPNN. *Adv. Civ. Eng.* **2019**, *2019*, 1–17. [CrossRef]
18. Shao, C.; Zheng, S.; Gu, C.; Hu, Y.; Qin, X. A novel outlier detection method for monitoring data in dam engineering. *Expert Syst. Appl.* **2022**, *193*, 116476. [CrossRef]
19. Hu, Y.; Gu, C.; Meng, Z.; Shao, C. Improve the model stability of Dam's displacement prediction using a numerical-statistical combined model. *IEEE Access* **2020**, *8*, 147482–147493. [CrossRef]
20. Shao, C.; Gu, C.; Meng, Z.; Hu, Y. A Data-Driven Approach Based on Multivariate Copulas for Quantitative Risk Assessment of Concrete Dam. *J. Mar. Sci. Eng.* **2019**, *7*, 353. [CrossRef]
21. Hu, Y.; Shao, C.; Gu, C.; Meng, Z. Concrete Dam Displacement Prediction Based on an ISODATA-GMM Clustering and Random Coefficient Model. *Water* **2019**, *11*, 714. [CrossRef]
22. Behnia, D.; Ahangari, K.; Noorzad, A.; Moeinossadat, S.R. Predicting crest settlement in concrete face rockfill dams using adaptive neuro-fuzzy inference system and gene expression programming intelligent methods. *J. Zhejiang Univ.–Sci. A Appl. Phys. Eng.* **2013**, *14*, 58–71. [CrossRef]
23. Zheng, S.; Gu, C.; Shao, C.; Hu, Y.; Xu, Y.; Huang, X. A Novel Prediction Model for Seawall Deformation Based on CPSO-WNN-LSTM. *Mathematics* **2023**, *11*, 3752. [CrossRef]
24. Su, H.; Chen, J.; Wen, Z.; Wang, F. Wavelet-fractal diagnosis model and its criterion for concrete dam crack status. *Trans. Inst. Meas. Control* **2018**, *40*, 1846–1853. [CrossRef]
25. Kao, C.; Loh, C. Monitoring of long-term static deformation data of Fei-Tsui arch dam using artificial neural networkbased approaches. *Struct. Control Health Monit.* **2011**, *20*, 282–303. [CrossRef]
26. Kim, Y.; Kim, B. Prediction of relative crest settlement of concrete-faced rockfill dams analyzed using an artificial neural network model. *Comput. Geotech.* **2008**, *35*, 313–322. [CrossRef]
27. Hamzic, A.; Avdagic, Z. Multilevel prediction of missing time series dam displacements data based on artificial neural networks voting evaluation. In Proceedings of the 2016 IEEE International Conference on Systems, Man, and Cybernetics (SMC), Budapest, Hungary, 9–12 October 2016; Volume 2016, pp. 3–6.
28. Smagulova, K.; James, A. A survey on LSTM memristive neural network architectures and applications. *Eur. Phys. J. Spec. Top.* **2019**, *228*, 2313–2324. [CrossRef]
29. Hu, Y.; Gu, C.; Meng, Z.; Shao, C.; Min, Z. Prediction for the settlement of concrete face rockfill dams using optimized LSTM model via correlated monitoring data. *Water* **2022**, *14*, 2157. [CrossRef]
30. Xu, Y.; Pan, P.; Xing, C. Dam settlement prediction based on random error extraction and multi-input LSTM network. *J. Surv. Eng.* **2022**, *148*, 04022006. [CrossRef]
31. Li, Y.; Yin, Q.; Zhang, Y.; Qiu, W. Prediction of long-term maximum settlement deformation of concrete face rockfill dams using hybrid support vector regression optimized with HHO algorithm. *J. Civ. Struct. Health Monit.* **2023**, *13*, 371–386. [CrossRef]
32. Su, H.; Chen, Z.; Wen, Z. Performance improvement method of support vector machine based model monitoring dam safety. *Struct. Control Health Monit.* **2016**, *23*, 252–266. [CrossRef]
33. Kang, F.; Liu, J.; Li, J.; Li, S. Concrete dam deformation prediction model for health monitoring based on extreme learning machine. *Struct. Control Health Monit.* **2017**, *24*, e1997. [CrossRef]
34. Bui, K.; Bui, D.; Zou, J.; Doan, C.; Revhaug, I. A novel hybrid artificial intelligent approach based on neural fuzzy inference model and particle swarm optimization for horizontal displacement modeling of hydropower dam. *Neural Comput. Appl.* **2018**, *29*, 1495–1506. [CrossRef]
35. Dai, B.; Gu, H.; Zhu, Y.; Chen, S.; Rodriguez, E.F. On the use of an improved artificial fish swarm algorithm-backpropagation neural network for predicting dam deformation behavior. *Complexity* **2020**, *2020*, 5463893. [CrossRef]
36. Dai, B.; Gu, C.; Zhao, E.; Qin, X. Statistical model optimized random forest regression model for concrete dam deformation monitoring. *Struct. Control Health Monit.* **2018**, *25*, e2170. [CrossRef]
37. Ma, Z.; Zhu, W.; Han, S.; Li, Z. Concrete crack genesis mining based on fuzzy neural network. *J. Hydraul. Eng.* **2010**, *37*, 33–35. (In Chinese)

8. Gu, H.; Yang, M.; Gu, C.; Huang, X. A factor mining model with optimized random forest for concrete dam deformation monitoring. *Water Sci. Eng.* **2021**, *14*, 330–336. [CrossRef]
9. Cao, E.; Bao, T.; Gu, C.; Li, H.; Liu, Y.; Hu, S. A novel hybrid decomposition-ensemble prediction model for dam deformation. *Appl. Sci.* **2020**, *10*, 5700. [CrossRef]
0. Yuan, R.; Su, C.; Cao, E.; Hu, S.; Zhang, H. Exploration of multi-scale reconstruction framework in dam deformation prediction. *Appl. Sci.* **2021**, *11*, 7334. [CrossRef]
1. Huang, L.; Chen, S.; Li, C.; Si, Z.; Zhang, F. Application of Gaussian process regression model based on wavelet denoising in settlement prediction of CFRD. *J. Water Resour. Water Eng.* **2023**, *34*, 144–150. (In Chinese)
2. Heidari, A.; Mirjalili, S.; Faris, H.; Aljarah, I.; Mafarja, M.; Chen, H. Harris hawks optimization: Algorithm and applications. *Future Gener. Comput. Syst.* **2019**, *97*, 849–872. [CrossRef]
3. Yu, J.; Kim, C.; Rhee, S. The comparison of lately proposed Harris Hawks optimization and Jaya optimization in solving directional overcurrent relays coordination problem. *Complexity* **2020**, *2020*, 3807653. [CrossRef]
4. Huang, Z.; Gu, C.; Peng, J.; Wu, Y.; Gu, H.; Shao, C.; Zheng, S.; Zhu, M. A statistical prediction model for sluice seepage based on MHHO-BiLSTM. *Water* **2024**, *16*, 191. [CrossRef]
5. Wang, S.; Yang, B.; Chen, H.; Fang, W.; Yu, T. LSTM-Based Deformation Prediction Model of the Embankment Dam of the Danjiangkou Hydropower Station. *Water* **2022**, *14*, 2464. [CrossRef]
6. Ranković, V.; Grujović, N.; Divac, D.; Milivojević, N. Development of support vector regression identification model for prediction of dam structural behavior. *Struct Saf.* **2014**, *48*, 33–39. [CrossRef]
7. He, P.; Wu, W. Levy flight-improved grey wolf optimizer algorithm-based support vector regression model for dam deformation prediction. *Front. Earth Sci.* **2023**, *11*, 1122937. [CrossRef]
8. Man, K.; Cao, Z.; Liu, X.; Song, Z.; Liu, R. Predicting TBM tunnelling parameters with HHO-LSTM-SVR model. *Railw. Stand. Des.* **2024**, *69*, 1–12. (In Chinese)

water

MDPI

Article

An Improved ResNet-Based Algorithm for Crack Detection of Concrete Dams Using Dynamic Knowledge Distillation

Jingying Zhang [1,2] and Tengfei Bao [1,2,3,*]

1 College of Water Conservancy and Hydropower Engineering, Hohai University, Nanjing 210098, China; 200402020006@hhu.edu.cn
2 The National Key Laboratory of Water Disaster Prevention, Hohai University, Nanjing 210098, China
3 College of Hydraulic and Environmental Engineering, China Three Gorges University, Yichang 443002, China
* Correspondence: baotf@hhu.edu.cn

Abstract: Crack detection is an important component of dam safety monitoring. Detection methods based on deep convolutional neural networks (DCNNs) are widely used for their high efficiency and safety. Most existing DCNNs with high accuracy are too complex for users to deploy for real-time detection. However, compressing models face the dilemma of sacrificing detection accuracy. To solve this dilemma, an improved residual neural network (ResNet)-based algorithm for concrete dam crack detection using dynamic knowledge distillation is proposed in this paper in order to obtain higher accuracy for small models. To see how well distillation works, preliminary experiments were carried out on mini-ImageNet. ResNet18 was trained by adding additional tasks to match soft targets generated by ResNet50 under dynamic high temperatures. Furthermore, these pre-trained teacher and student models were transferred to experiments on concrete crack detection. The results showed that the accuracy of the improved algorithm was up to 99.85%, an increase of 4.92%.

Keywords: residual neural network; knowledge distillation; transfer learning; concrete dam; crack detection

1. Introduction

Concrete structures are widely used in the construction of dams because of their high compressive strength. Under the influence of external forces [1] such as cyclic loads, temperature changes, and destructive earthquakes during long-term service, their surfaces are prone to cracks. In addition, cavitation [2–5] plays a significant role in dam damage. The development of cracks will destroy the integrity of concrete structures, which will cause leakage problems and affect the overall strength and stability of structures. Even worse, dams will end up in catastrophic accidents such as collapses and dam breaks. Therefore, crack detection is essential for the safety inspection of concrete dams to detect faults and repair them in a timely manner.

Historically, crack detection in concrete dams mainly depended on manual inspection. However, the results of manual detection are highly subjective, time-consuming, and laborious in the case of extensive fracture data. In addition, reservoir dams are primarily built in mountainous or hilly areas with complex terrain conditions, resulting in poor efficiency and safety in the detection process, especially in harsh environmental conditions such as underwater, where localization is even more prominent. In order to reduce the limitations of naked-eye observation and improve detection accuracy and efficiency, automatic detection technologies have emerged, such as vibrating wire sensors, fiber grating sensors, acoustic emission equipment, etc. However, these detection methods must determine where cracks are in advance, and their detection area is significantly small. In recent years, with the improvement of computer equipment performance and the rapid development of computer vision technology, crack detection methods based on artificial intelligence and deep learning have ushered in a tumultuous period of growth. Crack images of concrete structures are collected by image acquisition systems such as unmanned aerial vehicles (UAV) [6] and underwater detection robots and then processed by digital image processing techniques [7] or deep learning methods to classify and identify whether there are cracks in these images. Among them, image classification methods based on deep convolutional neural networks (DCNNs) are superior to digital image processing methods that rely on manual intervention and discrimination because they automatically extract image features. They have brought a series of breakthroughs in image classification and are widely used in crack detection.

Because of the complexity of computing, memory, and storage requirements, the training phase of the networks is performed on CPU or GPU clusters in a distributed computing environment. These networks, however, typically involve large models with numerous parameters. Once trained, a challenging aspect is the deployment of the trained models on resource-constrained inference systems such as portable devices, including underwater detection robots, as mentioned above, or sensor networks, and on applications in which real-time predictions are required. Performing inference on edge devices comes with severe memory, computing, and power constraints. We hope to deploy deep neural networks for real-time operations, which have more stringent requirements for latency and computing resources. We need to use smaller models to extract structures from less abundant data sets without losing accuracy.

Quantization using low-precision numerics [8–10] and model compression [11] have emerged as popular solutions for resource-constrained deployment scenarios. With quantization, a low-precision version of the network model is generated and deployed on the device. Operating in lower precision mode reduces computation, data movement, and storage requirements. However, the majority of existing works in low-precision DCNNs sacrifice accuracy over the baseline full-precision networks [12]. Therefore, we propose an improved DCNN crack detection algorithm using knowledge distillation. A smaller, low-memory footprint network is trained to mimic the behavior of the original complex network, "transferring" knowledge from the complex network to the smaller network, compressing the model without reducing accuracy, and allowing the smaller network to achieve accuracy equivalent to or slightly better than the original complex model.

The knowledge distillation system consists of a teacher network (large model) and a student network (small model). The category probability generated by the teacher model under high temperature is used as the "soft target" for training the student model, and knowledge is transferred from the large model in the training stage to the small model that is more suitable for deployment to obtain more accurate crack detection results on the small model [13,14]. It is necessary to select appropriate teacher networks and student networks for crack detection to ensure the benchmark accuracy of the knowledge distillation system. Common neural networks for crack detection include LeNet [15], AlexNet [16], VGG-Net [17], etc. These DCNNs improve model accuracy by increasing the depth of networks. However, as network layers continue to rise, training accuracy and testing accuracy rapidly decline when network layers increase to a certain number, indicating that the deeper these networks continue to grow, the more activation functions are introduced and the more discretely data are mapped, making it difficult to return to the origin space (also known as identity transformation), which means that DCNNs have a degradation problem. Therefore, residual neural networks (ResNet) are introduced as the basic framework of the improved crack detection algorithm. By adding a short-cut connection (also known as a residual connection) to nonlinear convolution layers, ResNet improves the efficiency of information dissemination and solves the problem of degradation. With ResNet50 used as the teacher model and ResNet18 used as the student model, knowledge distillation is performed to ensure basic accuracy.

On the other hand, scholars have continuously improved DCNNs used in crack detection, but they are all oriented toward enhancing their accuracy. Although the accuracy of these models has improved, there is still a large gap between the accuracy of training sets and validation sets, indicating that there is still considerable room for improvement in the generalization ability of networks. The improved algorithm has better model generalization ability, which is also a contribution to this paper.

In addition, the most critical parameter in knowledge distillation systems is distillation temperature [18–22]. However, current work generally uses fixed temperature parameters, which are usually set to 4. Although some research and experiments have verified this empirical value, it does not apply to all knowledge distillation tasks and models. For specific tasks, if one wants to match the soft targets of the teacher model and the student models' soft targets under the optimal temperature parameters, an exhausting search method is required, which poses a great challenge to computing resources and efficiency. In this context, we propose a dynamic temperature search method for knowledge distillation, adding an adversarial dynamic temperature module. As the main contribution of this paper, the proposed method makes it possible for the network to adjust temperature parameters by itself, and in each epoch of training, a suitable temperature can be found for distillation.

It is worth noting that the initial training samples of this algorithm were not concrete crack datasets because this type of dataset has a relatively small size and a single type, which are sometimes not easy to obtain. Applying this algorithm directly to concrete crack datasets for classification cannot guarantee a satisfactory training result. Therefore, preliminary experiments were carried out on mini-ImageNet and then transferred their training results to experiments on concrete crack detection to ensure more accurate detection of concrete cracks. Through experiments, the contribution of transfer learning (pretraining) to the improvement of model accuracy was proven.

This paper is organized as follows: Section 2 explains the proposed methodology, including knowledge distillation and transfer learning; Section 3 illustrates preliminary experiments conducted on mini-ImageNet and discusses the findings; Section 4 transfers the improved algorithm pre-trained in Section 3 to crack detection and discusses the results; and Section 5 applies the improved algorithm to the crack detection task of HKC Dam. Finally, Section 6 discusses the main conclusions.

2. Methodology

2.1. ResNet-Based Dynamic Knowledge Distillation Architecture

Large DCNNs have achieved remarkable success with good performance, especially in cases with large-scale data, because the over-parameterization improves the generalization performance when new data are considered [23–26]. With the significant breakthrough of DCNNs in image classification, much research has emerged that applies them to crack detection. However, deploying deep models on mobile devices and embedded systems is a great challenge due to the devices' limited computational capacity and memory. Model compression [11] was proposed to address this issue by transferring the information from a large model or an ensemble of models into training a small model without a significant drop in accuracy. The improved algorithm using dynamic knowledge distillation proposed in this paper regards large models as "teachers" and small models as "students". In this algorithm, a small student model is generally supervised by a large teacher model [27–29], and a temperature parameter [28] is introduced into the softmax function. Taking the probability distribution generated by the teacher model as "soft targets" for training the student model, it can be softened by adjusting the temperature. Soft targets usually have high entropy, providing much more information for each training case than hard targets, and the gradient variance between training cases is much smaller. If the student model is trained at the same temperature, it can learn such knowledge from the teacher model, achieving competitive and even superior performance.

A knowledge distillation system is composed of three key components: teacher-student architecture, knowledge, and distillation loss. A general framework for knowledge distillation is shown in Figure 1.

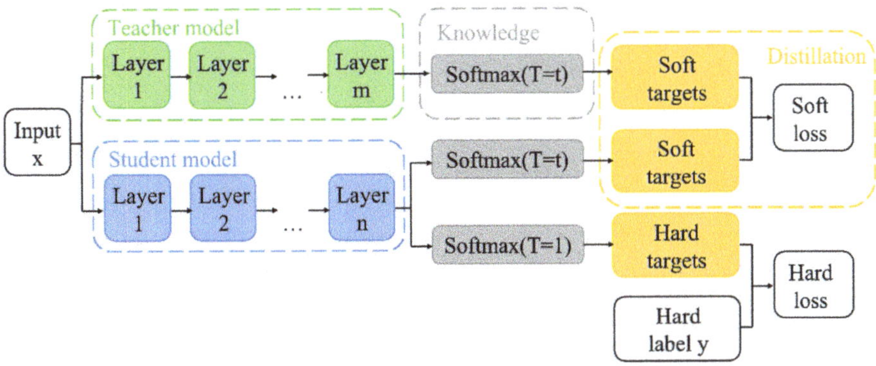

Figure 1. A generic framework for knowledge distillation.

As seen in Figure 1, in the process of knowledge distillation, a teacher model with a complex network structure needs to be trained on datasets at a high temperature in advance. Thereafter, a simplified student network must be guided to carry out the same task at the same temperature and at a low temperature for model migration. Neural networks typically use a converted softmax output layer to generate class probabilities. The difference among probability distributions is significantly small. Generally, softmax classifiers use an exponential function to amplify the gap between logits. After standardizing these logits, it will output a one-hot vector, which is also called a hard target. One variable of the one-hot vector equals 1, indicating that the input data belongs to a specific class, while the other variables are all 0, containing too little information, which is not conducive to the optimization of the network. Knowledge distillation provides a generalized softmax function to soften the output:

$$q_i = \frac{exp(z_i/T)}{\sum_1^N exp(z_i/T)} \quad (1)$$

where z_i represents logits generated by networks, q_i is a soft target, and T is a temperature usually set to 1. Using a higher T will produce a softer probability distribution over classes.

The temperature represents the softening degree of targets. When T approaches 0, the softmax function is still the standard softmax function and will still output a one-hot vector without the softening effect. When T approaches infinity, the output of the softmax function will be very soft, and more adequate information can be obtained from soft targets. Therefore, when training the teacher network, a higher T was set to make the output of softmax sufficiently soft target, and the output of the student network at the same temperature would be close to the teacher model. Finally, we trained the student network again at the normal temperature (T = 1) to output hard targets.

2.1.1. Selection of Appropriate Teacher and Student Models

DCNNs are continuously optimized by various means to improve detection accuracy, most of which are to deepen the networks. In a network without residual blocks, the deeper the network is, the stronger its learning ability should become. However, in practice, features learned from input data are often far from those we need because of gradient disappearance or explosion. The network needs to be improved to make it, once deepened, still extract more valuable features for training that are not worse than the shallower networks.

In this paper, two residual neural networks [30] with different depths were introduced as the teacher model and student model for knowledge distillation to solve the problem that the classification effect of convolutional neural networks deteriorates with the increase of depth and ensure the benchmark accuracy of subsequent experiments.

Taking a two-layer neural network as an example, as shown in Figure 2, the left part of the figure is an ordinary neural network, while the right one is a residual neural network. The main path of a common neural network is to input an x into the network. After two weight layers and one ReLU nonlinear activation, F(x) is outputted. Thereafter, F(x) is used as a new input to experience the next ReLU nonlinear activation. The difference between an ordinary neural network and a residual neural network is that there is an additional shortcut connection in the latter one. The input x of the residual neural network enters the deeper layer along with the output F(x) of its last layer, which acts together on the subsequent ReLU nonlinear activation function, forming a residual block. The curved arrow in Figure 2 is a "shortcut", which means that the input x should not only go through the main path of the ordinary neural network but also affect the deeper network with the output F(x). A convolutional neural network with such shortcut connections is a residual block, and several residual blocks constitute ResNet.

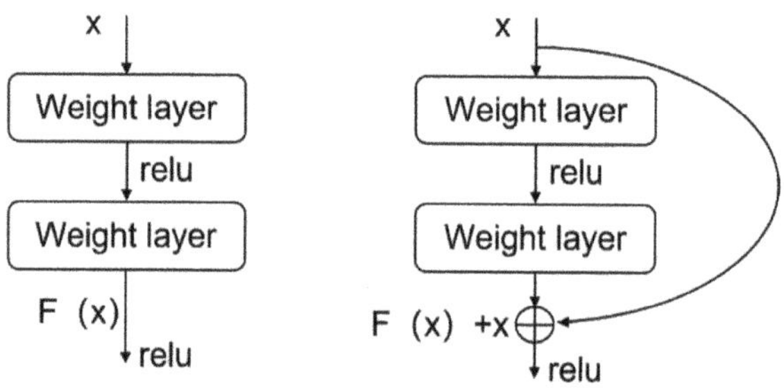

Figure 2. Structure of a residual block.

The structures of ResNet18 and ResNet50 are shown in Figure 3. They are divided into six stages. The first stage is convolution and maximum pooling once data are input. Data are convolved with residual blocks from the second to the fifth stage. Each stage has two different basic residual blocks: the Conv Block and the Identity Block (ID Block). The data are convolved three times in each residual block. Conv Blocks' function is to change the dimension of the network, so the input and output dimensions of these blocks are different and cannot be connected in series. In contrast, ID Blocks' input and output dimensions are the same, so they can be concatenated to deepen the network. In the sixth stage, data are pooled globally and flattened, and then data can be classified. The structure of ResNet18 is quite similar to that of ResNet50. The only difference is that the number of residual blocks differs from [2, 2, 2, 2] to [3, 4, 6, 3].

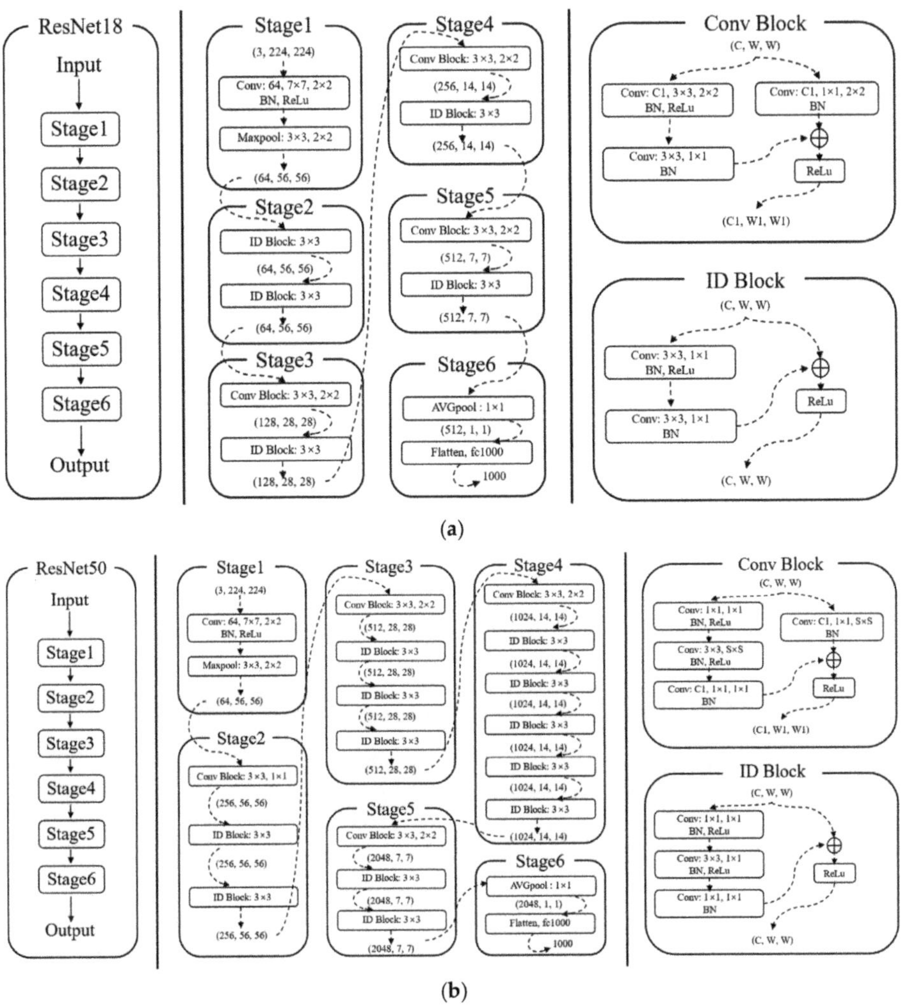

Figure 3. Structures of different residual neural networks: (**a**) ResNet18; (**b**) ResNet50.

2.1.2. Distillation Loss Function

Knowledge distillation aims to minimize the difference between the output of the student network and that of the teacher network. A distillation loss function can quantify this objective. As shown in Figure 1, the loss function in the distillation process consists of distillation loss (soft loss) and student loss (hard loss).

When the correct labels are known for all or some of the datasets, this method can be significantly improved by training the distilled model (the student model) to produce the correct labels. One way to achieve this is to use the correct labels to modify the soft targets, but we found that a better way is to simply use a weighted average of two different objective functions.

$$L = \alpha L_{\text{soft}} + (1 - \alpha) L_{\text{hard}} \tag{2}$$

here, L_{soft} is the first objective function, representing the cross entropy with the soft targets, and this cross-entropy is computed using the same high temperature in the softmax of the distilled model as was used for generating the soft targets from the cumbersome teacher model. The formula is as follows:

$$L_{\text{soft}} = -\sum_{1}^{N} t_i^T log\left(s_i^T\right) \tag{3}$$

where N represents the total number of labels/classes, and t_i^T and s_i^T are soft targets of the teacher and student models in class i when the temperature is T. The formulae to calculate t_i^T and s_i^T are as follows:

$$t_i^T = \frac{exp(u_i/T)}{\sum_1^N exp(u_i/T)} \tag{4}$$

$$s_i^T = \frac{exp(v_i/T)}{\sum_1^N exp(v_i/T)} \tag{5}$$

where u_i and v_i represent the logits of teacher and student networks, respectively.

The second objective function is L_{hard}. It represents the cross-entropy with the correct labels (hard labels). It is computed using exactly the same logits in the softmax function of the distilled model, but at a temperature of 1:

$$L_{\text{hard}} = -\sum_{1}^{N} c_i \frac{v_i}{\sum_1^N exp(v_i)} \tag{6}$$

where c_i represents the correct label on class j. In $c_i \in \{0,1\}$ 1 is the positive label and 0 is the negative one.

Since the magnitudes of the gradients produced by the soft targets scale as $1/T^2$, it is essential to multiply them by T^2 when considering soft loss. This ensures that the relative contributions of the hard and soft targets remain roughly unchanged if the temperature used for distillation is changed while experimenting with meta-parameters.

2.1.3. Dynamic Temperature for Knowledge Distillation

For a knowledge distillation system, the temperature parameter controls the smoothness of two prediction results and determines the distance between two probability distributions. The higher the temperature is, the smoother the probability distribution will be; otherwise, the closer the temperature is to zero, the sharper the probability distribution will be. At the same time, temperature affects how difficult it is for the student model to mimic the teacher model during the distillation process. Different distillation results will be produced at different temperatures. The common method in existing work is to use a fixed temperature parameter, generally set to 4.

However, the optimal temperature parameter for different distillation systems for various tasks is not necessarily equal to 4. Suppose we want to find the optimal temperature parameter for a specific task. In that case, we need to perform an exhaustive search, which will result in a large amount of computation, and the entire training process is very inefficient. At the same time, maintaining a static and fixed temperature parameter is not the best choice for student models. Based on curriculum learning [31], humans learn from simplicity to difficulty in the learning process. We also hope to form a step-by-step distillation difficulty model for the student during the distillation process.

Therefore, we propose to dynamically adjust the temperature during the training process by adding an antagonistic dynamic temperature module, making it possible for the network to automatically select a suitable temperature for distillation in each training epoch. After adding the dynamic temperature module, the overall architecture of knowledge distillation is shown in Figure 4.

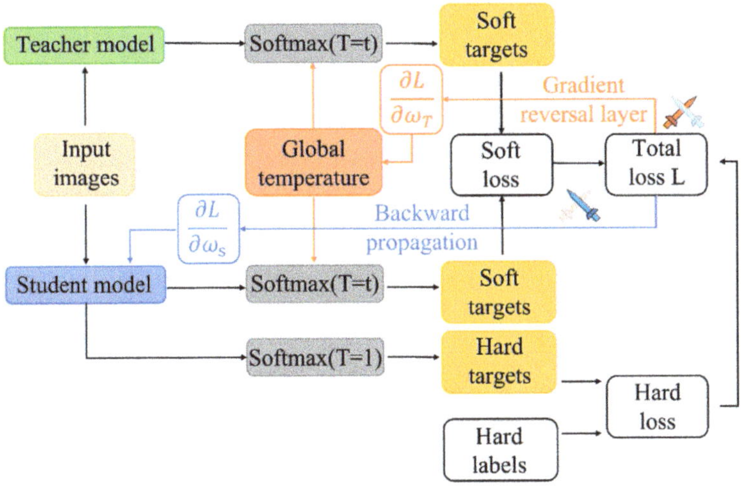

Figure 4. The architecture of dynamic knowledge distillation.

As shown in Figure 4, a dynamic temperature parameter is implemented through a nonparametric gradient reverse layer (GRL) to satisfy the need for confrontation: inserting a GRL between the softmax layer and the learnable temperature module and using the GRL to reverse the gradient of the learnable temperature parameter to amplify the distillation loss between the student and teacher model when the student network is trained to minimize distillation loss. The effect of confrontation can be achieved very directly in this case. The inverse gradient's weight rises with training progress, thereby increasing the learning difficulty.

After adding the dynamic temperature module, the target function of distillation loss is modified as follows:

$$\min_{\omega_s} \max_{\omega_T} L = \min_{\omega_s} \max_{\omega_T} \sum_{x \in D} \alpha L_{\text{soft}}(f^t(x;\omega_t), f^s(x;\omega_s), \omega_T) + (1-\alpha) L_{\text{hard}}(f^s(x;\omega_s), y) \quad (7)$$

The formula above can be solved by alternating algorithms: fixing one set of variables, solving another set of variables, and alternating between solving subproblems related to these two variables. The relevant pseudocode is shown in Algorithm 1.

Algorithm 1: Implementation of Dynamic Distillation

Input: Training dataset D; Total training epoch M; Pre-trained Teacher ω_t; Learnable Temperature Module ω_T
Output: Distilled Student ω_s
Initialize: Epoch m = 1; Randomly initialize: ω_T, ω_s
  while m $\leq$ M do
    for batch x in D do
      Forward propagation through ω_t and ω_s to obtain predictions $f^t(x;\omega_t), f^s(x;\omega_s)$
      Obtain temperature T by ω_T
      Calculate the loss L and update ω_s and ω_T by backward propagation
    end for
  m = m + 1;
  end while

In the dynamic temperature module, we use a global temperature, adapting the same temperature to all instances to be predicted. This efficient version does not incur additional computational costs for the distillation process because it involves only one learnable parameter.

2.2. Transfer Learning

In some deep learning scenarios, the cost of directly training the target task is too high, and the dataset provided by the target task is too small to support deep learning and achieve good results. For example, there had not yet been a set of crack images with rich samples for non-destructive crack detection of concrete structures, the research object of this chapter. Therefore, we expected that the model trained in advance on the task with sufficient samples could be directly applied to the new target task.

Transfer learning [32,33] is an effective method to solve training tasks without rich samples in computer vision. The training set and the test set do not need to meet the assumption of independent and identical distribution, so they can extract the common or essential features between two different but correlational tasks and use the common features to realize the mutual transfer of learning ability between these two tasks. The network in the target task also does not need to start training from scratch but extracts high-dimensional semantic features to complete the target task based on the common features mined in the source task. This shortens the training time and achieves considerable performance even with a small amount of data in the target task because it does not need a large number of training data to obtain features. In short, transfer learning can obtain common invariants from the source task and transfer them to the target task.

Transfer learning can be divided into sample transfer, feature transfer, and parameter/model transfer according to different learning objects [34]. The crack detection algorithm established in this paper mainly includes a backbone coding module and a classification module. The backbone coding module is primarily composed of multiple convolutional and pooling layers for feature coding. The classification module mainly consists of numerous fully connected layers whose primary function is to output the probability distribution of each category to achieve final results. The shallow convolutional layers near the input layer in the backbone coding module are suitable for extracting common features. In contrast, the deeper convolutional and classification layers are ideal for mining specific personality features. When the pre-trained network of the source task is transferred to the target task, the weight and bias parameters of the classification module need to be removed and redesigned to suit the target task. Therefore, the parameter/model transfer method was used in this chapter. Firstly, an improved image classification model on mini-ImageNet with excellent performance was trained. Secondly, its structure and parameters were frozen for invocation as the backbone coding module for feature extraction in the target task of crack detection. Thirdly, some important hyperparameters were fine-tuned, and the structure of the fully connected layer was modified to a two-classifier for crack detection. Thus, the multi-classification problem of mini-ImageNet was changed to the two-classification problem of crack detection, as shown in Figure 5. In other words, it was not necessary to start training the improved network used for crack detection from scratch but only to iterate and update parts of the parameters, which significantly reduced the training workload and shortened the training time.

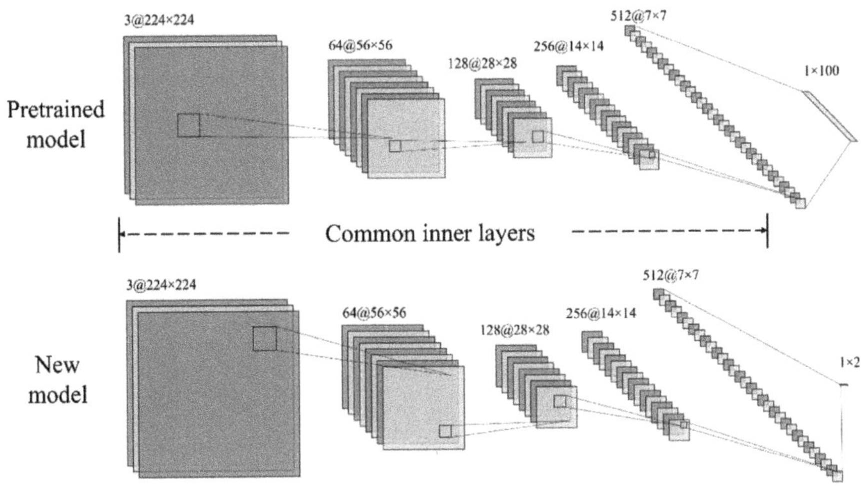

Figure 5. Steps of transfer learning.

3. Preliminary Experiments on Mini-ImageNet

ImageNet [35] is one of the most influential image recognition datasets in the world at present, with a total of 1,431,167 images in 1000 categories and their corresponding labels. Based on its large samples and diversity, ImageNet is the most popular dataset for pre-training various neural networks. The parameters are reliable and versatile after pre-training. However, training neural networks on ImageNet must consume a large amount of computing resources. Sixty thousand pictures (including 100 categories and 600 samples for each category) were extracted from ImageNet as the pretraining dataset of residual neural networks to save training time in this chapter.

3.1. Training Tricks

In deep learning, the learning rate is a vital hyperparameter that controls optimization algorithms' convergence speed. A high learning rate can lead to rapid convergence and cause the optimization algorithm to oscillate or converge to a suboptimal solution. A low learning rate, on the other hand, can ensure that the optimization algorithm converges to a good solution, but its convergence speed may be slow.

There are two strategies to solve this dilemma. The first strategy is to warm up the learning rate [36]. The parameters of ResNet18 and ResNet50 are randomly initialized at the initial stage of training. During this stage, if a relatively high learning rate is selected, it may cause oscillations in the networks. Warming up the learning rate can make the learning rate in the first few epochs of training relatively small. The networks can slowly become stable with a lower learning rate in the warm-up stage. When these two networks are relatively stable, the pre-set learning rate can be used for training. This strategy makes the convergence speed of the networks faster and their training effects better. In this paper, the steps of the warm-up stage were set to 1.

Another helpful trick is the decay strategy of the learning rate. The decay strategy used in this paper was cosine annealing [37,38]. When using gradient descent algorithms to optimize objective functions, the closer networks get to the global minimum loss, the smaller the learning rate should become to avoid networks missing this point. The cosine annealing decay strategy can reduce the learning rate through the cosine function. With the increase in input to the cosine function, the cosine value first slowly decreases, then accelerates to decline, and then slowly decreases again. This gradual decrease in learning rate helps to avoid overshooting the optimal solution and ensure that the optimization algorithm converges to a reasonable solution smoothly and stably.

3.2. Some Findings

To see how well distillation works, we trained two residual neural networks with different depths on mini-ImageNet, a large dataset with 60,000 samples. The accuracy of the teacher network ResNet50 was 81.79%, while the accuracy of the student network ResNet18 without knowledge distillation was 79.41%. However, if the smaller network was trained by adding the additional task of matching the soft targets produced by the large network at a higher temperature, empirically taking $T = 4$ and $\alpha = 0.6$ as an example, its accuracy reached 82.33%. It shows that soft targets can transfer a great deal of knowledge to the distilled model, including how to generalize what is learned from translated training data, even though the transfer set does not contain any translations. Table 1 shows the results of ablation experiments at different distillation temperatures.

Table 1. Results of ablation experiments.

Model	T	Epoch [1]	Train Acc [2] (%)	Hard Loss (%)	Best Acc [2] (%)	Val [3] Loss (%)
Student ResNet18	-	96	93.44	0.26	79.41	0.86
Teacher ResNet50	-	99	94.40	0.22	81.79	0.76
KD [4]	2	97	93.05	0.10	81.61	0.79
KD	3	93	92.53	0.11	81.95	0.78
KD	4	97	92.49	0.11	82.33	0.78
KD	5	94	91.99	0.12	82.04	0.77
KD(DT [5])	-	99	92.34	0.11	82.52	0.77

Note(s): [1] Epoch represents the number of training epochs with the highest accuracy. [2] Acc is the abbreviation for accuracy. [3] Val is the abbreviation for validation. [4] KD is the abbreviation for knowledge distillation, representing the distilled student network ResNet18. [5] DT is the abbreviation for dynamic temperature.

In addition, we find that knowledge distillation can also improve generalization by narrowing the gap between training and validation accuracy compared to that of the original student model and even the teacher model. For instance, the training accuracy and the validation accuracy of ResNet18 were 93.44% and 79.41%, which differed by 14.03%, and those of ResNet50 were 94.40% and 81.79%, which differed by 12.61%, while those of distilled ($T = 4$, $\alpha = 0.6$) ResNet18 were 92.49% and 82.33%, which differed by 10.16%. This is solid and decisive proof that the generalization ability of the improved algorithm is enhanced through knowledge distillation, which provides excellent support for transferring this algorithm to train crack datasets in the following experiments.

Furthermore, based on the empirical fixed temperature of 4, we set the variation range of dynamic temperature to [3, 5] and found that the model's accuracy using dynamic temperature parameters reached 82.52%. The final distillation temperature approached 3.95. Compared with the results of distillation systems under fixed temperature parameters, the distillation effect was significantly improved with a dynamic temperature module. The network itself can adjust the distillation temperature without repeating exhausting experiments to find the optimal fixed temperature or simply taking an empirical one. In the next experiment, we apply dynamic temperature to crack detection.

4. Experiments on Concrete Crack Detection

As mentioned above, the improved algorithm has achieved a more accurate image classification performance on mini-ImageNet. The next step is to transfer its training results to practice on crack datasets. The work to be carried out includes establishing crack datasets of concrete structures and transfer learning.

4.1. Concrete Crack Datasets

The original dataset [39,40] comprises 458 high-resolution images (4032 × 3024 pixels) of concrete cracks. Due to the limited number of samples, these images were cropped to generate 20,000 positive samples with cracks and 20,000 negative samples without cracks. Each sample is a 224 × 224 RGB image; thus, a dataset [41] of concrete cracks was established.

These images were randomly put into the training set and the validation set at a ratio of 8:2. There are two folders under the training set and the validation set: one for collecting positive samples and the other for collecting negative samples, respectively. Some samples with and without cracks are shown in Figure 6. It can be seen that the crack dataset established in this paper contains cracks of various shapes and widths and has good randomness and richness.

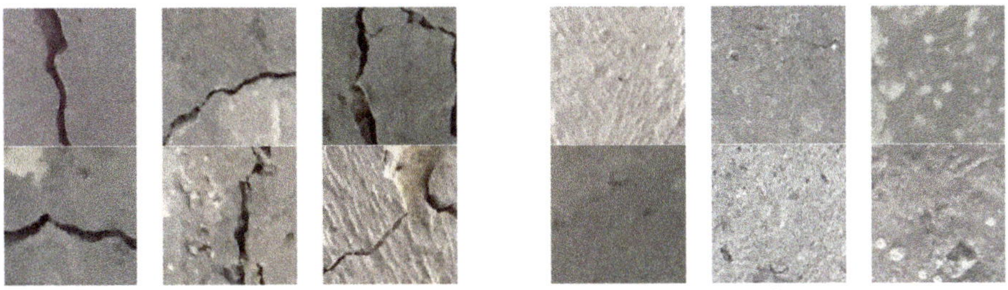
(a) (b)

Figure 6. Samples of the crack dataset: (**a**) samples with cracks; (**b**) samples without cracks.

4.2. Results

Taking the pre-trained network as the starting point and changing its fully connected layer, the multi-classification problem on mini-ImageNet was modified to a two-classification problem of concrete cracks (with or without cracks). Image enhancement measures such as random rotation and horizontal and vertical flipping were also taken to improve model generalization. In addition, the normalization of each channel was conducive to the training of networks, thus further enhancing the accuracy of the crack detection model. The network was trained for a total of 10 epochs. The performance of the improved algorithm on crack detection can be evaluated by the accuracy and loss of the training set and the validation set during each epoch of training, as shown in Figure 7.

It can be seen from Figure 7 that the improved ResNet-based algorithm using knowledge distillation performed exceptionally well on crack detection at the right beginning, taking the accuracy and loss of the validation set as the evaluation criteria. The high accuracy of the first epoch is the result of pre-training on mini-ImageNet and transfer learning. With the increase in training epochs, the improved algorithm performed better and better. When training to the eighth epoch, the algorithm achieved the best performance. Comparing the performance of ResNet18, ResNet50, and distilled ResNet18 on the task of crack detection, the detection ability of the teacher model is generally better than that of the student model. Additionally, the feature extraction ability of the distilled student model is enhanced, so its accuracy is higher than that of the original model, even exceeding that of the teacher model. Taking the optimal results as an example, the accuracy of the improved algorithm based on knowledge distillation applied to the crack detection of concrete structures was as high as 99.85%. Compared with the original algorithm, the accuracy was increased by 3.78%. With six images randomly selected from the crack dataset, correct detection results were obtained, taking the trained model of this epoch as the ultimate one for detection, as shown in Figure 8.

Meanwhile, we also carried out ablation experiments from different angles. For example, we compared the crack detection results of student and teacher networks with and without pretraining, respectively, as well as other improved methods of ResNets other than knowledge distillation, such as the insertion of convolutional block attention module (CBAM) and some other DCNNs applied to crack detection, such as AlexNet and VGG-Net. The results are shown in Table 2.

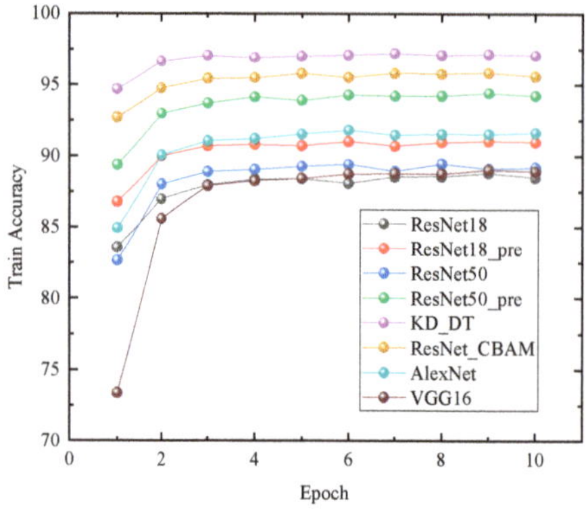

(a)

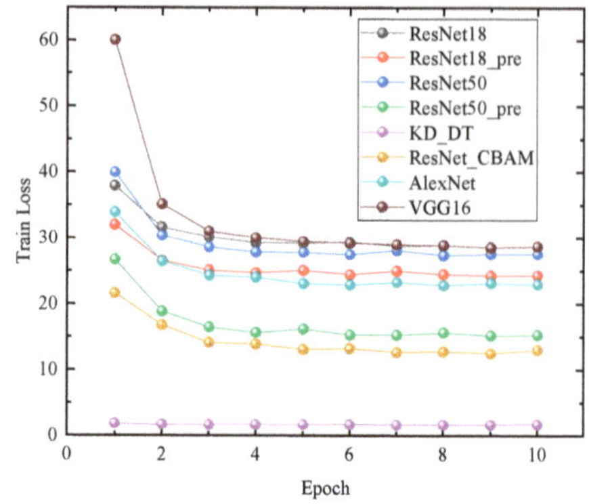

(b)

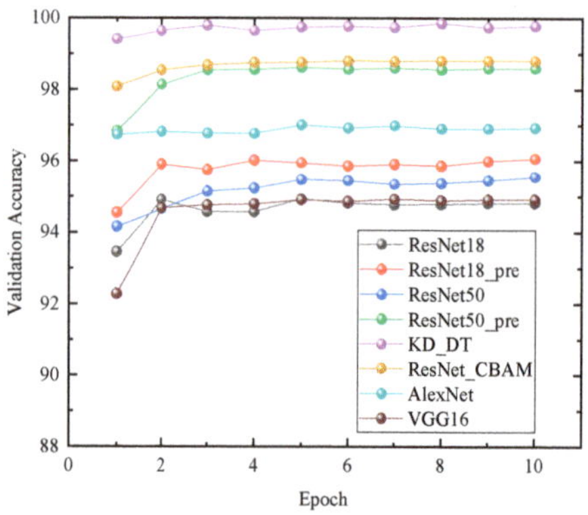

(c)

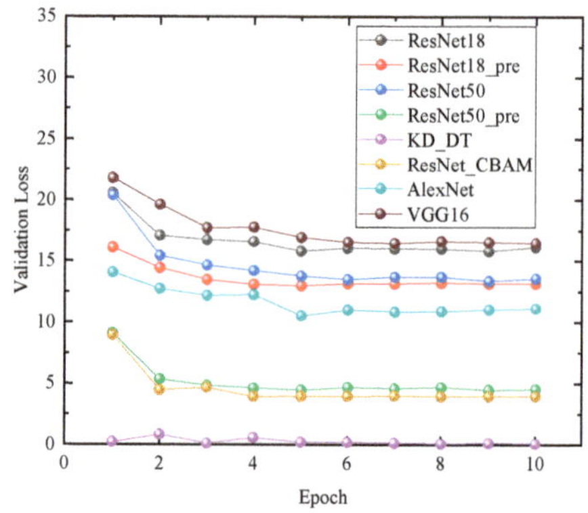

(d)

Figure 7. Performance of different crack detection algorithms: (**a**) training accuracy; (**b**) training loss; (**c**) validation accuracy; and (**d**) validation loss.

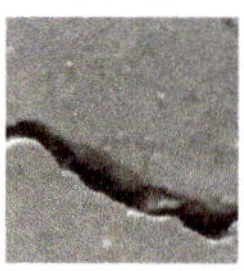
(a)

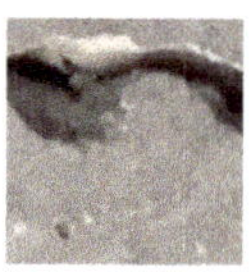
(b)

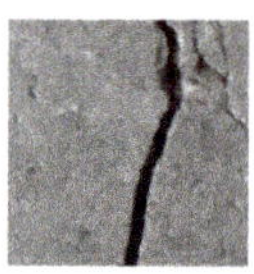
(c)

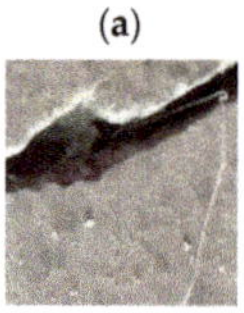
(d)

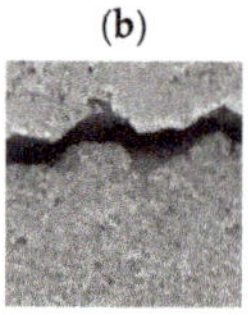
(e)

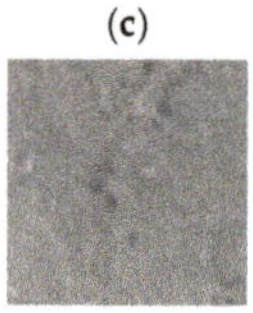
(f)

Figure 8. Crack detection results: (**a**) positive; (**b**) positive; (**c**) positive; (**d**) positive; (**e**) positive; and (**f**) negative.

From Table 2, it can be seen that the improved pre-trained network based on dynamic knowledge distillation presented an improvement of 4.92% in accuracy compared to the original student network without pretraining. The accuracy was 1.05% higher than that of networks with CBAM. Compared with other DCNNs, such as AlexNet and VGG-Net, the accuracy was improved by 2.83% and 4.92%, respectively. This algorithm is well prepared for application to crack detection in concrete dams. In the subsequent case study, we take the HKC Dam as an example to conduct crack detection on the concrete structures of its spillway.

Table 2. Crack detection results for different networks.

Model	ResNet18 *		ResNet50 *		KD	CBAM	AlexNet	VGG-Net
Best Acc	94.93	96.07 (+1.14)	95.56 (+0.63) (−0.51)	98.62 (+3.69) (+2.55)	99.85 (+4.92) (+3.78)	98.80 (+3.87) (+2.73)	97.02 (+2.09) (+0.95)	94.93 (+0.00) (−1.14)

Note(s): * The student and teacher models have two columns of data, which are the crack detection results of pre-trained models and those without pretraining. The values in parentheses in the second row are the degrees of improvement compared to the accuracy of the student model without pre-training, and the values in parentheses in the third row are the degrees of improvement compared to that of the pre-trained student model.

5. Case Study

Long-term and periodic changes in water pressure and temperature, structural settlement, geological disasters, and other factors can easily lead to the evolution of small cracks in concrete dams into more severe damage. Timely crack detection is necessary to ensure the safety and stability of concrete dams. Furthermore, accurately identifying whether there are cracks in a specific area of a dam is an essential requirement for crack detection. Since the evolution of cracks in concrete dams is quite a slow process, lasting for several years to several decades, few crack images are available in the short term. It is necessary to conduct research on crack detection in the case of a small number of samples. Taking the HKC Dam as an example, the improved ResNet-based algorithm using knowledge distillation mentioned above in this paper and already trained on crack datasets was applied to identify the cracks in the concrete dam.

5.1. Case Description—HKC Dam

HKC Dam is located at the exit of the Qin River's last section, the Yellow River's primary tributary. The basin area controlled by the reservoir is 9223 km^2, accounting for 68.2% of the basin area of the Qin River. The design flood control standard has a 500-year return period, and the check standard has a 2000-year return period. With a total storage capacity of 317 million cubic meters, the dam is a hydraulic project focusing on flood control, water supply, and the comprehensive utilization of irrigation and power generation. The main structures include a concrete face rockfill dam, spillway, and power generation system, as shown in Figure 9. We used UAVs to acquire images of the dam, and it was found that the spillway of the dam had many cracks due to long-term operation, as shown in Figure 10.

Figure 9. Realistic view of HKC Dam.

Figure 10. Crack samples of the spillway.

5.2. Application to HKC Dam

When applying the improved ResNet-based algorithm to crack detection at HKC Dam, the first step was to divide crack images into several blocks, the size of which is determined by that of the crack images in the above chapter (224 × 224 pixels). Thereafter, we classified each block using the previously trained model, resulting in an accuracy rate of 98.39%. A threshold segmentation method, a kind of digital image processing technique, was also used to detect cracks for comparison. The results of these two methods are shown in Table 3.

Table 3. Crack detection results for HKC Dam.

Original Images	Original Feature Maps	Segmented Images	Detection Results
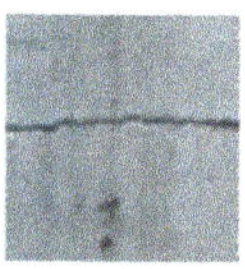	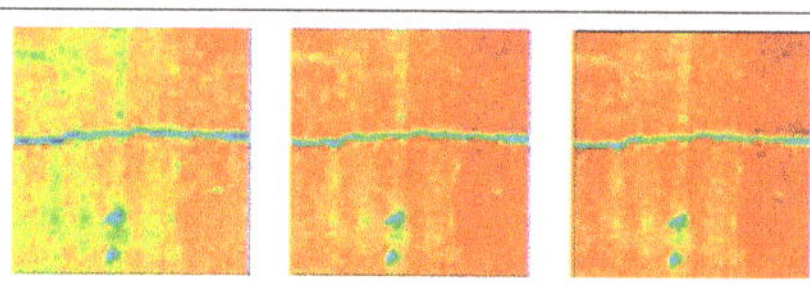	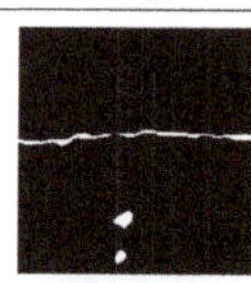	positive
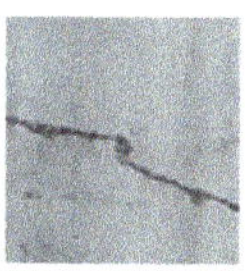	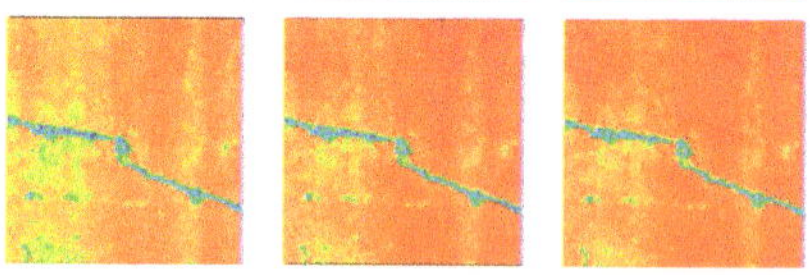	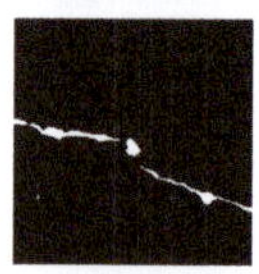	positive
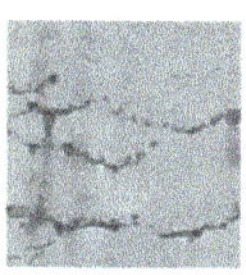	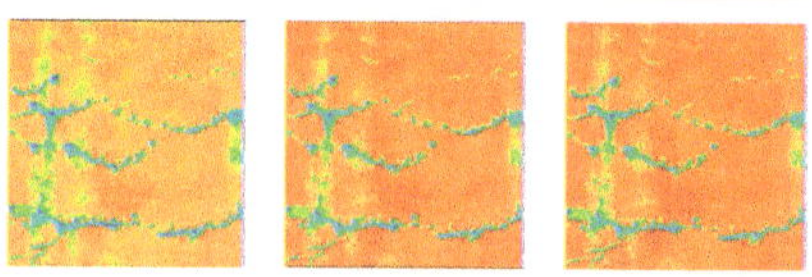	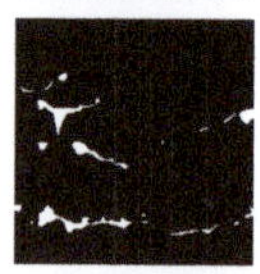	positive
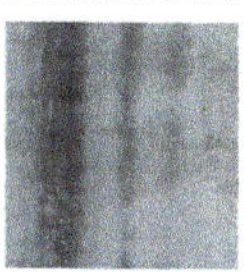	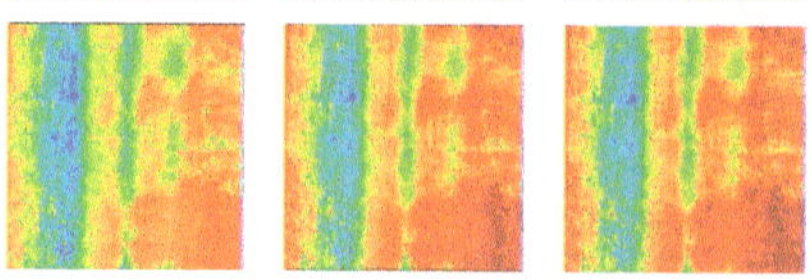	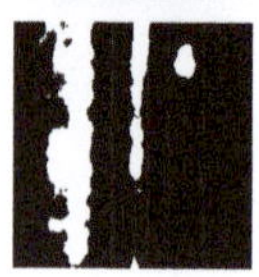	negative

Taking the four samples shown in Table 3 as an example, it was determined that they were all samples with cracks using the threshold segmentation method, but there was one without cracks. The improved ResNet-based algorithm using knowledge distillation still had good performance in the case of crack detection in concrete dams. It should be noted that the above algorithm cannot directly extract cracks from the image at the pixel level. This crack detection algorithm based on deep learning still needs to be combined with image segmentation technology to complete the pixel-level extraction of cracks. However, compared with the traditional digital image processing techniques, the improved algorithm proposed in this paper does not require any pre-processing of the original images, such as manually filtering background and noise in images. As long as all the samples are classified according to their categories, the algorithm can perform well in crack detection, which proves that it has a certain progressiveness.

6. Discussion

In this paper, we present an improved ResNet-based algorithm for crack detection in concrete dams using dynamic knowledge distillation. The improved algorithm eliminates the incompatibility of high accuracy and model compression of DCNNs applied to crack detection tasks, such as AlexNet, VGG-Net, ResNet50, and other improved versions using CBAM, enabling student models to have superior crack recognition capabilities compared to their teacher and facilitating their deployment on devices with limited resources.

Meanwhile, our method has abolished the practice of using fixed distillation temperatures in previous knowledge distillation systems and proposes to dynamically adjust the temperature during the training process by adding an antagonistic dynamic temperature module, making it possible for the network to select a suitable temperature for distillation automatically in each training epoch. This measure also greatly benefits the improvement of model accuracy.

Besides, we used transfer learning to solve the problem of insufficient samples in the crack detection task. Preliminary experiments were carried out on mini-ImageNet, and the pre-trained model was transferred to the target task. Through experiments, we have come to the following conclusions:

1. Preliminary experiments on mini-ImageNet prove soft targets can transfer a great deal of knowledge to the distilled model. If the smaller network is trained by adding the additional task of matching the soft targets produced by the large network at a higher temperature ($T = 4$), its accuracy will increase from 79.41% to 82.33%. We also find that knowledge distillation can improve generalization by narrowing the gap between training and validation accuracy compared to the original student model (from 14.03% to 10.16%) and even the teacher model (from 12.61% to 10.16%). In addition, the distillation effect is further improved with a dynamic temperature module (from 82.33% to 82.52%), compared with the results of distillation systems under fixed temperature parameters. The final distillation temperature approaches 3.95, which is different from the empiric value of 4, which confirms our previous conjecture.
2. Experiments on concrete crack detection prove the improved pre-trained network based on dynamic knowledge distillation has an improvement of 4.92% compared to the original student network without pretraining, with an accuracy of 99.85%. The accuracy is 1.05% higher than that of networks with CBAM. Compared with other DCNNs, such as AlexNet and VGG, the accuracy is improved by 2.83% and 4.92%, respectively. Experimental results demonstrate that the proposed dynamic distillation and transfer learning are highly beneficial for crack detection tasks and can satisfy the dual requirements of high accuracy and model compression. It is particularly true for tasks with insufficient samples, such as the application of HKC Dam.
3. When common feature encoders obtained from concrete cracks with rich features were applied to crack detection in HKC Dam through transfer learning, its accuracy reached 98.39%, making it easy to draw sweeping conclusions: dynamic distillation

and transfer learning can help networks improve the ability to extract common features such as texture and contour of cracks and alleviate the overfitting problem of datasets involving unrich samples.

Author Contributions: Conceptualization, J.Z.; methodology, J.Z.; software, J.Z.; validation, J.Z.; formal analysis, J.Z.; investigation, J.Z.; resources, T.B.; data curation, J.Z.; writing—original draft, J.Z.; writing—review and editing, T.B.; visualization, J.Z.; supervision, T.B.; project administration, T.B.; funding acquisition, T.B. All authors have read and agreed to the published version of the manuscript.

Funding: This research was funded by the National Key Research and Development Program of China, grant number 2018YFC1508603, and the National Natural Science Foundation of China, grant number 51739003.

Data Availability Statement: Publicly available datasets were analyzed in this study. This data can be found here: [https://drive.google.com/drive/folders/1Td3qB6WCJMzKMOb-YdTict5302S8RqF1 (accessed on 8 November 2022); https://data.mendeley.com/datasets/5y9wdsg2zt/2] (accessed on 11 February 2023).

Conflicts of Interest: The authors declare no conflict of interest.

References

1. Rezaiee-Pajand, M.; Kazemiyan, M.S.; Aftabi Sani, A. A Literature Review on Dynamic Analysis of Concrete Gravity and Arch Dams. *Arch. Comput. Methods Eng.* **2021**, *28*, 4357–4372. [CrossRef]
2. Lee, Y.-H.; Ryu, J.-H.; Heo, J.; Shim, J.-W.; Lee, D.-W. Stability Improvement Method for Embankment Dam with Respect to Conduit Cracks. *Appl. Sci.* **2022**, *12*, 567. [CrossRef]
3. Ge, M.; Petkovšek, M.; Zhang, G.; Jacobs, D.; Coutier-Delgosha, O. Cavitation Dynamics and Thermodynamic Effects at Elevated Temperatures in a Small Venturi Channel. *Int. J. Heat Mass Transf.* **2021**, *170*, 120970. [CrossRef]
4. Ge, M.; Sun, C.; Zhang, G.; Coutier-Delgosha, O.; Fan, D. Combined Suppression Effects on Hydrodynamic Cavitation Performance in Venturi-Type Reactor for Process Intensification. *Ultrason. Sonochem.* **2022**, *86*, 106035. [CrossRef]
5. Ge, M.; Manikkam, P.; Ghossein, J.; Kumar Subramanian, R.; Coutier-Delgosha, O.; Zhang, G. Dynamic Mode Decomposition to Classify Cavitating Flow Regimes Induced by Thermodynamic Effects. *Energy* **2022**, *254*, 124426. [CrossRef]
6. Feng, C.; Zhang, H.; Wang, H.; Wang, S.; Li, Y. Automatic Pixel-Level Crack Detection on Dam Surface Using Deep Convolutional Network. *Sensors* **2020**, *20*, 2069. [CrossRef]
7. Mohan, A.; Poobal, S. Crack Detection Using Image Processing: A Critical Review and Analysis. *Alex. Eng. J.* **2018**, *57*, 787–798. [CrossRef]
8. Vanhoucke, V.; Senior, A.; Mao, M. Improving the Speed of Neural Networks on CPUs. 2011; pp. 1–8. Available online: https://www.semanticscholar.org/paper/Improving-the-speed-of-neural-networks-on-CPUs-Vanhoucke-Senior/fbeaa499e10e98515f7e1c4ad89165e8c0677427#citing-papers (accessed on 3 August 2023).
9. Venkatesh, G.; Nurvitadhi, E.; Marr, D. Accelerating Deep Convolutional Networks Using Low-Precision and Sparsity. In Proceedings of the 2017 IEEE International Conference on Acoustics, Speech and Signal Processing (ICASSP), New Orleans, LA, USA, 5–9 March 2017; pp. 2861–2865.
10. Zhou, A.; Yao, A.; Guo, Y.; Xu, L.; Chen, Y. Incremental Network Quantization: Towards Lossless CNNs with Low-Precision Weights. *arXiv* **2017**, arXiv:1702.03044.
11. Buciluă, C.; Caruana, R.; Niculescu-Mizil, A. Model Compression. In Proceedings of the 12th ACM SIGKDD International Conference on Knowledge Discovery and Data Mining, Philadelphia, PA, USA, 20–23 August 2006; Association for Computing Machinery: New York, NY, USA, 2006; pp. 535–541.
12. Hong, Y.-W.; Leu, J.-S.; Faisal, M.; Prakosa, S.W. Analysis of Model Compression Using Knowledge Distillation. *IEEE Access* **2022**, *10*, 85095–85105. [CrossRef]
13. Gou, J.; Yu, B.; Maybank, S.J.; Tao, D. Knowledge Distillation: A Survey. *Int. J. Comput. Vis.* **2021**, *129*, 1789–1819. [CrossRef]
14. Mishra, A.; Marr, D. Apprentice: Using Knowledge Distillation Techniques to Improve Low-Precision Network Accuracy. *arXiv* **2017**, arXiv:1711.05852.
15. LeCun, Y.; Bottou, L.; Bengio, Y.; Haffner, P. Gradient-Based Learning Applied to Document Recognition. *Proc. IEEE* **1998**, *86*, 2278–2323. [CrossRef]
16. Krizhevsky, A.; Sutskever, I.; Hinton, G.E. ImageNet Classification with Deep Convolutional Neural Networks. *Commun. ACM* **2017**, *60*, 84–90. [CrossRef]
17. Simonyan, K.; Zisserman, A. Very Deep Convolutional Networks for Large-Scale Image Recognition. *arXiv* **2014**, arXiv:1409.1556.
18. Sarfraz, F.; Arani, E.; Zonooz, B. Knowledge Distillation Beyond Model Compression. In Proceedings of the 2020 25th International Conference on Pattern Recognition (ICPR), Milan, Italy, 10–15 January 2021; pp. 6136–6143.

19. Wang, J.; Bao, W.; Sun, L.; Zhu, X.; Cao, B.; Yu, P.S. Private Model Compression via Knowledge Distillation. In Proceedings of the AAAI Conference on Artificial Intelligence, Honolulu, HI, USA, 27 January–1 February 2019; Volume 33, pp. 1190–1197. [CrossRef]
20. Jafari, A.; Rezagholizadeh, M.; Sharma, P.; Ghodsi, A. Annealing Knowledge Distillation. In Proceedings of the 16th Conference of the European Chapter of the Association for Computational Linguistics: Main Volume, Online, 19–23 April 2021; pp. 2493–2504.
21. Walawalkar, D.; Shen, Z.; Savvides, M. Online Ensemble Model Compression Using Knowledge Distillation. In *Proceedings of the Computer Vision–ECCV 2020: 16th European Conference, Glasgow, UK, August 23–28, 2020*; Vedaldi, A., Bischof, H., Brox, T., Frahm, J.-M., Eds.; Springer International Publishing: Cham, Switzerland, 2020; pp. 18–35.
22. Sun, S.; Cheng, Y.; Gan, Z.; Liu, J. Patient Knowledge Distillation for BERT Model Compression. *arXiv* **2019**, arXiv:1908.09355.
23. Allen-Zhu, Z.; Li, Y.; Liang, Y. Learning and Generalization in Overparameterized Neural Networks, Going Beyond Two Layers. *arXiv* **2018**, arXiv:1811.04918.
24. Arora, S.; Cohen, N.; Hazan, E. On the Optimization of Deep Networks: Implicit Acceleration by Overparameterization. In Proceedings of the 35th International Conference on Machine Learning, Stockholm, Sweden, 10–15 July 2018.
25. Brutzkus, A.; Globerson, A. Why Do Larger Models Generalize Better? A Theoretical Perspective via the XOR Problem. In Proceedings of the International Conference on Machine Learning, Long Beach, CA, USA, 9–15 June 2019.
26. Tu, Z.; He, F.; Tao, D. Understanding Generalization in Recurrent Neural Networks. In Proceedings of the International Conference on Learning Representations, New Orleans, LA, USA, 6–9 March 2020.
27. Ba, L.J.; Caruana, R. Do Deep Nets Really Need to Be Deep? In Proceedings of the 27th International Conference on Neural Information Processing Systems, Montreal, QC, Canada, 8–13 December 2014.
28. Hinton, G.; Vinyals, O.; Dean, J. Distilling the Knowledge in a Neural Network. *arXiv* **2015**, arXiv:1503.02531.
29. Urban, G.; Geras, K.J.; Kahou, S.E.; Aslan, O.; Wang, S.; Caruana, R.; Mohamed, A.; Philipose, M.; Richardson, M. Do Deep Convolutional Nets Really Need to Be Deep and Convolutional? *arXiv* **2016**, arXiv:1603.05691.
30. He, K.; Zhang, X.; Ren, S.; Sun, J. Deep Residual Learning for Image Recognition. In Proceedings of the IEEE Conference on Computer Vision and Pattern Recognition, Boston, MA, USA, 7–12 June 2015.
31. Li, Z.; Li, X.; Yang, L.; Zhao, B.; Song, R.; Luo, L.; Li, J.; Yang, J. Curriculum Temperature for Knowledge Distillation. In Proceedings of the AAAI Conference on Artificial Intelligence, Washington, DC, USA, 7–14 February 2023.
32. Romero, A.; Ballas, N.; Kahou, S.E.; Chassang, A.; Gatta, C.; Bengio, Y. FitNets: Hints for Thin Deep Nets. *arXiv* **2014**, arXiv:1412.6550.
33. Yim, J.; Joo, D.; Bae, J.; Kim, J. A Gift from Knowledge Distillation: Fast Optimization, Network Minimization and Transfer Learning. In Proceedings of the 2017 IEEE Conference on Computer Vision and Pattern Recognition (CVPR), Honolulu, HI, USA, 21–26 July 2017; pp. 7130–7138.
34. Pan, S.J.; Yang, Q. A Survey on Transfer Learning. *IEEE Trans. Knowl. Data Eng.* **2010**, *22*, 1345–1359. [CrossRef]
35. Deng, J.; Dong, W.; Socher, R.; Li, L.J.; Li, K.; Fei-Fei, L. ImageNet: A Large-Scale Hierarchical Image Database. In Proceedings of the 2009 IEEE Conference on Computer Vision and Pattern Recognition, Miami, FL, USA, 20–25 June 2009; pp. 248–255. [CrossRef]
36. Goyal, P.; Dollar, P.; Girshick, R.; Noordhuis, P.; Wesolowski, L.; Kyrola, A.; Tulloch, A.; Jia, Y.; He, K. Accurate, Large Minibatch SGD: Training ImageNet in 1 Hour. *arXiv* **2017**, arXiv:1706.02677.
37. Loshchilov, I.; Hutter, F. SGDR: Stochastic Gradient Descent with Warm Restarts. *arXiv* **2016**, arXiv:1608.03983.
38. He, T.; Zhang, Z.; Zhang, H.; Zhang, Z.; Xie, J.; Li, M. Bag of Tricks for Image Classification with Convolutional Neural Networks. In Proceedings of the 2019 IEEE/CVF Conference on Computer Vision and Pattern Recognition (CVPR), Long Beach, CA, USA, 15–20 June 2019; pp. 558–567.
39. Özgenel, Ç.; Sorguc, A. Performance Comparison of Pretrained Convolutional Neural Networks on Crack Detection in Buildings. In Proceedings of the International Symposium on Automation and Robotics in Construction, Berlin, Germany, 20–25 July 2018.
40. Zhang, L.; Yang, F.; Daniel Zhang, Y.; Zhu, Y.J. Road Crack Detection Using Deep Convolutional Neural Network. In Proceedings of the 2016 IEEE International Conference on Image Processing (ICIP), Phoenix, AZ, USA, 25–28 September 2016; pp. 3708–3712.
41. Özgenel, Ç.F. Concrete Crack Images for Classification; Mendeley Data, V2; 2019. Available online: https://data.mendeley.com/datasets/5y9wdsg2zt/2 (accessed on 3 August 2023).

water

MDPI

Article

Prediction for the Sluice Deformation Based on SOA-LSTM-Weighted Markov Model

Jianhe Peng [1], Wei Xie [2,3,4], Yan Wu [5], Xiaoran Sun [1], Chunlin Zhang [6], Hao Gu [2,3,4], Mingyuan Zhu [2,3,4,5] and Sen Zheng [2,3,4,*]

1 Anhui and Huaihe River Institute of Hydraulic Research (Anhui Provincial Water Conservancy Engineering Quality Testing Center Station), Hefei 230088, China; pjh@ahwrri.org.cn (J.P.); sxr@ahwrri.org.cn (X.S.)
2 The National Key Laboratory of Water Disaster Prevention, Hohai University, Nanjing 210098, China; hhu0829@126.com (W.X.); zhumingyuan@hhu.edu.cn (M.Z.)
3 College of Water Conservancy and Hydropower Engineering, Hohai University, Nanjing 210098, China
4 National Engineering Research Center of Water Resources Efficient Utilization and Engineering Safety, Hohai University, Nanjing 210098, China
5 Xinjiang Institute of Water Resources and Hydropower Research, Urumqi 830049, China; xjskywy@126.com
6 Anhui Huaihe River Management Bureau, Bengbu 233099, China
* Correspondence: zheng_sen@hhu.edu.cn

Abstract: Increasingly, deformation prediction has become an essential research topic in sluice safety control, which requires significant attention. However, there is still a lack of practical and efficient prediction modeling for sluice deformation. In order to address the limitations in mining the deep features of long-time data series of the traditional statistical model, in this paper, an improved long short-term memory (LSTM) model and weighted Markov model are introduced to predict sluice deformation. In the method, the seagull optimization algorithm (SOA) is utilized to optimize the hyper-parameters of the neural network structure in LSTM primarily to improve the model. Subsequently, the relevant error sequences of the fitting results of SOA-LSTM model are classified and the Markovity of the state sequence is examined. Then, the autocorrelation coefficients and weights of each order are calculated and the weighted and maximum probability values are applied to predict the future random state of the sluice deformation. Afterwards, the prediction model of sluice deformation on the SOA-LSTM-weighted Markov model is proposed. Ultimately, the presented model is used to predict the settlement characteristics of an actual sluice project in China. The analysis results demonstrate that the proposed model possesses the highest values of R^2 and the smallest values of *RMSE* and absolute relative errors for the monitoring data of four monitoring points. Consequently, it concluded that the proposed method shows better prediction ability and accuracy than the SOA-LSTM model and the stepwise regression model.

Keywords: deformation prediction model; sluice; long short-term memory; seagull optimization algorithm; weighted Markov model

1. Introduction

There is no doubt that sluices have played a significant role in flood control, irrigation, shipping, hydroelectric power and water supply for a very long time [1,2]. Therefore, research on the safety properties of sluices has become a subject of intense interest [3]. Sluice deformation as the most intuitive index to reflect the sluice structural characteristics has brought about widespread attention. Many failure incidents of sluices have already taken place caused by extreme deformations, which have seriously affected the safety of people's lives and property. Hence, it is becoming increasingly difficult to ignore the research on monitoring techniques, prediction and assessment methods of sluice deformation.

In order to monitor the sluice deformation in real-time to access the structural health conditions of sluices, besides traditional monitoring techniques, GPS/GNSS [4,5], geodetic methods [6,7], laser scanning devices [8–10], interferometric synthetic aperture radar (InSAR) [11,12], etc. are all viable alternatives. Among them, GPS is the Global Positioning System, which is designed as an all-weather space-based navigation system for both military and civilian navigation needs. GNSS is the Global Navigation Satellite System, which serves as a popular means of deformation monitoring technique with the advantages of functioning in all weather conditions, is fully automated, requires no need for visibility conditions, etc. The geodetic monitoring methods usually utilize different optical and mechanical instruments, combined with other monitoring methods, such as the visual line method, lead line method and forward intersection method. Laser scanning devices and InSAR are both the latest emerging monitoring technology, which have been applied universally in deformation monitoring due to their high monitoring precision. Yang et al. [13] applied interferometric synthetic aperture radar (InSAR) in sluice deformation monitoring. Zhang et al. [14] introduced a digital photography and measurement robot into the monitoring of sluice health. As important means of deformation monitoring techniques, the above methods possess many merits such as high accuracy and efficiency [15] and have been widely applied. The monitoring results are able to provide reliable information of structural characteristics of hydraulic engineering.

In view of extreme deformations always leading to failure cases of hydraulic structure engineering, establishing deformation prediction models based on monitoring data series is of great importance [16,17]. For example, mathematical statistics, structural analysis and artificial intelligence algorithms have been utilized in the studies of variation law, early warning and risk analysis related to the deformation of dams for decades [18–20]. Recently, with the rapid development of artificial intelligence algorithms, artificial neural networks [21–23], grey system models [24–26], clustering algorithms [27–29] and intelligent optimization algorithms [30–32] have been widely applied in the deformation prediction of hydraulic structure engineering. These algorithms are able to overcome the shortcomings of traditional prediction models in terms of multidimensional input, model adaptive learning and overfitting. Given that deformation monitoring data series of hydraulic structure engineering possess apparent variations in regulation in the time dimension, the recurrent neural network (RNN) has excellent development and application prospects in learning sequenced data. However, RNNs are prone to the problems of gradient vanishing and gradient explosion during long sequence training. Therefore, as an RNN, LSTM is introduced to establish the more sensible deformation prediction model construction due to its advantage in learning from experience to classify, processing and predicting long time series [33–35]. Song et al. [36] and Hu et al. [37] optimized the hyperparameters of the LSTM model to establish deformation prediction models of concrete dam and concrete faced rockfill dam, respectively. The application results verified the good analysis accuracy and efficiency of the models, which demonstrates the practical application value of LSTM in engineering safety monitoring.

In recent decades, some researchers carried out monitoring, laboratory tests and numerical analyses of sluice deformation. Si et al. analyzed the overall strength and deformation of complex sluice chamber structures based on the finite element method [38]. Ding et al. [39] established the horizontal deformation prediction model of sluice based on the back propagation neural network weighted Markov model. The sluice deformation prediction model serves as a useful tool which can effectively analyze sluice performance in real-time and prevent potential risks. Unfortunately, the existing studies focus primarily on the dam deformation prediction. So far, there are only a few studies concerning sluice deformation prediction model establishment in the literature.

Above all, there is an urgent need to propose an efficient prediction model of sluice deformation. Considering the obvious merits of the LSTM, this paper aims to present a novel prediction model for sluice deformation. First, the deformation prediction method of sluice based on a statistical model is introduced in Section 2. Subsequently, the principle of the LSTM network optimized by the seagull optimization algorithm (SOA) is studied in Section 3. Then, the weighted Markov model is applied to adjust the prediction residuals of the SOA-LSTM model. Accordingly, the sluice deformation prediction model is proposed by combining the SOA, LSTM and weighted Markov model in Section 4. Subsequently, a practical sluice project in China is introduced as a case study to verify the prediction ability and accuracy of the proposed model in Section 5. The fitted and predicted results are demonstrated and discussed in Section 6. Finally, the conclusions are drawn in Section 7.

2. Statistical Model of Sluice Deformation

The statistical model is a practical tool for hydraulic structure deformation prediction, which assumes that the expectation of the disturbance term is zero in the monitoring data, and approximately obeys the normal distribution. However, the monitoring data in practical engineering scenarios may not necessarily satisfy this condition. The prediction accuracy of the statistical model is generally more complex and depends on the selection of influencing factors. In this paper, the contribution of three components of hydraulic structure deformation is mainly analyzed, such as water pressure, temperature, and time effect factors, to the final results. Among all the factors, the time effect component has many influencing factors, which makes it difficult to formulate an explicit expression. Therefore, in this paper, the expression of deformation can be adopted as follows.

$$\delta = a_0 + \delta_H + \delta_T + \delta_\theta \tag{1}$$

where δ_H denotes the water pressure component; δ_T is the temperature component, which can be expressed as a harmonic function; δ_θ denotes the time effect component. The deformation expression can be converted to:

$$\delta = a_0 + \sum_{i=1}^{3} a_i H^i + \sum_{i=1}^{2} (b_{1i} \sin \frac{2\pi it}{365} - b_{2i} \cos \frac{2\pi it}{365}) + c_1\theta + c_2 \ln \theta \tag{2}$$

where δ is the amount of deformation; H^i is the i-th power of the upstream water depth, m; t is the number of days from the observation date to the beginning of the monitoring day; θ is equal to the number of days from the observation date to the beginning of the monitoring day divided by 100; a_0, a_i, b_{1i}, b_{2i}, c_1 and c_1 are the constant terms and the regression coefficient, respectively, which can be obtained by the stepwise regression analysis method.

3. SOA-LSTM Network Model

3.1. Principles of (Long Short-Term Memory) LSTM Neural Network

LSTM is a special kind of recurrent neural network (RNN), which effectively solves the problems of gradient explosion, gradient disappearance, and long-term dependence of traditional recurrent neural networks in the process of adaptation and prediction by introducing the concepts of cell state and gating. As a result, it has excellent time-series prediction performance. The structure of the LSTM unit mainly consists of a cell state, forgetting gate, input gate, and output gate as depicted in Figure 1. The cell state is also called the memory unit, which provides a channel for information transmission. The cell state is also known as the memory cell, which provides a channel for the transmission of information. The cell state update is controlled by the forgetting gate, input gate, and output gate.

The forgetting gate decides with a certain probability whether to retain the cell state of the previous moment and selects the weight of the information to be retained. Meanwhile, it reads the output of the hidden layer of the previous moment h_{t-1} and the input of the current moment x_t to obtain the output of the forgetting gate f_t through a sigmoid activation function:

$$f_t = \sigma\left[\mathbf{W}_f \cdot (h_{t-1}, x_t) + b_f\right] \tag{3}$$

where $\mathbf{W}_f$ represents the weight matrix of the forgetting gate; b_f denotes the bias term; σ is the sigmoid activation function, $\sigma(x) = 1/(1-e^{-x})$; the output of the forgetting gate f_t controls the degree of forgetting the information of the cell state in the previous moment, which takes the range of $[0,1]$. When $f_t = 1$, f_t indicates complete retention, when $f_t = 0$, f_t indicates complete forgetting.

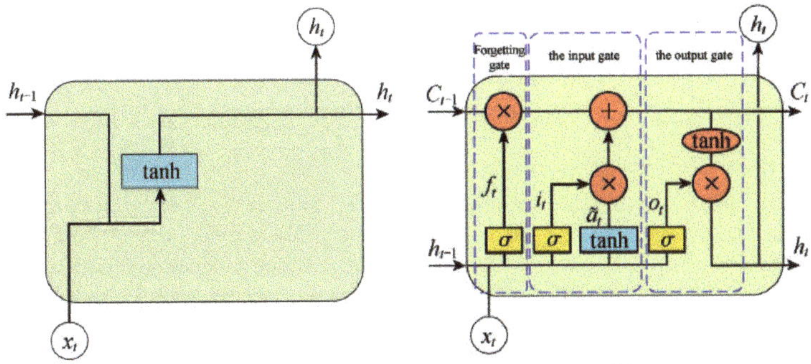

Figure 1. (**a**) Cell structure of RNN; (**b**) cell structure of LSTM.

The role of the input gate is to filter the new information being input at the current moment for storage in the cell state. The introduction of new information is determined by the sigmoid layer in conjunction with the tanh layer, where the sigmoid layer determines the extent i_t to which new information is selected, and the tanh layer is applied to generate candidates $\widetilde{a}_t$ for storing the learned new information.

$$i_t = \sigma[\mathbf{W}_i \cdot (h_{t-1}, x_t) + b_i] \tag{4}$$

$$\widetilde{a}_t = \tanh[\mathbf{W}_c \cdot (h_{t-1}, x_t) + b_c] \tag{5}$$

where $\mathbf{W}_i$, $\mathbf{W}_c$ are the weight matrices of the sigmoid layer and the tanh layer, respectively; b_i, b_c are the bias terms of the sigmoid layer and the tanh layer, respectively; tanh is the hyperbolic tangent function, $\tanh x = (1-e^{-x})/(1+e^{-x})$.

Based on the selection of the cell state at the previous moment by the forgetting gate and the input information at the current moment by the input gate, the cell state can be updated as follows.

$$C_t = f_t C_{t-1} + i_t a_t \tag{6}$$

The output gate extracts valid information from the current cell state, which is used to generate a new hidden layer. Firstly, the output portion of the current cell state is determined by the sigmoid function. Subsequently, the current cell state is processed by the tanh function. Finally, a new hidden layer h_t is generated as follows.

$$h_t = o_t \tanh C_t \tag{7}$$

where $o_t = \sigma[\mathbf{W}_o \cdot (h_{t-1}, x_t) + b_o]$ is the output of the hidden layer; $\mathbf{W}_o$ is the weight matrix of the output gate; b_o is the bias term.

In summary, the hidden layer output h_t and cell state C_t of the LSTM at the current moment is determined by the hidden layer output h_{t-1} and cell state C_{t-1} of the previous moment in conjunction with the input x_t of the current moment.

When establishing the sluice deformation prediction model based on LSTM, the number of hidden layers plays an essential role in the training accuracy and efficiency of the model. Theoretically, although the fitting accuracy of the LSTM model will improve with the increase in the number of hidden layers, too many hidden layers will lead to a decrease in the training efficiency and even cause a decrease in the fitting and prediction accuracy of the model. Consequently, in this paper, the network structure with two hidden layers of LSTM is adopted.

According to Equation (2), it can be observed that the number of input nodes provided to the LSTM network is nine, and the number of output nodes is one. When selecting the number of nodes in the hidden layer, the initial number of nodes is usually calculated using the empirical formula, and then the number of nodes is incrementally increased or decreased in an attempt to find out the number of nodes of the more optimal hidden layer. Furthermore, the network training utilizes the dropout technique to avoid overfitting, i.e., the model training effect is good but the prediction performance is poor, and the batch processing technique is used to improve the training efficiency of the model [36].

3.2. Principles of Seagull Optimization Algorithm (SOA)

SOA is a novel bio-inspired algorithm proposed by Dhiman et al. [40]. Compared to traditional algorithms like particle swarm optimization (PSO) and the genetic algorithm (GA), SOA has advantages in terms of simplicity, ease of implementation, and better search performance [41]. SOA treats the migration and aggressive behaviors of seagulls in nature as an optimization process. In this algorithm, each seagull represents an individual searching entity in the search space, and the position of a seagull represents a potential solution to the optimization problem. The migration and aggressive behaviors of seagulls are illustrated in Figure 2.

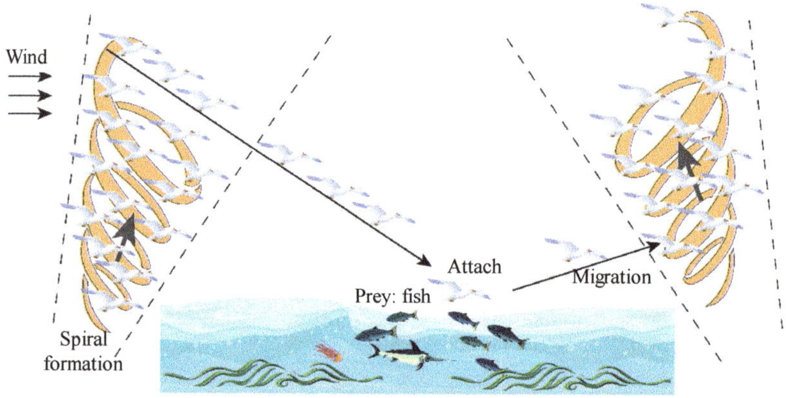

Figure 2. The migration and aggressive behaviors of seagulls.

The implementation process of the algorithm is expressed as follows.

(1) Migration behavior (global search) During the seagull migration process, a seagull moves from one position to another. The above migration behavior should satisfy the following three conditions:

  (1) Avoiding collision. To avoid collisions between adjacent seagulls, an additional variable A is introduced to update the position of the seagull during the iterative calculation, as shown in Figure 3a.

$$\vec{N_s} = A \cdot \vec{P_s}(i) \tag{8}$$

  where $\vec{N_s}$ is the new position; $\vec{P_s}(i)$ is the current position of the seagull; i is the number of the current iteration; A is the additional variable representing the movement of seagull in the given search space. The calculation method of A is as follows.

$$A = f_c - i \cdot \left(\frac{f_c}{\text{Max}_{iteration}} \right) \tag{9}$$

where f_c is the control frequency of variable A, its range is $[0, f_c]$. Typically, f_c is set to 2; i is the number of current iterations, $i = 0, 1, 2 \cdots \text{Max}_{iteration}$; $\text{Max}_{iteration}$ is the maximum number of iterations.

(2) Moving towards the best neighbor. After avoiding collisions between adjacent seagulls, a seagull moves towards the direction of the best neighbor, as illustrated in Figure 3b.

$$\vec{B_S} = B \cdot \left(\vec{P_{gS}}(i) - \vec{P_s}(i) \right) \tag{10}$$

where $\vec{B_S}$ represents the direction in which the seagull moves from its original position $\vec{P_s}(i)$ to the position of the best neighboring seagull $\vec{P_{gS}}(i)$; B is a random parameter used to balance the global and local search, and its calculation formula can be denoted as:

$$B = 2 \cdot A^2 \cdot rd \tag{11}$$

where rd is a random value between 0 and 1.

(3) Moving towards the best position. Finally, the seagull updates its own position based on the best position, as depicted in Figure 3c.

$$\vec{D_s} = \left| \vec{N_s}(i) + \vec{B_s}(i) \right| \tag{12}$$

where $\vec{D_s}$ represents the distance between the current position and the global best position.

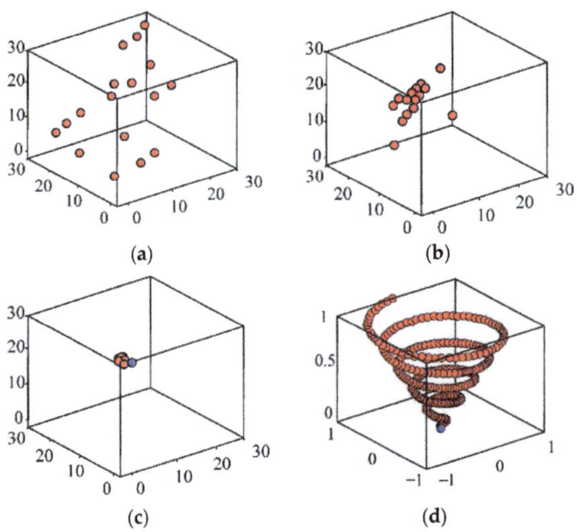

Figure 3. Schematic diagram of migration and aggressive behavior of SOA: (**a**) avoiding collision, (**b**) moving towards the best neighbor, (**c**) moving towards the best position, (**d**) the spiral aggressive behavior.

(2) Aggressive behavior (local search) When a seagull needs to attack its prey during flight, it forms a spiral formation in the air, as shown in Figure 3d. This behavior can be described in the xyz three-dimensional plane as follows.

$$x = r \cdot \sin k \tag{13}$$

$$y = r \cdot \cos k \tag{14}$$

$$z = r \cdot k \tag{15}$$

$$r = u \cdot e^{k \cdot v} \tag{16}$$

where r denotes the spiral radius formed by the seagull swarm; k is a random angle within the range $[0, 2\pi]$; v and u are constants that determine the spiral shape; e denotes the base of the natural logarithm. Considering both seagull migration and aggressive behaviors, the calculation formula of seagull location updating can be obtained based on Equation (12) to Equation (16) as follows.

$$\vec{P_s}(i) = x \cdot y \cdot z \cdot \vec{D_s} + \vec{P_{gs}}(i) \tag{17}$$

where $\vec{P_s}(i)$ is the attack position of the seagull, which is denoted as its final location is updated.

3.3. SOA Optimized LSTM

Since LSTM has many hyper-parameters that need to be adjusted manually, such as the number of neurons in the hidden layer, the initial learning rate, the maximum number of iterations, the minimum number of batches, the time step, etc., these hyper-parameters have a certain impact on the prediction results, and it takes a lot of time to adjust these parameters. Determining how to find the hyper-parameters that can make the whole prediction process faster and more accurate is a problem that needs to be solved.

The seagull optimization algorithm, as a biomimetic global optimization algorithm, can be used to optimize the hyper-parameters of the neural network.

In the prediction of sluice deformation, the factors that can affect the amount of deformation are mainly obtained as the characteristic variables for the prediction of deformation and are normalized as the input data. Simultaneously, sluice deformation monitoring data are normalized to ensure the prediction accuracy. By adjusting and designing the LSTM network structure and using SOA to optimize and solve the hyper-parameters of the LSTM model, the optimized parameters are directly used as the parameter values of the LSTM model to predict the sluice deformation. Finally, the prediction value of sluice deformation is obtained. The specific modeling and prediction ideas are as follows.

(1) Data processing. First, the data set is normalized, and 80% of the data in the monitoring sequence is selected as the training set, and 20% of the data in the monitoring sequence is selected as the test set.
(2) Single LSTM model structure design. The LSTM model results are designed in terms of the number of network layers, the number of neurons in the input and hidden layers, the optimizer, and the loss function of the LSTM, and the multivariate multidimensional single-step LSTM network model with an initial input of nine dimensions, a time step of one, and an output of one dimension is constructed.
(3) Construction of SOA LSTM model. The LSTM hyper-parameters to be optimized, such as the number of neurons in the hidden layer, the initial learning rate, the maximum number of iterations, the minimum number of batches, the time step, and the regularization parameter, are used as the solution objectives of the seagull optimization algorithm, and the root mean square error (*RMSE*) on the training set is taken as the fitness function to obtain the SOA-LSTM model. The computational formula is given by:

$$RMSE = \sqrt{\frac{1}{n}\sum_{i=1}^{n}(y_i - \hat{y}_i)^2} \tag{18}$$

where n is the number of data in the training set; y_i is the measured value of data in training set; $\hat{y}_i$ is the fitted value of data in training set.

(4) Model training and prediction. The training data are applied to train the SOA-LSTM model, the test data are input into the trained model, and finally obtain the predicted values.
(5) Model comparison. The prediction results are utilized to calculate the prediction accuracy evaluation index to test the model effect.

The optimization process of LSTM based on SOA is exhibited in Figure 4.

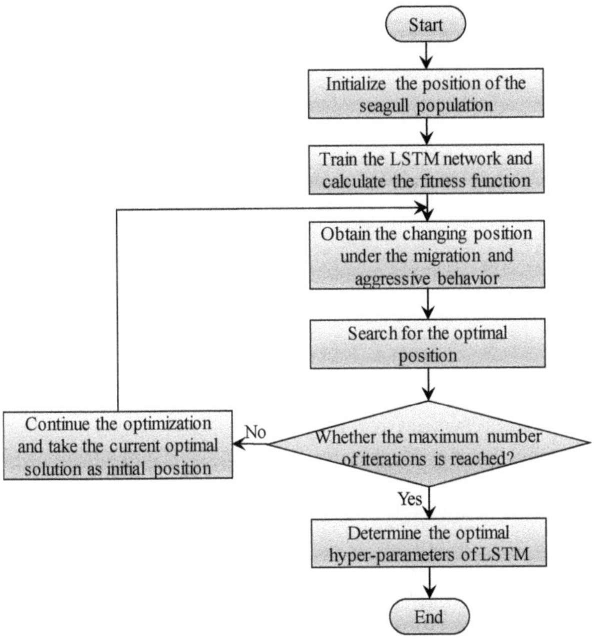

Figure 4. The flow chart of optimization of LSTM based on SOA.

4. SOA-LSTM-Weighted Markov Model for Sluice Deformation Prediction

4.1. Weighted Markov Model-Based Sluice Deformation Prediction

The measured data of sluice deformation has a certain degree of volatility. Due to the advantage of Markov chain in terms of dealing with data with high random volatility, it has been applied in predicting the deformation of hydraulic structures. There are three common Markov models: the Markov model based on absolute distribution, the superposition Markov model, and the weighted Markov model. In this paper, the weighted Markov model is introduced, which is not restricted by the time chi-square of the Markov chain, and takes into account the strength of dependence of different lags, fully exploiting the information contained in the sample data. Furthermore, the accuracy of the weighted Markov model is generally higher than the other two models, and the steps of the weighted Markov model building are as follows:

(1) State classification. The mean square deviation method is proposed to classify the values of random sequence indicator into five state intervals.
(2) Markovianity test. For discrete sluice deformation sequences, the chi-square statistic can be constructed to test for Markovianity.
(3) State transition probability matrix. The one-step state transition frequency matrix $f_{ij}^{(k)}$ with different lags (k is the step size) and the one-step state transition probability matrix $p_{ij}^{(k)}$ should be calculated, separately.

(4) Autocorrelation coefficients and transfer weights of each order. After selecting the lag time, the autocorrelation coefficient r_k and transfer weight w_k of each order should be calculated, separately.
(5) Prediction table preparation. The sluice deformation of several years before the prediction year should be taken as the initial state, and then the sluice deformation should be combined with the transfer probability matrix of the corresponding lag time to obtain the probability of each order state in the prediction year. Afterwards, the predicted state probabilities of each year of the same state are weighted and summed, which is the average probability of being in each state in the prediction year. The above calculations can be compiled into a prediction table, and the state corresponding to the maximum probability value in the prediction table is taken and predicted by using the weighted Markov model.

4.2. Sluice Deformation Prediction Procedure of SOA-LSTM-Weighted Markov Model

In view of the fact that the LSTM network can deal well with nonlinear problems, the Markov chain is suitable for predicting random sequences with large volatility. Therefore, the SOA-LSTM network-weighted Markov model synthesizes the above advantages, which can be applied to sluice deformation prediction. The prediction procedure of the model is as follows:

(1) Based on the training set, the SOA-LSTM network model is established.
(2) After comparing the measured data with the fitted value, the error sequence can be obtained.
(3) The weighted Markov model is constructed according to the steps in Section 3.1.
(4) The sluice deformation prediction model based on the SOA-LSTM-weighted Markov model can be established.

The computational flow of the SOA-LSTM-weighted Markov model is exhibited in Figure 5.

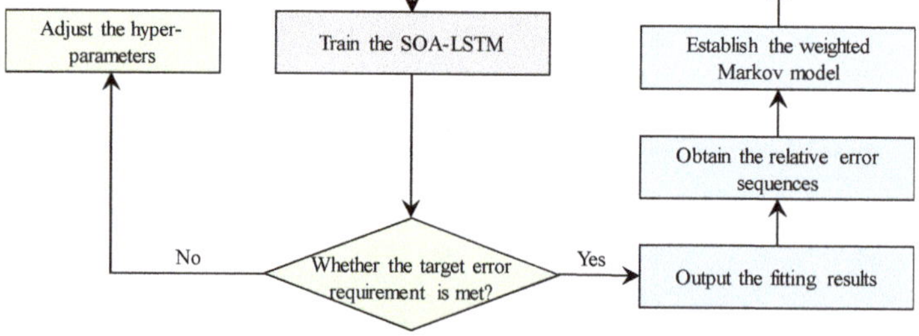

Figure 5. The establishment process of SOA-LSTM-weighted Markov model.

5. Project Overview

The Bengbu sluice project is located in the middle reaches of the Huaihe River. It now consists of a regulating sluice with 12 holes, a regulating sluice with 28 holes, a hydropower plant, the ship lock, the floodway, embankment and so on. The regulating sluice with 12 holes is composed of six bottom boards, each bottom board with one monitoring point at each corner. One settlement monitoring point at each downstream of the left-side and right-side piers. Six settlement monitoring points at each symmetrical position of upstream and downstream on the left and right bank wing walls. A total of 50 monitoring points were installed in the regulating sluice with 12 holes. Figure 6 illustrates the layout of settlement monitoring points of the regulating sluice with 12 holes of the Bengbu sluice.

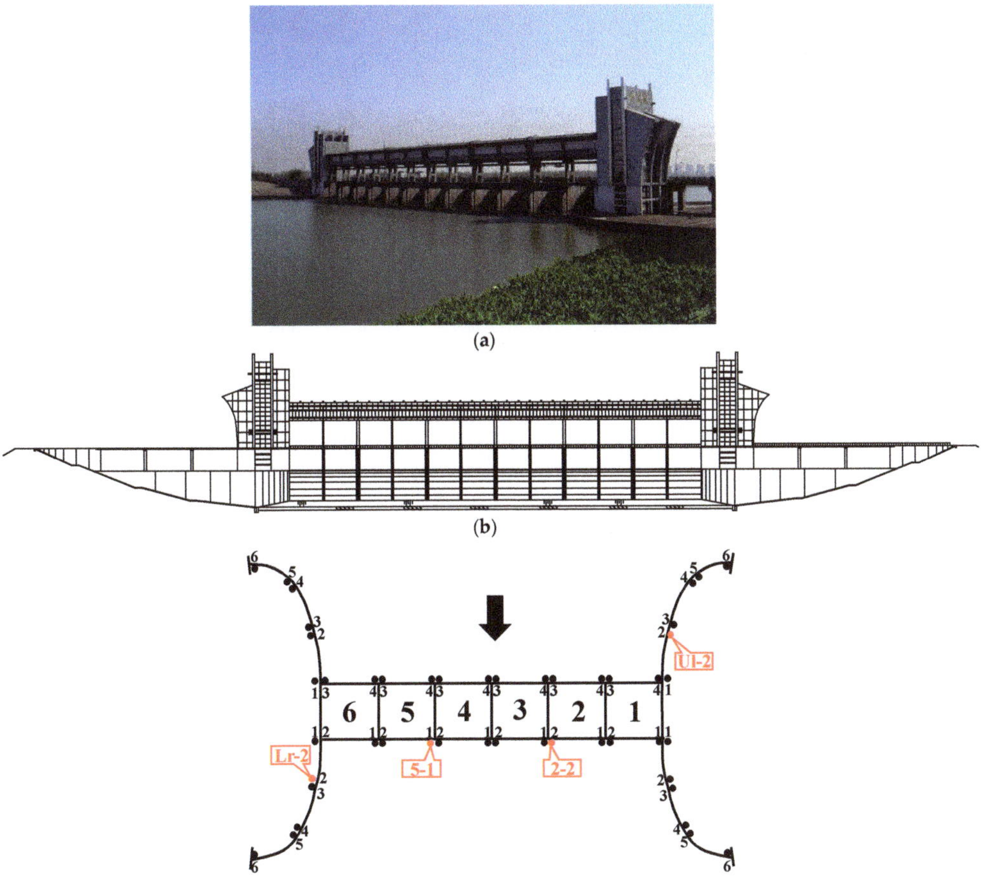

Figure 6. (**a**) Engineering scene of the regulating sluice with 12 holes at Bengbu sluice; (**b**) elevation map of the regulating sluice with 12 holes at Bengbu sluice; (**c**) layout of settlement monitoring points of the regulating sluice with 12 holes at Bengbu sluice.

The pre-flood settlement monitoring of the regulating sluice of Bengbu has been in progress since 10 March 2021 and the monitoring data are recorded every 30 days. According to the settlement monitoring results, the largest cumulative average variation in settlement of the regulating sluice with 12 holes appeared at the lower right-wing wall at monitoring point Lr-2, with a value of −18.11 mm. The largest cumulative variation in settlement was appeared of monitoring point 3–4, with the largest difference value of 31.59 mm at the upper sluice pier. The largest cumulative settlement appeared at monitoring point 5–1, with the cumulative settlement of −35.75 mm. The settlement eigenvalues of the regulating sluice with 12 holes are counted in Table 1.

Table 1. The settlement eigenvalues of the regulating sluice with 12 holes.

Project Site	Maximum Average Variation (mm)	Minimum Settlement (mm)		Maximum Settlement (mm)		Maximum Settlement Variation (mm)
		Point	Settlement	Point	Settlement	
Upper left-wing wall	−14.99	Ul-6	−12.71	Ul-2	−16.33	3.62
Lower left-wing wall	−14.95	Ll-6	−9.80	Ll-4	−17.82	8.02
Upper right-wing Wall	−16.80	Ur-1	−14.14	Ur-2	−18.37	4.23
Lower right-wing wall	−18.11	Lr-2	−8.03	Lr-5	−28.71	20.68
Upper sluice pier	−13.52	1-3	−0.21	3-4	−31.80	31.59
Lower sluice pier	−14.30	2-2	−4.19	5–1	−35.75	31.56
Abutment pier	−8.45	Right pier	−7.59	Left pier	−9.32	1.73

Note: In the table, "−" means the sinking settlement.

In this study, we selected the settlement data (the vertically downward is positive, the vertically upward is negative) of four monitoring points (2-2, Lr-2, Ul-2, and 5–1) for analysis. The monitoring points 2-2, Lr-2, Ul-2, and 5–1 are marked with red dots in Figure 6. After preprocessing the settlement data, the measured data of each monitoring point can be obtained in Figure 7. It is noted that the vertical settlements of all monitoring points have similar variation rules with time; they increase slowly, and tend to converge without obvious periodic variations.

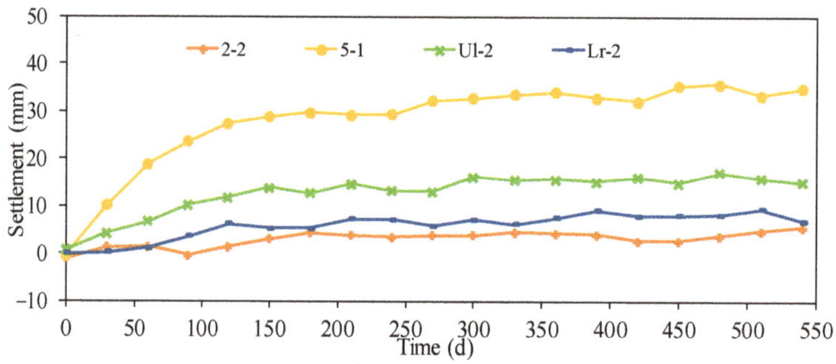

Figure 7. Settlement measured data series of four selected monitoring points.

In this paper, the recent settlement monitoring data of monitoring point 5–1 of the sluice was taken as an example to illustrate the implementation process of the proposed method. Eighty (80)% of the monitoring data were utilized to establish the stepwise regression model, the SOA-LSTM model, and the SOA-LSTM-weighted Markov model. Based on the modeling results, the relative errors of the test set were compared to verify the reliability of the proposed model. The specific settlement monitoring data of point 5–1 are illustrated in Table 2.

Table 2. Settlement monitoring data of monitoring point 5–1.

Time (d)	Monitoring Data (mm)	Time (d)	Monitoring Data (mm)	Time (d)	Monitoring Data (mm)
0	1.43	210	29.14	420	34.71
30	11.72	240	30.83	450	33.20
60	17.74	270	29.32	480	33.43
90	23.85	300	29.72	510	36.25
120	26.79	330	32.21	540	35.75
150	28.95	360	34.41		
180	28.56	390	33.06		

6. Results and Discussions

6.1. Fitted and Predicted Results of the Stepwise Regression Model

The measured values, calculated values and relative errors obtained by the stepwise regression model are listed in Table 3. The fitted and predicted results of the stepwise regression model are exhibited in Figure 8. The predicted values are exhibited with a blue background in Figure 8.

Table 3. The fitted and predicted values of the stepwise regression model.

Time (d)	Monitoring Data	Predicted Value (mm)	Relative Error (%)	Time	Monitoring Data	Predicted Value (mm)	Relative Error (%)
0	1.43	1.431	0	300	29.72	29.17	−1.85
30	11.72	10.87	−7.23	330	32.21	29.66	−7.91
60	17.74	17.99	1.46	360	34.41	30.10	−12.53
90	23.85	21.93	−8.01	390	33.06	30.51	−7.69
120	26.79	24.48	−8.61	420	34.71	30.89	−11.00
150	28.95	25.63	−11.48	450	33.20	31.24	−5.89
180	28.56	26.56	−7.02	480	33.43	31.57	−5.56
210	29.14	27.35	−6.16	510	36.25	31.88	−12.04
240	30.83	28.03	−9.09	540	35.75	32.18	−10.01
270	29.32	28.63	−2.33				

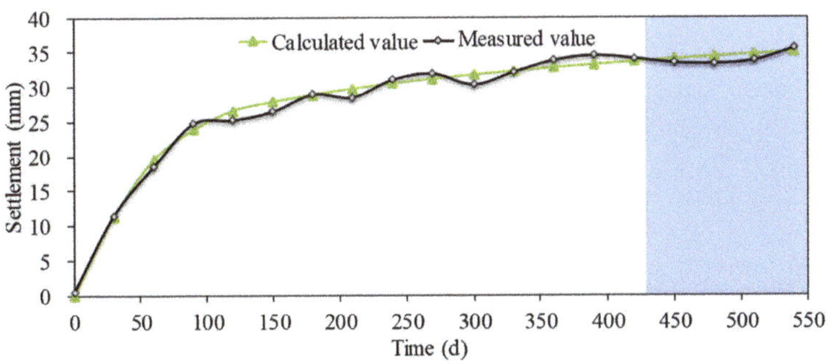

Figure 8. Fitted and predicted values of the stepwise regression model.

The calculation formula of the relative error of each calculated datapoint can be written as:

$$RE = \left(\dot{y}_i - y_i\right)/y_i \times 100\% \quad (19)$$

where RE denotes the relative error; $\dot{y}_i$ is the fitted or predicted value; y_i is the measured value of sluice settlement.

6.2. Fitted and Predicted Results of the SOA-LSTM Network Model

In total, 80% of the sluice settlement monitoring data were used as training set to establish the SOA-LSTM model. The output and input of the LSTM model are the sluice settlement monitoring data and the influence factors of the sluice settlement. Therefore, the number of nodes in the input layer and output layer are nine and one, respectively. The influence of nodes number on the model performance is similar to the mechanism of the influence of the number of hidden layers on the model performance.

In this work, we adopt the SOA to optimize the hyper-parameters the LSTM model. In SOA, the population size was selected as 20, the number of iterations was taken as 30. Figure 9 illustrates the objective function value of the seagull at each iteration step, and the best objective function value can be determined as 5.326 at the 13th iterations by SOA.

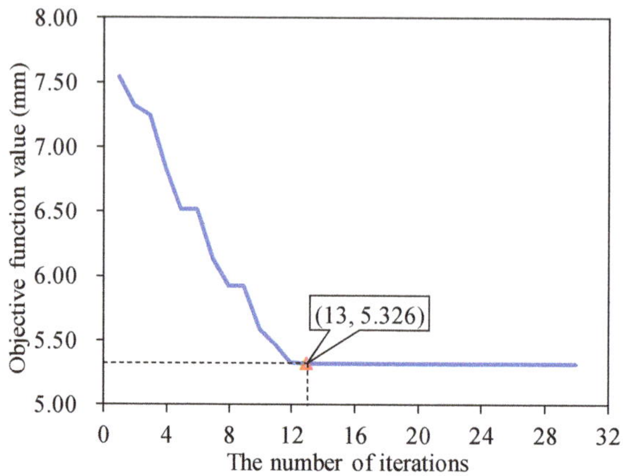

Figure 9. Plot of the objective function values versus the number of iterations.

The set of upper bound and lower bound of the search area for are exhibited in Table 4. After calculation, the results of the optimized hyperparameters of LSTM model are presented in Table 4.

Table 4. Upper bound and lower bound of the parameters set in SOA, and the optimized results of the LSTM model.

Time (d)	Hidden Layers	Nodes in First Layer	Nodes in Second Layer	Learning Rate
Upper bound	10	200	200	0.1
Lower bound	1	10	10	0.001
Optimized parameters	2	120	60	0.005

Figure 10 depicts the relationship curve between the number of nodes in the hidden layer and the average relative error of the training samples. The initial number of nodes in the hidden layer is 10, and the trial calculation is carried out by incrementing 10 nodes one by one, and the average relative error of the training samples is minimized as the criterion. As can be seen in Figure 10, after the trial calculation, in the two hidden layers of the LSTM adopted in this paper, the minimum value of the relative error can be taken when the node number of the first hidden layer is 120. The minimum value of the relative error can be taken when the number node of the second hidden layer is 60. The batch size and the maximum number of iterations can be set to 50 and 400, respectively. The target error of the model training can be set to 0.001.

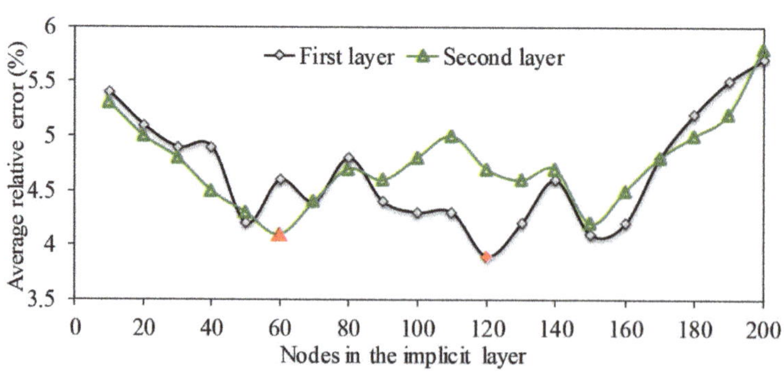

Figure 10. The average relative error with the number of nodes in the hidden layer.

Figure 11 exhibits the process line of the fitted and predicted values of the sluice settlement of the monitoring point 5–1 based on the SOA-LSTM network model.

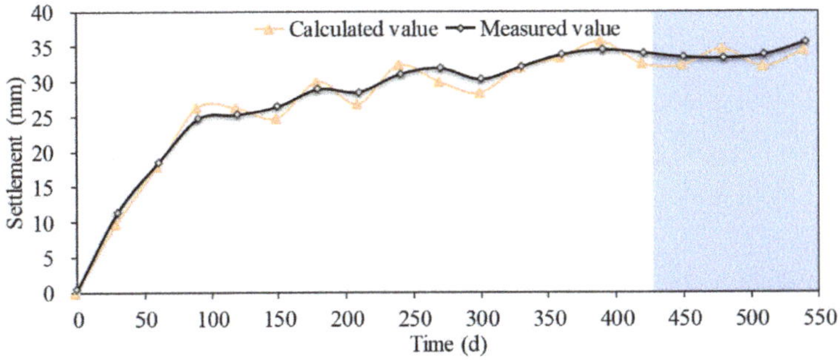

Figure 11. Fitted and predicted values of the SOA-LSTM model.

As can be seen from Figure 11, except for a few curve peaks with relatively large data jumps, the fitted value of the LSTM network model has the same trend as the measured value. The average absolute value of the relative error is 5.72%, which verifies the better fitted effect of the model.

6.3. Fitted and Predicted Results of the SOA-LSTM-Weighted Markov Model

To further improve the prediction accuracy of the SOA-LSTM-weighted Markov model proposed in this work, the errors of the SOA-LSTM model were corrected by the weighted Markov model. The following computational steps and results are described as follows.

(1) State classification. The relative error series has a mean of 75% and a root mean square error of 7.64%, which is classified into five states corresponding to the intervals of (−22.44%, −9.16%), (−9.16%, −4.58%), (−4.58%, 3.07%), (3.07%, 7.65%) and (7.65%, 24.20%), respectively.

(2) Markovianity test. The one-step transfer frequency matrix f_{ij}, one-step transfer probability matrix $p^{(1)}$, and marginal probability and chi-square statistics were calculated, and the calculation results are described in Table 5. The value of the calculated statistic $\chi^2 = 2\sum\limits_{i=1}^{m}\sum\limits_{i=1}^{m} f_{ij}\left|\ln\frac{P_{ij}}{P_{.j}}\right|$ is 109.9, and given the significance level, $\alpha = 0.01$. Then $\chi^2_\alpha = \left[(m-1)^2\right] = \chi^2_{0.01}(16) = 32.0$. Because $\chi^2 > \chi^2_{0.01}(16)$, the random sequence is highly significant in terms of Markovianity, and can be predicted by the weighted Markov model.

Table 5. Calculation results of chi-square statistics.

Status	Marginal Probability	$f_{i1}\left\|\ln\frac{P_{i1}}{P_{,1}}\right\|$	$f_{i2}\left\|\ln\frac{P_{i2}}{P_{,2}}\right\|$	$f_{i3}\left\|\ln\frac{P_{i3}}{P_{,3}}\right\|$	$f_{i4}\left\|\ln\frac{P_{i4}}{P_{,4}}\right\|$	$f_{i5}\left\|\ln\frac{P_{i5}}{P_{,5}}\right\|$	Total
1	0.127	10.768	5.410	1.94	0.961	0.000	19.079
2	0.158	0.575	0.062	2.010	0.867	0.851	4.365
3	0.436	2.083	0.972	7.210	0.439	1.836	12.54
4	0.157	0.000	0.087	1.521	0.247	1.957	3.842
5	0.122	0.723	0.000	2.513	4.984	6.745	14.965
Total	1	14.149	6.531	15.194	7.498	11.389	54.761

(3) The state transition probability matrix determination. The relative errors of the fitted values of the SOA-LSTM model were taken as a random sequence, and the one-step state transfer probability matrixes with the step size of 2, 3, 4, and 5, respectively.

(4) Calculation of the autocorrelation coefficients and transfer weights of each order. The autocorrelation coefficients (r_k) and transmission weights (w_k) of each order are shown in Table 6.

Table 6. Autocorrelation coefficients and transfer weights of each order.

Project	Order				
	1	2	3	4	5
r_k	0.568	0.354	0.248	0.126	0.139
w_k	0.416	0.245	0.168	0.081	0.105

(5) Prediction table preparation. According to the relative errors of the SOA-LSTM model, the corresponding state transfer probability is weighted and calculated to predict the relative error state of test set. Table 7 illustrates a subset of the weighted prediction results of the relative error state of the test set. The state S with the maximum probability value in the prediction table was determined, E was taken as the median value of the corresponding state interval. x_{SL} was selected as the predicted value of the SOA-LSTM network model. x_{SIM} was selected as the predicted value of the SOA-LSTM-weighted Markov model. Therefore, the corrected predicted value can be expressed as:

$$x_{BM} = x_{BP}/(1+E) \tag{20}$$

Table 7. Some state-weighted predictions of relative error for settlement of the monitoring point.

Initial Time (d)	Status	Lag Time	Weights	Status				
				1	2	3	4	5
60	3	1	0.4106	0.0771	0.1328	0.5587	0.1528	0.0759
120	3	2	0.2514	0.0979	0.1757	0.4892	0.1363	0.0978
180	3	3	0.1659	0.1000	0.2000	0.4785	0.1000	0.1207
240	4	4	0.0824	0.1581	0.0000	0.6296	0.1043	0.1062
300	5	5	0.1021	0.0657	0.0657	0.4000	0.2678	0.1985
	P_j (Weighted sum)			0.0922	0.1315	0.5249	0.0901	0.1447

Taking the calculated results of the stepwise regression model, the SOA-LSTM model as the basis for comparison, the fitted results, predicted results and the relative errors of three models are illustrated in Figures 12 and 13. Among them, Figure 12 shows the comparison of the predicted effects of three models. Figure 13 exhibits some of the absolute values of the relative errors of the settlement predicted value of three models.

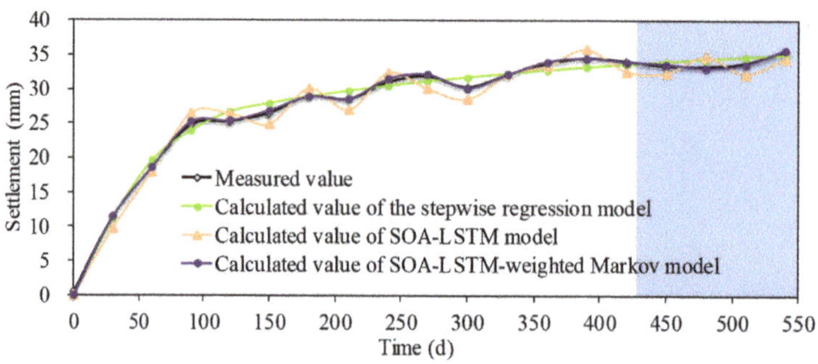

Figure 12. Comparison of the fitted and predicted values of three models.

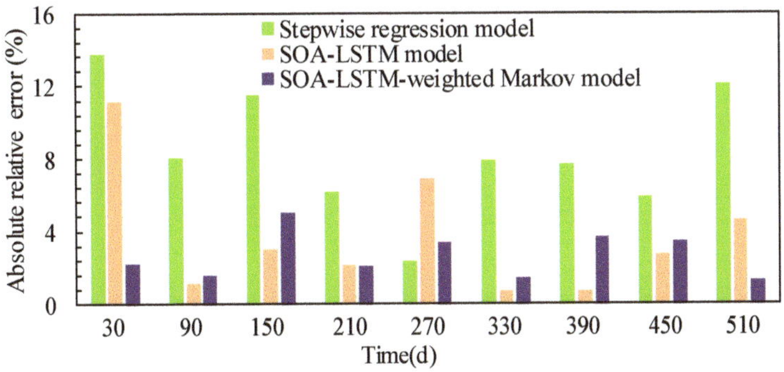

Figure 13. Some of the absolute relative errors of three models.

In Figure 12, it can be seen that the measured values of sluice settlement at monitoring point 5–1 are nonlinear with time, and the settlement predicted by the SOA-LSTM model and the SOA-LSTM-weighted Markov model have the same trend with the measured data. However, the settlement predicted by the stepwise regression model shows a linear change different from the measured values. The predicted value of the stepwise regression model cannot reflect the nonlinear change characteristics of the measured data, and its prediction result is not as good as that of the SOA-LSTM model and the SOA-LSTM-weighted Markov model in dealing with nonlinear data. As can be seen from Figure 13, the relative errors of SOA-LSTM-weighted Markov model are generally the smallest, while the relative errors of stepwise regression model are the largest overall. The average absolute values of the relative errors of the stepwise regression model, the SOA-LSTM model, and the SOA-LSTM-weighted Markov model are 7.55%, 3.12%, and 1.57%, respectively. It illustrates that the accuracy of the proposed model is better than the other two models.

With the aim of verifying the superiority of the SOA-LSTM-weighted Markov model in fitting and predicting sluice deformation, the measured data of monitoring points 2-2, 5–1, Lr-2 and Ul-2 were utilized to establish the prediction models based on the stepwise regression model, SOA-LSTM model, and SOA-LSTM-weighted Markov model, respectively. The calculated results are compared in Figure 14. Figure 15 shows the coefficient of determination (R^2) of the training set and the root mean square error (*RMSE*) of the test set of each model. The calculation formula of R^2 can be given by:

$$R^2 = \frac{\sum_{i=1}^{N}(\hat{\delta}_i - \overline{\delta})}{\sum_{i=1}^{N}(\delta_i - \overline{\delta})} \tag{21}$$

where δ_i represents the measured data of training set; $\overline{\delta}$ denotes the average value of the measured data of training set; $\hat{\delta}_i$ is the fitted value of the training set; N means the amount of data in the training set.

It can be observed in Figure 14 that although the models can fit the measured data reasonably, the proposed model possesses not only the best fitting effects, but also the best forecast effects. In Figure 15, the R^2 and *RMSE* of the stepwise regression model vary from 0.8001 to 0.9114 and from 1.6705 to 3.8577, respectively. The R^2 and *RMSE* of the stepwise regression model vary from 0.7801 to 0.8514 and from 2.0705 to 4.9561, respectively. The R^2 and *RMSE* of the stepwise regression model vary from 0.9301 to 0.9951 and from 0.725 to 1.517, respectively. Among all four monitoring points, the proposed SOA-LSTM-weighted Markov model has the smallest *RMSE*, which indicates that the proposed model is feasible and has better prediction accuracy overall.

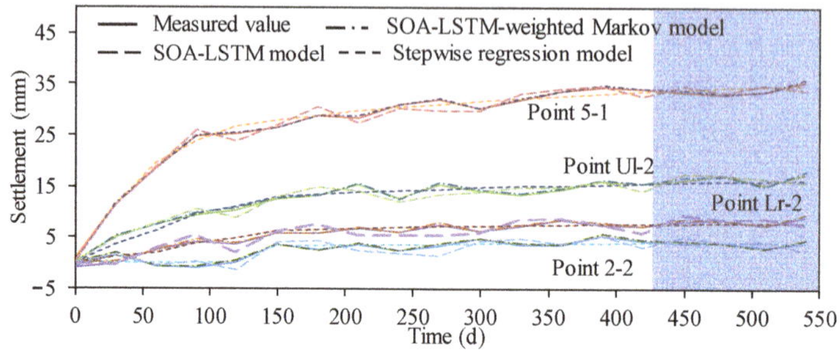

Figure 14. Comparison results of the fitting and prediction results of different models based on measured data of four monitoring points.

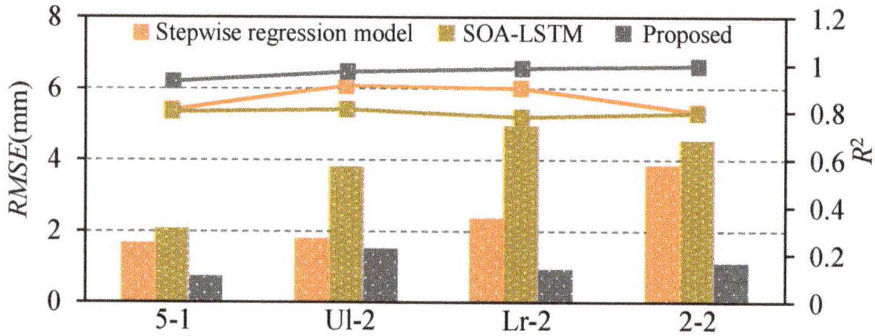

Figure 15. R^2 and *RMSE* of different prediction models.

7. Conclusions

In this paper, the relative error random sequence is obtained by adopting the fitting results of the training set of the SOA-LSTM model and the mean square deviation method is used to classify the states and test the Markovianity of the random sequence. In this case, the SOA-LSTM-weighted Markov model is established. Subsequently, the calculated results of the stepwise regression model, the SOA-LSTM model and the SOA-LSTM-weighted Markov model were compared to verify the fitted and predicted effects of the proposed model. On the basis of the prediction results and the relative errors, the main conclusions can be drawn as follows:

(1) The stepwise regression model is more suitable for dealing with linear problems; it has some limitations in fitting and predicting nonlinear and fluctuating monitoring sequences. By contrast, the proposed model shows obvious superiority in dealing with monitoring data with fluctuation.

(2) The SOA improves the training efficiency of the neural network hyper-parameters of the LSTM model. The SOA-LSTM model can more accurately reflect the nonlinear change rule of the sluice settlement, and the weighted Markov model takes into account the strength of the dependency relationship between different lags, which can fully utilize the information of the training set, significantly reduce the model prediction error, and improve the model prediction accuracy. As listed in Table 1, the maximum settlement variation ranges from 1.73 mm to 31.59 mm, which indicates the apparent fluctuation of sluice settlement monitoring data. On the basis of the fitted results, predicted results and absolute relative errors, the proposed model demonstrates the largest values of R^2 and the smallest values of *RMSE* for the monitoring data of four monitoring points. Therefore, the SOA-LSTM-weighted Markov model is especially suitable for fitting and predicting the larger fluctuation of the sluice settlement, it is also especially suitable for dealing with the more volatile settlement monitoring data.

(3) Due to the better regularity of the actual data series selected in this paper, the prediction accuracy of the SOA-LSTM model and the SOA-LSTM-weighted Markov model is higher. When more monitoring data are obtained, the prediction model should be trained and adjusted in time, and then the subsequent dynamic prediction of the sluice settlement can be carried out, which will be able to better describe the settlement law of the sluice. To our knowledge, the proposed model expresses the stable and accurate results in predicting long sequenced monitoring data. In view of the practical application performance in the case study, the proposed model is supposed to applied extensively in deformation prediction of more sluice projects to improve safety monitoring efficiency, which provides a new aid for structural engineers involved in the real-time health assessment and safety monitoring in sluices. The study in this paper focuses mainly on the data-driven sluice deformation prediction model. A further study should pay more attention to investigating the physics-data double-driven intelligent prediction model, and the proposed method will have prospective applications in sluice structural behavior safety monitoring.

Author Contributions: Conceptualization, J.P. and W.X.; methodology, Y.W. and X.S.; software, S.Z., C.Z. and H.G.; validation, J.P. and X.S.; formal analysis, M.Z.; investigation, S.Z.; resources, J.P.; data curation, X.S.; writing—original draft preparation, W.X. and S.Z.; writing—review and editing, S.Z. and H.G.; supervision, J.P.; project administration, J.P., X.S., C.Z., Y.W. and M.Z.; funding acquisition, J.P., Y.W. and H.G. All authors have read and agreed to the published version of the manuscript.

Funding: This research was funded by Anhui Provincial Natural Science Foundation: "Water Sciences" Joint Fund (2208085US17), National Natural Science Foundation of China (Grant No. U2243223), the Fundamental Research Funds for the Central Universities (Grant No. B230201011), Jiangsu Young Science and Technological Talents Support Project (Grant No.TJ-2022-076), Fund of Water Conservancy Technology of Xinjiang Province (XSKJ-2023-23), and Water Conservancy Science and Technology Project of Jiangsu (Grant No. 2022024).

Data Availability Statement: Not applicable.

Conflicts of Interest: The authors declare no conflict of interest.

References

1. Gu, Z.; Cao, X.; Liu, G.; Lu, W. Optimizing operation rules of sluices in river networks based on knowledge-driven and data-driven mechanism. *Water Resour. Manag.* **2014**, *28*, 3455–3469. [CrossRef]
2. He, X.; Wang, Y.P.; Zhu, Q.; Zhang, Y.; Zhang, D.; Zhang, J.; Yang, Y.; Gao, J. Simulation of sedimentary dynamics in a small-scale estuary: The role of human activities. *Environ. Earth Sci.* **2015**, *74*, 869–878. [CrossRef]
3. Zhu, Y.; Tang, H. Automatic damage detection and diagnosis for hydraulic structures using drones and artificial intelligence techniques. *Remote Sens.* **2023**, *15*, 615. [CrossRef]
4. Komac, M.; Holley, R.; Mahapatra, P.; Marel, H.; Bavec, M. Coupling of GPS/GNSS and radar interferometric data for a 3D surface displacement monitoring of landslides. *Landslides* **2015**, *12*, 241–257. [CrossRef]
5. Tu, R.; Liu, J.; Lu, C.; Zhang, R.; Zhang, P.; Lu, X. Cooperating the BDS, GPS, GLONASS and strong-motion observations for real-time deformation monitoring. *Geophys. J. Int.* **2017**, *209*, 1408–1417. [CrossRef]
6. De Lacy, M.C.; Ramos, M.I.; Gil, A.J.; Franco, Ó.D.; Herrera, A.M.; Avilés, M.; Domínguez, A.; Chical, J.C. Monitoring of vertical deformations by means high-precision geodetic levelling. Test case: The Arenoso dam (South of Spain). *J. Appl. Geod.* **2017**, *11*, 31–41. [CrossRef]
7. Milillo, P.; Bürgmann, R.; Lundgren, P.; Salzer, J.; Perissin, D.; Fielding, E.; Biondi, F.; Milillo, G. Space geodetic monitoring of engineered structures: The ongoing destabilization of the Mosul dam, Iraq. *Sci. Rep.* **2016**, *6*, 37408. [CrossRef] [PubMed]
8. Zhang, L.; Liu, P.; Yan, X.; Zhao, X. Middle displacement monitoring of medium–small span bridges based on laser technology. *Struct. Control Health Monit.* **2020**, *27*, e2509. [CrossRef]
9. Zhao, X.; Liu, H.; Yu, Y.; Xu, X.; Hu, W.; Li, M.; Ou, J. Bridge displacement monitoring method based on laser projection-sensing technology. *Sensors* **2015**, *15*, 8444–8463. [CrossRef]
10. Rodriguez, A.; Comeron, A.; Mesalles, E. Performance assessment of a homodyne laser radar for surface displacement monitoring. *Opt. Eng.* **2001**, *40*, 2812–2821. [CrossRef]
11. Li, Z.; Cao, Y.; Wei, J.; Duan, M.; Wu, L.; Hou, J.; Zhu, J. Time-series InSAR ground deformation monitoring: Atmospheric delay modeling and estimating. *Earth Sci. Rev.* **2019**, *192*, 258–284. [CrossRef]
12. Yang, Z.; Li, Z.; Zhu, J.; Wang, Y.; Wu, L. Use of SAR/InSAR in mining deformation monitoring, parameter inversion, and forward predictions: A review. *IEEE Geosci. Remote Sens. Mag.* **2020**, *8*, 71–90. [CrossRef]

13. Yang, X.; Wang, D.; Xu, Y.; Hou, M.; Wang, Z. Performance assessment of Insar-based vertical displacement monitoring of sluices in coastal soft soil area. *KSCE J. Civ. Eng.* **2022**, *26*, 371–380. [CrossRef]
14. Zhang, G.; Yu, C.; Guo, G.; Li, L.; Zhao, Y.; Li, H.; Gong, Y. Monitoring sluice health in vibration by monocular digital photography and a measurement robot. *KSCE J. Civ. Eng.* **2019**, *23*, 2666–2678. [CrossRef]
15. Ferretti, A.; Fumagallim, A.; Novali, F.; Prati, C.; Rocca, F.; Rucci, A. A new algorithm for processing interferometric data-stacks SqueeSAR. *IEEE Trans. Geosci. Remote Sens.* **2011**, *49*, 3460–3470. [CrossRef]
16. Peng, M.; Zhang, L.M. Analysis of human risks due to dam break floods—Part 2: Application to Tangjiashan landslide dam failure. *Nat. Hazards* **2012**, *64*, 1899–1923. [CrossRef]
17. Kalkan, Y.; Potts, L.V.; Bilgi, S. Assessment of vertical deformation of the Atatürk dam using geodetic observations. *J. Surv. Eng.* **2016**, *142*, 04015011. [CrossRef]
18. Li, W.; Ye, Y.; Hu, N.; Wang, X.; Wang, Q. Real-time warning and risk assessment of tailings dam disaster status based on dynamic hierarchy-grey relation analysis. *Complexity* **2019**, *2019*, 5873420. [CrossRef]
19. Shao, C.; Gu, C.; Yang, M.; Xu, Y.; Su, H. A novel model of dam displacement based on panel data. *Struct. Control Health Monit.* **2018**, *25*, e2037. [CrossRef]
20. Zhu, Y.; Xie, M.; Zhang, K.; Li, Z. A dam deformation residual correction method for high arch dams using phase space reconstruction and an optimized long short-term memory network. *Mathematics* **2023**, *11*, 2010. [CrossRef]
21. Zheng, S.; Gu, C.; Shao, C.; Hu, Y.; Xu, Y.; Huang, X. A novel prediction model for seawall deformation based on CPSO-WNN-LSTM. *Mathematics* **2023**, *11*, 3752. [CrossRef]
22. Li, M.; Wang, J. An empirical comparison of multiple linear regression and artificial neural network for concrete dam deformation modelling. *Math. Prob. Eng.* **2019**, *2019*, 7620948. [CrossRef]
23. Wu, X.; Zheng, D.J.; Liu, Y.T.; Chen, Z.Y.; Chen, X.Q. Temporal convolution network-based time frequency domain integrated model of multiple arch dam deformation and quantification of the load impact. *Struct. Control Health Monit.* **2022**, *29*, e3090. [CrossRef]
24. Wang, L.; Zhang, S.C.; Li, Y.H. Application of dynamic gray forecast model in dam deformation monitoring and forecast. *J. Xi'an Univ. Sci. Technol.* **2005**, *3*, 014.
25. Huang, C.; Zhou, L.; Liu, F.; Cao, Y.; Liu, Z.; Xue, Y. Deformation prediction of dam based on optimized grey verhulst model. *Mathematics* **2023**, *11*, 1729. [CrossRef]
26. Chu, C.; Xu, G. Application of grey deformation prediction model optimized by double coefficient for tailings dam. *Math. Prob. Eng.* **2022**, *2022*, 6103860. [CrossRef]
27. Hu, Y.; Shao, C.; Gu, C.; Meng, Z. Concrete dam displacement prediction based on an ISODATA-GMM clustering and random coefficient model. *Water* **2019**, *11*, 714. [CrossRef]
28. Chen, W.; Wang, X.; Cai, Z.; Liu, C.; Zhu, Y.; Lin, W. DP-GMM clustering-based ensemble learning prediction methodology for dam deformation considering spatiotemporal differentiation. *Knowl.-Based Syst.* **2021**, *222*, 106964. [CrossRef]
29. Hu, J.; Ma, F. Zoned deformation prediction model for super high arch dams using hierarchical clustering and panel data. *Eng. Comput.* **2020**, *37*, 2999–3021. [CrossRef]
30. Lin, C.; Weng, K.; Lin, Y.; Zhang, T.; He, Q.; Su, Y. Time series prediction of dam deformation using a hybrid STL–CNN–GRU model based on sparrow search algorithm optimization. *Appl. Sci.* **2022**, *12*, 11951. [CrossRef]
31. Shao, C.; Zheng, S.; Gu, C.; Hu, Y.; Qin, X. A novel outlier detection method for monitoring data in dam engineering. *Expert Syst. Appl.* **2022**, *193*, 116476. [CrossRef]
32. Xing, Y.; Chen, Y.; Huang, S.; Wang, P.; Xiang, Y. Research on dam deformation prediction model based on optimized SVM. *Processes* **2022**, *10*, 1842. [CrossRef]
33. Hochreiter, S.; Schmidhuber, J. Long short-term memory. *Neural Comput.* **1997**, *9*, 1735–1780. [CrossRef] [PubMed]
34. Abdel-Nasser, M.; Mahmoud, K. Accurate photovoltaic power forecasting models using deep LSTM-RNN. *Neural Comput. Appl.* **2019**, *31*, 2727–2740. [CrossRef]
35. Altan, A.; Karasu, S.; Zio, E. A new hybrid model for wind speed forecasting combining long short-term memory neural network, decomposition methods and grey wolf optimizer. *Appl. Soft Comput.* **2021**, *100*, 106996. [CrossRef]
36. Song, S.; Zhou, Q.; Zhang, T.; Hu, Y. Automatic concrete dam deformation prediction model based on TPE-STL-LSTM. *Water* **2023**, *15*, 2090. [CrossRef]
37. Hu, Y.; Gu, C.; Meng, Z.; Shao, C.; Min, Z. Prediction for the settlement of concrete face rockfill dams using optimized LSTM model via correlated monitoring data. *Water* **2022**, *14*, 2157. [CrossRef]
38. Si, H.Y.; Chen, M. Three dimensional structural analysis of complex sluice chamber structures. *Appl. Mech. Mater.* **2012**, *170*, 1971–1976. [CrossRef]
39. Ding, Q.; Huang, Y.; Xie, T.; Li, F.; Gao, L. Prediction of horizontal displacement of sluice gate based on BP neural network weighted Markov model. *J. Water Resour. Water* **2020**, *31*, 187–193.

40. Dhiman, G.; Kumar, V. Seagull optimization algorithm: Theory and its applications for large-scale industrial engineering problems. *Knowl.-Based Syst.* **2019**, *165*, 169–196. [CrossRef]
41. Cao, Y.; Li, Y.; Zhang, G.; Jermsittiparsert, K.; Razmjooy, N. Experimental modeling of PEM fuel cells using a new improved seagull optimization algorithm. *Energy Rep.* **2019**, *5*, 1616–1625. [CrossRef]

water

MDPI

Article

The Use of UAV for Measuring the Morphology of Ice Cover on the Surface of a River: A Case Study of the Low Head Dam and Fishway Inlet Area in the Odra River

Jan Błotnicki [1], Paweł Jarzembowski [2], Maciej Gruszczyński [1,*] and Marcin Popczyk [3]

[1] Institute of Environmental Engineering, Wrocław University of Environmental and Life Sciences, Grunwaldzki Sq. 24, 50-363 Wrocław, Poland; jan.blotnicki@upwr.edu.pl
[2] Institute of Environmental Biology, Wrocław University of Environmental and Life Sciences, Kożuchowska St. 7a, 51-631 Wrocław, Poland; pawel.jarzembowski@upwr.edu.pl
[3] Faculty of Mining, Safety Engineering and Industrial Automation, Silesian University of Technology, Akademicka St. 2a, 44-100 Gliwice, Poland
* Correspondence: maciej.gruszczynski@upwr.edu.pl

Abstract: The application of UAV to acquire data on the morphometry of frazil ice floe in motion is demonstrated in the measurements conducted in the area of the Wrocław Water Junction at the Opatowice weir on the Odra River (Poland). Image processing was performed using open-source software dedicated to image analysis. The methodology presented in the publication offers a cost-effective and low-overhead technique for describing ice phenomena in lowland rivers. The focus of the methodology was on measuring the area, average size, perimeter, and circularity of frazil ice floe. The measurements were carried out for individual frames captured by a UAV, and the results were analyzed using statistical techniques. In prior research, the team effectively assessed ice velocity on an identical test sample. Deriving the average velocity, surface area, and fundamental morphological traits of frazil ice facilitates the automated segmentation, classification, and prediction of potential risks related to ice blockages on water routes. These risks encompass potential waterway obstructions, as well as infrastructure impairments, and may pose a danger to human safety.

Keywords: UAV; ice jam; frazil ice floe; remote sensing; frazil ice jam; teledetection; ice morphometry; data processing

1. Introduction

The subject of this article is an analysis of the morphometry of drifting ice on the water surface. In the literature review, a significant research gap is noted, where researchers primarily focus on analyzing the processes of the appearance, disappearance, concentration, structure, and thickness of ice in studies utilizing remote measurement methods. However, an analysis of morphometrical parameters such as average size, area, perimeter, and circularity allows for the prediction of the direction of ice phenomena's development and potential threats to existing infrastructure.

The primary objective of this study is to develop a comprehensive methodology for assessing the morphometrical characteristics of frazil ice floes in lowland river systems. The presented work is part of a broader project focused on designing a remote sensing system for the detection, evaluation, and prediction of ice-related phenomena in a central European lowland river [1]. In this publication, we demonstrate the capabilities of determining the morphometrical traits of each observed ice element within the frame. Subsequently, our next step involves evaluating changes in morphometrical features based on the occurrence period of ice-related events. By employing UAV-based remote sensing techniques, we aim to advance our understanding of frazil ice dynamics and contribute to improved river management strategies. This research addresses the need for effective monitoring and predictive capabilities in mitigating the potential risks associated with ice-related incidents in river environments.

The study thoroughly examines the morphometrical characteristics of each drifting ice floe. This approach offers a fresh perspective in ice flow research. We posit that these described features may also play a crucial role in potential ice blockages. At present, data regarding the attributes of ice floes for predictive learning are not available. The study's hypothesis proposes that a shift in the morphological traits of ice floes would occur in the event of a blockage, enabling the prediction of its timing. Therefore, the continuous observation of ice traits stands as a central objective for the research team.

Traditional observation methods do not capture the specific data presented in our publication. The only comparable parameter is the extent of shore ice. However, our primary interest lay in changes to ice cover rather than the presence of shore ice. As such, direct comparisons between data acquired via UAVs and traditional methods were not applicable to the objectives of this study. Ice phenomena measurements pose a challenge due to their inherent difficulty and the subjective nature of the results, which are often presented descriptively. These measurements primarily serve as operational monitoring for various services involved in flood protection, the mitigation of river blockage events, and the management of water structures. The main objective is to exercise control and perform necessary actions in these areas.

Inland water bodies experience the formation of ice covers when the water temperature cools down to near freezing point. In polish conditions, ice typically forms during the transition from autumn to winter due to the strong radiation heat loss from the water surface and subsequent cooling. The formation of ice can lead to blockages, causing disruptions to traffic flow and posing risks to structures such as dams. The occurrence and characteristics of ice phenomena are largely influenced by climatic conditions, including their intensity and frequency [2]. The primary factor governing the formation of ice phenomena on rivers is the air temperature. The course of atmospheric circulation plays a crucial role in shaping the development of ice phenomena by affecting water thermal variability and, consequently, the various forms of ice formation on the river [3]. Designing a water infrastructure in rivers and lakes in northern regions requires careful consideration of ice-related issues. These issues encompass various phenomena such as floating ice, anchor ice, ice jams in rivers, and frazil ice. Among these, the detrimental impact of frazil ice primarily manifests in the obstruction of water passages in engineering structures [4].

The accumulation of frazil ice at water intakes poses significant challenges as it can obstruct the flow passages and potentially completely block the inflow. The consequences of such blockages range from minor temporary inconveniences to severe impacts, particularly in cases where industrial or nuclear plants are compelled to halt operations or when municipalities experience a depletion of their fire-fighting reserves [5].

Documenting the progression and causes of frazil ice blockages is a complex task due to several factors. Firstly, these blockages occur underwater, often during dark and cold winter nights, making direct observation challenging. Additionally, each intake design possesses unique characteristics that influence the behavior of frazil ice accumulation. Moreover, the hydraulic, hydrological, and cryological processes associated with each river vary, further complicating the understanding of frazil ice blockages. Furthermore, frazil ice can manifest in diverse forms, as described by Daly [6], and its occurrence is highly dependent on localized flow patterns and weather conditions.

The intricate nature of frazil ice blockages necessitates comprehensive research and analysis to better comprehend their development and mitigative measures. This understanding is crucial for the effective design and operation of water intake systems, particularly in regions prone to frazil ice formation. By investigating the underlying processes and considering site-specific factors, strategies can be devised to minimize the impacts of frazil ice blockages on critical infrastructure and ensure the uninterrupted functioning of vital facilities [4,7].

Various factors contribute to the formation and characteristics of frazil ice. (1) Turbulent conditions in the water, such as those caused by tides, currents, and the passage of ships, can significantly influence the production and movement of frazil ice. The presence of turbulence promotes the mixing and dispersion of ice crystals within the water column. (2) The presence of impurities or suspended particles in the water can affect the formation and growth of frazil ice. Contaminants can serve as nucleation sites for ice crystals, influencing their size and distribution. (3) The depth of the water body plays a role in frazil ice formation. Shallow waters allow for a greater interaction between the water surface and the atmosphere, leading to faster cooling and an increased likelihood of frazil ice formation. (4) The presence of existing ice cover on the water surface can influence the development of frazil ice. Ice cover can provide a surface for frazil ice crystals to adhere to, leading to the formation of thicker ice layers or ice jams. (5) Wind speed and direction can impact the movement and dispersion of frazil ice. Strong winds can cause ice crystals to accumulate in certain areas or be transported downstream, affecting the spatial distribution and concentration of frazil ice. (6) The ambient air and water temperatures play a fundamental role in frazil ice formation. Cold temperatures below the freezing point of water are necessary for the nucleation and growth of ice crystals. Temperature fluctuations and variations in thermal conditions influence the rate and extent of frazil ice production. These factors interact with each other in complex ways, and their combined effects determine the extent and characteristics of frazil ice floe phenomena in natural water bodies. Understanding these factors is crucial for accurately predicting and managing the risks associated with frazil ice in various environmental and engineering contexts.

2. Materials and Methods

2.1. Formation of Frazil Ice

The formation of frazil ice involves the development of small crystalline particles resulting from the supercooling of water and the presence of turbulence. This natural phenomenon is commonly observed in both river and ocean environments [8] In the context of freshwater systems, such as northern rivers, individual frazil ice particles are transported by turbulent flow and undergo collision events. Through these collisions and the cohesive properties of the ice, the particles agglomerate, forming clusters known as frazil flocs. This process is referred to as flocculation [9]. The production mechanism of frazil ice follows a consistent pattern, which can be illustrated by a typical test curve depicted in Figure 1 (e.g., [4,10–13]). Furthermore, a vital notion is worth considering—it is important to note that this relationship holds true only in the absence of snowfall.

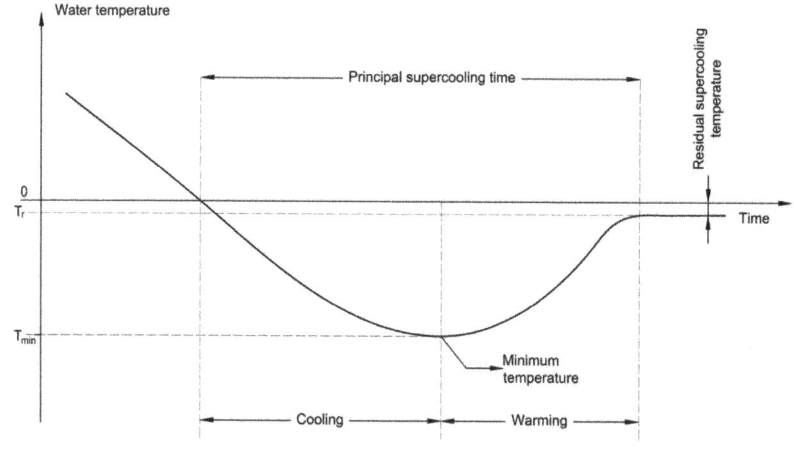

Figure 1. Temperature of water during frazil ice production, acc. to [4,10–13].

In the first stage, the water temperature, initially above zero degrees Celsius, gradually decreases at a rate determined by external atmospheric conditions. Over a short time period, during which atmospheric conditions remain constant, the cooling rate remains relatively stable. As the temperature approaches 0 degrees Celsius, it continues to drop slightly below zero degrees at the same rate. Subsequently, the rate of cooling begins to rapidly decrease until it reaches zero.

At this point, very small particles suddenly emerge and disperse uniformly throughout the water mass. These particles are too minuscule to be discerned with the naked eye but can be observed during nighttime due to the reflection they generate when illuminated (Figure 2a). One of the subsequent stages involves the transformation of frazil ice into frazil ice floes, which are visible in Figure 2b. The photograph was taken at the same time, below the surveyed river cross-section.

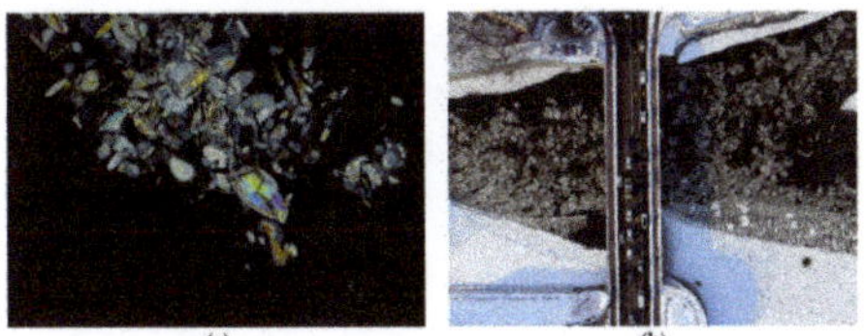

(a) (b)

Figure 2. (**a**) Raw digital image showing frazil particles and flocs for freshwater; source: [9]. (**b**) Frazil ice floe in Odra river on 12 February 2021, with a different object of observation (Wrocław, Poland).

Active frazil ice poses a significant hazard due to its inherent properties and behavior. In the process of frazil ice formation, molecules of supercooled water with temperatures ranging from approximately −0.00 °C to −0.030 °C actively seek nucleation sites to initiate crystallization. These sites can include contaminants present in the water or even on the surface, such as pieces of steel. As the supercooled water molecules come into contact with these nucleation sites, they rapidly freeze and form a slushy and adhesive ice known as active frazil. The freezing process can also extend to the surrounding water molecules, leading to the further development of ice. The presence of seed crystals or solid surfaces facilitates the nucleation and growth of frazil ice, enhancing the potential for ice formation and accumulation. This phenomenon highlights the dangerous nature of active frazil ice, as its stickiness and ability to adhere to objects can contribute to the formation of ice jams, blockages, and other flow obstructions, posing risks to navigation, infrastructure, and overall river dynamics (Figure 2b).

In the presence of an ice cover, freshwater systems experience an interaction between frazil ice and the existing ice layer. The ice cover serves as thermal insulation, reducing the rate of heat exchange between the water and the atmosphere. As a result, the formation of frazil ice is suppressed, leading to a decreased presence of passive frazil ice in the water.

However, it is important to note that the occurrence of active frazil ice is still possible, albeit to a lesser extent, even in the presence of an ice cover. Various contributing factors, such as turbulent flow patterns, tidal influences, and other hydrodynamic processes, can still generate conditions conducive to the formation of active frazil ice. These factors create the potential for supercooled water molecules to find nucleation sites, initiate crystallization, and form sticky and slushy ice.

The concentration of ice in the water column plays a significant role in determining the extent of frazil ice formation. The exact concentration of ice present in the water is an unknown factor that can vary depending on the specific conditions, including temperature, flow velocity, and the history of ice formation and breakup. This uncertainty highlights the complexity of predicting and quantifying the occurrence of active frazil ice in freshwater systems with an existing ice cover.

Understanding the interplay between ice cover, passive frazil ice, and the possibility of active frazil ice formation is crucial for assessing the overall dynamics and hazards associated with frazil ice in freshwater environments. Further research and monitoring efforts are necessary to improve our understanding of these complex interactions and their implications for freshwater systems with ice cover.

2.2. Interaction of Ice with Infrastructure

The majority of instances involving interactions between ice and structures can be categorized into four distinct classes [14]. The first class dynamic ice forces, which originate from floating ice sheets and floes propelled by streamflow, currents or wind. This mode of action is typically crucial for structures situated in riverine environments and may hold significance in specific lacustrine scenarios. When the structure's surface facing the ice impact is vertical, the ice generally undergoes failure through crushing and splitting, resulting in predominantly horizontal forces. In instances where the structure's surface is adequately inclined, the ice experiences failure through bending and shear, giving rise to the development of both horizontal and vertical force components. The second class is static ice forces, which arise from a relatively intact ice sheet experiencing thermal expansion and contraction, or enduring steady pressure from wind or currents. Typically, the ice does not fail but rather undergoes plastic deformation around the structure. The third class comprises forces from broken pack or "rubble" ice, which result from a mass of broken pack or "rubble" ice being driven against a structure, which commonly occurs in situations like a river ice jam or along lake shores. The behavior of the ice pack is somewhat analogous to granular soil. Pile-ups on the shore and forces exerted on ice booms, bridges, and other structures fall within this category. The fourth class is uplift and drawdown forces, which are linked to the adherence of floating ice to piles, among other elements. Forces are generated when static water pressures undergo changes due to factors like tides, reservoir operation, and fluctuations in lake levels.

Accurately predicting ice forces on river piers is of paramount significance. Overestimation can lead to substantial construction expenses, while underestimation poses a risk of structural failure, entailing both financial and safety concerns. Nonetheless, achieving precise estimates of ice forces is a formidable challenge due to the fluctuating nature of ice characteristics and hydraulic conditions [15].

Water-control structures, encompassing dams and barrages, are susceptible to various ice-related challenges. These encompass static ice loads on the dam and associated structures, complications in operating gates or stoplogs due to freezing, impacts from ice floes, the expansion of ice within closed structures like gates, and potential damage to the reservoir shoreline's armour stones. Bridges also face distinct ice-related issues, including ice loads on bridge piers or the structure itself, arising from factors like ice floe impacts, static ice loads, ice jamming, and fluctuations in water levels. Additionally, there may be ice-induced rises in water levels leading to flooding, as well as hydraulic scour caused by ice. Ferry crossings are affected by high accumulations of drifting ice that may exceed the capacity of ice-breaking ferries. Ice jams and runs can cause damage to the ferry landing infrastructure. Furthermore, managing ice in the channel and shoreline access points through measures like bubblers is crucial to ensure navigability. Navigation infrastructure may suffer damage due to the presence of artificial islands around piers in shipping channels. These islands are intended to protect against ship impacts and regulate the formation of ice cover. Water intakes are susceptible to blockage by frazil, both active and inactive. Shallow intakes may experience ice buildup and pileup from the surface downwards. Severe ice runs and jams can lead to significant damage [16]. Infrastructure situated across and along rivers with winter ice cover may encounter significant loads from the ice. Specifically, ice runs have the potential to induce critical ice interactions [15]. Bank protection works face challenges such as ice "bulldozing" along the shoreline, ice-induced hydraulic scour, and potential harm to armour stones. Flood protection works, including levees, may experience reduced ice and flow conveyance capacity, which could potentially lead to overtopping. Additionally, the armour stones used in these structures may be subject to damage [16].

While dams or weirs are not typically designed exclusively for ice control, they stand out as the most efficient type of ice management structure. These constructions, by elevating water levels and moderating water velocity, facilitate the formation of a thermal ice sheet and the aggregation of frazil floes originating from the swifter upstream river sections. The reservoir formed by a dam or weir acts as a reservoir for frazil ice that has been transported from the more open stretches upstream, preventing its accumulation at downstream sites that are prone to freeze-up jams. By averting these freeze-up jams, the subsequent breakup run can navigate through conventional problematic areas without hindrance [17].

Furthermore, there is a notable lack of understanding regarding the impact of in-stream structures on the river ice regime, and there is a scarcity of design guidelines that are accessible for both designers and regulatory bodies [18].

In a river channel, the mere presence of stationary ice leads to elevated water levels for a given discharge in comparison to open-water conditions. This is attributed to the ice occupying a section of the channel's cross-section and introducing a roughness factor that retards the average flow velocity and modifies the velocity profile [16].

2.3. Utilized Research Methods and Instruments

The observation and study of frazil ice floe and ice phenomena in natural watercourses and reservoirs are challenging due to the random nature of these processes and the diversity of natural conditions. Therefore, to gain a better understanding of phenomena related to frazil ice, laboratory studies are conducted, allowing for the precise control of operating devices (such as pumps and turbines) by regulating flow conditions and temperature [13]. Additionally, these studies enable the more effective observation of occurring processes, utilizing a variety of measurement tools and methods, such as shadow photography [19], Laser Doppler Velocimeter optical system [20], a self-constructed electro-resistant instrument [21], filters [6], optical instruments [22], shallow water ice profile sonar (SWIPS) (e.g., [23–25]), acoustic instruments like ADCP (e.g., [26–30]), and machine learning models [31]. Table 1 presents selected scientific articles conducted in both field and laboratory settings, characterizing the research problem, instruments that were utilized, and the location of measurements.

Table 1. Examples of selected studies on ice phenomena.

Area of Study	Research Instrument	Location	Issues	Key References
Field study	Satellite observation	River	Detecting ice sheet, detecting forms of ice	[32]
	Satellite observation	Sea	Ice observations, predicting ice formation	[27]
	UAV observation	River	Ice monitoring	[33]
	Camera on buoys	Sea	Parameterizing ice	[34]
	Acoustic Zooplankton Fish Profiler (AZFP)	Area of land-fast sea ice	Estimating diameters of frazil ice crystals	[35]
	N/D	River	Measuring amount and time of supercooling, radiative heat balance	[36]
	Ground-penetrating radar (GPR), video camera	River	Ice thickness, ice temperature profile, density and crystal structure of ice	[37]
	Satellite observation–Synthetic Aperture Radar (SAR)	River	The extent of river ice	[38]
	Acoustic Doppler Current Profiler (ADCP), Ice-Profiling Sonar (IPS), conductivity-temperature recorder	Sea	Study of underwater frazil ice formation	[39]
Laboratory experiment	observation of frazil ice model (polyethylene material)	Flume	Waved-shape accumulation of ice jam	[40,41]
	Laser Doppler Velocimeter (LDV)	Water tank	Concentration of frazil ice	[20]
	constant temperature anemometer (CTA), monochrome cameras,	Counter-rotating flume	Effects of turbulence intensity on frazil ice formation and evolution	[5]
	Resistance Temperature Detector (RTD); Optical Gauge Sensors (OGS); Acoustic Sonars; Acoustic Doppler Velocimeter (ADV), sampling of frazil ice	Flume	Reproducing natural flow conditions	[42]

2.4. Research Site

The investigation was conducted on 12 February 2021 at the Opatowice weir, situated at km 245 + 035 along the course of the Odra River in Wrocław, Poland (Figure 3). This weir is part of the Bartoszowice-Opatowice barrage, which comprises two weirs and two locks. The Bartoszowice-Opatowice barrage plays a significant role in the canalized Odra River, serving three primary functions: (1) maintaining controlled water levels for navigation purposes, (2) stabilizing soil and water conditions for agricultural and forestry activities, and (3) regulating the riverbed. Within this barrage, the Bartoszowice weir and the Opatowice weir enable the division of water flow into the City Odra River and the Flood Canal, while also supplying water to the Bartoszowice-Zacisze Navigation Canal.

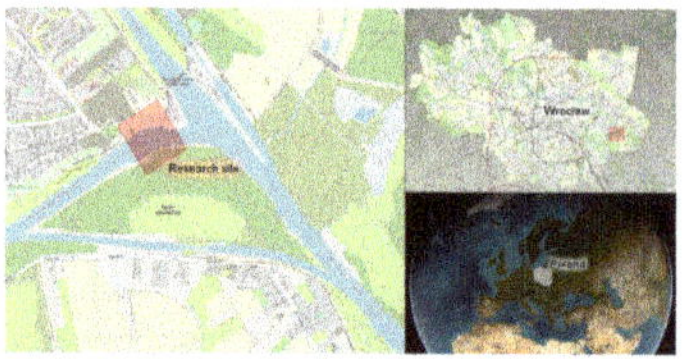

Figure 3. Research site.

The Odra River in the vicinity of Wrocław represents a prototypical example of a medium-sized, lowland river in Central Europe. The hydrotechnical infrastructure, including the weir, exemplifies a conventional water management system with all the associated components. The choice of this specific location was deliberate, as we believe it can be regarded as representative, and consequently, the findings can be extrapolated to other lowland rivers of similar characteristics. This selection facilitates the identification of fundamental patterns and processes applicable to broader contexts within this category of river systems.

2.5. *UAV Setup*

In order to obtain optimal results of aerial photography, the DJI Mavic Air 2 UAV equipped with a 12-megapixel camera and a 1/2″ CMOS sensor flight was conducted at an altitude of 69 m above ground level (AGL), which falls within the upper limit of allowable flight elevation according to the Polish Civil Aviation Authority. When this altitude was reached, orthorectification was performed, ensuring that the camera matrix was parallel to the imaging plane. The frame was meticulously adjusted to encompass the entire study area, containing the river, its banks and reference points. Then, the flight was stabilized, and the images were captured during a period of minimal wind movement.

The aerial photographs were taken on 12 February 2021 at approximately 15:30 (UTC +2), at a temperature of −7.4 degrees Celsius. The research was conducted using the UAV DJI Mavic Air 2, which is equipped with an RGB camera. Table 2 provides precise specifications of the quadcopter and camera.

Table 2. Technical specifications of the UAV unit.

Main unit	Dimensions	183 × 253 × 77 mm
	Weight	570 g
RGB Camera	Camera type	12MP, 1/2″ CMOS
	Image size	4000 × 3000 pixel
	Shutter release	Rolling shutter
	Lens	f/2.8; 4.44 mm (equiv. 24 mm), FOV 84°
	Sensor dimensions	6.40 × 4.80 mm
	Crop factor	5.41
	Additional features	Internal motion unit (IMU), magnetometer
Video properties	Video Resolution	1920 × 1080
	Video Frame Rates	29.97 fps
	Color Profiles	Normal
	Ground Sampling Distance (GSD)	5.18 cm/px

2.6. The Applied Research Methodology

The presented work is an extension of the methodology presented in previous work [1], where we utilized the open-source Fiji (version 1.53t) in conjunction with the TrackMare plugin [43] to assess the flow velocity of frazil ice floes. This approach enabled us to generate moving trajectories, calculate distances covered and determine preference indices in two-choice assays. The analysis was extended to include ice morphometry, i.e., shape detection and identification, surface area, perimeter, circularity. In the study, the open-source Fiji version 1.53t [44] was employed which is costless, multiplatform and under an open license. It is widely used in image analysis. It has options for importing ready-made scripts. To determine the selected morphometrical parameters of the ice, a stabilized UAV video recording was used. To extract frames from the recording, images from 0.5 s, 1.5 s, 2 s, 2.5 s, 3 s, 3.5 s, 4 s, 4.5 s, 5 s, 5.5 s, 6 s, 6.4 s were recorded for a total of 12 images. The files were then saved in the lossless tiff format. The 12 images prepared in this way were transformed to an 8-bit gray-scale. The batch files that were prepared in this way were subjected to a thresholding process that allows for the selection of bright fragments—in this case, a flowing floe against the background of dark, unfrozen water. This process can be performed manually for each frame. To avoid an arbitrary selection being made by the operator, the Renya Entropa [45] algorithm was used. Thanks to this, the same thresholding conditions were achieved for all batch files. The results of the Renya Entropa algorithm are binary files (black and white). In order to eliminate minor errors, the close option was used—adding one pixel (dilation) from the perimeter of the floe and then subtracting (erosion) one pixel [46]. The files were converted to grayscale and then to 8 bits. The closure function smoothes objects and fills small holes (less than 26.83 cm^2), which may be the result of noise generated by camera matrices. The watershed function was used to automatically separate or cut apart particles in contact (e.g., aggregated ice fields). This is a fast and flexible algorithm for computing watersheds in digital gray-scale images. The algorithm is based on an immersion process analogy, in which the flooding of the water in the picture is efficiently simulated using a queue of pixels. This is commonly used in applications like medical image analysis, geospatial image processing, and object recognition, where the precise segmentation of objects in images is crucial [47]. All measurements were performed automatically. Subsequently, the processed images were analyzed using the 'Analyse Particles' function, which computed the ice area, perimeter, and circularity using the formula $4\,\pi$ (area/perimeter^2). A circularity value of 1.0 indicates a perfect circle, while a value approaching 0.0 signifies an increasingly elongated polygon. Images that extended beyond the defined reading area were excluded from the analysis. The workflow diagram applied in this study is depicted in Figure 4.

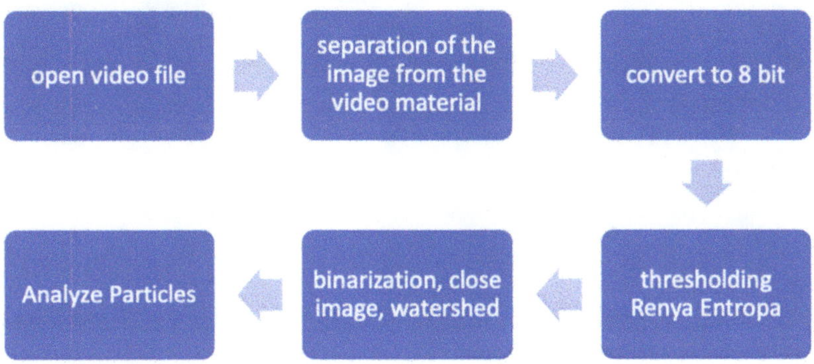

Figure 4. Workflow diagram of Fiji.

3. Results

By processing a total of 12 images, we successfully identified 3444 pieces of drifting ice. It is important to note that any fragments of frazil ice floes that extended partially beyond the designated selection area were excluded from the analysis. On average, each image contained 287 pieces of ice. The calculated average size of the drifting ice was 3.379 square meters, with a perimeter of 6.369 m and a circularity of 0.743. The aim of the work was to develop a new methodology to determine the basic morphometrical parameters of the ice floes. Therefore, it was decided to use basic statistics such as mean, media, modal, and standard deviation to check the homogeneity of readings from individual frames of the film. Table 3 provides a comprehensive presentation of the analytical results that were obtained.

Table 3. Basic statistics on ice morphometry.

Variable	Sample Size (-)	Total Area (m^2)	Average Size (m^2)	Perimeter (m)	Circularity (-)
Min.	282	957.9	3.292	6.253	0.735
1st Qu.	283	962.2	3.362	6.329	0.742
Median	287	966.7	3.373	6.382	0.743
Mean	287.5	971.4	3.379	6.369	0.743
3rd Qu.	286	973.4	3.392	6.396	0.744
Max.	295	1011.1	3.523	6.524	0.751

In Figure 5, the image analysis process is outlined. It begins with a grayscale image in the upper left corner, followed by a thresholded image in the upper right corner. The catchment feature is displayed in the lower left corner, and the count mask image is shown in the lower right corner. This comprehensive illustration provides a clear visual representation of the analytical steps involved in image processing. The colors were randomly assigned by the program for better data visualization.

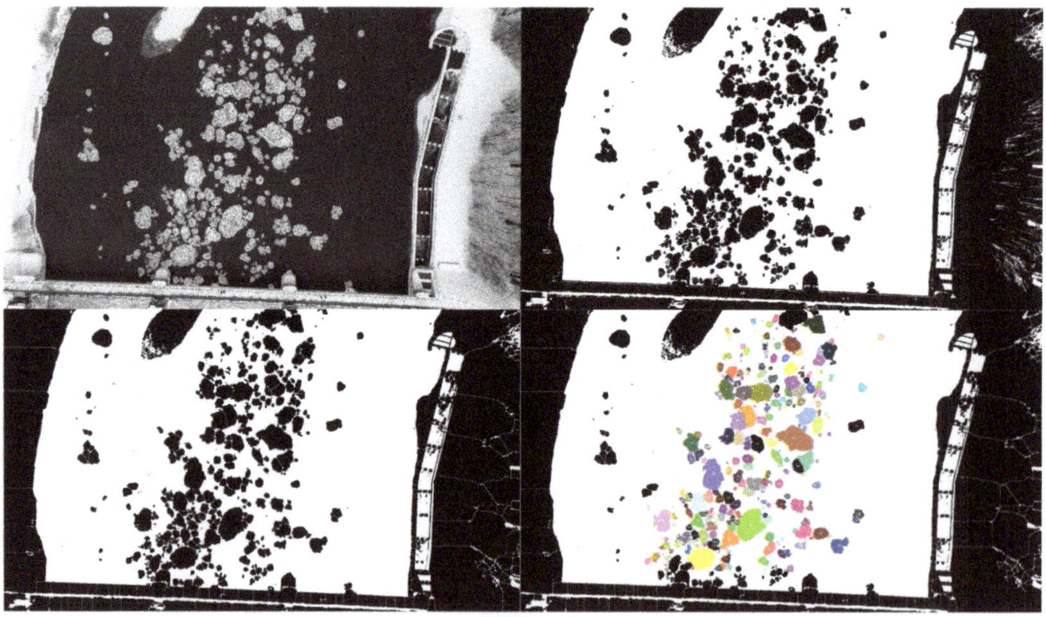

Figure 5. Image breakup process. Grayscale image in the **upper left** corner; image after thresholding in the **upper right** corner; watershed function in the **lower left** corner; image with counting mask in the **lower right** corner.

The fundamental morphometrical parameters were described using standard statistical measures, including minimum, maximum, first and third quartiles, median, and mean. The examination of individual frames within the recording demonstrated the method's high efficacy, with results exhibiting only minor deviations from the average. Figures 6–8 illustrate boxplots depicting the morphometrical characteristics of frazil ice floes, specifically surface area, perimeter, and circularity. These visuals affirm the method's robustness and consistency in parameter readings.

Furthermore, Figures 6–8 offer a statistical insight into the stability of the solution, emphasizing the limited variability among readings across individual frames. This striking uniformity suggests a high level of consistency in the acquired data, with minimal fluctuations discernible from one moment to the next. While this finding is noteworthy, additional analysis is warranted to elucidate the underlying factors contributing to this observed low variability and to evaluate its implications for our study.

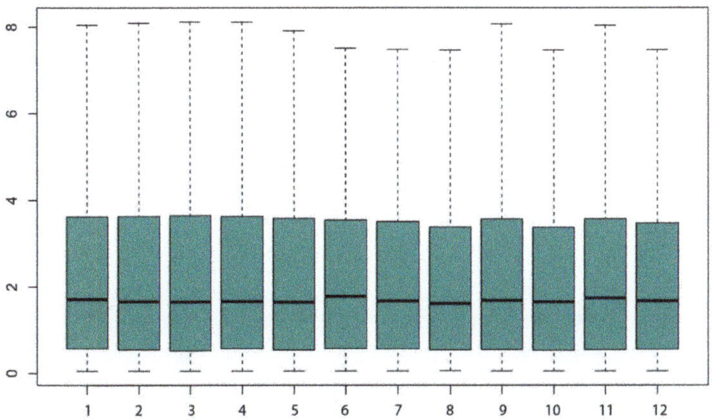

Figure 6. Boxplots of area (m^2) of frazil ice, divided into individual film frames (outliers removed).

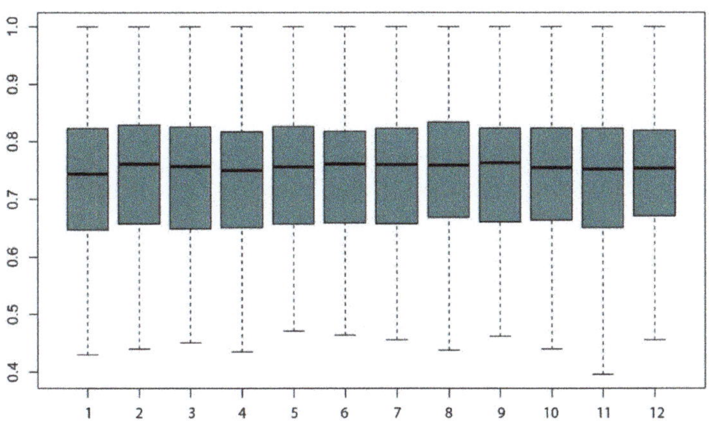

Figure 7. Boxplots of circularity (dimensionless) of drifting ice, divided into individual film frames (outliers removed).

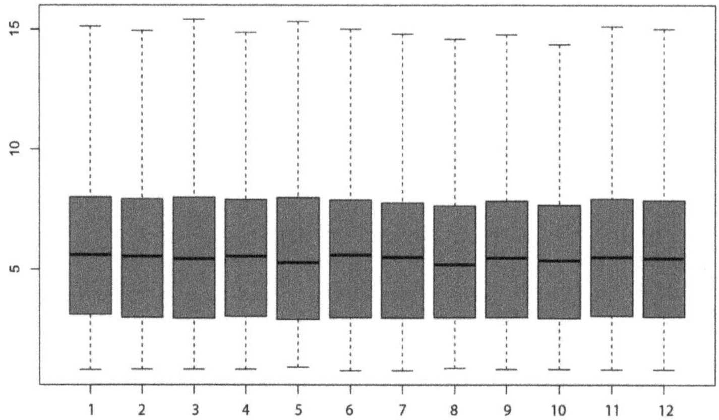

Figure 8. Boxplots of perimeter (m) of drifting ice, divided into individual film frames (outliers removed).

4. Discussion

A systematic approach to the observation of ice phenomena is implemented across all regions where ice phenomena occur, due to the significant impact of ice on human activities and the utilization of waterways. Conventional observation methods have been associated with considerable time and labor demands, coupled with limited quantitative and qualitative data acquisition pertaining to ice phenomena within the observed area. An alternative or substantial complement to traditional observation techniques lies in the application of remote sensing for image analysis. The potential of employing remote sensing for ice phenomena observation has been recognized in several academic institutions, as summarized in Table 1. Considering a thorough review of the literature and drawing from our own experiences, as presented in the publication [1], investigations were conducted to ascertain the morphological parameters of frazil ice floes.

The research has demonstrated the feasibility of determining morphometrical parameters of floes through UAV observations. Through an analysis of the potential and capabilities of UAVs and remote sensing to assess the occurrence of ice phenomena and, consequently, in determining hazards to hydrotechnical infrastructure and adjacent riverine structures, it can be affirmed that this constitutes a promising and innovative tool, representing a valuable avenue for future research.

The very process of image processing, as well as the binarization, segmentation and analysis of components, depends on the computer's computing power; however, with standard desktop PC hardware, it should not take more than 5 s using macro due to the larger dataset, making it possible to investigate how morphometrical parameters can be utilized in predicting ice blockages. The usage of a drone along with image processing software under an open license potentially provides an excellent tool for monitoring ice flow and the early detection of threats stemming from blockages in inland waterways.

When utilizing UAVs to capture images of frazil ice floes in motion, several challenges and limitations were encountered. One of the challenges is maintaining flight stability while recording video. Keeping the camera steady is very difficult, even with the use of advanced stabilizers. During post-production stabilization using automatic algorithms (in Davinci Resolve Studio 18.1 software), we encountered a problem with stabilizing relative to moving objects, meaning that the algorithm took the moving ice floes as reference points. As a result, the 'stabilized image' moved along with the floes. We had to use a different approach by defining points for stabilization.

The images acquired during the flight were of high quality, captured under favorable lighting conditions and in good weather conditions (no rain or snow, absence of shadows, and no wind). This is in line with our objective of observing natural, pre-existing conditions in the field, rather than using a controlled laboratory setting. Adverse atmospheric conditions pose a potential threat to the possibility of obtaining reliable data. It is worth noting that even human observers would face challenges under adverse atmospheric conditions, which can impede research endeavors.

This presents another possible direction for future research, involving an assessment of the influence of photographic material quality on the obtained results. This entails examining how the resolution of the captured image, noise originating from the sensor, exposure time, and the occurrence of atmospheric precipitation can affect the quality of results. The acquired image was taken in an orthogonal view, which is optimal for minimizing distortions caused by the perspective effect. However, studies (e.g., [33,34,48–50]) indicate that this is not a necessary condition, and it is possible to obtain reliable results without orthorectification. Utilizing cameras permanently installed in the technical infrastructure of a facility is possible. However, it is highly unlikely that they will be positioned at the same altitude as a drone's flight, which, in our study, was at 69 m. In such a scenario, it would be necessary to employ a lens with a shorter focal length, thereby increasing the field of view. However, this adjustment would also lead to optical distortions and could decrease the resolving power. As a consequence, it may impact the quality of the obtained results, necessitating distortion correction.

It is important to underscore that this study serves as an initial foray into exploring the potential of remote sensing for monitoring ice conditions. The acknowledgment of potential inaccuracies stemming from alterations in survey height is meant to flag an avenue for future refinements and advancement of the methodology.

The advantages of our remote sensing technique include wide coverage, enabling the analysis of large areas that may be difficult or inaccessible for field observers. Additionally, the reproducibility of measurements allows for changes to be monitored over time, which is extremely useful for analyzing dynamic processes such as ice phenomena. Modern technologies also provide a high spatial resolution, enabling the capture of very detailed images. Moreover, remote sensing enhances safety and time efficiency by allowing for the remote analysis of areas, eliminating the need to send individuals to the field or perform risky operations. The collection of multispectral data is facilitated, enabling the analysis of different aspects of the study area, including features related to frazil ice floe. Furthermore, remote data can be easily integrated with other spatial data through Geographic Information Systems (GIS), enabling a more comprehensive analysis and interpretation. Remote sensing proves particularly effective in studying large-scale ice phenomena, allowing for the analysis of regional or even global patterns and trends. The availability of UAV technology further enhances the technique's flexibility and allows for high-resolution measurements in areas that may be challenging for traditional methods. Finally, remote sensing ensures environmental safety by enabling the monitoring and analysis of ice phenomena in a way that minimizes impact on the environment and natural ecosystems. The significance of field measurements for method validation is acknowledged. However, given the expanse of our study area and the complexities posed by the proximity to the dam, conducting extensive field studies comparable to the scope of our remote sensing approach was not feasible. Safety concerns and logistical challenges were particularly pronounced in this large river environment.

The intricate hydrodynamics and potential hazards in our study area, situated on a major river near a dam, presented substantial obstacles to manual floe measurements. As a result, direct field measurements were precluded. Consequently, the utilization of remote sensing technology emerged as the most viable approach.

5. Conclusions

Our method, as described in this paper, for calculating frazil ice floe from digital images is not only user-friendly, but also efficient, fast and consistent. Both beginners and expert ImageJ users are able to process and analyze images without difficulty.

Obtaining data is essential for developing an effective ice jam reporting program. With access to relevant information such as river flow rates, and ice thickness, it becomes straightforward to create a comprehensive system for monitoring and reporting ice jams. This data-driven approach enables accurate predictions and timely responses to mitigate the risks associated with ice blockages in waterways. There are active plans to broaden and fine-tune this methodology into a comprehensive monitoring system. Collaborative efforts with river management authorities are underway to implement a prototype at select sites.

The research team from the Wrocław University of Environmental and Life Science and the Silesian University of Technology initiated a study in 2021 utilizing UAVs and image analysis software to identify morphometry that can be utilized to predict ice-related hazards. We focused investigations primarily in our home region (Wrocław), enabling a swift response and the collection of reliable data, as ice phenomena in the middle Europe climatic zone, particularly in the context of ongoing climate change, occur infrequently and dissipate rapidly. The employed technique proved effective, motivating the research team to attempt a more detailed characterization of the ice properties at an individual fragment level. While manual description of frazil caps based on aerial photographs is feasible, the authors of the publication focused on automating this process to facilitate the commercial application of image analysis for winter river risk management. The utilization of statistical parameters during the analysis of research results demonstrated that the method is of high quality and exhibits high repeatability and result stability across successive frames of the examined sample. According to the authors' assessment, the research findings indicate the potential implementation of the developed methodology as an automated monitoring system for ice phenomena on lowland rivers. This is particularly crucial considering the hazards associated with the formation of frazil ice in rivers, which can have significant consequences. Hazards described in the introduction emphasize the need for the monitoring and management of frazil ice on rivers, particularly in areas where the risk of blockages is high. The regular monitoring of ice conditions, implementation of appropriate warning systems, and proactive measures can help mitigate the risks associated with frazil ice and ensure the safety of both human activities and the natural environment. By employing a drone under suitable atmospheric and lighting conditions, coupled with the utilization of open-source software analysis, it becomes possible to accurately determine the morphological parameters of frazil ice floes, particularly in the context of their breakup into distinct clusters.

We opted for the use of drones for several compelling reasons. Firstly, drones allow for us to obtain high-quality photographic data in a highly interpretable format, particularly when captured orthogonally. Drones provide us with the flexibility to conduct observations in any selected hydrotechnic structure without the need for prior infrastructure setup for stationary observation systems. This not only saves time but also offers a broader scope of potential observation sites. Furthermore, employing UAVs significantly enhances safety for researchers. It eliminates the necessity for researchers to physically interact with potentially hazardous environments, such as wide rivers, weirs, and areas with significant turbulence, where conducting similar studies would be perilous and logistically challenging.

The need for further research and refinement is imperative to optimize the methodology for varying environmental conditions and scenarios. In the future, more extensive studies will be conducted, incorporating diverse heights and conditions in order to validate and enhance the accuracy of the approach. This study lays the groundwork for future investigations in this domain, with a commitment to remedying this limitation in ongoing research. In the publication, we present only a portion of the analyzed parameters of ice phenomena collected using UAVs and remote sensing. In our opinion, it is too early to assess the impact of the research on waterway management. Our efforts aim to predict the emergence, development, decay, and negative effects of ice phenomena.

Author Contributions: Conceptualization, P.J. and M.G.; data curation, J.B.; formal analysis, J.B., P.J., M.G. and M.P.; funding acquisition, M.G. and M.P.; investigation, J.B.; methodology, P.J. and M.G.; project administration, J.B. and M.G.; resources, J.B.; software, P.J.; supervision, J.B. and M.G.; validation, P.J.; visualization, J.B.; writing—original draft, J.B.; writing—review and editing, J.B. All authors have read and agreed to the published version of the manuscript.

Funding: The APC was funded by Wroclaw University of Environmental and Life Sciences and the Silesian University of Technology.

Data Availability Statement: The data presented in this study are openly available in Base of Knowledge at DOI:10.57755/dhpp-sc05, reference number UPWR5ea2b2899b6f4671b45e51a69c496a0e.

Acknowledgments: Special thanks to the language editor—Joanna Barańska.

Conflicts of Interest: The authors declare no conflict of interest. The funders had no role in the design of the study; in the collection, analyses, or interpretation of data; in the writing of the manuscript; or in the decision to publish the results.

References

1. Błotnicki, J.; Gruszczyński, M.; Jarzembowski, P.; Popczyk, M. Application of Close-Range Remote Sensing for Automatic Identification of Ice Jams in Rivers in the Area of the Inlet to the Fishway. *Zesz. Nauk./Akad. Morska Szczecinie* **2023**, *75*, 97–106. [CrossRef]
2. Łukaszewicz, T.J.; Jawgiel, K. Zmienność Częstości i Charakteru Przebiegu Zjawisk Lodowych Na Rzece Parsęcie w Aspekcie Zmian Klimatycznych. *BADANIA Fizjogr. Geogr. Fiz.* **2017**, *68*, 79–98. [CrossRef]
3. Ptak, M.; Choiński, A.; Kirviel, J. Long-Term Water Temperature Fluctuations in Coastal Rivers (Southern Baltic) in Poland. *Bull. Geogr. Phys. Geogr. Ser.* **2016**, *11*, 35–42. [CrossRef]
4. Michel, B. Theory of Formation and Deposit of Frazil Ice. In Proceedings of the Eastern Show Conference, the 1963 Annual Meeting, Quebec City, QC, Canada, 14–15 February 1963; pp. 130–148.
5. Clark, S.P.; Doering, J.C. Frazil Flocculation and Secondary Nucleation in a Counter-Rotating Flume. *Cold Reg. Sci. Technol.* **2009**, *55*, 221–229. [CrossRef]
6. Daly, S.F.; Rand, J.H. Development of an Underwater Frazil-Ice Detector. *Cold Reg. Sci. Technol.* **1990**, *18*, 77–82. [CrossRef]
7. Richard, M.; Morse, B. Multiple Frazil Ice Blockages at a Water Intake in the St. Lawrence River. *Cold Reg. Sci. Technol.* **2008**, *53*, 131–149. [CrossRef]
8. Martin, S.; Kauffman, P. A Field and Laboratory Study of Wave Damping by Grease Ice. *J. Glaciol.* **1981**, *27*, 283–313. [CrossRef]
9. Schneck, C.C.; Ghobrial, T.R.; Loewen, M.R. Laboratory Study of the Properties of Frazil Ice Particles and Flocs in Water of Different Salinities. *Cryosphere* **2019**, *13*, 2751–2769. [CrossRef]
10. Ye, S.Q.; Doering, J.; Shen, H.T. A Laboratory Study of Frazil Evolution in a Counter-Rotating Flume. *Can. J. Civ. Eng.* **2004**, *31*, 899–914. [CrossRef]
11. McFarlane, V.; Loewen, M.; Hicks, F. Measurements of the Evolution of Frazil Ice Particle Size Distributions. *Cold Reg. Sci. Technol.* **2015**, *120*, 45–55. [CrossRef]
12. Clark, S.; Doering, J. Laboratory Experiments on Frazil-Size Characteristics in a Counterrotating Flume. *J. Hydraul. Eng.* **2006**, *132*, 94–101. [CrossRef]
13. Chen, Y.; Lian, J.; Zhao, X.; Guo, Q.; Yang, D. Advances in Frazil Ice Evolution Mechanisms and Numerical Modelling in Rivers and Channels in Cold Regions. *Water* **2023**, *15*, 2582. [CrossRef]
14. US Army Corps of Engineers. *Engineering and Design: Ice Engineering (Engineer Manual 1110-2-1612)*; US Army Corps of Engineers: Washington, DC, USA, 1999.
15. Rødtang, E.; Alfredsen, K.; Høyland, K.; Lia, L. Review of River Ice Force Calculation Methods. *Cold Reg. Sci. Technol.* **2023**, *209*, 103809. [CrossRef]
16. Burrell, B.C.; Turcotte, B.; Comfort, G.; Groeneveld, J. Infrastructure and River Ice: An Overview. In Proceedings of the 21st Workshop on the Hydraulics of Ice Covered Rivers, Saskatoon, SK, Canada, 29 August–1 September 2021.
17. Daly, S.F. Frazil Ice. In *River Ice Formation. Committee on River Ice Processes and the Environment*; Beltaos, S., Ed.; CRIPE: Edmonton, AB, Canada, 2013; pp. 107–134.
18. Tuthill, A.M. *Ice Jam at the Rio Blanco Diversion Weir on the White River in Colorado Cold Regions Research*; Defense Technical Information Center: Washington, DC, USA, 2008.
19. Arakawa, K.; Higuchi, K. Studies on the Freezing of Water. (I). *J. Fac. Sci. Hokkaido Univ. II* **1952**, *4*, 201–209.
20. Schmidt, C.C.; Glover, J.R. A Frazil Ice Concentration Measuring System Using a Laser Doppler Velocimeter. *J. Hydraul. Res.* **1975**, *13*, 1975. [CrossRef]
21. Tsang, G. An Instrument for Measuring Frazil Concentration. *Cold Reg. Sci. Technol.* **1985**, *10*, 235–249. [CrossRef]
22. Pegau, W.S.; Paulson, C.A.; Zaneveld, J.R.V. Optical Measurements of Frazil Concentration. *Cold Reg. Sci. Technol.* **1996**, *24*, 341–353. [CrossRef]

23. Marko, J.R.; Topham, D.R. Laboratory Measurements of Acoustic Backscattering from Polystyrene Pseudo-Ice Particles as a Basi for Quantitative Characterization of Frazil Ice. *Cold Reg. Sci. Technol.* **2015**, *112*, 66–86. [CrossRef]
24. Marko, J.R.; Jasek, M. Sonar Detection and Measurements of Ice in a Freezing River I: Methods and Data Characteristics. *Cold Reg Sci. Technol.* **2010**, *63*, 121–134. [CrossRef]
25. Marko, J.R.; Jasek, M. Sonar Detection and Measurement of Ice in a Freezing River II: Observations and Results on Frazil Ice. *Col Reg. Sci. Technol.* **2010**, *63*, 135–153. [CrossRef]
26. Dmitrenko, I.A.; Wegner, C.; Kassens, H.; Kirillov, S.A.; Krumpen, T.; Heinemann, G.; Helbig, A.; Schröder, D.; Hölemann, J.A Klagge, T.; et al. Observations of Supercooling and Frazil Ice Formation in the Laptev Sea Coastal Polynya. *J. Geophys. Res. Ocear* **2010**, *115*, 1–9. [CrossRef]
27. Ito, M.; Ohshima, K.I.; Fukamachi, Y.; Simizu, D.; Iwamoto, K.; Matsumura, Y.; Mahoney, A.R.; Eicken, H. Observations o Supercooled Water and Frazil Ice Formation in an Arctic Coastal Polynya from Moorings and Satellite Imagery. *Ann. Glacio* **2015**, *56*, 307–314. [CrossRef]
28. Ito, M.; Ohshima, K.I.; Fukamachi, Y.; Mizuta, G.; Kusumoto, Y.; Nishioka, J. Observations of Frazil Ice Formation and Upwarc Sediment Transport in the Sea of Okhotsk: A Possible Mechanism of Iron Supply to Sea Ice. *J. Geophys. Res. Ocean.* **2017**, *122* 788–802. [CrossRef]
29. Ito, M.; Ohshima, K.I.; Fukamachi, Y.; Mizuta, G.; Kusumoto, Y.; Kikuchi, T. Underwater Frazil Ice and Its Suspension Deptl Detected from ADCP Backscatter Data around Sea Ice Edge in the Sea of Okhotsk. *Cold Reg. Sci. Technol.* **2021**, *192*, 103382 [CrossRef]
30. Leonard, G.H.; Purdie, C.R.; Langhorne, P.J.; Haskell, T.G.; Williams, M.J.M.; Frew, R.D. Observations of Platelet Ice Growth anc Oceanographic Conditions during the Winter of 2003 in McMurdo Sound, Antarctica. *J. Geophys. Res. Ocean.* **2006**, *111*, 1–14 [CrossRef]
31. Koo, Y.; Xie, H.; Kurtz, N.T.; Ackley, S.F.; Wang, W. Sea Ice Surface Type Classification of ICESat-2 ATL07 Data by Using Data-Driven Machine Learning Model: Ross Sea, Antarctic as an Example. *Remote Sens. Environ.* **2023**, *296*, 113726. [CrossRef]
32. Pawłowski, B.; Łoś, H.; Osińska-Skotak, K. Próba Klasyfikacji Typów Pokrywy Lodowej Na Dolnej Wiśle Na Podstawie Satelitarnych Zobrazowań Radarowych Pod Kątem Oceny Lodowego Wypełnienia Koryta. In *Nowoczesne Metody i Rozwiązania w Hydrologii i Gospodarce Wodnej*; Absalon, D., Matysik, M., Ruman, M., Eds.; Komisja Hydrologiczna Polskiego Towarzystwa Geograficznego: Warszawa, Poland, 2015; Volume 3, pp. 313–324. ISBN 9788578110796.
33. Zhang, X.; Zhou, Y.; Jin, J.; Wang, Y.; Fan, M.; Wang, N.; Zhang, Y. ICENETv2: A Fine-Grained River Ice Semantic Segmentation Network Based on UAV Images. *Remote Sens.* **2021**, *13*, 633. [CrossRef]
34. Rogers, W.E.; Thomson, J.; Shen, H.H.; Doble, M.J.; Wadhams, P.; Cheng, S. Dissipation of Wind Waves by Pancake and Frazil Ice in the Autumn Beaufort Sea. *J. Geophys. Res. Ocean.* **2016**, *121*, 7991–8007. [CrossRef]
35. Frazer, E.K.; Langhorne, P.J.; Leonard, G.H.; Robinson, N.J.; Schumayer, D. Observations of the Size Distribution of Frazil Ice in an Ice Shelf Water Plume. *Geophys. Res. Lett.* **2020**, *47*, 2020gl090498. [CrossRef]
36. Gilfilian, R.; Kline, W.; Osterkamp, T.; Benson, C. Ice Formation in a Small Alaskan Stream. In Proceedings of the Symposia on the Role of Snow and Ice in Hydrology: Properties and Processes, Banff, AB, Canada, September 1972; pp. 505–512.
37. Li, Z.; Li, C.; Yang, Y.; Zhang, B.; Deng, Y.; Li, G. Physical Mechanism and Parameterization for Correcting Radar Wave Velocity in Yellow River Ice with Air Temperature and Ice Thickness. *Remote Sens.* **2023**, *15*, 1121. [CrossRef]
38. Stonevicius, E.; Uselis, G.; Grendaite, D. Ice Detection with Sentinel-1 SAR Backscatter Threshold in Long Sections of Temperate Climate Rivers. *Remote Sens.* **2022**, *14*, 1627. [CrossRef]
39. Ito, M.; Fukamachi, Y.; Ohshima, K.I.; Shirasawa, K. Observational Evidence of Supercooling and Frazil Ice Formation throughout the Water Column in a Coastal Polynya in the Sea of Okhotsk. *Cont. Shelf Res.* **2020**, *196*, 104072. [CrossRef]
40. Hou, Z.; Wang, J.; Sui, J.; Song, F.; Li, Z. Impact of Local Scour around a Bridge Pier on Migration of Waved-Shape Accumulation of Ice Particles under an Ice Cover. *Water* **2022**, *14*, 2193. [CrossRef]
41. Chen, P.; Sui, J.; Cao, G.; Cheng, T. Waved-Shape Accumulation of Ice Jam—Analysis and Experimental Study. *Water* **2022**, *14* 3945. [CrossRef]
42. Richard, M.; Cornett, A. NRC Frazil Ice Research Facility. In Proceedings of the Annual Conference of the Canadian Society of Civil Engineering, Montreal, QC, Canada, 12–15 June 2019; pp. 1–10.
43. Ershov, D.; Phan, M.-S.; Pylvänäinen, J.W.; Rigaud, S.U.; Le Blanc, L.; Charles-Orszag, A.; Conway, J.R.W.; Laine, R.F.; Roy, N.H. Bonazzi, D.; et al. Bringing TrackMate in the Era of Machine-Learning and Deep-Learning. *bioRxiv* **2021**. [CrossRef]
44. Schindelin, J.; Arganda-Carreras, I.; Frise, E.; Kaynig, V.; Longair, M.; Pietzsch, T.; Preibisch, S.; Rueden, C.; Saalfeld, S.; Schmid B.; et al. Fiji: An Open-Source Platform for Biological-Image Analysis. *Nat. Methods* **2012**, *9*, 676–682. [CrossRef] [PubMed]
45. Kapur, J.N.; Sahoo, P.K.; Wong, A.K.C. A New Method for Gray-Level Picture Thresholding Using the Entropy of the Histogram *Comput. Vis. Graph. Image Process.* **1985**, *29*, 273–285. [CrossRef]
46. Hartig, S.M. Basic Image Analysis and Manipulation in ImageJ. *Curr. Protoc. Mol. Biol.* **2013**, *102*, 14.15.1–14.15.12. [CrossRef]
47. Vincent, L.; Soille, P. Watersheds in Digital Spaces: An Efficient Algorithm Based on Immersion Simulations. *IEEE Trans. Pattern Anal. Mach. Intell.* **1991**, *13*, 583–598. [CrossRef]
48. Qiao, G.; Yang, M.; Wang, H. A Detection Approach for Floating Debris Using Ground Images Based on Deep Learning. *Remote Sens.* **2022**, *14*, 4161. [CrossRef]

49. Lin, F.; Hou, T.; Jin, Q.; You, A. Improved Yolo Based Detection Algorithm for Floating Debris in Waterway. *Entropy* **2021**, *23*, 1111. [CrossRef] [PubMed]
50. Li, S.; Liu, S.; Cai, Z.; Liu, Y.; Chen, G.; Tu, G. TC-YOLOv5: Rapid Detection of Floating Debris on Raspberry Pi 4B. *J. Real-Time Image Process.* **2023**, *20*, 17. [CrossRef]

water

MDPI

Article

Research on Temperature Control Index for High Concrete Dams Based on Information Entropy and Cloud Model from the View of Spatial Field

Guang Yang [1,†], Jin Sun [2,*,†], Jianwei Zhang [1], Jingtai Niu [3,*], Bowen Luan [1], Zhendong Huang [1] and Ahui Zhao [1]

1 School of Water Conservancy, North China University of Water Resources and Electric Power, Zhengzhou 450045, China; yangguang2019@ncwu.edu.cn (G.Y.); zhangjianwei@ncwu.edu.cn (J.Z.); bowenluan@outlook.com (B.L.); hhzd493@outlook.com (Z.H.); zhaoahui1998@outlook.com (A.Z.)
2 College of Surveying and Geo-Informatics, North China University of Water Resources and Electric Power, Zhengzhou 450045, China
3 School of Hydraulic and Ecological Engineering, Nanchang Institute of Technology, Nanchang 330099, China
* Correspondence: sunjin@ncwu.edu.cn (J.S.); niujingtai@163.com (J.N.)
† These authors contributed equally to this work.

Abstract: It is significant to adopt scientific temperature control criteria for high concrete dams in the construction period according to practical experience and theoretical calculation. This work synthetically uses information entropy and a cloud model and develops novel in situ observation data-based temperature control indexes from the view of a spatial field. The order degree and the disorder degree of observation values are defined according to the probability principle. Information entropy and weight parameters are combined to describe the distribution characteristics of the temperature field. Weight parameters are optimized via projection pursuit analysis (PPA), and then temperature field entropy (TFE) is constructed. Based on the above work, multi-level temperature control indexes are set up via a cloud model. Finally, a case study is conducted to verify the performance of the proposed method. According to the calculation results, the change law of TFEs agrees with actual situations, indicating that the established TFE is reasonable, the application conditions of the cloud model are wider than those of the typical small probability method, and the determined temperature control indexes improve the safety management level of high concrete dams. Research results offer scientific reference and technical support for temperature control standards adopted at other similar projects.

Keywords: high concrete dam; in situ observation data; temperature field; temperature control index; information entropy; cloud model; dam safety

1. Introduction

Several high concrete dams have been constructed in China over the years, e.g., the Xiluodu arch dam (285.50 m) and the Xiangjiaba gravity dam (162.00 m), as shown in Figure 1a,b [1–8]. These huge projects have tremendously contributed to the economic growth and social development of China. However, they also pose serious security risks, which could cause unpredictable and disastrous consequences. Historically, severe concrete dam failure accidents occurred and caused vast economic losses and causalities, such as the St. Francis dam, the Malpasset dam, and the Vajont dam [9–11]. Therefore, governments of the world set strict requirements on the safe construction and operation of concrete dams [12], especially for high concrete dams. Instead of a sudden occurrence, a dam accident is a gradual process from quantitative to qualitative changes. In situ observation and risk assessment are essential for making informed decisions about the construction safety of high concrete dams, with the ultimate goal of protecting both people and property [13]. If robust and reliable real-time safety monitoring strategies are conducted, then most hidden dangers can be found and catastrophic accidents can be avoided [14–16].

(a) (b)

Figure 1. Typical high concrete dams: (**a**) Xiluodu arch dam and (**b**) Xiangjiaba gravity dam.

Concrete temperature is an important in situ observation indicator for concrete dams [17–19]. Several factors, e.g., cement hydration heat, atmospheric temperature, water temperature, etc., play an important role in concrete temperature variation. Concrete materials have poor temperature conductivity. The influence of temperature is mainly reflected in temperature stress. If temperature stress exceeds the threshold, then it can cause dam cracks. Adopting scientific temperature control methods [20–22] can effectively prevent dam cracks, which is crucial to guarantee dam safety. Conventionally, researchers focus on investigating single-point control indexes. However, with the continuous construction of ultra-high concrete dams, single-point indexes cannot meet practical demands. There is an urgent need to construct temperature control indexes from the perspective of the spatial field. Chen et al. proposed a temperature control approach by using the finite element (FE) method, and on this basis, the temperature field evaluation and the thermal-induced stress of the discussed roller-compacted-concrete (RCC) dam were analyzed [23]. Chen et al. used the AdaBoost-ANN algorithm to reconstruct the temperature field of concrete arch dams. The calculation results showed that the proposed algorithm is convenient for dam safety assessment [24]. Pan et al. used the convolutional neural network to analyze the temperature field distribution of the Xiaowan dam, and this method can accurately predict possible thermal cracks on the dam surface [25]. Hou et al. used the FE method to invert the thermal parameters of dam concrete. The calculation results efficiently reflect the temperature field distribution of the discussed ultra-high arch dam [4]. Zhou et al. applied the distributed temperature sensing technology to monitor the temperature field distribution of high-arch dams under construction [26]. In China, cold areas play an important role in further hydropower development. Compared with regions with a mild climate, cold areas where cold waves appear frequently have a large day-night temperature difference, a low annual average temperature, etc., and many scientific problems need to be solved urgently. Although there has been some progress in research related to the spatial temperature field, in-depth research remains rare, especially for those using in situ observation data.

Information entropy, also known as Shannon entropy, is a measure of uncertainty degree caused by insufficient or chaotic information in a system. The greater the information entropy, the higher the chaotic degree and uncertainty degree the studied system shows. Many applications have been employed in the field of dam safety monitoring. Chen et al. used information entropy theory to analyze the seepage characteristics of complex rock mass dams [27]. Su et al. comprehensively applied information entropy theory and the FE method to analyze the stress and deformation of concrete-faced rockfill dams [28]. Cui et al. proposed the multi-time-scale mutual information entropy method to reveal the cyclic synchronous changes of drought-flood abrupt alternation in the Three Gorges reservoir area [29]. Lei et al. employed information entropy to characterize the spatial deformation distribution of high concrete dams. On this basis, the multi-level early-warning indexes were constructed by applying the typical small probability method [30]. Zheng et al. used Copula entropy and mutual information theory to analyze and process in situ observation data of dams, effectively reducing the influence of correlations among different factors [31]. Huang et al. constructed the viscoelastic control indexes for monitoring the deformation safety of gravity dams in alpine regions by comprehensively using the typical small probability method and the maximum entropy method [32]. Zhang et al. used the maximum entropy method to determine operational control loads of concrete dams according to deformation observation data, and the analysis results showed that the proposed method, compared with traditional approaches, has better performances [33]. Qin et al. used weighted fusion entropy and Hurst exponent difference to evaluate the crack evolution of concrete dams [34]. Information entropy, as an intelligent and information-based mathematical method, not only lays a theoretical foundation for measuring uncertainty in random events, but also possesses broad universality and scalability to describe system state evolution. Further research needs to be conducted to characterize the evolution laws of high concrete dam temperature fields based on information entropy theory and then to construct reasonable control indexes for ensuring dam safety.

The cloud model algorithm is a mathematical model using probability theory to describe and quantify uncertain information. It consists of two parts, i.e., a cloud generator and a cloud transformation. Forward and reverse cloud generators can generate cloud droplets and determine mathematical features, respectively. Cloud transformation converts cloud droplets into a probability distribution used for further analysis. The cloud model possesses strong abilities in handling uncertain and imprecise information, which has been applied in various fields, e.g., data mining, decision-making, and pattern recognition. For dam safety monitoring, the cloud model has also achieved successful applications. Han et al. proposed the cloud model-Monte Carlo coupling model that can accurately evaluate risk degree in dam safety management [35]. Wu et al. integrated the cloud model and Dempster-Shafer evidence theory to assess the safety state of tailing dams. The developed approach has strong robustness and a fault-tolerant capacity [36]. Li et al. used the cloud model to analyze in situ observation data of dams, and the proposed method has good performance in discovering abnormal values [37]. Sang et al. developed the extended cloud model and the extended analytic hierarchy process method to address uncertainty problems in the safety assessment of dams [38]. Guo et al. used a cloud model and modal strain energy to identify uncertain damage problems caused by noise [39]. According to the above analyses, the cloud model algorithm can achieve the conversion of qualitative concepts into quantitative data, which provides an effective way to construct temperature control indexes for the multi-level risk management of high concrete dams.

Consequently, by synthetically using information entropy and a cloud model, this study develops temperature control indexes based on in situ observation data. The paper is organized as follows. In Section 2, temperature field entropy (TFE) for describing the evolution characteristics of temperature fields is established, and projection pursuit analysis (PPA) is explored to optimize weight parameters. Section 3 investigates the principle of the cloud model, and then the cloud model-based temperature control indexes are constructed. Section 4 conducts a case study to verify the proposed method. In Section 5, the main conclusions of the research are summarized, and some suggestions are provided.

2. Temperature Field Characterization Using Information Entropy

2.1. TFE Construction

The evaluation process of dam temperature can be roughly divided into two stages, i.e., the construction period and the operation period [40–42]. During the construction period, the temperature field of the dam body is affected by multiple factors, such as cement hydration heat, artificial cooling, and ambient temperature. In the stage of operation, cement hydration heat has basically dissipated, and the temperature field is affected by ambient temperature, e.g., air temperature and water temperature [43–46]. Concrete temperature has a significant impact on the stability of the dam during the construction period, but its effects during regular operation afterward are essentially insignificant. This work mainly focuses on the construction period, and information entropy is used to characterize the temperature field evaluation of high concrete dams.

Here, δ_{ij} denotes the observation value of the ith observation point on the jth observation date, the corresponding order degree is recorded as u_{ij}, ω_i denotes the weight parameter of the ith point, $i = 1, 2, \cdots, n$ and n is the total number of observation points. Under the condition of temperature increase, u_{ij} is defined as follows:

$$u_{ij} = F_i(\delta_{ij}) = \int_{-\infty}^{\delta_{ij}} f_i(\delta)\mathrm{d}\delta, \tag{1}$$

where $f_i(\delta)$ is the probability density function of δ_{ij}, and $F_i(\delta)$ denotes the probability distribution function. For temperature decrease, u_{ij} is built as follows:

$$u_{ij} = 1 - F(\delta_{ij}) = \int_{\delta_{ij}}^{+\infty} f_i(\delta)\mathrm{d}\delta. \tag{2}$$

According to the definition of information entropy, the temperature entropy (TE) of δ_{ij} is expressed by S_{ij} as follows:

$$S_{ij} = -\left[u_{ij}\ln u_{ij} + (1 - u_{ij})\ln(1 - u_{ij})\right] = -\sum_{k=1}^{2} u_{ij}^{p}\ln u_{ij}^{p}, \tag{3}$$

where p is a parameter with a value of one or two, and $u_{ij}^1 = u_{ij}$ and $u_{ij}^2 = 1 - u_{ij}$ are the order degree and the disorder degree of δ_{ij}, respectively. Obviously, the disorder degree decreases while the order degree increases and vice versa.

This work regards the temperature field of high concrete dams as an interactional system in which characteristic points have different contribution degrees. ω_i is used to describe the contribution degree of the ith point. The weight entropy (WE) S_ω is established as follows:

$$S_\omega = -\sum_{i=1}^{n} \omega_i \ln \omega_i. \tag{4}$$

Information entropy and weight parameters are combined to describe the distribution characteristics of the temperature field. Then, TFE can be constructed as follows:

$$\begin{aligned} S_j &= -\sum_{i=1}^{n}\sum_{p=1}^{2} \omega_i u_{ij}^{p} \ln\left(\omega_i u_{ij}^{p}\right) \\ &= -\sum_{i=1}^{n}\sum_{p=1}^{2} \omega_i u_{ij}^{p}\left(\ln\omega_i + \ln u_{ij}^{p}\right) \\ &= -\sum_{i=1}^{n}\sum_{p=1}^{2} \omega_i u_{ij}^{p} \ln\omega_i - \sum_{i=1}^{n}\sum_{p=1}^{2} \omega_i u_{ij}^{p} \ln u_{ij}^{p} \\ &= -\sum_{i=1}^{n} \omega_i \ln\omega_i \sum_{p=1}^{2} u_{ij}^{p} - \sum_{i=1}^{n} \omega_i \sum_{p=1}^{2} u_{ij}^{p} \ln u_{ij}^{p} \\ &= S_\omega + \sum_{i=1}^{n} \omega_i S_{ij} \end{aligned} . \tag{5}$$

Equation (5) reveals that TFE involves WE and the weighted TE. The lower the quantity of TFE, the more dangerous the temperature status demonstrates.

2.2. Weight Optimization

PPA, as a high-dimensional data analysis method, can identify data structures and characteristics by projecting high-dimensional monitoring data into a low-dimensional space. PPA [47–50] is used to optimize the weight parameters. The calculation steps are given as follows.

Calculate S_{ij} by using Equations (1)–(3). The matrix S is constructed as follows:

$$S = \{S_{ij}\} = \begin{Bmatrix} S_{11} & S_{12} & \cdots & S_{1m} \\ S_{21} & S_{22} & \cdots & S_{2m} \\ \vdots & \vdots & \ddots & \vdots \\ S_{n1} & S_{n2} & \cdots & S_{nm} \end{Bmatrix}, \tag{6}$$

where m is the total number of observation dates, and n denotes the total number of observation points. Project $\boldsymbol{S}$ to the unit direction $\boldsymbol{P} = \{p_1, p_2, \cdots, p_j, \cdots, p_n\}$, where ${p_1}^2 + {p_2}^2 + \cdots + {p_n}^2 = 1$. The projection function $G(i)$ is expressed as follows:

$$G(i) = \sum_{j=1}^{m} p_j \cdot S_{ij}. \tag{7}$$

The best projection direction $\boldsymbol{P}^*$ is determined by solving the following optimization problem, namely:

$$\text{Max} : H(p) = S_G \cdot Q_G \tag{8}$$

$$\text{Constraint} : \sum_{j=1}^{m} {p_j}^2 = 1. \tag{9}$$

In Equation (8), S_G and Q_G are expressed as follows:

$$S_G = \sqrt{\frac{\sum_{i=1}^{n} [G(i) - \overline{g}(i)]^2}{n-1}} \tag{10}$$

$$Q_G = \sum_{i=1}^{n} \sum_{j=1}^{m} (R - r_{ij}) \cdot f(R - r_{ij}), \tag{11}$$

where $\overline{g}(i)$ is the mean value of $G(i)$, $R = 0.1S_G$, $r_{ij} = |G(i) - G(j)|$ is the distance between $G(i)$ and $G(j)$, and $f(t)$ is a unit step function equal to 1 if $t > 0$; otherwise, it is 0. Substitute $\boldsymbol{P}^*$ into Equation (7) to compute $G^*(i)$. Then, calculate the individual contribution ω_i by using the following:

$$\omega_i = \frac{G^*(i)}{\sum_{i=1}^{n} G^*(i)}. \tag{12}$$

3. Temperature Control Index Construction Based on Cloud Model

3.1. Cloud Model Algorithm

Here, U denotes a domain, C is a qualitative concept, $x \in U$ is a variable, and the certainty degree of x is recorded as u, a random number in the range of [0, 1]. The distribution of x on U is called a cloud, and the point (x, u) is a cloud droplet.

The mathematical features of cloud model, i.e., the expectation E_x, the entropy E_n, and the hyper entropy H_e, are calculated by using reverse cloud generator. For a data set $x = \{x_1, x_2, \cdots, x_l, \cdots, x_k\}$, k is its data size, and the mean value $\overline{x}$ and the variance σ_x are calculated as follows:

$$\overline{x} = \frac{1}{k} \sum_{l=1}^{k} x_l \tag{13}$$

$$\sigma_x = \sqrt{\frac{1}{k-1} \sum_{l=1}^{k} (x_l - \overline{x})^2}. \tag{14}$$

The calculation expressions of E_x, E_n, and H_e are as follows:

$$E_x = \overline{x}, \tag{15}$$

$$E_n = \left(\frac{\pi}{2}\right)^{1/2} \frac{1}{k} \sum_{l=1}^{k} |x_l - E_x|, \tag{16}$$

$$H_e = \sqrt{\sigma_x{}^2 - E_n{}^2}. \tag{17}$$

Cloud droplets are generated by using a forward cloud generator. The algorithm steps are as follows [51–55]. Generate the normal random number E'_{n_i} with the mean value of E_n and the variance of $H_e{}^2$, expressed as follows:

$$E'_{n_i} = NORM\left(E_n, H_e{}^2\right). \tag{18}$$

Build the normal random number x_i with the mean value of E_x and the variance of E'_{n_i}, recorded as follows:

$$x_i = NORM(E_x, E'_{n_i}). \tag{19}$$

The certainty degree u_i of x_i is calculated as follows:

$$u_i = e^{-\frac{(x_i - E_x)^2}{2E'_{n_i}}}. \tag{20}$$

Repeat the above steps, and all preset cloud droplets can be generated. The upper and lower bounds of cloud droplets are expressed by $y^u(x)$ and $y^l(x)$ as follows:

$$y^u(x) = e^{-\frac{(x-E_x)^2}{2(E_n+3H_e)^2}} \tag{21}$$

$$y^l(x) = e^{-\frac{(x-E_x)^2}{2(E_n-3H_e)^2}}. \tag{22}$$

The three mathematical features have different impacts on the morphology of artificial clouds. If artificial clouds have the same values of E_n and H_e but different E_x, then their dispersion degrees, distribution ranges, and cloud thickness are basically the same. However, the center areas are different, as shown in Figure 2a,b. If artificial clouds have the same values of H_e and E_x but different E_n, then their center areas and cloud thickness are similar, while the dispersion degrees and distribution ranges are different, as shown in Figure 2a,c. Moreover, it can be seen from Figure 2a,d that same E_x, same E_n, and different H_e result in similar center areas, similar dispersion degrees, similar distribution ranges, but different cloud thickness. The greater the value of H_e, the thicker the artificial cloud exhibits.

3.2. Temperature Control Index

Table 1 shows the contribution degrees of different domains, which are important references to establish the temperature control index. Considering practical situations of high concrete dams, TFEs under unfavorable conditions are taken as samples, and the data set S' is constructed as follows:

$$S' = \{S_1, S_2, \cdots, S_k\}. \tag{23}$$

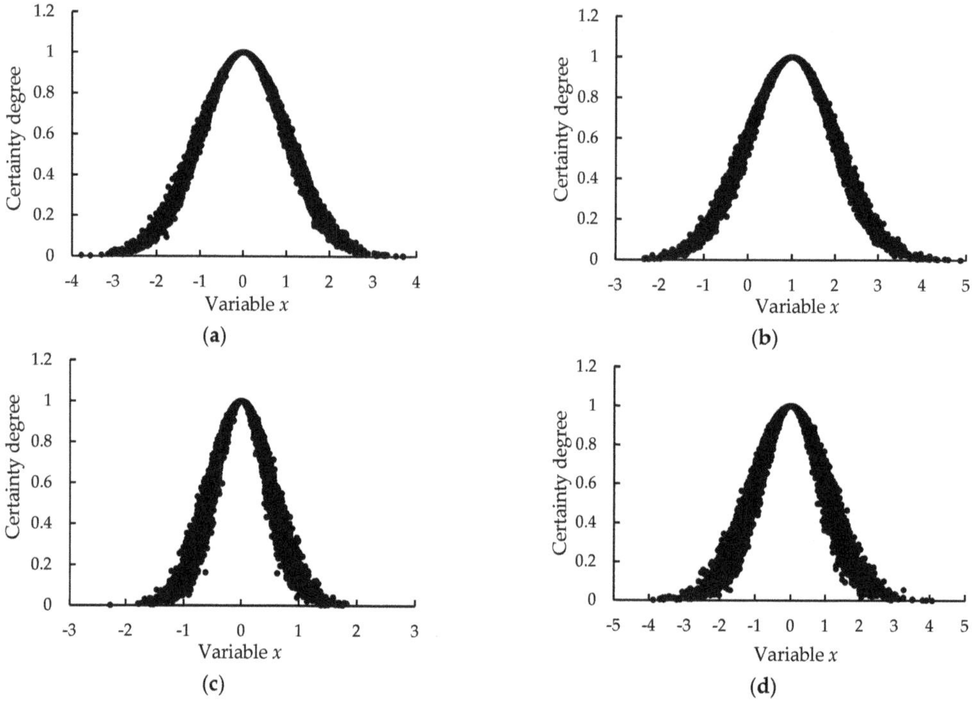

Figure 2. Cloud droplet distributions with different mathematical features: (**a**) $E_x = 0$, $E_n = 1$, and $H_e = 0.05$; (**b**) $E_x = 1$, $E_n = 1$, and $H_e = 0.05$; (**c**) $E_x = 0$, $E_n = 0.5$; and $H_e = 0.05$; and (**d**) $E_x = 0$, $E_n = 1$, and $H_e = 0.1$.

Table 1. The ΔD and β values of ΔU.

Domain ΔU	Contribution Degree ΔD	Value of β
$[E_x - 0.67E_k, E_x + 0.67E_k]$	50.0%	0.67
$[E_x - E_k, E_x + E_k]$	68.3%	1.00
$[E_x - 1.96E_k, E_x + 1.96E_k]$	95.0%	1.96
$[E_x - 2E_k, E_x + 2E_k]$	95.5%	2.00
$[E_x - 2.58E_k, E_x + 2.58E_k]$	99.0%	2.58
$[E_x - 3E_k, E_x + 3E_k]$	99.7%	3.00

E_x, E_n, and H_e are calculated by employing the reverse cloud generator, and cloud droplets are generated using the forward cloud generator. α denotes the confidence level. The relationship between the confidence level α and the contribution degree ΔD is $\alpha = 1 - \Delta D$. According to the principle of small probability event, if small probability events occur, then the studied dam is in different degrees of abnormality. The corresponding temperature control index $S_{m,\alpha}$ is constructed as follows:

$$S_{m,\alpha} = E_x - \beta E_n, \tag{24}$$

where β is a parameter related to α. According to Table 1, if α is set as 0.01, then β equals 2.58. Dam security can be evaluated by comparing S_j and $S_{m,\alpha}$. If $S_j < S_{m,\alpha}$, then the studied dam may be in an abnormal state on the jth observation date. Appropriate measures should be taken to improve the temperature state of the studied dam.

In actual applications, several confidence levels $\alpha_1, \alpha_2, \cdots, \alpha_k$ ($\alpha_1 > \alpha_2 > \cdots > \alpha_k$) are set, and the corresponding $S_{m,\alpha_1}, S_{m,\alpha_2}, \cdots, S_{m,\alpha_k}$ ($S_{m,\alpha_1} < S_{m,\alpha_2} < \cdots < S_{m,\alpha_k}$) are used for multi-level dam risk management. It should be noted that α is determined by considering various factors, e.g., engineering and build grades. This parameter should be carefully selected.

The algorithm flowchart is shown in Figure 3.

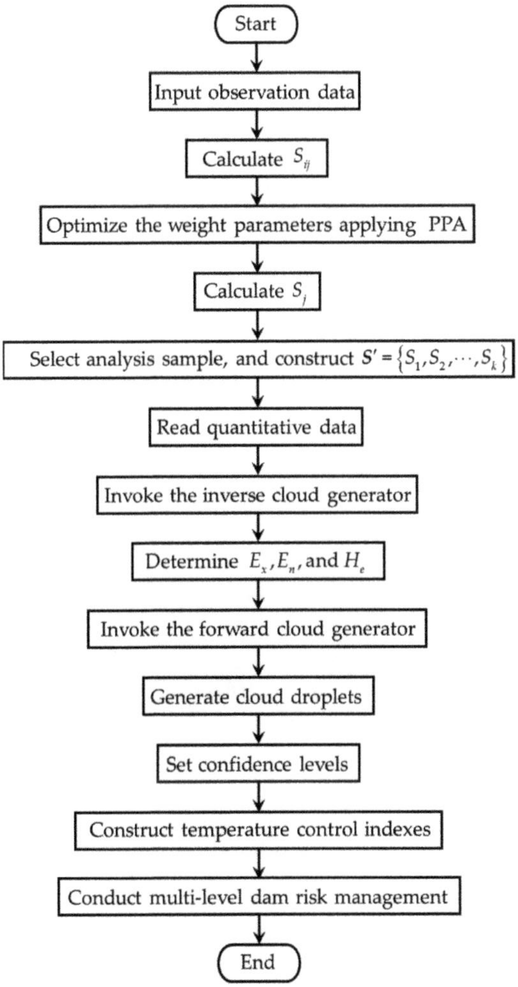

Figure 3. Algorithm flowchart.

4. Case Study

4.1. In Situ Observation Data

The studied dam is a RCC gravity dam, and it is located on the Jinsha River in southwestern China. This dam has a crest elevation of 1424.00 m, a maximum dam height of 160.00 m, and a crest length of 640.00 m. The storage capacity of the reservoir is 9.13×10^9 m^3. The normal and checked water levels are 1418.00 m and 1421.07 m, respectively. The proposed methodology is adopted for the #6 dam section. Thirty-three observation points are installed to monitor temperature variations, as shown in Figure 4. Figure 5 gives the time curves of air temperature that show a periodic change pattern. Taking November 2011 as an example, the average daily temperature is 14.2 °C, the maximum temperature is 30.7 °C, and the minimum temperature is 4.7 °C. The average, maximum, and minimum temperature differences between day and night are 17.4 °C, 22.5 °C, and 7.4 °C, respectively.

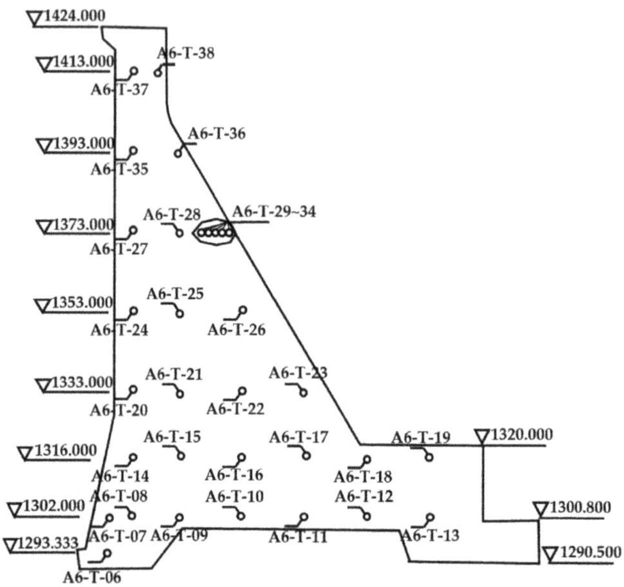

Figure 4. Distribution of the 33 observation points.

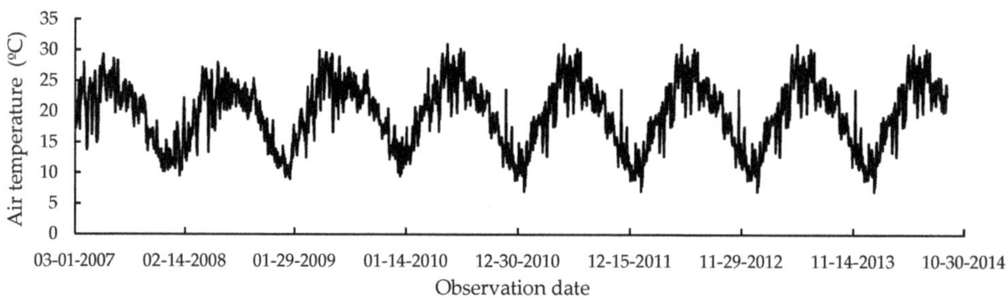

Figure 5. Time curves of air temperature.

The time curve of the concrete temperature observed at A6-T-21 is shown in Figure 6. In the early stage of concrete pouring, the concrete temperature gradually increases due to cement hydration heat. As cement hydration heat gradually dissipates, the concrete temperature exhibits a decreasing trend. Figure 7a,b show the temperature field distributions before and after water storage, respectively. After the upstream water level reaches 1397.83 m, the temperature change on the upstream surface is mainly affected by air temperature and water temperature. Compared with the downstream surface, the change amplitude of the upstream surface is relatively smaller. It can be seen from Table 2 that the maximum and minimum values are 49.4 °C and 12.4 °C, respectively, and that the maximum annual variation is 22.0 °C, which appeared at A6-T-37 in 2009.

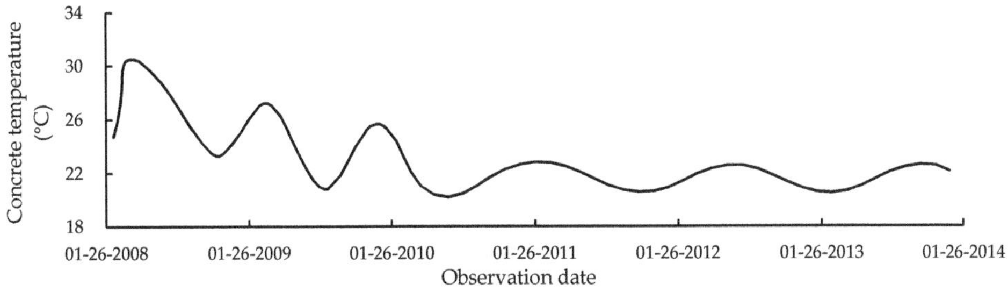

Figure 6. Time curve of concrete temperature observed at A6-T-21.

Table 2. Maximum value, minimum value, and maximum annual variation.

Observation Point	Point Number	Maximum Value/ Date (mm-dd-yyyy)	Minimal Value/ Date (mm-dd-yyyy)	Maximum Annual Variation/Year (yyyy)
A6-T-06	1	31.1/09-11-2007	17.0/08-30-2007	14.2/2007
A6-T-07	2	31.9/10-22-2008	20.0/01-30-2012	10.8/2008
A6-T-08	3	34.0/09-05-2008	23.0/02-01-2010	10.9/2008
A6-T-09	4	33.0/09-10-2008	17.9/03-11-2008	15.1/2008
A6-T-10	5	30.8/05-12-2008	17.4/03-11-2008	13.5/2008
A6-T-11	6	34.6/10-07-2008	19.9/02-22-2012	11.3/2008
A6-T-12	7	40.6/05-19-2008	22.2/03-22-2008	18.4/2008
A6-T-13	8	33.8/09-02-2008	18.1/02-01-2010	15.0/2008
A6-T-14	9	30.4/06-06-2008	18.1/08-19-2010	10.3/2008
A6-T-15	10	30.5/07-04-2008	17.4/01-03-2011	10.3/2008
A6-T-16	11	32.3/05-23-2008	19.9/08-11-2011	11.8/2008
A6-T-17	12	31.7/05-16-2008	18.5/03-18-2011	11.8/2008
A6-T-18	13	31.5/05-16-2008	20.7/12-28-2011	10.9/2008
A6-T-19	14	31.7/05-16-2008	19.6/04-19-2008	12.2/2008
A6-T-20	15	37.7/05-17-2008	14.7/12-28-2011	18.9/2008
A6-T-21	16	37.1/08-15-2008	17.9/05-05-2008	19.2/2008
A6-T-22	17	35.5/08-29-2008	18.1/05-05-2008	17.4/2008
A6-T-23	18	32.1/06-06-2008	18.6/05-05-2008	13.5/2008
A6-T-24	19	38.0/07-24-2008	19.0/07-03-2008	19.0/2008
A6-T-25	20	34.2/09-05-2008	19.0/07-03-2008-	15.3/2008
A6-T-26	21	30.5/02-11-2010	20.0/07-03-2008	10.5/2008
A6-T-27	22	31.7/01-14-2009	12.4/03-10-2011	10.6/2008
A6-T-28	23	28.5/12-17-2010	18.3/12-03-2008	5.9/2008
A6-T-29	24	28.4/08-11-2010	19.0/12-03-2008	4.9/2008
A6-T-30	25	40.2/11-05-2009	18.1/12-03-2008	14.3/2009
A6-T-31	26	32.0/12-17-2010	18.6/12-03-2008	5.8/2010
A6-T-32	27	31.8/12-17-2010	18.0/12-03-2008	6.3/2008
A6-T-33	28	27.9/12-17-2010	17.8/12-03-2008	6.9/2008
A6-T-34	29	30.5/08-26-2010	14.0/12-04-2008	7.8/2010
A6-T-35	30	49.4/05-28-2009	17.3/07-22-2011	21.2/2009
A6-T-36	31	39.6/05-28-2009	17.5/01-20-2011	20.6/2009
A6-T-37	32	48.5/10-04-2009	19.0/04-06-2011	22.0/2009
A6-T-38	33	45.3/09-28-2009	22.4/02-10-2012	16.5/2009

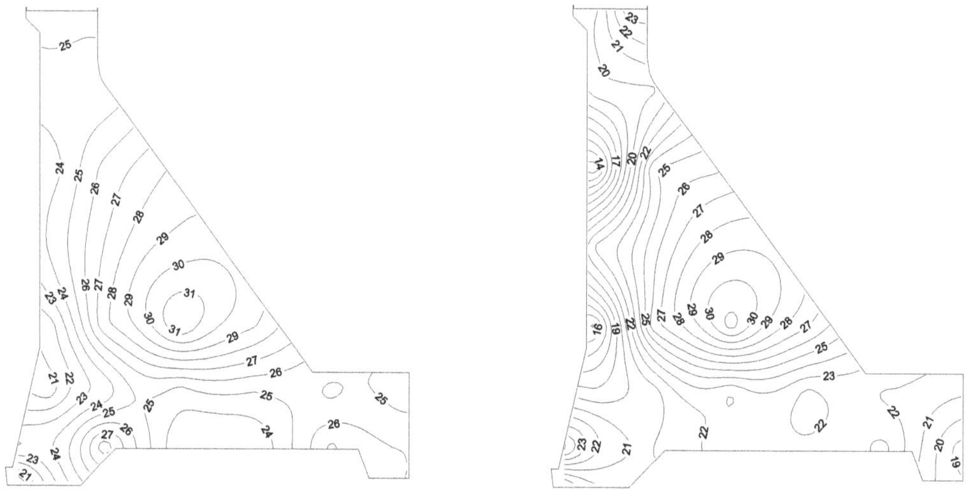

(a) (b)

Figure 7. Temperature field distributions, (**a**) before water storage and (**b**) after water storage.

4.2. Weight Distribution

The concrete pouring of the #6 dam section began on 29 July 2007. The construction was completed on 16 November 2009. The dam foundation elevation is 1288.00 m, and the dam crest elevation is 1424.00 m. The average pouring temperature is approximately 16 °C. The analysis period is divided into four stages according to the pouring process. Stage 1 is from 26 January 2008 to 6 March 2008. The corresponding pouring elevation is 1311.00 m. Stage 2 is from 7 March 2008 to 10 August 2008 with a pouring elevation of 1360.00 m. Stage 3 is from 11 August 2008 to 26 November 2009, and the pouring elevation is 1422.50 m. Stage 4 is from 27 November 2009 to 20 January 2014. At this stage, the dam construction has been completed. The weight distribution is optimized by PPA, and the results are shown in Figure 8. The point numbers are given in Table 2. To verify the reasonability of the weight distribution, the temperature fields of the four stages are simulated using the FE method. The numerical simulation considers the comprehensive effects of cement hydration heat, ambient temperature, and cooling water pipes. The initial temperature of cooling water is approximately 11.4 °C. Figure 9a,b show the FE models of the dam body, foundation, and cooling water pipe. The ABAQUS 2021 software is used to conduct numerical simulations in this work. The temperature distributions are presented in Figure 10a–d. Apparently, observation points with higher calculation values have higher weight parameters, indicating that the weight distribution patterns are in line with actual situations.

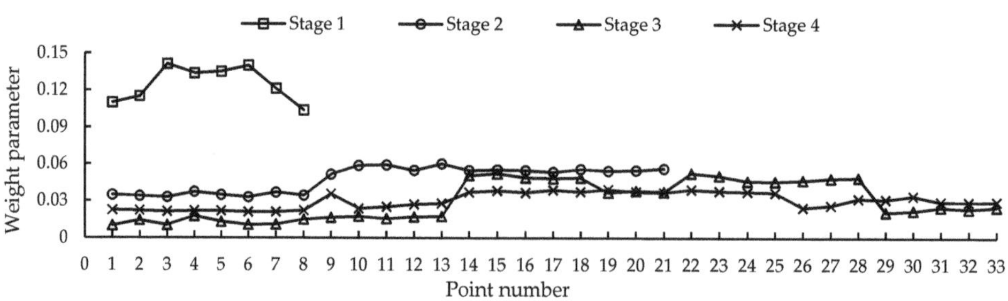

Figure 8. Weight distribution.

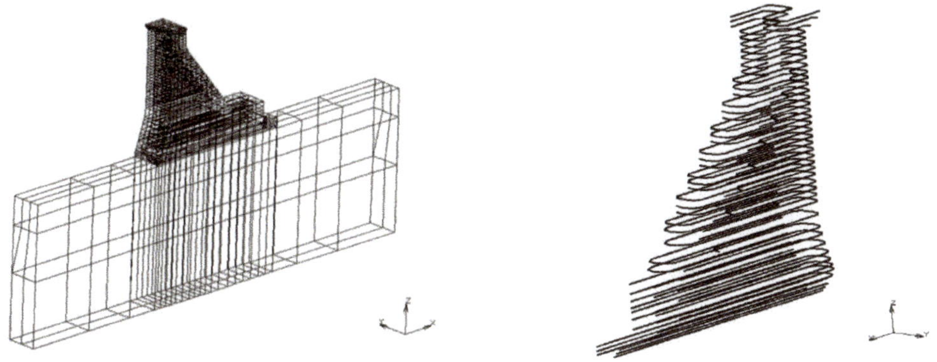

(a) (b)

Figure 9. The established FE models. (**a**) The dam body and foundation, and (**b**) the cooling water pipe.

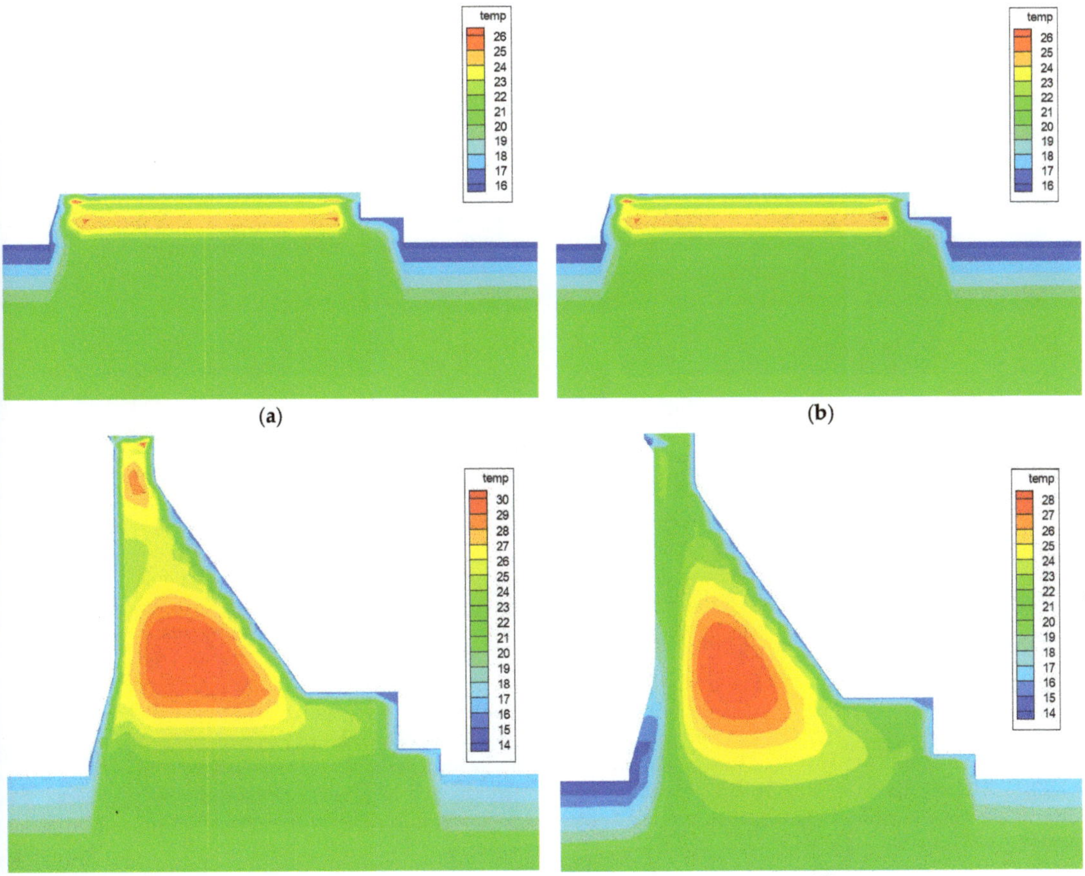

Figure 10. Simulation results of temperature fields: (**a**) pouring elevation at 1311.00 m; (**b**) pouring elevation at 1360.00 m; (**c**) pouring elevation at 1422.50 m; and (**d**) 4 years after completion.

4.3. Comparative Analysis

At an early stage of the construction period, concrete materials release abundant heat, and dam temperature quickly increases, leading to a decrease in TFEs. Thereafter, the concrete temperature of the dam body gradually decreases due to artificial cooling and heat dissipation, leading to an increase in TFEs. Since the end of Stage 3, the concrete temperature of the dam body has started to decrease and tends to stabilize gradually. Due to periodic changes in ambient temperature, TFEs show periodic fluctuations at Stage 4, as demonstrated in Figure 11. The regularities of TFEs agree with actual situations, and Equation (5) can effectively characterize the evolution characteristics of a high concrete dam temperature field.

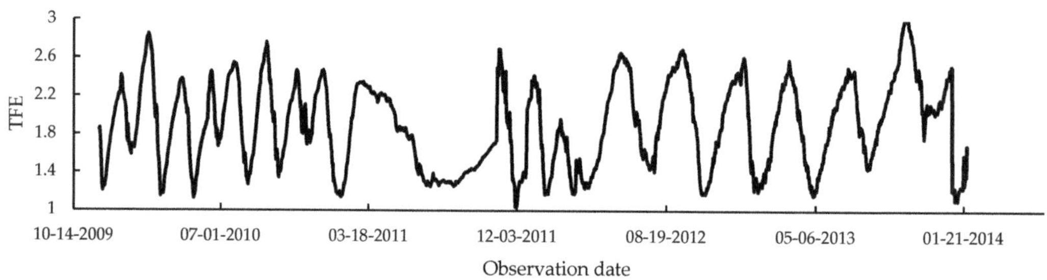

Figure 11. Time curve of TFEs for the #6 dam section.

In this work, the typical small probability method is compared with the cloud model. The analysis process is as follows. The probability distribution of $S' = \{S_1, S_2, \cdots, S_k\}$ is tested by the Kolmogorov–Smirnov (K–S) method [56,57]. Then, its probability density function $f(S)$ and distribution function $F(S)$ are determined. According to the confidence level α, its abnormal probability is expressed by $P(S \leq S_{m,\alpha})$ as follows:

$$P(S \leq S_{m,\alpha}) = \int_{-\infty}^{S_{m,\alpha}} f(S)\mathrm{d}x. \tag{25}$$

Then, the control index value $S_{m,\alpha}$ is established as follows:

$$S_{m,\alpha} = F^{-1}(\overline{S}, \sigma_S, \alpha), \tag{26}$$

where $\overline{S}$ and σ_S denote the mean value and the variance of S', respectively. The two parameters are calculated by Equations (13) and (14). The annual extreme values of S_j are listed in Table 3. The results of the K–S test are shown in Table 4, and the most reasonable probability distribution is the normal distribution with $\overline{S} = 1.14$ and $\sigma_S = 0.0589$. For the cloud model, the values of E_x, E_n, and H_e are 1.1400, 0.0501, and 0.0070, respectively, and the values of β are 1.96 ($\alpha = 5\%$) and 2.58 ($\alpha = 1\%$), respectively. The distribution of cloud droplets is shown in Figure 12. This paper sets two control levels according to different dangerous situations, i.e., Level 1 ($\alpha = 5\%$) and Level 2 ($\alpha = 1\%$). The calculated results of the typical small probability method and the cloud model are listed in Table 5. Their results are relatively consistent, indicating that the constructed indexes are reasonable and credible. Moreover, the cloud model, compared with the typical small probability method, also has the advantage that it can still obtain high-precision control indexes in the case of small samples.

Table 3. Annual extreme values.

Year	2008	2009	2010	2011	2012	2013
Minimum value of TFEs	0.723	0.968	1.126	1.034	1.170	1.160

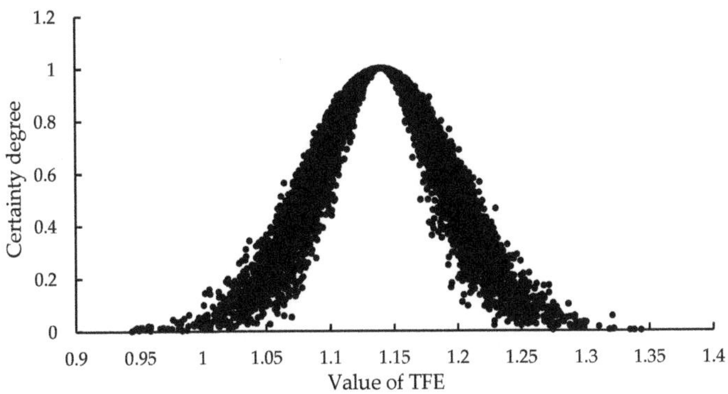

Figure 12. Distribution of cloud droplets.

Table 4. K–S test results.

Probability Distribution	Confidence Level
Lognormal distribution	0.23
Normal distribution	0.15
Uniform distribution	0.64
Triangular distribution	0.33
Exponential distribution	0.41
γ distribution	0.87
β distribution	0.78

Table 5. Temperature control indexes of the two methods.

Method	Temperature Control Index	
Typical small probability method	$\alpha = 5\%$ 1.0431	$\alpha = 1\%$ 1.0031
Cloud model	$\alpha = 5\%$ 1.0418	$\alpha = 1\%$ 1.0107

According to the above analyses, the established control indexes are appropriate for global monitoring of temperature fields, whereas conventional indexes are appropriate for local monitoring. Given that global monitoring is more important than local monitoring, the established indexes have greater significance than conventional indexes. This research offers scientific reference and technical support for the temperature control of other similar high concrete dam projects.

5. Conclusions and Suggestions

This work establishes the temperature control indexes for high concrete dams from the perspective of the spatial field. Information entropy theory is applied to describe the evaluation regularity of the temperature field, and then the expression of TFE that considers the contribution degrees of different observation points is established. Using TFE, the cloud model is employed to build the temperature control indexes. A case study is conducted in which the cloud model is compared with the typical small probability method. According to the calculation results, the variation law of TFEs agrees with actual situations, and the determined temperature control indexes can improve the safety management level of high concrete dams. Based on actual situations, different temperature control measures can be taken, e.g., pre-cooling methods such as lowering the temperature of concrete-forming components or post-cooling methods like running cold water through a pipe embedded in dam concrete.

Through contrastive analysis, the following conclusions can be drawn. When using the typical small probability method, the probability distribution of dangerous observation values needs to be determined. It is not applicable when the observation time is short. If the analysis samples do not conform to a definite distribution, then the applicability of this method is greatly limited. The cloud model directly obtains the mathematical features using the reverse cloud generator and generates cloud droplets by employing the forward cloud generator. As a result, this model can effectively represent the characteristics of a pan-normal distribution, and it has wider application conditions than the typical small probability method. Moreover, the cloud model considers the randomness and fuzziness of information, and no strict requirements are put on the sample size. Therefore, there is no need to test the probability distribution of the analysis example. In addition, it should be noted that this case study sets two confidence levels, i.e., $\alpha = 1\%$ and 5%. In fact, there is no unified standard for selecting the values of α. The determination of α should be given full consideration to practical experience and theoretical calculation, and detailed arguments need to be conducted.

Author Contributions: Detailed author contributions of this research are listed as follows. Conceptualization, G.Y., B.L. and J.S.; methodology, G.Y., J.S. and B.L.; software, B.L. and J.Z.; investigation, Z.H. and A.Z.; formal analysis, J.Z.; resources, Z.H. and A.Z.; writing-original draft preparation, G.Y. and J.S.; writing-review and editing, J.S. and J.N. All authors have read and agreed to the published version of the manuscript.

Funding: This research is supported by the National Natural Science Foundation of China (grant numbers 52109155 and 52279133), the Scientific Research Foundation of high-level talents of the North China University of Water Resources and Electric Power (grant number 202005017), and the Belt and Road Special Foundation of the State Key Laboratory of Hydrology-Water Resources and Hydraulic Engineering (grant number 2019492211).

Data Availability Statement: The data used to support the findings of this study are available from the corresponding authors upon request.

Acknowledgments: I am grateful to Hohai University. In addition, I thank Chongshi Gu and Hao Gu for their precious suggestions.

Conflicts of Interest: The authors declare that they have no known competing financial interest or personal relationship that could have appeared to influence the work reported in this paper.

Nomenclature and Abbreviations

Nomenclature

i	an observation point
j	an observation date
δ_{ij}	the observation value of the ith point on the jth date
u_{ij}	a parameter related to the order degree
p	a parameter related to u_{ij}^1 and u_{ij}^2
u_{ij}^1	the order degree of δ_{ij}
u_{ij}^2	the disorder degree of δ_{ij}
$f_i(\delta)$	the probability density function of δ_{ij}
$F_i(\delta)$	the probability distribution function of δ_{ij}
S_{ij}	the temperature entropy of δ_{ij}
ω_i	the weight of the ith point
S_j	the TFE on the jth date
S	a matrix
m	the total number of observation dates
n	the total number of observation points
$\mathbf{P}$	the unit projection direction
$G(i)$	the projection function
$\mathbf{P}^*$	the best projection direction
$\overline{g}(i)$	the mean value of $G(i)$

S_G	a parameter related to PPA
Q_G	a parameter related to PPA
$G^*(i)$	a parameter related to $\boldsymbol{P}^*$
R	a parameter related to PPA
r_{ij}	the distance between $G(i)$ and $G(j)$
$f(t)$	a unit step function
t	a parameter related to $f(t)$
U	the domain
C	a qualitative concept
x	a variable
u	the certainty degree of x
E_x	the expectation of the cloud model
E_n	the entropy of the cloud model
H_e	The hyper entropy of the cloud model
$\boldsymbol{x}$	a data set
$\overline{x}$	the mean value of $\boldsymbol{x}$
σ_x	the variance of $\boldsymbol{x}$
$\boldsymbol{S}'$	a data set
$f(S)$	the probability density function of $\boldsymbol{S}'$
$F(S)$	the distribution function of $\boldsymbol{S}'$
P	the abnormal probability
$\overline{S}$	the mean value of $\boldsymbol{S}'$
σ_S	the variance of $\boldsymbol{S}'$
k	the size of the data sets $\boldsymbol{x}$ and $\boldsymbol{S}'$
y^u	the upper bounds of cloud droplets
y^l	the lower bounds of cloud droplets
ΔU	a micro-interval in the domain U
ΔD	the contribution degree of ΔU
$U_{\Delta U}$	the certainty degree of ΔU
α	the confidence level
$S_{m,\alpha}$	the temperature control index with a confidence level of α
β	a parameter related to α
Abbreviations	
PPA	projection pursuit analysis
TFE	temperature field entropy
FE	finite element
RCC	roller compacted concrete
TE	temperature entropy
WE	weight entropy
K-S	Kolmogorov–Smirnov

References

1. Sevieri, G.; De Falco, A. Dynamic structural health monitoring for concrete gravity dams based on the Bayesian inference. *J. Civ. Struct. Health Monit.* **2020**, *10*, 235–250. [CrossRef]
2. Shao, C.; Gu, C.; Yang, M.; Xu, Y.; Su, H. A novel model of dam displacement based on panel data. *Struct. Control Health Monit.* **2018**, *25*, e2037. [CrossRef]
3. Zhou, Y.; Bao, T.; Shu, X.; Li, Y.; Li, Y. BIM and ontology-based knowledge management for dam safety monitoring. *Automat. Constr.* **2023**, *145*, 104649. [CrossRef]
4. Hou, C.; Chai, D.; Cheng, H.; Ning, S.; Yang, B.; Zhou, Y. Simulation Feedback of Temperature Field of Super-High Arch Dam during Operation and Its Difference with Design Temperature. *Water* **2022**, *14*, 4028. [CrossRef]
5. Peng, H.; Lin, P.; Xiang, Y.; Hu, J.; Yang, Z. Effects of Carbon Thin Film on Low-Heat Cement Hydration, Temperature and Strength of the Wudongde Dam Concrete. *Buildings* **2022**, *12*, 717. [CrossRef]
6. Qiu, H.; Hu, T.; Zhang, S.; Xiao, Y. Deriving Operating Rules of Hydropower Reservoirs Using Multi-Strategy Ensemble Henry Gas Solubility Optimization-Driven Support Vector Machine. *Water* **2023**, *15*, 437. [CrossRef]
7. Lou, S.; Mo, L.; Zhou, J.; Wang, Y.; He, W. Supply and Demand Forecasting of Water Resource Coupling System in Upstream Yangtze River under Changing Environmental Conditions. *Water* **2021**, *13*, 640. [CrossRef]
8. Pan, J.; Wang, J. Effect of abutment movements on nonlinear seismic response of an arch dam. *Struct. Infrastruct. Eng.* **2019**, *16*, 1106–1120. [CrossRef]

9. Brown, E.T. Reducing risks in the investigation, design and construction of large concrete dams. *J. Rock. Mech. Geotech.* **2017**, *9*, 197–209. [CrossRef]
10. Ahmadi, S.M.; Yamamoto, Y. A New Dam-Break Outflow-Rate Concept and Its Installation to a Hydro-Morphodynamics Simulation Model Based on FDM (An Example on Amagase Dam of Japan). *Water* **2021**, *13*, 1759. [CrossRef]
11. Xu, W.; Zhou, Q.; Dong, X. SPH–DEM coupling method based on GPU and its application to the landslide tsunami. Part II: Reproduction of the Vajont landslide tsunami. *Acta Geotech.* **2021**, *17*, 2121–2137. [CrossRef]
12. Li, B.; Zhang, Z.; Wang, X.; Liu, X. Investigation on the Debonding Failure Model of Anchored Polyurea Coating under a High-Velocity Water Flow and Its Application. *Sustainability* **2019**, *11*, 1261. [CrossRef]
13. Hariri-Ardebili, M.A.; Mahdavi, G.; Nuss, L.K.; Lall, U. The role of artificial intelligence and digital technologies in dam engineering: Narrative review and outlook. *Eng. Appl. Artif. Intel.* **2023**, *126*, 106813. [CrossRef]
14. Ma, K.; Xu, N.; Liang, Z. Stability Assessment of the Excavated Rock Slope at the Dagangshan Hydropower Station in China Based on Microseismic Monitoring. *Adv. Civ. Eng.* **2018**, *2018*, 4567258. [CrossRef]
15. Ma, K.; Sun, X.; Zhang, Z.; Hu, J.; Wang, Z. Intelligent Location of Microseismic Events Based on a Fully Convolutional Neural Network (FCNN). *Rock. Mech. Rock. Eng.* **2022**, *55*, 4801–4817. [CrossRef]
16. Prakash, G.; Sadhu, A.; Narasimhan, S.; Brehe, J.-M. Initial service life data towards structural health monitoring of a concrete arch dam. *Struct. Control Health Monit.* **2018**, *25*, e2036. [CrossRef]
17. Yosef, T.Y.; Song, C.R.; Chang, K.-T. Hydro-thermal coupled analysis for health monitoring of embankment dams. *Acta Geotech.* **2018**, *13*, 447–455. [CrossRef]
18. Yun, T.; Butler, K.E.; MacQuarrie, K.T.B. Investigation of seepage near the interface between an embankment dam and a concrete structure: Monitoring and modelling of seasonal temperature trends. *Can. Geotech. J.* **2023**, *60*, 453–470. [CrossRef]
19. Song, C.R.; Yosef, T.Y. Seepage Monitoring of an Embankment Dam Based on Hydro-Thermal Coupled Analysis. *J. Eng. Mater. Technol.* **2017**, *139*, 021024. [CrossRef]
20. Din, Z.U.; Ali, A.; De la Sen, M.; Zaman, G. Entropy generation from convective–radiative moving exponential porous fins with variable thermal conductivity and internal heat generations. *Sci. Rep.* **2022**, *12*, 1791. [CrossRef]
21. Xie, Z.; Yu, T. Determination of Monitoring Control Value for Concrete Gravity Dam Spatial Deformation Based on POT Model. *Cmes-Comp. Model. Eng.* **2023**, *135*, 2119–2135. [CrossRef]
22. Zhou, X.; Huang, X.; Zhao, H.; Ma, K. Development of a revised method for indicators of hydrologic alteration for analyzing the cumulative impacts of cascading reservoirs on flow regime. *Hydrol. Earth Syst. Sci.* **2020**, *24*, 4091–4107. [CrossRef]
23. Chen, H.; Liu, Z. Temperature control and thermal-induced stress field analysis of GongGuoQiao RCC dam. *J. Therm. Anal. Calorim.* **2018**, *135*, 2019–2029. [CrossRef]
24. Chen, Z.; Zheng, D.; Li, J.; Wu, X.; Qiu, J.; Nguyen Thanh, N. Temperature Field Online Reconstruction for In-Service Concrete Arch Dam Based on Limited Temperature Observation Data Using AdaBoost-ANN Algorithm. *Math. Probl. Eng.* **2021**, *2021*, 9979994. [CrossRef]
25. Pan, J.; Liu, W.; Wang, J.; Jin, F.; Chi, F. A novel reconstruction method of temperature field for thermomechanical stress analysis of arch dams. *Measurement* **2022**, *188*, 110585. [CrossRef]
26. Zhou, H.; Pan, Z.; Liang, Z.; Zhao, C.; Zhou, Y.; Wang, F. Temperature Field Reconstruction of Concrete Dams based on Distributed Optical Fiber Monitoring Data. *Ksce. J. Civ. Eng.* **2019**, *23*, 1911–1922. [CrossRef]
27. Chen, X.; Chen, J.; Wang, T.; Zhou, H.; Liu, L. Characterization of Seepage Velocity beneath a Complex Rock Mass Dam Based on Entropy Theory. *Entropy* **2016**, *18*, 293. [CrossRef]
28. Su, H.; Hu, J.; Wen, Z. Structure analysis for concrete-faced rockfill dams based on information entropy theory and finite element method. *Int. J. Numer. Anal. Meth. Geomech.* **2011**, *36*, 1041–1055. [CrossRef]
29. Cui, H.; Jiang, S.; Ren, L.; Xiao, W.; Yuan, F.; Wang, M.; Wei, L. Dynamics and potential synchronization of regional precipitation concentration and drought-flood abrupt alternation under the influence of reservoir climate. *J. Hydrol. Reg. Stud.* **2022**, *42*, 101147. [CrossRef]
30. Lei, P.; Chang, X.; Xiao, F.; Zhang, G.; Su, H. Study on early warning index of spatial deformation for high concrete dam. *Sci. China Technol. Sci.* **2011**, *54*, 1607–1614. [CrossRef]
31. Zheng, D.; Li, X.; Yang, M.; Su, H.; Gu, C. Copula entropy and information diffusion theory-based new prediction method for high dam monitoring. *Earthq. Struct.* **2018**, *14*, 143–153. [CrossRef]
32. Huang, Y.; Wan, Z. Study on Viscoelastic Deformation Monitoring Index of an RCC Gravity Dam in an Alpine Region Using Orthogonal Test Design. *Math. Probl. Eng.* **2018**, *2018*, 8743505. [CrossRef]
33. Zhang, J.; Gu, C. Maximum Entropy Method for Operational Loads Feedback Using Concrete Dam Displacement. *Entropy* **2015**, *17*, 2958–2972. [CrossRef]
34. Qin, X.; Gu, H.; Yuan, D.; Guo, J.; Chen, X.; Xu, B.; Wu, B.; Sohn, H. An Evaluation Method of Crack Variation on Structural Performance of Concrete Dams with Fusion Entropy Based on Observation and Simulation. *Struct. Control Health Monit.* **2023**, *2023*, 4040761. [CrossRef]
35. Han, L.; Chen, M.; Sun, Z.; Si, J.; Ma, L.; Ji, W.; Zhang, H. Stability analysis of slopes based on cloud model-Monte Carlo coupling. *Front. Earth Sci.* **2023**, *11*, 1196677. [CrossRef]
36. Wu, X.; Duan, J.; Zhang, L.; AbouRizk, S.M. A hybrid information fusion approach to safety risk perception using sensor data under uncertainty. *Stoch. Environ. Res. Risk Assess.* **2017**, *32*, 105–122. [CrossRef]

37. Li, H.; Li, Z.; Ma, F.; Liu, C. Similarity analysis of dam behavior characterized by multi-monitoring points based on Cloud model. *Int. J. Distrib. Sens. Netw.* **2020**, *16*, 1550147720920226. [CrossRef]
38. Sang, L.; Wang, J.; Sui, J.; Dziedzic, M. A New Approach for Dam Safety Assessment Using the Extended Cloud Model. *Water Resour. Manag.* **2022**, *36*, 5785–5798. [CrossRef]
39. Guo, H.; Yuan, H.; Huang, Q. Structural damage identification based on gray cloud rule generator algorithm. *Adv. Mech. Eng.* **2019**, *11*, 1–13. [CrossRef]
40. Su, H.; Li, J.; Wen, Z. Evaluation of Various Temperature Control Schemes for Crack Prevention in RCC Arch Dams During Construction. *Arab. J. Sci. Eng.* **2014**, *39*, 3559–3569. [CrossRef]
41. Salazar, F.; Vicente, D.J.; Irazábal, J.; de-Pouplana, I.; San Mauro, J. A Review on Thermo-mechanical Modelling of Arch Dams During Construction and Operation: Effect of the Reference Temperature on the Stress Field. *Arch. Comput. Method. Eng.* **2020**, *27*, 1681–1707. [CrossRef]
42. Shi, N.; Chen, Y.; Li, Z. Crack Risk Evaluation of Early Age Concrete Based on the Distributed Optical Fiber Temperature Sensing. *Adv. Mater. Sci. Eng.* **2016**, *2016*, 4082926. [CrossRef]
43. An, G.; Yang, N.; Li, Q.; Hu, Y.; Yang, H. A Simplified Method for Real-Time Prediction of Temperature in Mass Concrete at Early Age. *Appl. Sci.* **2020**, *10*, 4451. [CrossRef]
44. Li, M.; Si, W.; Du, S.; Zhang, M.; Ren, Q.; Shen, Y. Thermal deformation coordination analysis of CC-RCC combined dam structure during construction and operation periods. *Eng. Struct.* **2020**, *213*, 110587. [CrossRef]
45. de Oliveira, J.R.S.; Pettres, R. A thermal analysis of concrete structures performed by layers using boundary element formulation and dual reciprocity. *Eng. Anal. Bound. Elem.* **2023**, *150*, 542–554. [CrossRef]
46. Wang, L.; Yang, H.; Zhou, S.; Chen, E.; Tang, S. Mechanical properties, long-term hydration heat, shinkage behavior and crack resistance of dam concrete designed with low heat Portland (LHP) cement and fly ash. *Constr. Build. Mater.* **2018**, *187*, 1073–1091. [CrossRef]
47. Cho, H.; Lee, E.-K. Tree-Structured Regression Model Using a Projection Pursuit Approach. *Appl. Sci.* **2021**, *11*, 9885. [CrossRef]
48. Lee, E.-K. PPtreeViz: An R Package for Visualizing Projection Pursuit Classification Trees. *J. Stat. Softw.* **2018**, *83*, 1–30. [CrossRef]
49. Cao, C.; Song, S.; Chen, J.; Zheng, L.; Kong, Y. An Approach to Predict Debris Flow Average Velocity. *Water* **2017**, *9*, 205. [CrossRef]
50. An, H.; Yang, W.; Huang, J.; Huang, A.; Wan, Z.; An, M. Identify and Assess Hydropower Project's Multidimensional Social Impacts with Rough Set and Projection Pursuit Model. *Complexity* **2020**, *2020*, 9394639. [CrossRef]
51. Jin, Y.; Sun, Y.; Ma, H. A Developed Artificial Bee Colony Algorithm Based on Cloud Model. *Mathematics* **2018**, *6*, 61. [CrossRef]
52. Gao, H.; Zhang, X.; Liu, Y.; Li, D. Cloud Model Approach for Lateral Control of Intelligent Vehicle Systems. *Sci. Program.* **2016**, *2016*, 6842891. [CrossRef]
53. Jiao, S.; Pan, Z.; Chen, Y.; Li, Y. Cloud Annealing: A Novel Simulated Annealing Algorithm Based on Cloud Model. *IEICE Trans. Inf. Syst.* **2020**, *E103-D*, 85–92. [CrossRef]
54. Khatri, P.; Hayasaka, T.; Holben, B.N.; Singh, R.P.; Letu, H.; Tripathi, S.N. Increased aerosols can reverse Twomey effect in water clouds through radiative pathway. *Sci. Rep.* **2022**, *12*, 20666. [CrossRef]
55. Zhu, F. Mathematical Modeling Method of the Improved Genetic Algorithm for Random Power Fluctuation. *Int. T. Electr. Energy Syst.* **2022**, *2022*, 1889952. [CrossRef]
56. Liu, Y.; Chen, X.; Wang, H. The fused Kolmogorov–Smirnov screening for ultra-high dimensional semi-competing risks data. *Appl. Math. Model.* **2021**, *98*, 109–120. [CrossRef]
57. Lee, S.-H.; Lee, J.-W.; Kim, M.-K.; Park, H.-M. An Analysis on the Effectiveness of Nitrogen Oxide Reduction from Applying Titanium Dioxide on Urban Roads Using a Statistical Method. *Atmosphere* **2021**, *12*, 972. [CrossRef]

water

MDPI

Article

A Statistical Prediction Model for Sluice Seepage Based on MHHO-BiLSTM

Zihui Huang [1,2,3], Chongshi Gu [1,2,3,*], Jianhe Peng [4], Yan Wu [5], Hao Gu [1,2,3,*], Chenfei Shao [1,2,3], Sen Zheng [1,2,3] and Mingyuan Zhu [1,2,3,5]

1 The National Key Laboratory of Water Disaster Prevention, Hohai University, Nanjing 210098, China; 15616528050@163.com (Z.H.)
2 College of Water Conservancy and Hydropower Engineering, Hohai University, Nanjing 210098, China
3 National Engineering Research Center of Water Resources Efficient Utilization and Engineering Safety, Hohai University, Nanjing 210098, China
4 Anhui and Huaihe River Institute of Hydraulic Research (Anhui Provincial Water Conservancy Engineering Quality Testing Center Station), Hefei 230088, China
5 Xinjiang Institute of Water Resources and Hydropower Research, Urumqi 830049, China
* Correspondence: csgu@hhu.edu.cn (C.G.); ghao@hhu.edu.cn (H.G.)

Abstract: The current seepage prediction model of the sluice gate is rarely used. To solve the problem, this paper selects the bidirectional long and short-term neural network (BiLSTM) with high information integration and accuracy, which can well understand and capture the temporal pattern and dependency relationship in the sequence and uses the multi-strategy improved Harris Hawks optimization algorithm (MHHO) to analyze its two hyperparameters: By optimizing the number of forward and backward neurons, the overfitting and long-term dependence problems of the neural network are solved, and the convergence rate is accelerated. Based on this, the MHHO-BiLSTM statistical prediction model of sluice seepage is established in this paper. To begin with, the prediction model uses water pressure, rainfall, and aging effects as input data. Afterward, the bidirectional long short-term memory neural network parameters are optimized using the multi-strategy improved Harris Hawks optimization algorithm. Then, the statistical prediction model based on the optimization algorithm proposed in this paper for sluice seepage is proposed. Finally, the seepage data of a sluice and its influencing factors are used for empirical analysis. The calculation and analysis results indicate that the optimization algorithm proposed in this paper can better search the optimal parameters of the bidirectional long short-term memory neural network compared with the original Harris Eagle optimization algorithm, optimizing the bidirectional long short-term memory neural network (HHO-BiLSTM) and the original bidirectional long short-term memory neural network (BiLSTM). Meanwhile, the bidirectional long and short-term neural network (BiLSTM) model shows higher prediction accuracy and robustness.

Keywords: sluice seepage; prediction model; MHHO; BiLSTM

1. Introduction

The hydraulic industry has a long history of improving human life, ensuring national security, and promoting economic development. Sluice is a common hydraulic engineering, and its main role is to regulate the flow of water. A low-head water retaining structure can be used to generate electricity, control flooding, provide irrigation, or provide water [1], and ensure the rational use of water resources and water environment protection [2]. Once the sluice has an accident, the damage to the traffic and urban infrastructure and the casualties caused by the accident are far more serious than the consequences caused by general public facilities [3]. In the overall safety of sluice, seepage safety occupies an important position. Within the data, it can be seen that many geological disasters and accidents are related to seepage damage [4]. Many sluices have many defects in technology and equipment, which makes the problem of seepage increasingly prominent. It can not only affect the flood protection and flood discharge of the sluice, but also affect the anti-sliding stability of the sluice base and sluice room [5]. The study of seepage problems mainly includes the numerical simulation of seepage, mathematical modeling of seepage, forward and inversion of the seepage field, engineering applications, calculation methods of seepage, quantitative analytical mathematical method, etc. Among them, many researchers have conducted numerical simulations of seepage over the years [6,7], for example, the finite element method, boundary element method, finite analysis method [8], finite integral method, finite difference method, numerical manifold method, and so on, have all been used. Compared with the dam, due to the complexity of the sluice environment, the monitoring data are relatively scarce; at present, the structure behavior analysis experience is very limited. However, the importance of sluices makes their operating state analysis the subject of widespread attention in recent decades.

Since seepage is generally considered a key factor in the design, construction, and safe operation of floodgates, the seepage prediction analysis of floodgates has attracted more and more attention. Since the development of finite element software has accelerated in recent years, many scholars have used different finite element software to simulate, analyze, and calculate seepage flow in hydraulic structure engineering, for example, PLAXIS [9], COMSOL Multiphysics [10–12], GeoStudio [13], FLAC3D [14,15], SEEP/W [16], ABAQUS [17,18], FEFLOW [19], FLUENT [20], FLOW3D [21], and so on.

At the same time, the latest development of machine learning also provides a new method for the accurate analysis and prediction of many seepage problems of hydraulic structure projects. To detect abnormal seepage in hydraulic structure engineering in time, it is of great significance to use machine learning methods and historical monitoring data to predict the development trend of seepage characteristics in hydraulic structure engineering. Therefore, seepage prediction models for various hydraulic structures have been proposed and widely used. Hou et al. [22] proposed an integrated multi-target prediction model (MPM) that combines two deep learning methods into a single framework. This model captures both the variation pattern of measurement values with changing load and time, as well as the spatial distribution relationship of these values. Zhao et al. [23] combined the rough set theory (RS) and a long short-term memory network model (LSTM) to develop an RS-LSTM model written in Python by taking advantage of the timeliness of monitoring value changes and the lag of external influences. To optimize XGBoost hyperparameters, Zhang et al. [24] used hybrid Gray Wolf optimization (HGWO), which combines differential evolution (DE) and Gray Wolf optimization (GWO), as well as quad cross-validation, and then used it to predict seepage. Ishfaque et al. [25] built a deep learning model combining recurrent neural networks (RNN) and long short-term memory (LSTM) to predict the seepage degree of hydraulic structure engineering. The influencing factors mainly considered in this method are time series.

Multidimensional inputs present significant challenges to traditional statistical models, deterministic models, mixed models, model adaptive learning, and analyzing complex nonlinear relations. In recent years, neural networks have been widely used in hydraulic structures in deformation prediction [26]. Multilayer neural networks have been widely used in time series forecasting because of their good nonlinear fitting ability. To address the limitation of traditional multilayer neural networks in learning and remembering long-term dependent information, the long short-term memory (LSTM) architecture has been proposed [27]. LSTM neural networks have gained significant popularity and are extensively used in seepage prediction for various hydraulic structures. Compared with the traditional prediction model, it has good prediction performance [28,29]. However, LSTM can only use the previous input information to capture context information, and the dependence on subsequent input is relatively weak. This unidirectional dependency modeling may not capture the temporal patterns and dependencies completely and cannot make full use of the global information in the sequence to solve the gross error problem. For sequences with complex gross errors, the predictive power of LSTM may be limited.

Compared with ordinary feedforward neural networks, LSTM is more difficult to train. This may lead to problems such as gradient disappearance or gradient explosion in the process of model training, and it is necessary to use weight initialization techniques and gradient clipping methods to train stability. In recent years, some scholars have optimized LSTM. For example, Girsang A S [30] combined improved Easy Data Augmentation (EDA) with reverse translation methods to improve the macro accuracy and F1 of the BiLSTM model. Based on this, this paper selects a bidirectional long short-term memory neural network (BiLSTM) with stronger context awareness, bidirectional dependent modeling, and stronger gross error solving ability, and adopts a multi-strategy optimization Harris Eagle optimization algorithm (MHHO) with better optimization accuracy and stability to optimize the parameters of this model. This can improve the overfitting and long dependence problems of the BiLSTM model and accelerate the convergence rate of the model.

In conclusion, according to the working characteristics and surrounding environment of the sluice, this paper preliminarily analyzes the possible influencing factors of the seepage of the sluice and selects three influencing factors, namely water pressure, rainfall, and aging, which have a great influence on the seepage of the sluice. As a means of improving accuracy and stability, the parameters of BiLSTM were optimized by using the multi-decision improved Harris Hawks optimization algorithm (MHHO). Therefore, based on the identified main factors, a prediction model for the seepage of slugs is developed according to the BiLSTM of Section 2.2. Then, the MHHO algorithm is used to improve the performance of the BiLSTM model. Section 3 of the paper shows that the model is valid by looking at a case study of sluice engineering in China. A comparison is made between HHO-based models and BiLSTM- and HHO-BiLSTM-based models, and the accuracy and combined prediction effect of this method are verified. Finally, conclusions are drawn in Section 4.

2. Principles of the MHHO-BiLSTM Model

2.1. Statistical Model of Sluice Seepage

The statistical model refers to the regression equation obtained by mathematical statistics based on the cause quantity and the concurrent effect quantity obtained by long-term observation data. It can reflect the change in monitoring effect size and be used in extensional forecasting (including missing measurement interpolation) and operation monitoring [31].

Sluice is a kind of low-head hydraulic structure that uses a sluice gate to hold back and discharge water, and seepage is the main factor that interferes with the sluice structure's stability. Seepage may cause seepage deformation of the foundation, and serious seepage deformation will cause the foundation to be damaged and even lead to the sluice crashing. The size of the seepage flow reflects the anti-seepage effect of the main body of the sluice, the foundation curtain, and the impermeable parts of the sluice, such as the cover of the sluice, the sheet pile, and the bottom plate, which is an important basis for evaluating the operation safety of the sluice. Rainfall, water levels, and time are the primary factors affecting sluice seepage. The water pressure component, rainfall component, and time effect component are mainly analyzed, and the deformation expression adopted is as follows:

$$Q = Q_h + Q_p + Q_\theta = a_0 + a_1 \int_{-\infty}^{0} \frac{1}{\alpha_1} \frac{1}{\sqrt{2\pi x_2}} e^{-\frac{(t-x_1)^2}{2{x_2}^2}} H(t)dt + b_1 \int_{-\infty}^{0} \frac{1}{\alpha_2} \frac{1}{\sqrt{2\pi x_4}} e^{-\frac{(t-x_3)^2}{2{x_4}^2}} p(t)dt + c_1\theta + c_2 ln\theta \quad (1)$$

where Q is the flow of seepage; Q_h is the component of water level component; Q_p is the component of rainfall; Q_θ is the component of time; a_0, a_1, b_1, c_1, and c_2 are regression coefficients for each component (i = 1, 2); α_1 is the adjustment parameter related to the water level weight distribution function, at a fixed observation time $t = t_0$, it is a constant; α_2 is the adjustment parameter related to rainfall weight distribution function, at a fixed observation time $t = t_0$, it is a constant; x_1 is the water level's lag hours of; x_2 is the number of hours affected by water level; $H(t)$ is the time t's water level; x_3 is the rainfall factor's lag hours; x_4 is the number of hours affected by the rainfall factor; $p(t)$ is rainfall factor values at time t extracted by factor analysis; and θ is divided by 2400, which is the number of hours since the monitoring began until the initial observation began 1.

2.2. BiLSTM

2.2.1. Principles of LSTM

Depending on the length of the time series, the recurrent neural network (RNN) model can be applied. However, when the standard RNN model processes a time series with a long time span, the earliest information will lose its effectiveness along with the model information passing, and the RNN cannot establish long-range structural connections. A vanishing gradient (gradient explosion) problem can be improved by setting hyperparameters, modifying activation functions, dropout pruning, and so on. However, these methods cannot fundamentally solve the problem of gradient disappearance of RNN, so the accuracy improvement effect is poor. In order to solve the problem of long-term dependence of time series, Sepp Hochreiter et al. [27] proposed a long short-term memory network (LSTM) in 1997. It is an important branch of the recurrent neural network model (RNN), which has the advantages of RNN and is improved by adding an internal gate control mechanism to maintain the long-term preservation of information on the basis of RNN. Compared with RNN, LSTM has better performance in long-time series.

LSTM consists of a forgotten gate, input gate, and output gate, and its neuron structure diagram is shown in Figure 1.

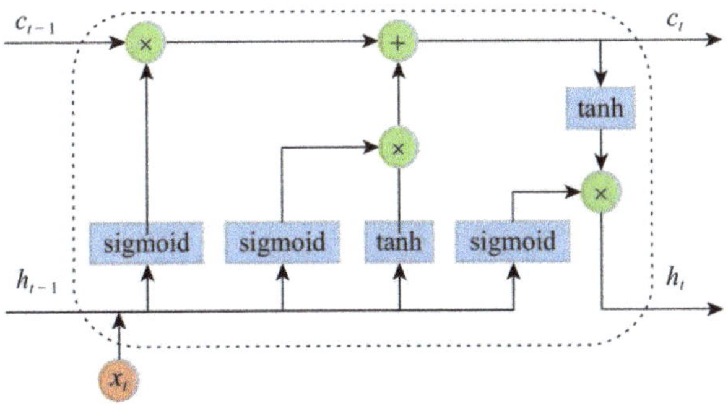

Figure 1. The network structure of LSTM [27].

In Figure 1, from left to right, they are, respectively, the forgetting gate, input gate, cell status update, and output gate.

RNN also exploits sequence dependencies via internal state transfers. It incorporates a gating mechanism to overcome the defects of RNN gradient updates within LSTM. An LSTM gate consists of three parts: forgotten, input, and output, and a state unit coordinates the process.

2.2.2. Principles of BiLSTM

LSTM is used as a basis for the development of BiLSTM, which is a bidirectional long-short-term memory network. LSTM's learn time series only of past moments, not future moments because cell state is transmitted one-way from front to back. BiLSTM has two unit state conveyor belts, which enables the BiLSTM model to learn the characteristics of future seepage information and perform recursion and feedback on it while using the seepage data information in the past. Therefore, BiLSTM can efficiently utilize the time characteristics of time series and improve the accuracy of model prediction so that the prediction results are more accurate than LSTM.

The BiLSTM network structure is shown in Figure 2.

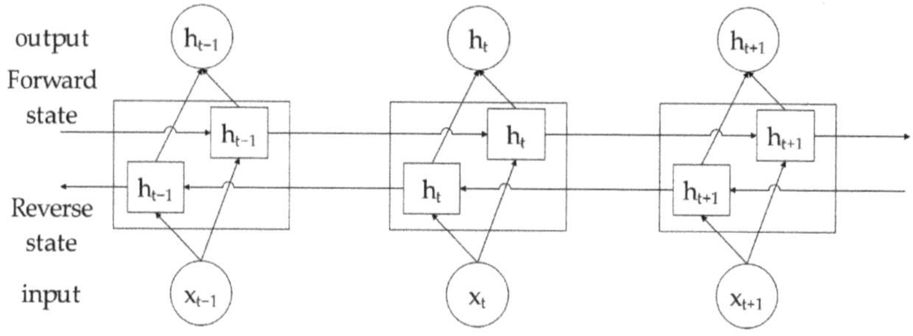

Figure 2. The network structure of BiLSTM.

Suppose $\overrightarrow{h}_t$ is the time t hidden layer state of the forward LSTM network, which is calculated using the following formula. It can be regarded as a single-layer LSTM network, the process of calculating the state at time t from the state at time $t-1$, and x_t is the input at time t.

$$\overrightarrow{h}_t = LSTM\left(x_t, \overrightarrow{h}_{t-1}\right) \tag{2}$$

where $\overrightarrow{h}_t$ is the time t hidden layer state of forward LSTM network; $LSTM$ is the LSTM unit; x_t is the t-time input; and $\overrightarrow{h}_{t-1}$ is time $t-1$ hidden layer state of forward LSTM network.

Similarly, if $\overrightarrow{h}_t$ is time t hidden layer state of the forward LSTM network, its calculation formula is as follows:

$$\overleftarrow{h}_t = LSTM\left(x_t, \overleftarrow{h}_{t-1}\right) \tag{3}$$

where $\overleftarrow{h}_t$ is time t hidden layer state of forward LSTM network; $LSTM$ is the LSTM unit; x_t is the t-time input; and $\overleftarrow{h}_{t-1}$ is time $t-1$ hidden layer state of forward LSTM network.

Combined with hidden layer states $\overrightarrow{h}_t$ and $\overleftarrow{h}_t$, the BiLSTM network produces a hidden state h_t as its output.

2.3. *MHHO Optimizes BiLSTM*

2.3.1. MHHO

Harris Hawks Optimization (HHO)

The Harris Hawks Optimization (HHO) is a nature-inspired algorithm proposed by Heidari et al. [32] in 2019, which uses mathematical formulas to simulate the strategies of Harris hawks capturing prey under different mechanisms in reality. The algorithm is divided into two stages of exploration and development, and the transformation is mainly carried out according to the escape energy E; it is used to measure the state of each individual in the group, representing the quality of the individual's current position. A high escape energy means that an individual is in a not-so-good position, at which time more exploration is needed. Individuals with high escape energies are more likely to explore in a better direction to find a better solution. Its calculation formula is as follows:

$$E = 2E_0\left(1 - \frac{t}{T}\right), E_0 = 2r - 1 \tag{4}$$

where E_0 is the initial state of prey energy; t is the current iteration count; T is the maximum number of iterations; and r is the random number between [0, 1].

(a) Exploration phase

When escaping energy $|E| \geq 1$, the algorithm enters the exploration phase and performs extensive search operations. The location update formula is as follows:

$$X(t+1) = \begin{cases} X_k(t) - r_1|X_k(t) - 2r_2X(t)| & q \geq 0.5 \\ (X_r(t) - X_m(t)) - r_3(lb - r_4(ub - lb)) & q < 0.5 \end{cases} \quad (5)$$

$$X_m(t) = \frac{1}{N}\sum_{i=1}^{N} X_i(t) \quad (6)$$

where $X(t+1)$ is the position of the Harris Hawks on the t + 1 iteration; $X(t)$ is the position of the Harris Hawks on the t + 1 iteration; $X_k(t)$ is a randomly selected individual in a population; $X_r(t)$ is the prey location, which is the current best individual; $X_m(t)$ is the current population average position; N is the number of individuals in a population; r_1, r_2, r_3, r_4 are four independent uniformly distributed random numbers with the values [0, 1]; and lb and ub are the lower and upper bounds of the problem, respectively.

(b) Development phase

Case 1: Soft surround. When the escaping energy $|E| \geq 0.5$ and $r \geq 0.5$, the soft bounding strategy is adopted to update the position; the method is as follows:

$$X(t+1) = \Delta X(t) - E|JX_r(t) - X(t)| \quad (7)$$

$$\Delta X(t) = X_r(t) - X(t) \quad (8)$$

where J is the jump energy, which is a random number with the value [0, 2], and $\Delta X(t)$ is the distance between the best individual and the current individual position in the t-th iteration.

Case 2: Hard surround. When the escaping energy $|E| < 0.5$ and $r \geq 0.5$, prey escape failure, surrounded by a hard strategy is adopted to improve the position update at this time, the formula is as follows:

$$X(t+1) = X_r(t) - E|\Delta X(t)| \quad (9)$$

Case 3: Progressive fast dive soft surround. When the escaping energy $|E| \geq 0.5$ and $r < 0.5$, the algorithm is updated by a progressive strategy of rapid dive and soft encircling. If the fitness is not improved after updating with position Y, position Z is updated; if the fitness value is not improved, the original position is retained. The specific formula is as follows:

$$X(t+1) = \begin{cases} Y, & if\ F(Y) < F(X(t)) \\ Z, & if\ F(Z) < F(X(t)) \end{cases} \quad (10)$$

$$Y = X_r(t) - E|JX_r(t) - X(t)| \quad (11)$$

$$Z = Y + S \times L_F(x) \quad (12)$$

$$L_F(x) = 0.01 \times \frac{uv}{|v|^{\frac{1}{\beta}}} \quad (13)$$

$$\sigma = \left[\Gamma(1+\beta) \times \sin\left(\frac{\pi\beta}{2}\right) / \Gamma\left(\frac{1+\beta}{2}\right) \times \beta \times 2^{\frac{\beta-1}{2}}\right]^{\frac{1}{\beta}} \quad (14)$$

where D is the dimension of a problem; S is the random vector of size $1 \times D$; Leavy is the Leavy flight function; u and v are random numbers that are evenly distributed between (0, 1); β is the default constant, set to 1.5; and Γ is the Gamma function.

Case 4: Progressive fast dive hard encircle. When the escaping energy $|E| < 0.5$ and $r < 0.5$, the progressive fast dive and hard encircling strategy are adopted to update the position. Currently, the optimal position continues to shrink from the average position for optimization as the population grows. The formula is as follows:

$$Y = X_r(t) - E|JX_r(t) - X_m(t)| \tag{15}$$

If the fitness is worth improving after the use of position Y, the position is updated to Y; otherwise, the Z position is used, and the original position is returned if it fails.

Multi-Decision Improved Harris Hawks Optimization Algorithm (MHHO)

While the existing enhancement strategies do enhance the exploration and optimization performance of the HHO algorithm to some extent, they typically only focus on improving specific update strategies or altering the energy reduction method. However, it does not effectively address the issue of global and local optimization blindness. In addition, other optimization algorithms also have their own shortcomings, and many scholars have adopted other optimization algorithms to improve the original algorithm, hoping to achieve better results. For example, Ahmad Sharafati et al. [33] proposed several novel hybrid adaptive neuro-fuzzy inference system (ANFIS) methods called ANFIS-PSO (particle swarm optimization), ANFIS-ACO (ant colony optimization), ANFIS-DE (differential evolution), and ANFIS-GA (genetic algorithm) as predictive models to estimate scour depth downstream of a sluice gate, thereby adopting a new stochastic model based on the integration of Group Method of Data Handling (GMDH) and Generalized Likelihood Uncertainty Estimation (GLUE) to predict scour depth around piers in cohesive soils [34]. This study shows the application of an adaptive neuro-fuzzy inference system (ANFIS) incorporated with particle swarm optimization (ANFIS-PSO), ant colony (ANFIS-ACCO), differential evolution (ANFIS-DE), and genetic algorithm (ANFIS-GA) and assesses the scour depth prediction performance and associated uncertainty in different scour conditions, including live-bed and clear-water [35]. Based on this, to solve the above problems, this paper uses the Multi-Strategy Harris Hawks Optimization algorithm (MHHO) [36], which improves the HHO algorithm from three aspects: Cauchy mutation, random shrinkage exponential function, and adaptive weight. Firstly, the Cauchy distribution function is introduced to improve the global search abilities of the HHO algorithm by mutating the Harris Hawks' position information. Secondly, the energy-decreasing mechanism is modified by the random contraction index function to effectively adjust the transition. Finally, the HHO algorithm's local search ability is further improved by introducing a weight factor that is adaptive.

(a) Cauchy variation

Using the HHO optimization, it is difficult to find a local optimum value; by using the Cauchy distribution function, the Harris Hawks algorithm possesses the ability to enhance population diversity, extend the search space, and improve its capacity for global search. In this paper, a global optimal object is optimized by using the Cauchy operator by utilizing both ends of the distribution function. Cauchy distributions can be formulated as follows:

$$f(x) = \frac{1}{\pi}\left(\frac{1}{x^2+1}\right) \tag{16}$$

Cauchy's function peaks at relatively low values. Following the Cauchy mutation, Harris Hawks will spend less time mining the local interval, searching for the global optimum value. Furthermore, the Cauchy function exhibits a gradual decrease from the peak toward both sides. This property becomes significant when updating the position by implementing Cauchy mutations. Harris Hawks have the capability to escape local optima and minimize the presence of local optima constraints. With the Cauchy mutation, the global optimal solution can be obtained, and the optimal solution can be updated using the following formula:

$$X'_{best} = X_{best} + X_{best} \times Cauchy(0,\ 1) \tag{17}$$

(b) The random shrinking exponential function

In the HHO algorithm, the energy magnitude of the prey, represented by E, plays a crucial role in the regulation and transition between global exploration and local exploitation phases. As E becomes smaller, the HHO algorithm is more inclined to perform local mining. As E becomes larger, the algorithm is more inclined to perform global mining. Nevertheless, the traditional HHO algorithm's energy equation represents E as a linear decrease from its maximum value to the minimum value, which fails to accurately capture the natural process of Harris Hawks hunting prey. Mathematicians simulate the predator-prey interaction and draw the conclusion that the random contraction exponential function is more suitable for expressing the energy change when the prey flees. Therefore, this optimization algorithm proposes to modify the linear decreasing energy regulation mechanism and integrate the random contraction index function into the decreasing process of prey energy E, and the corresponding energy equation is as follows:

$$E = E_0\left(2rand \times \exp\left(-\left(\frac{\pi}{2} \times \frac{t}{T}\right)\right)\right) \tag{18}$$

(c) The adaptive weight

It is important to consider the inertia weighting factor, which determines how much time the algorithm spends on global searches when it is high. The algorithm uses relatively more time for local search when the inertia weighting factor is small, and it is capable of fine-tuning the search for optimal results. The prey position represents the current optimal solution of the HHO algorithm, and the position of the Harris Hawks habitat is inversely guided. Its iterative update directly determines the optimization performance of the algorithm. Therefore, in order to improve the local exploitation ability of the algorithm, it is necessary to re-update the domain of the prey position to find a better solution.

Considering the above, this optimization algorithm introduces an adaptive weight method. During the four hunting mechanisms, when Harris Hawks are searching for prey, the prey adjusts its position using a reduced adaptive weight, enhancing the local optimization capabilities of the algorithm. The adaptive weight formula and prey position update can be expressed as follows:

$$\omega = \sin\left(\frac{\pi t}{2T} + \pi\right) + 1 \tag{19}$$

$$X'_r = \omega \times X_r \tag{20}$$

where T is the maximum number of iterations, and t is the current iteration count. The integration of the adaptive weighting factor ω enhances the local exploration capabilities of the MHHO algorithm, resulting in an improved ability to mine local optima.

2.3.2. The Establishment Process of the Sluice Seepage Prediction Model Based on MHHO-BiLSTM

The establishment process of the sluice seepage prediction model based on MHHO-BiLSTM is divided into the following steps:

Step 1: Entering the data. Input the original sluice seepage data required for MHHO-BiLSTM model training.

Step 2: Normalization. The normalization method is used to process the data, and the data range is limited between positive and negative 1, which can fix the order of magnitude of the data to improve the training speed and facilitate data analysis.

Step 3: Dividing the normalized processed data into the training set and test set. The data is partitioned with a 70% to 30% ratio, where 70% of the data is allocated for training purposes, and the remaining 30% is reserved as the test set. The training set is trained by the BiLSTM model and continuously back-replaced, and the test set is predicted by the BiLSTM model for the future period of sluice seepage data.

Step 4: Based on the BiLSTM model's results, the maximum correntropy test is performed to determine if the data meets the requirements.

Step 5: The MHHO algorithm optimizes the parameters. BiLSTM uses the MHHO algorithm to optimize the number of hidden layer nodes, the initial learning rate and the regularization coefficient, and values that are not accurate after iterations are cycled.

Step 6: The maximum correntropy test is carried out again on the calculated data, and the predicted data and test index can be output if the requirements are met. If the requirements are not met, the fifth step is carried out until the optimal solution and fitness value are obtained.

The modeling process of MHHO-BiLSTM is illustrated in Figure 3.

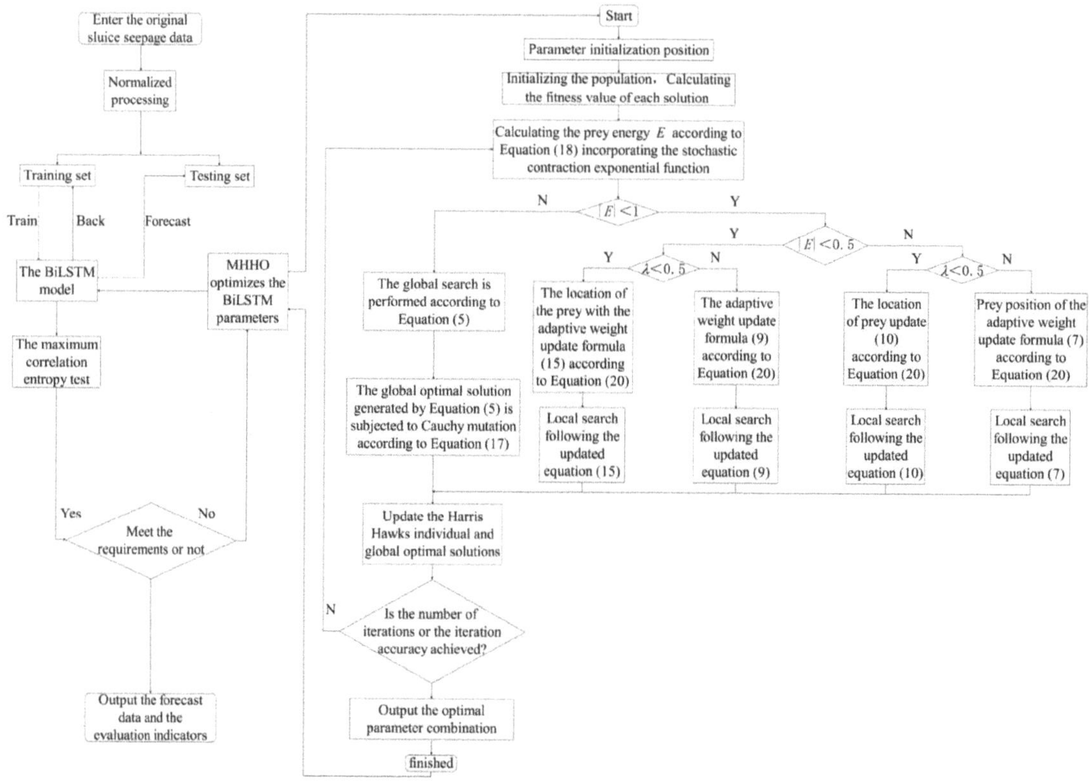

Figure 3. The modeling process of MHHO-BiLSTM model.

2.4. Evaluation Method for the Effectiveness of the Prediction Model

The effect of time series prediction can be comprehensively considered from two aspects: the subjective level and objective level. Taking a subjective view of the chart, one can see how the predicted experimental data differ from the standard data; that is, the quality of the predicted results can be directly judged by the observation method. However, subjective analysis alone will lack a scientific basis, so it is necessary to supplement objective analysis. From the objective level, we can import the results data of the prediction model into the calculation formula for verification and obtain the numerical error results. In this paper, four types of evaluation indicators are selected, and the calculation formulas of different evaluation indicators are different, as shown in the following formula:

Index 1: Mean Absolute Error (*MAE*), which reflects the average value of the absolute value of the deviation between the forecast data and the average. It is calculated as follows:

$$MAE = \frac{1}{m}\sum_{i=1}^{m}\left|y_{test,i} - \overline{y}_{test,i}\right| \tag{21}$$

Index 2: Mean Square Error (*MSE*), which reflects the average value of the square deviation between the predicted data and the measured data. The calculation formula can be written as follows:

$$MSE = \frac{1}{m}\sum_{i=1}^{m}\left(y_{test,i} - \hat{y}_{test,i}\right)^2 \tag{22}$$

Index 3: Root Mean Square Error (*RMSE*), which is calculated by dividing the predicted data by the measured data and multiplying the ratio of measurements by the number of observations. The calculation formula is given by the following:

$$RMSE = \sqrt{\frac{1}{m}\sum_{i=1}^{m}\left(y_{test,i} - \hat{y}_{test,i}\right)^2} \tag{23}$$

Index 4: The coefficient of determination (R-square, R^2); this index reflects the ratio relationship between the sum of the squared differences in the real value and the predicted value and the sum of the squared differences in the real value and the mean value. The accuracy of the model is judged according to the value of R^2, which ranges from [0, 1]. The closer the value is to 1, the more accurate the model is. The calculation formula is as follows:

$$R^2 = 1 - \frac{\sum_{i=1}^{m}\left(y_{test,i} - \hat{y}_{test,i}\right)^2}{\sum_{i=1}^{m}\left(y_{test,i} - \overline{y}_{test,i}\right)^2} \tag{24}$$

3. Case Study

This paper selected Bengbu sluice monitoring data for empirical analysis. The position of its pressure-measuring tube is shown in Figure 4. A total of 72 groups of seepage data from 2021 to 2022 were selected for the model test. Among them, the observed seepage value from January 2021 to June 2022 is the training sample, and the observed seepage value from June 2022 to December 2022 is the test sample (July and August are the local flood seasons, and the rainfall is the maximum in a year. The measured seepage data are introduced into the model as rainfall characteristics).

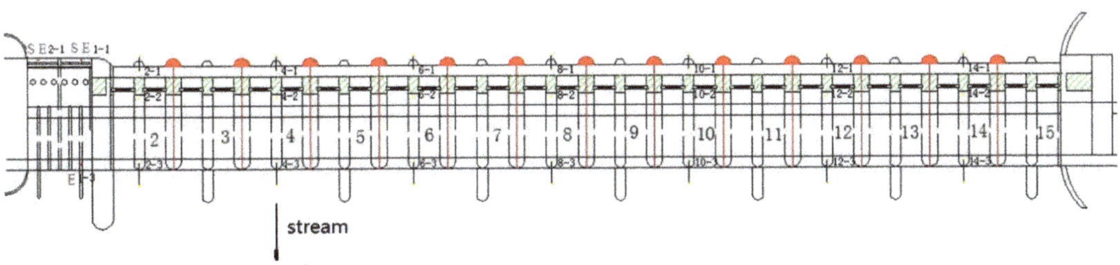

Figure 4. The position diagram of pressure monitoring pipe in Bengbu Gate 28-hole control gate.

3.1. The Seepage Prediction of Sluice Based on Statistical Model

In this paper, the measured seepage data of a monitoring point in Bengbu Gate in 2022 are used as test samples, and the BiLSTM network model and MHHO-BiLSTM model are, respectively, used to predict, combined with the prediction results of the stepwise regression statistical model, and the relative errors of the three models are compared. The reliability of the MHHO-BiLSTM model established in this paper is verified. The specific seepage measurement results are shown in Table 1.

Table 1. The analysis table of measured seepage data of Bengbu Gate.

Time (d)	The Height of Pressure Tube (m)	Time (d)	The Height of Pressure Tube (m)	Time (d)	The Height of Pressure Tube (m)
515	4.2	585	5.08	655	3.74
525	3.74	595	4.16	665	3.56
535	3.56	605	4.14	675	3.34
545	3.9	615	3.84	685	3.84
555	4.68	625	3.8	695	3.88
565	4.92	635	3.66	705	3.94
575	5.96	645	4.18	715	3.92

Using stepwise regression fitting, Table 2 presents predicted values and relative errors.

Table 2. Stepwise regression fitting predicted values and relative errors of monitoring points.

Time (d)	Measured Value (m)	Predicted Value (m)	Relative Error (%)	Time (d)	Measured Value (m)	Predicted Value (m)	Relative Error (%)
515	4.20	3.92975	0.27025	625	3.80	3.50992	0.29008
525	3.74	3.40368	0.33632	635	3.66	3.23059	0.42941
535	3.56	3.24000	0.32000	645	4.18	3.80737	0.37263
545	3.90	3.64363	0.25637	655	3.74	3.47025	0.26975
555	4.68	4.34023	0.33977	665	3.56	3.24419	0.31581
565	4.92	4.69632	0.22368	675	3.34	3.01331	0.32669
575	5.96	5.60482	0.35518	685	3.84	3.56232	0.27768
585	5.08	4.72975	0.35025	695	3.88	3.56771	0.31229
595	4.16	3.85467	0.30533	705	3.94	3.61926	0.32074
605	4.14	3.83909	0.30091	715	3.92	3.68754	0.23246
615	3.84	3.52096	0.31904				

The seepage prediction calculated by the stepwise regression method is shown in Figure 5.

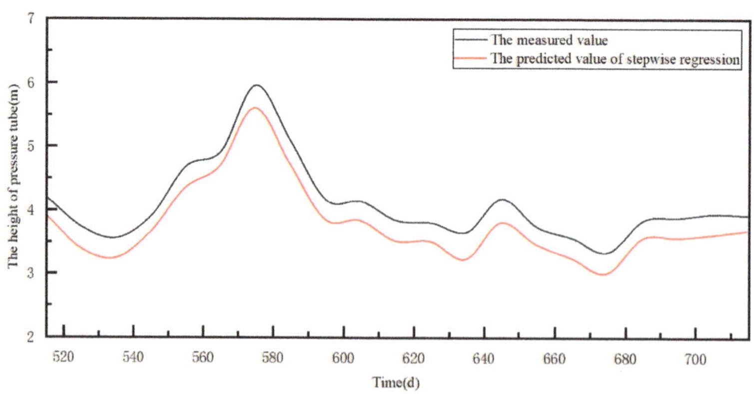

Figure 5. Seepage prediction diagram calculated by stepwise regression model.

3.2. The BiLSTM Network Model Training and Prediction

The measured seepage value of the No. 2-1 monitoring point on the upstream side of the middle pier on the 2# bottom plate of the Bengbu Gate from January 2021 to June 2022 is used as the training sample. According to the prediction principle of the exponential curve, the LSTM model's outputs and inputs are the seepage value and influence factor parameters, respectively. Thus, the 10 and 1 nodes are in the input and output layers, respectively. The output item is the No. 2-1 monitoring point on the upstream side of the middle pier of the 2# bottom plate of the Bengbu Gate. The predicted samples are measured seepage values from January 2022 to December 2022 at monitoring points No. 2-1 on the upstream side of the middle pier of the 2# bottom slab of the Bengbu Gate.

Initially, there are 10 nodes in the hidden layer, and those 10 nodes are increased one by one for trial calculations. An evaluation criterion is determined based on the minimum average relative error of the training sample. This paper uses the BiLSTM hidden layer, which, as shown in Figure 2, has a minimum relative error of 4.31% when there are 100 hidden layer nodes. Construction training is conducted with 50 batches and 500 iterations, with a 0.001 target error. Figure 6 describes the number of hidden layer nodes and the training sample average relative error of the curve, and Figure 7 is the process line between the measured seepage value and the fitting value of the LSTM network model at the monitoring point No. 2-1 on the upstream side of the middle pier of the 2# bottom plate.

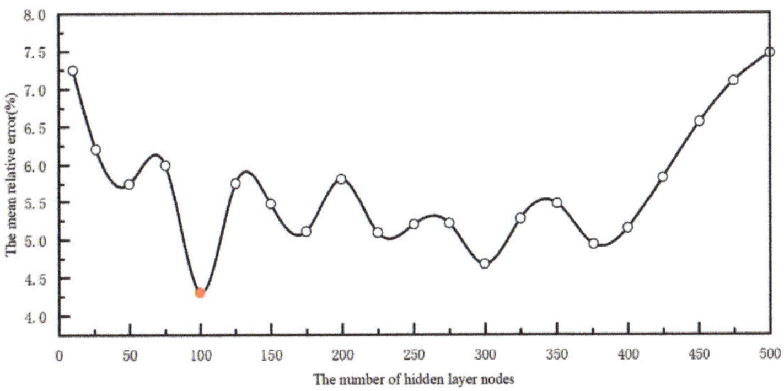

Figure 6. The number of hidden layer nodes and the training sample average relative error of curve (The white dots are the number of hidden layer nodes, and the red dots are the number of optimal hidden layer nodes).

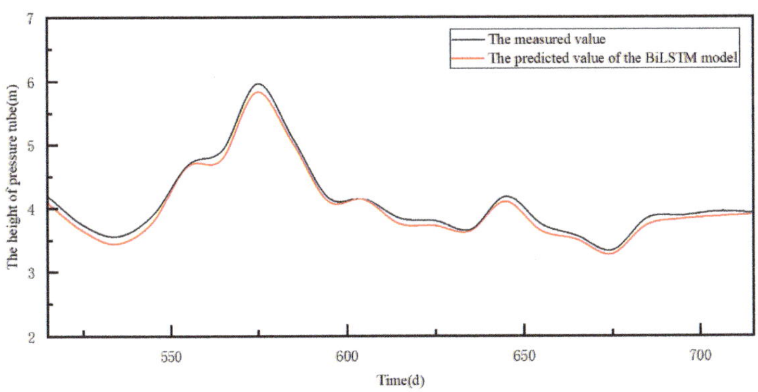

Figure 7. Process line between measured seepage value and the BiLSTM model fitting value (June 2022~December 2022).

It can be seen from Figure 7 that the fitting value of the BiLSTM network model is consistent with the measured value, and the fitting effect of the model is good, except for several cusps of the curve with relatively large data runout. As can be seen from Table 3, the absolute relative error of the BiLSTM network model in predicting seepage at the No. 2-1 monitoring point on the upstream side of the middle pier of Bengbu Gate 2# bottom floor ranges from 0.35% to 14.96%, and the average absolute value of relative error is 7.41%, indicating a good prediction effect.

Table 3. Comparison between the predicted results of the BiLSTM model and the measured values.

Time (d)	Measured Value (m)	Predicted Value (m)	Relative Error (%)	Time (d)	Measured Value (m)	Predicted Value (m)	Relative Error (%)
515	4.2	4.10652	0.09348	625	3.8	3.72805	0.07195
525	3.74	3.66147	0.07853	635	3.66	3.63839	0.02161
535	3.56	3.45042	0.10958	645	4.18	4.10113	0.07887
545	3.9	3.7796	0.1204	655	3.74	3.64533	0.09467
555	4.68	4.65414	0.02586	665	3.56	3.50907	0.05093
565	4.92	4.77195	0.14805	675	3.34	3.28215	0.05785
575	5.96	5.83144	0.12856	685	3.84	3.73229	0.10771
585	5.08	5.03003	0.04997	695	3.88	3.82805	0.05195
595	4.16	4.11445	0.04555	705	3.94	3.86686	0.07314
605	4.14	4.13654	0.00346	715	3.92	3.89562	0.02438
615	3.84	3.75722	0.08278				

3.3. The Establishment and Prediction of the MHHO-BiLSTM Model

The utilization of the MHHO optimization algorithm aims to enhance the prediction accuracy of BiLSTM network models even further. Using the fitting values of the BiLSTM network model from June 2022 to December 2022 in Section 3.2, the relative error sequence can be obtained. The subsequent calculation procedures and outcomes are outlined as follows.

The multi-strategy Improved Harris Hawks Optimization Algorithm (MHHO) is used to optimize three parameters (the number of hidden layer nodes, initial learning rate, and regularization coefficient) of BiLSTM. The initial parameters were set as follows: the population number N = 36; the maximum number of iterations T = 500; and the boundary ranges of the three parameters are $[1 \times 10^{-5}, 1 \times 10^{-2}]$, [0.0001, 0.002], and [10, 100], respectively. Iterative optimization using Harris Hawks with multi-strategy improved methods yields the optimal parameter combination as follows: there are 100 nodes in the hidden layer, the initial learning rate is 0.002, and the regularization coefficient is 1.52×10^{-5}. The convergence curve of MHHO-BiLSTM is shown in Figure 8.

In order to further test the effectiveness and superiority of the MHHO-BiLSTM model in sluice seepage prediction, three models, namely MHHO-BiLSTM, BiLSTM, and step-by-step regression, were selected for comparison. In order to compare the prediction effects of MHHO-BiLSTM, BiLSTM, and stepwise regression, the mean absolute error, mean square error, root mean square error, and determination coefficient are taken as the measurement standards, and the evaluation index data of the three models are shown in Table 4. The errors of the test set and the training set are shown in Figures 9 and 10, respectively.

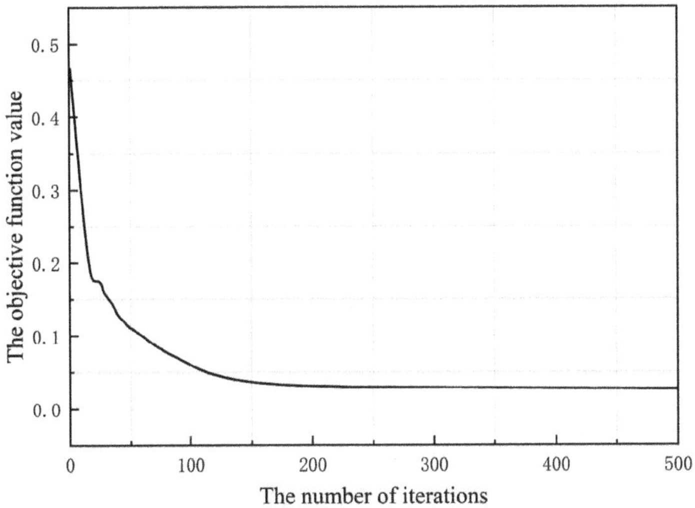

Figure 8. The convergence of the MHHO-BiLSTM.

Table 4. The comparison of evaluation indexes of the three models.

	Stepwise Regression	The Model of the BiLSTM	The Model of the MHHO-BiLSTM
MAE	0.5351	0.4645	0.4631
MSE	0.0582	0.0120	0.0028
RMSE	0.2412	0.1095	0.0530
R^2	0.8923	0.9747	0.9942

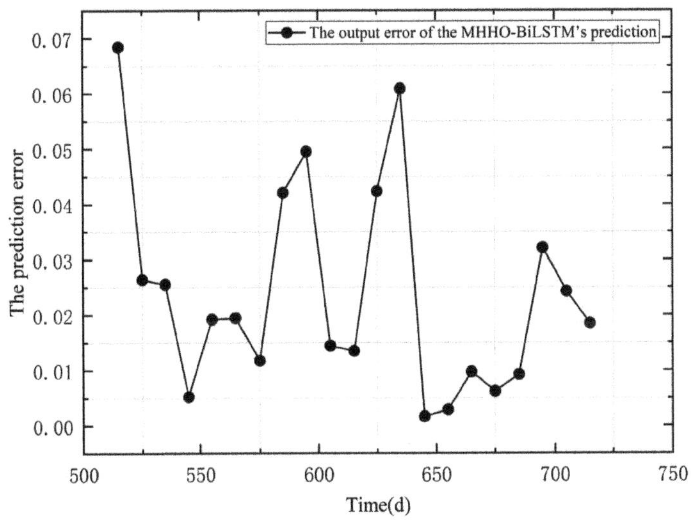

Figure 9. The prediction error of testing sets.

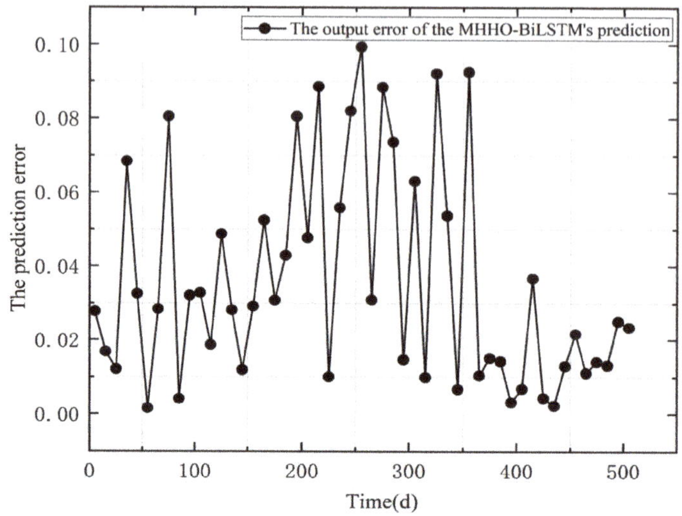

Figure 10. The prediction error of training sets.

3.4. *The Comparison of Model Accuracy*

According to the calculation results of seepage flow at the No. 2-1 monitoring point on the upstream side of the middle pier of the 2# bottom plate of the Bengbu Gate in Sections 3.2 and 3.3, combined with the stepwise regression statistical model's predictions, the measured values from January 2022 to December 2022 are used to make prediction comparisons. Table 4 and Figures 10 and 11 show the prediction results of the three models, as well as relative errors for each. In Figure 11, the predicted seepage value is compared to the measured value at the pier measurement point; Figure 12 is the absolute relative error of the predicted seepage value at the pier measurement point; Table 5 is the comparison of the relative error of the predicted seepage value at the pier measurement point; M-B model represents the MHHO-BiLSTM model.

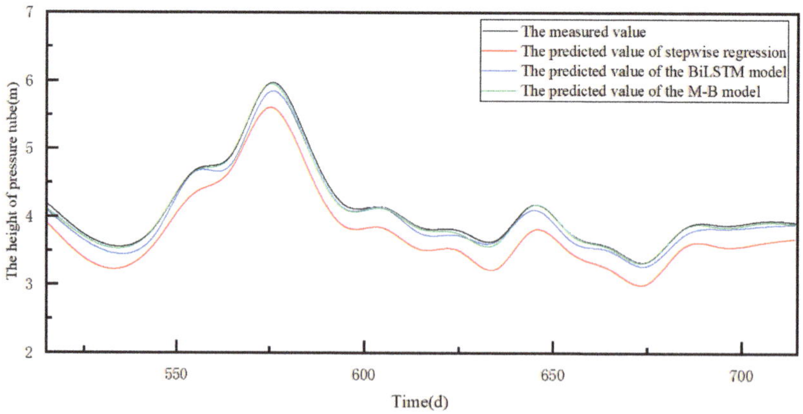

Figure 11. The measured values of the pier and the predicted seepage values of the three models.

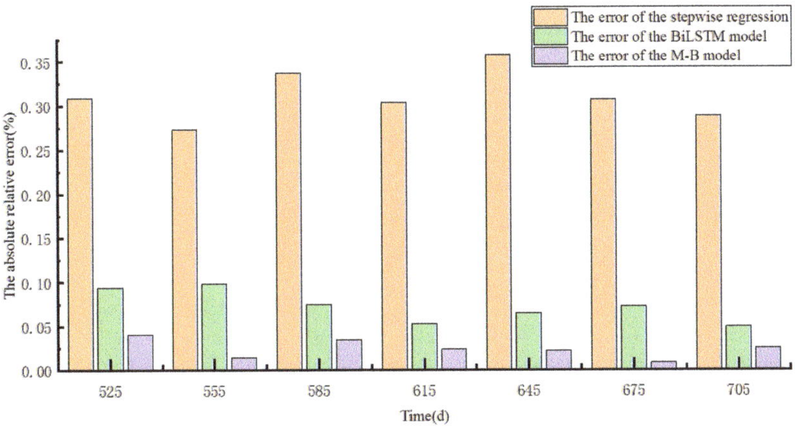

Figure 12. The relative error absolute values of the three models predicting seepage at the pier monitoring points.

Table 5. The comparison table of relative errors predicted by three models at the pier monitoring points.

Time (d)	Measured Value (m)	Stepwise Regression's Forecast (m)	Stepwise Regression's Error (%)	BiLSTM's Forecast (m)	BiLSTM's Error (%)	M-B's Forecast (m)	M-B's Error (%)
525	4.56	4.28	0.2766	4.48	0.0820	4.55	0.0107
555	4.49	4.22	0.2724	4.43	0.0595	4.48	0.0126
585	5.87	5.57	0.2980	5.76	0.1080	5.86	0.0149
615	4.48	4.16	0.3233	4.42	0.0612	4.45	0.0325
645	3.97	3.66	0.3131	3.89	0.0836	3.95	0.0191
675	5.03	4.72	0.3086	4.94	0.0930	5.00	0.0307
705	3.90	3.61	0.2858	3.83	0.0643	3.87	0.0210

As shown in Figure 11, the measured seepage value at the No. 2-1 monitoring point on the upstream side of the middle pier of floor 2# changes with time non-linearly. In addition, the predicted seepage by the BiLSTM network model and the MHHO-BiLSTM model corresponds to the measured data change trend. In dealing with nonlinear data, the prediction effect of the stepwise regression statistical model is inferior to the BiLSTM network model and MHHO-BiLSTM model. As can be seen from Figure 12 and Table 5, in terms of the relative error of model prediction in the 12 forecasting months, the MHHO-BiLSTM model is the smallest on the whole, while the stepwise regression statistical model is the largest on the whole. Statistically, the stepped-regression model, the BiLSTM network model, and the MHHO-BiLSTM model have mean absolute relative errors of 30.38%, 7.41%, and 1.99%, respectively. It can be seen that the accuracy of the BiLSTM network model is better than the stepped-regression statistical model, and the accuracy of the MHHO-BiLSTM model is better than the BiLSTM network model.

4. Conclusions

Via the main research contents, the following conclusions can be drawn:

(1) According to the results of the main influencing factors of seepage, MHHO was introduced to optimize the BiLSTM model and establish the seepage model of the sluice. The model increases the correlation coefficient R^2 from 0.8923 to 0.9942 and decreases the root-mean-square error RMSE from 0.2412 to 0.0530. Compared with the BiLSTM model and stepwise regression model, the R^2 value and RMSE value of the predicted results of this method are the largest and the smallest. Case study

analysis shows that the MHHO-BiLSTM model has good predictive performance, which indicates the good predictive ability of long-term data series.

(2) The MHHO optimization algorithm can help the BiLSTM model find the optimal parameter combination: the number of forward neurons and the number of backward neurons, so as to improve the performance of the BiLSTM model. The global search capability of the MHHO optimization algorithm can help regulate the complexity of the BiLSTM model to reduce the risk of overfitting, thereby improving the generalization ability of the BiLSTM model on previously unseen data. MHHO optimization algorithm has a strong global search ability, which can help accelerate the convergence process of the BiLSTM model and reduce the training time and calculation cost. MHHO optimization algorithm may help to improve the performance of the BiLSTM model when dealing with long-term dependence and strengthen its modeling ability for long-term dependence by optimizing parameters or model structure.

(3) Sluice seepage problems involve time series data, and BiLSTM is able to consider both the current moment input and the previous and subsequent input information. This enables BiLSTM to more comprehensively capture the time before and after information and timing patterns in the sluice seepage data, which helps to better understand and model the seepage behavior. BiLSTM has strong modeling capabilities and can handle complex nonlinear relationships and timing dependencies. The seepage problem of sluice often involves many influencing factors, and the relationship between these factors may be complicated. BiLSTM is able to build a model and make predictions by learning the temporal patterns in the data. BiLSTM's bidirectional structure allows the entire time series data to be processed at once. This means that BiLSTM is able to use global information to make predictions rather than just local information at the current moment. For the sluice seepage problem, global dependency modeling is helpful to better understand the correlation and influence of different times in the sluice system. In the seepage problem of the sluice gate, there may be a correlation between several variables, such as temperature, water pressure, aging, and so on. BiLSTM can process multiple input variables at the same time, building predictive models by learning the interactions between them. This gives BiLSTM an advantage in solving multivariable problems.

(4) However, BiLSTM models typically require large amounts of data to be trained in order to effectively capture and learn patterns and trends in time series data. If the amount of data available is small, the BiLSTM model may not be adequately trained, resulting in degraded model performance. These disadvantages should be considered comprehensively according to the specific situation, and the use of the BiLSTM model in predicting the seepage problem of sluice should be weighed. Issues such as the amount of data may need to be fully considered to achieve better predictive performance.

This paper provides a new idea for the seepage prediction of sluice. In addition, the robustness of the MHHO optimization algorithm should be evaluated in more engineering projects and in different scenarios, and it should be extensively validated to verify the applicability and universality of the proposed model.

Author Contributions: Conceptualization, Z.H., C.G. and S.Z.; methodology, Z.H., H.G. and C.S.; software, Z.H., J.P. and C.S.; validation, S.Z. and Y.W.; formal analysis, M.Z.; investigation, H.G.; resources, C.G., J.P. and Y.W.; data curation, Z.H. and C.S.; writing—original draft preparation, Z.H. and S.Z.; writing—review and editing, C.G. and H.G.; supervision, C.G.; project administration, C.G., H.G., J.P., Y.W. and M.Z.; funding acquisition, C.G., J.P., Y.W., H.G. and C.S. All authors have read and agreed to the published version of the manuscript.

Funding: This research was funded by the National Natural Science Foundation of China (Grant Nos. U2243223 and 52209159), the Fundamental Research Funds for the Central Universities (Grant No. B230201011), the Anhui Provincial Natural Science Foundation "Water Sciences" Joint Fund (2208085US17), the Jiangsu Young Science and Technological Talents Support Project (Grant No. TJ-2022-076), the Fund of Water Conservancy Technology of Xinjiang Province (XSKJ-2023-23), the Water Conservancy Science and Technology Project of Jiangsu (Grant No. 2022024), and the China Postdoctoral Science Foundation (2023M730934).

Data Availability Statement: Restrictions apply to the availability of these data. Data was obtained from the Anhui Provincial Natural Science Foundation "Water Sciences" Joint Fund and are available from the author with the permission of the Anhui Provincial Natural Science Foundation "Water Sciences" Joint Fund.

Conflicts of Interest: The authors declare no conflicts of interest.

References

1. Fan, X.; Wu, Z.; Liu, L.; Wen, Y.; Yu, S.; Zhao, Z.; Li, Z. Analysis of Sluice Foundation Seepage Using Monitoring Data and Numerical Simulation. *Adv. Civ. Eng.* **2019**, *2019*, 2850916. [CrossRef]
2. Hong, P.; Cao, B.; Ai, D. Based on the exploration of sluice engineering construction technology in hydraulic engineering. In Proceedings of the Guangzhou Sub-Forum of 2023 Smart City Construction Forum, Guangzhou, China, 26 March 2023. (In Chinese)
3. Zhou, W. Safety Problems and treatment measures of sluice operation. *Sci. Life* **2012**, 186. (In Chinese)
4. Li, H. Study on Seepage Flow of the Sluice Foundation Based on AutoBANK Software. Master's Thesis, Ocean University of China, Qingdao, China, 2015. (In Chinese)
5. Zhao, Y.; Cai, T.; Li, J.; Gou, J. Study on seepage problem of sluice. *Dwelling* **2019**, *1*, 157. (In Chinese)
6. Zhou, X.; Gu, X.; Yu, M.; Qian, Q. Seismic bearing capacity of shallow foundations resting on rock masses subjected to seismic loads. *KSCE J. Civ. Eng.* **2016**, *20*, 216–228. [CrossRef]
7. Yue, N.; Zhang, D.; Chen, J.; Song, P.; Wang, S.; Qiu, S.; Su, G.H.; Zhang, Y. The development and validation of the inter-wrapper flow model in sodium-cooled fast reactors. *Prog. Nucl. Energy* **2018**, *108*, 54–65. [CrossRef]
8. Pokharel, G.; Honjo, Y. Mapped infinite elements in multilayered seepage analysis. In *Computer Methods and Advances in Geomechanics*; Balkma: Rotterdam, The Netherlands, 1994; pp. 1243–1248.
9. Pongsivasathit, S.; Petchgate, W.; Horpibulsuk, S.; Piyaphipat, S. Composite contiguous pile wall and deep mixing column wall as a dam—Design, construction and performance. *Case Stud. Constr. Mater.* **2022**, *16*, e00904. [CrossRef]
10. Zhang, K.; Xu, L.; Qiu, L.; Tan, J.; Yang, C.; Zhang, K. Numerical Analysis of Concrete Gravity Dam Seepage Characteristics Evolution considering the Calcium Leaching Effect. *Adv. Civ. Eng.* **2021**, *2021*, 9042863. [CrossRef]
11. Jia, Y.; Ding, Y.; Wang, X.; Zhang, J.; Chen, X. A Numerical Analysis of the Leakage Characteristics of an Embankment Dam Slope with Internal Erosion. *Front. Earth Sci.* **2022**, *10*, 866238. [CrossRef]
12. Wang, R.; Zhu, Y. Finite element analysis of seepage of earth-rock dams in dry and rainy seasons. *IOP Conf. Ser. Earth Environ. Sci.* **2019**, *344*, 012112. [CrossRef]
13. Sun, Y.; Li, Z.; Yang, K.; Wang, G.; Hu, R. Analysis of the Influence of Water Level Change on the Seepage Field and Stability of a Slope Based on a Numerical Simulation Method. *Water* **2023**, *15*, 216. [CrossRef]
14. Zhang, M.; Yao, D.; Lu, H.; Wang, H. Solution of seepage field in different soil layers of concrete dam foundation by flow net method. *IOP Conf. Ser. Earth Environ. Sci.* **2020**, *546*, 052053. [CrossRef]
15. Bensmaine, A.; Benmebarek, N.; Bensmebarek, S. Numerical Analysis of Seepage Failure Modes of Sandy Soils within a Cylindrical Cofferdam. *Civ. Eng. J.* **2022**, *8*, 1388–1405. [CrossRef]
16. Norouzi, R.; Salmasi, F.; Arvanaghi, H. Uplift pressure and hydraulic gradient in Sabalan Dam. *Appl. Water Sci.* **2020**, *10*, 111. [CrossRef]
17. Chen, Q.; Zhang, P.; Ding, H.; Cong, H.; Yun, L. Study on Seepage Characteristics of Composite Bucket Foundation under Eccentric Load. *China Ocean. Eng.* **2021**, *35*, 123–134. [CrossRef]
18. Liu, B.; Li, J.; Liu, Q.; Liu, X. Analysis of Damage and Permeability Evolution for Mudstone Material under Coupled Stress-Seepage. *Materials* **2020**, *13*, 3755. [CrossRef] [PubMed]
19. Guo, Q.; Huang, J. Analysis of the influence of the interlayer staggered zone in the basalt of Jinsha River Basin on the main buildings. *IOP Conf. Ser. Earth Environ. Sci.* **2018**, *113*, 012099. [CrossRef]
20. Bai, L.; Che, W.; Wang, S. Study on the Influence of Groundwater Seepage on the form of the Layout of Soil Source Heat Pump. *Procedia Eng.* **2016**, *146*, 445–449. [CrossRef]
21. Safari Ghaleh, R.; Aminoroayaie Yamini, O.; Mousavi, S.H.; Kavianpour, M.R. Numerical Modeling of Failure Mechanisms in Articulated Concrete Block Mattress as a Sustainable Coastal Protection Structure. *Sustainability* **2021**, *13*, 12794. [CrossRef]
22. Hou, W.; Wen, Y.; Deng, G.; Zhang, Y.; Wang, X. A multi-target prediction model for dam seepage field. *Front. Earth Sci.* **2023**, *11*, 1156114. [CrossRef]

23. Zhao, M.; Jiang, H.; Chen, S.; Bie, Y. Prediction of Seepage Pressure Based on Memory Cells and Significance Analysis o Influencing Factors. *Complexity* **2021**, *2021*, 5576148. [CrossRef]
24. Zhang, K.; Gu, C.; Zhu, Y.; Chen, S.; Dai, B.; Li, Y.; Shu, X. A Novel Seepage Behavior Prediction and Lag Process Identificatior Method for Concrete Dams Using HGWO-XGBoost Model. *IEEE Access* **2021**, *9*, 23311–23325. [CrossRef]
25. Ishfaque, M.; Dai, Q.; Haq, N.U.; Jadoon, K.; Shahzad, S.M.; Janjuhah, H.T. Use of Recurrent Neural Network with Long Short-Term Memory for Seepage Prediction at Tarbela Dam, KP, Pakistan. *Energies* **2022**, *15*, 3123. [CrossRef]
26. Zheng, S.; Gu, C.; Shao, C.; Hu, Y.; Xu, Y.; Huang, X. A Novel Prediction Model for Seawall Deformation Based on CPSO-WNN LSTM. *Mathematics* **2023**, *11*, 3752. [CrossRef]
27. Hochreiter, S.; Schmidhuber, J. Long short-term memory. *Neural Comput.* **1997**, *9*, 1735–1780. [CrossRef] [PubMed]
28. Hu, Y.; Gu, C.; Meng, Z.; Shao, C.; Min, Z. Prediction for the Settlement of Concrete Face Rockfill Dams Using Optimized LSTM Model via Correlated Monitoring Data. *Water* **2022**, *14*, 2157. [CrossRef]
29. Bolboacă, R.; Haller, P. Performance Analysis of Long Short-Term Memory Predictive Neural Networks on Time Series Data *Mathematics* **2023**, *11*, 1432. [CrossRef]
30. Girsang, A.S. Modified EDA and backtranslation augmentation in deep learning models for Indonesian aspect-based sentimen analysis. *Emerg. Sci. J.* **2022**, *7*, 256–272.
31. Sheng, J.; Liu, J.; Li, J. Statistical mathematical model of seepage flow in earth-rock dam. *Sci. Res. Water Conserv. Water Transp* **1995**, *4*, 435–443. (In Chinese)
32. Heidari, A.A.; Mirjalili, S.; Faris, H.; Aljarah, I.; Mafarja, M.; Chen, H. Harris hawks optimization: Algorithm and applications *Futur. Gener. Comput. Syst.* **2019**, *97*, 849–872. [CrossRef]
33. Sharafati, A.; Tafarojnoruz, A.; Shourian, M.; Yaseen, Z.M. Simulation of the depth scouring downstream sluice gate: The validation of newly developed data-intelligent models. *J. Hydro-Environ. Res.* **2019**, *29*, 20–30. [CrossRef]
34. Sharafati, A.; Tafarojnoruz, A.; Yaseen, Z.M. New stochastic modeling strategy on the prediction enhancement of pier scour depth in cohesive bed materials. *J. Hydroinform.* **2020**, *22*, 457–472. [CrossRef]
35. Sharafati, A.; Tafarojnoruz, A.; Motta, D.; Yaseen, Z.M. Application of nature-inspired optimization algorithms to ANFIS model to predict wave-induced scour depth around pipelines. *J. Hydroinform.* **2020**, *22*, 1425–1451. [CrossRef]
36. Guo, Y.; Liu, S.; Gao, W.; Zhang, L. Improved harris hawks optimization algorithm with multiple strategies. *Microelectron. Comput* **2021**, *38*, 18–24. (In Chinese)

water

Article

A Method for Identifying Gross Errors in Dam Monitoring Data

Liqiu Chen [1,2,3], Chongshi Gu [1,2,3,*], Sen Zheng [1,2,3] and Yanbo Wang [1,2,3]

1 The National Key Laboratory of Water Disaster Prevention, Hohai University, Nanjing 210098, China; clq139466@gmail.com (L.C.); zheng_sen@hhu.edu.cn (S.Z.); a19617230662022@163.com (Y.W.)
2 College of Water Conservancy and Hydropower Engineering, Hohai University, Nanjing 210098, China
3 National Engineering Research Center of Water Resources Efficient Utilization and Engineering Safety, Hohai University, Nanjing 210098, China
* Correspondence: csgu@hhu.edu.cn

Abstract: Real and effective monitoring data are crucial in assessing the structural safety of dams. Gross errors, resulting from manual mismeasurement, instrument failure, or other factors, can significantly impact the evaluation process. It is imperative to eliminate such anomalous data. However, existing methods for detecting gross errors in concrete dam deformation often focus on analyzing a single monitoring effect quantity. This can lead to sudden jumps in values of effect quantity caused by changes in environmental variables being mistakenly identified as gross error. Therefore, a method based on Fuzzy C-Means clustering algorithm (FCM) partitioning and density clustering algorithm (Ordering Points To Identify the Clustering Structure, OPTICS) combined with Local Outlier Factor (LOF) algorithm for gross error identification is proposed. Firstly, the FCM algorithm is used to achieve the division of measurement point areas. Then, the OPTICS and LOF algorithms are jointly utilized to determine the gross errors. Finally, the real gross errors are identified by comparing the time of occurrence of the gross errors at measurement points in the same area. Through the case study, the results indicate that the method can effectively identify spurious, gross errors in the monitoring effect quantity caused by environmental mutations. The accuracy of gross error detection is significantly improved, and the rate of misjudgment of gross errors is reduced.

Keywords: dam monitoring data; gross errors; environmental change; FCM algorithm; OPTICS algorithm; LOF algorithm

1. Introduction

To avoid sudden dam diseases in the service process, which may lead to serious engineering accidents and cause significant loss of life and property, it is essential to continuously monitor the operational status of the dam body in real-time. Regular analysis of monitoring data, including deformation, seepage, and uplift pressure, allows for a comprehensive understanding of the dam's health status and ensures its long-term stable operation [1–3]. Accurate monitoring data are an important prerequisite and fundamental guarantee for effective dam safety monitoring. Therefore, the preprocessing of dam monitoring data is necessary.

The original data from concrete dam safety monitoring often includes gross errors resulting from reading errors, instrument failures, and other factors. These errors represent unreasonably inaccurate data points that can significantly distort dam monitoring data. If not effectively addressed, they can compromise the accuracy of safety evaluations, early warnings, and forecasts for the dam. Consequently, eliminating gross errors is an important part of preliminary processing [4–6]. To identify gross errors, many scholars have carried out related research. Traditional gross error identification methods are generally based on statistical theories, such as 3σ criterion, Grubbs criterion, Romanowski criterion, and Dixon criterion, which have been widely employed for gross error identification in dam monitoring data [7–10]. However, when the structure of dam monitoring data is complex, problems of missing judgment and misjudgment are easy to occur. As a result, many scholars have endeavored to enhance traditional methods, aiming to improve the detection rate. For example, Zhao et al. [11] proposed an improved 3σ criterion based on the Minimum Covariance Determinant. Li et al. [12] introduced an enhanced Pauta criterion based on M estimation. With the development of big data analysis technology, machine learning has also been gradually applied to gross error recognition [13]. Examples include Neural Networks [14,15], Support Vector Machines [16,17], Decision Trees [18,19], and others. Song et al. [20] proposed a gross error detection method of singular spectrum analysis (SSA) combined with nonlinear autoregression (NAR). Qi et al. [21,22] used full convolutional networks to learn from artificially labeled datasets and realize the recognition of gross error data. Hu et al. [23] proposed a method for identifying anomalies by combining dynamic time warping with a local outlier factor. Song et al. [24] developed an analysis method for detecting outliers in dam deformation data based on multivariable panel data and K-means clustering theory. Shao et al. [25] converted monitoring data into images and processed outliers in combination with the cuckoo-search algorithm. Bao et al. [26] visualized time series data and fed them into a deep neural network to detect anomalous data [27–29]. Zhang et al. [30] integrated multiple learners and proposed an anomaly diagnosis method using an anomaly index matrix updated with real-time data. Gu et al. [31] used an improved IGG method and an extreme learning machine to identify gross errors in deformation monitoring data. Li et al. [32] proposed an outlier identification method based on a BP neural network. Liu et al. [33] utilized the wavelet transform to identify outliers in time series data. While these methods have shown success in identifying gross errors, many of them are computationally complex. Moreover, several primarily rely on time series change rules and fail to fully consider the influence of environmental changes on dam behavior. Typically, significant fluctuations in environmental factors such as water levels and temperature can lead to corresponding variations in key parameters like deformation. These variations represent the actual behavior of the dam, providing valuable data reflecting its true state. However, they are prone to being misinterpreted as gross errors.

To address the aforementioned issues, the local density difference between normal data and data containing gross errors is considered in this paper, and a method for identifying gross errors in dam safety monitoring data is presented. This method combines FCM clustering, OPTICS, and LOF algorithms. Firstly, the FCM algorithm is applied to partition the dam displacement measurement points, followed by the use of an enhanced OPTICS algorithm for preliminary gross error identification. Subsequently, the LOF value of each data point in the preliminary gross error dataset is calculated. If the LOF value exceeds a predefined threshold, the data is flagged as a gross error. Finally, potential misjudgments resulting from changes in environmental quantities are mitigated by comparing the data with other measurement points within the same cluster, ultimately determining the final classification of gross errors. The combination of OPTICS and LOF algorithms significantly enhances the accuracy and sensitivity of gross error detection while also reducing the rate of misjudgments to some extent. Because the proposed method only calculates LOF values for a subset of the data, it can effectively reduce computational complexity.

2. Fundamentals of a Gross Error Identification Algorithm for Dam Monitoring Data

2.1. FCM Algorithm

The Fuzzy C-means (FCM) clustering algorithm introduces fuzzy set theory into cluster analysis. Instead of categorizing sample data into a specific partition with 100% certainty, it calculates the membership degree of each data sample corresponding to each class by optimizing the objective function. This approach achieves optimal data clustering, effectively enhancing the algorithm's resistance to noise and its overall fuzziness [34,35]. To obtain accurate clustering results, it is necessary to comprehensively analyze the change trend of the measured value series. Therefore, the comprehensive distance, a weighted combination of absolute distance, incremental distance, and growth rate distance, is adopted as the similarity index [36]. The indicators are explained as follows.

(1) The absolute distance between the measuring point i and the measuring point j, denoted $d_{ij}(AD)$.

$$d_{ij}(AD) = \left[\sum_{t=1}^{T}(\delta_{it} - \delta_{jt})^2\right]^{\frac{1}{2}} \quad (1)$$

where δ_{it} is the deformation value of the measuring point i at time t; δ_{jt} is the deformation value of the measuring point j at time t.

(2) The incremental distance between the measuring point i and the measuring point j is denoted as $d_{ij}(KD)$.

$$d_{ij}(KD) = \left[\sum_{t=1}^{T}(\Delta\delta_{it} - \Delta\delta_{jt})^2\right]^{\frac{1}{2}} \quad (2)$$

where $\Delta\delta_{it} = \delta_{it} - \delta_{it-1}$; $\Delta\delta_{jt} = \delta_{jt} - \delta_{jt-1}$.

(3) The growth rate distance between the measuring point i and the measuring point j is denoted as $d_{ij}(GRD)$.

$$d_{ij}(GRD) = \left[\sum_{t=1}^{T}\left(\frac{\Delta\delta_{it}}{\delta_{it-1}} - \frac{\Delta\delta_{jt}}{\delta_{jt-1}}\right)^2\right]^{\frac{1}{2}} \quad (3)$$

(4) The comprehensive distance between the measuring point i and the measuring point j is denoted $d_{ij}(CED)$.

$$d_{ij}(CED) = \omega_1 d_{ij}(AD) + \omega_2 d_{ij}(KD) + \omega_3 d_{ij}(GRD) \quad (4)$$

where ω_1, ω_2 and ω_3 represent the weight coefficients of the three distances, which meet the requirement that $\omega_1 + \omega_2 + \omega_3 = 1$. This paper adopts the entropy weight method [37] for their calculation.

Assuming that the concrete dam has n deformation measurement points, $x_1, x_2, \cdots, x_n$, and the number of clusters is c, then the objective function of FCM is as follows:

$$\min J_m(\mathbf{U}, c) = \sum_{j=1}^{c}\sum_{i=1}^{n} u_{ij}^m d_{x_i v_j}(CED) \quad (5)$$

where $\mathbf{U}$ is the membership matrix, $\mathbf{U} = [u_{ij}]_{c \times n}$; u_{ij} is the membership degree of the measuring point x_i and cluster center v_j; m is the fuzzy index and the best value interval is [1.5, 2.5]; $d_{x_i v_j}(CED)$ is the comprehensive distance between the measuring point x_i and cluster center v_j.

The specific steps of the FCM algorithm are as follows:

Step 1: Set the fuzzy index m and cluster number c, initialize the membership matrix $\mathbf{U}^0$, set the convergence accuracy $\varepsilon > 0$, and calculate the comprehensive distance between the measurement points.

Step 2: Calculate cluster centers v_j:

$$v_j = \sum_{i=1}^{n} u_{ij}^m x_i / \sum_{i=1}^{n} u_{ij}^m \quad (6)$$

Step 3: Update the membership matrix $\mathbf{U}$:

$$u_{ij} = \left[\sum_{k=1}^{c}\left(\frac{d_{x_i v_j}}{d_{x_i v_k}}\right)^{2/(m-1)}\right]^{-1} \quad (7)$$

Step 4: Repeat steps 2 and 3 until the difference in membership matrices between consecutive iterations is less than the set threshold ε.

In this paper, the Silhouette Coefficient method is employed to determine the optimal number of clusters (c). Various values of c are utilized for clustering, and the corresponding silhouette coefficients are calculated. The silhouette coefficient (S) ranges between [−1, 1], and the closer it is to 1, the better the clustering effect. The silhouette coefficient S_i for a single measurement point x_i is defined as follows:

$$S_i = \frac{p(x_i) - q(x_i)}{\max\{q(x_i), p(x_i)\}} \quad (8)$$

where $p(x_i)$ is the average distance between measuring point x_i and other measuring points within the same cluster; $q(x_i)$ is the minimum value of the average distance between measuring point x_i and the samples of other clusters.

2.2. OPTICS Algorithm and Its Improvement

2.2.1. OPTICS Algorithm

Ordering Points To Identify the Clustering Structure (OPTICS) is a density-based clustering algorithm, which is an improvement of DBSCAN [38]. Compared with DBSCAN, although the neighborhood radius (ε) and the minimum number of neighborhood points ($MinPts$) must be set, OPTICS achieves clustering results that are less dependent on the selection of ε and $MinPts$. It obtains a series of parameter-set density-based clusterings by introducing the core-distance and reachability-distance [39,40]. The OPTICS algorithm is capable of effectively handling datasets with non-uniform density, unlike other algorithms that require all clusters to possess similar densities. The fundamental concept behind OPTICS is to initiate a random selection of data and expand towards the densest area. Each data point has a reachability-distance to represent its density relationship with other data points. Based on the reachability-distance of each data point, an accessible distance map is generated, reflecting the density of the data. The relevant definition of the algorithm is as follows:

Definition 1. *Neighborhood radius (ε): Given a data set D, for any point p in D, the neighborhood radius is defined as a circular region with the point p as the center and ε as the radius.*

Definition 2. *Minimum number of neighborhood points ($MinPts$): The minimum number of points within the neighborhood radius required for a point to be considered a core point.*

Definition 3. *Core point: The core point has at least $MinPts$ data points in the neighborhood radius.*

Definition 4. *Core-Distance: The minimum neighborhood radius of a point that makes it a core point, that is, the distance from it to the $MinPts$ closest point in its neighborhood. If a point is not a core point, its core-distance is undefined.*

Definition 5. *Reachability-Distance: The reachability-distance from one point q to another core point p is $Max(coreDist(p), d(p, q))$, which is the greater value between the distance from q to p and the core-distance of p. If p is not a core point, then the reachability-distance from q to p is undefined. $d(p, q)$ represents the distance between p and q.*

Definition 6. *Direct Density Reachability: If the reachability-distance from a point q to another core point p is less than or equal to ε, and q is in the neighborhood radius of p, then it is said to have direct density reachability from p to q.*

The core-distance and reachability-distance are shown in Figure 1. Given the neighborhood radius $\varepsilon = 6$ and the minimum number of neighborhood points $MinPts = 5$, then the core-distance of p is 3, the reachability-distance from q to p is $\max(coreDist(p), d(p,q)) = 7$, and the reachability-distance from r to p is $\max(coreDist(p), d(p,r)) = 3$.

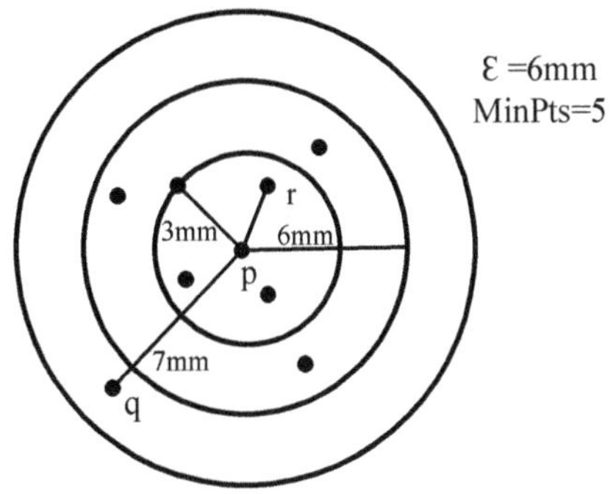

Figure 1. Schematic diagram of the OPTICS algorithm.

In the clustering process, OPTICS uses two queues: an ordered seed queue and a result queue. The orderly seed queue is used to temporarily store the points to be processed and is arranged in ascending order according to the reachability-distance. The data points with the smallest reachability-distance are preferentially selected for processing, and the data points are expanded to the densest area of data. In this way, data objects with high density can be quickly found. The result queue is used to store the output data points and is arranged in descending order of reachability-distance to form a decision graph. On the decision graph, the regions of steep rises and steep falls can be identified, representing the anomalous points and the points inside the cluster, respectively.

2.2.2. Improved OPTICS Algorithm

Although the clustering results are minimally affected by the setting of the parameters, and the number of clusters does not need to be specified in advance, the time complexity is quite high. When data is selected from the ordered seed queue for expansion, the core-distance and reachability-distance are calculated for any point in its neighborhood. If a data point is already in the seed sequence and the reachability-distance is greater than the calculated result, the original data is replaced and reordered. If it does not belong to the seed sequence, it is added to the corresponding position of the seed sequence according to its reachability-distance. The fastest sorting algorithm is used to perform the above sorting, and the time complexity of each data added or updated in the seed sequence is $O(n \log n)$.

Because of the high time complexity of the OPTICS algorithm, a non-sorted seed sequence is adopted to improve the efficiency of the algorithm [41]. A temporary variable repository is created to store data points with the minimum reachability-distance in the seed sequence, which is no longer sorted by the reachability-distance. When the seed sequence adds or updates a point, simply compare the size of the reachability-distance with the temporary variable and replace the temporary variable if it is smaller. When a point needs to be extracted from the seed sequence, it is only necessary to extract the point from the temporary variable repository. Subsequently, identify the new minimum reachability-distance point from the seed queue and store it in the temporary variable repository. Compared with sorting, this method of traversing the seed queue to find the minimum point greatly improves efficiency, and the time complexity of traversing a seed sequence is $O(n)$. The specific implementation steps of the algorithm are as follows:

Step 1: Create a seed queue, a result queue, and a temporary variable repository. The seed queue is designed to store a large number of samples to be processed, and the result queue is intended for storing samples after processing. The temporary variable repository is utilized to store the next sample to be processed.

Step 2: An unprocessed core point is randomly selected and placed into the result queue. At the same time, the reachability-distance of sample points in its neighborhood is calculated. Add these sample points to the seed queue, selecting the sample with the minimum reachability distance and placing it in the temporary variable repository.

Step 3: If the seed queue is empty, return to step 2 (reselect the data). Otherwise, extract the sample p from the temporary variable repository. If p is not a core point, the fourth step is performed. If p is a core point, expand any unexpanded data q in its neighborhood ε and calculate its reachability-distance. If q is already in the seed queue and the new reachability-distance is less than the original value, update the reachability-distance. If q is not in the seed queue, it is placed in the seed queue.

Step 4: The p is written into the result queue, and the sample points with the smallest reachability-distance are re-found in the seed queue and put into the temporary variable repository. Repeat step 3 until all points in the data set are processed. Then, the algorithm ends, and the ordered sample points in the result queue are output.

2.3. LOF Algorithm

The LOF algorithm is a method for determining whether a sample is abnormal according to the ratio of the local density of a sample to that of its neighbors [23]. The outlier degree of data is determined by calculating the outlier factor of each data sample. The greater the outlier factor value, the higher the likelihood that the sample is abnormal data. Moreover, the algorithm is capable of computing the LOF values for specified data points without the need to traverse all data points, thus significantly reducing computational complexity. The related concepts of the algorithm are as follows:

Definition 7. *The distance $d_b(a)$ between object a and object b: the distance between point a and its b-th nearest neighbor.*

Definition 8. *The b-th distance neighborhood of an object a (denoted as $N_b(a)$): the set of data points whose distance from a is less than or equal to $d_b(a)$.*

Definition 9. *Reachability-Distance: The reachability-distance from object a to object o is the greater of the $d_b(o)$ and the direct distance between object o and a, i.e.,:*

$$reach - \text{distance}_b(a,o) = \max\{d_b(o), d(a,o)\} \tag{9}$$

Definition 10. *Locally Reachable Density: The locally reachable density of an object a is defined as the reciprocal of the average reachability-distance from each point in the $N_b(a)$ to a, i.e.,:*

$$lrd_b(a) = 1 \Bigg/ \left[\frac{\sum\limits_{o \in N_b(a)} reach - dist_b(a,o)}{|N_b(a)|} \right] \tag{10}$$

Definition 11. *Local Outlier Factor (LOF): The local outlier factor of an object a is defined as the average of the ratio of the locally reachable density of $N_b(a)$ to the locally reachable density of a, i.e.,*

$$LOF_b(a) = \frac{\sum\limits_{o \in N_b(a)} \frac{lrd_b(o)}{lrd_b(a)}}{|N_b(a)|} \tag{11}$$

Figure 2 shows a schematic diagram of the LOF algorithm identifying a few outliers in the sample.

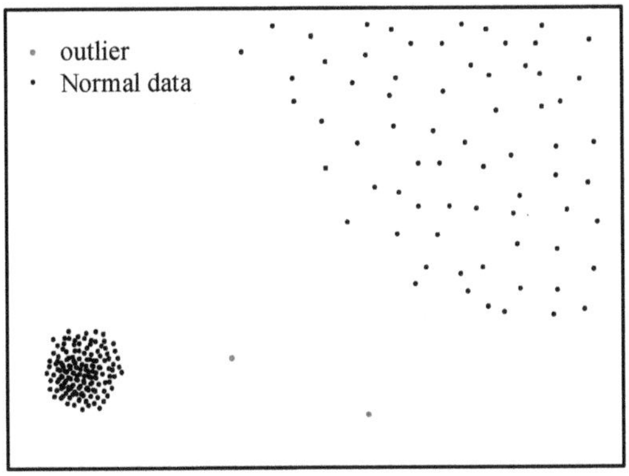

Figure 2. Schematic diagram of the LOF algorithm for identifying outliers.

3. Gross Error Identification Method for Dam Safety Monitoring Based on FCM-OPTICS-LOF Algorithm

The dam body undergoes displacement due to water pressure, temperature, and aging effects. While the displacement magnitude varies at different measuring points, there is a certain commonality in the load received. As a result, several measuring points may exhibit similar deformation patterns. Therefore, the FCM clustering algorithm is employed to partition the dam measuring points based on deformation, ensuring that measuring points within each region share similar deformation process lines. Subsequently, the OPTICS clustering algorithm, combined with the LOF algorithm, is utilized to identify gross errors in the monitoring data.

In the reachability-distance diagram of OPTICS, the abscissa represents the order of the output data points, and the ordinate represents the reachability-distance. A point with a larger reachability-distance indicates lower data density and a higher likelihood of gross error. Typically, gross errors constitute only a small portion of the overall data. Therefore, data at the top ρ% of the reachability-distance (where ρ is the selected percentage) are treated as preliminary gross errors. Following this, the LOF value of each data point in the preliminary gross errors is calculated, and data with an LOF value greater than the threshold are identified as gross errors. On the basis of identifying gross errors with the OPTICS algorithm, the LOF algorithm is further used to detect local abnormalities in data. This multi-level gross error identification method can be applied to various types of data sets and reduce the missed detection rate.

However, data points with low densities may also be true values characterizing the normal operating conditions of the dam due to changes in environmental quantities (e.g., water level). To address this, it is necessary to identify the gross errors of other measurement points located in the same zone but on different vertical lines, and the appearance time of the gross errors should be compared. If a measurement point is detected with gross errors at a certain moment while other measurement points in the same area at this moment are valid, it is judged as a gross error. If two measurement points exhibit gross errors simultaneously, then it is considered that the anomaly is the real value caused by the change in environmental quantity. Thus reducing the misjudgment rate. The specific implementation process of the method is shown in Figure 3.

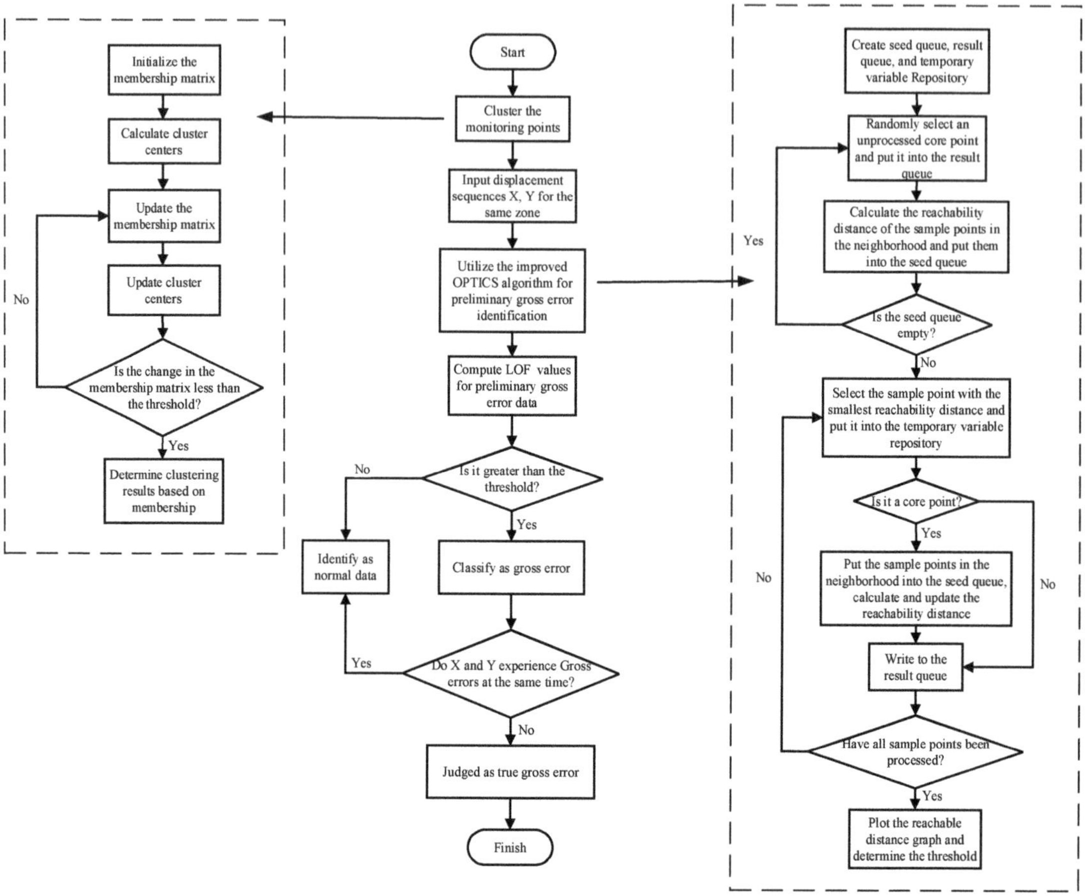

Figure 3. The flow chart of gross error identification.

4. Case Study

4.1. Project Overview

A certain concrete hyperbolic arch dam is located in Sichuan Province, China, with a maximum dam height of 305 m. The dam crest elevation is 1885 m, and the lowest elevation of the foundation is 1580 m. A total of 29 vertical displacement monitoring points are arranged in the dam sections #5, #9, #11, #13, #16, and #19 to monitor the deformation of the dam body. To validate the effectiveness of the proposed gross error identification method in this paper, displacement monitoring data from the dam are selected for analysis for the period from 1 September 2015, to 31 December 2018. Figure 4 shows a photograph of the hyperbolic arch dam.

Figure 4. The photograph of the hyperbolic arch dam.

4.2. Clustering Partition

The FCM clustering algorithm is employed to assess the similarity in the deformation of the 29 measurement points. The range of clusters (c) is set from 3 to 10, and the silhouette coefficient is computed for each cluster. As illustrated in Figure 5, when the number of clusters (c) is set to 5, the silhouette coefficient reaches its maximum value of 0.64. At this point, the samples within the cluster demonstrate high aggregation, and there is significant separation between samples belonging to different clusters. Consequently, we select $c = 5$ as the optimal number of clusters. The dam partitioning results are displayed in Figure 6.

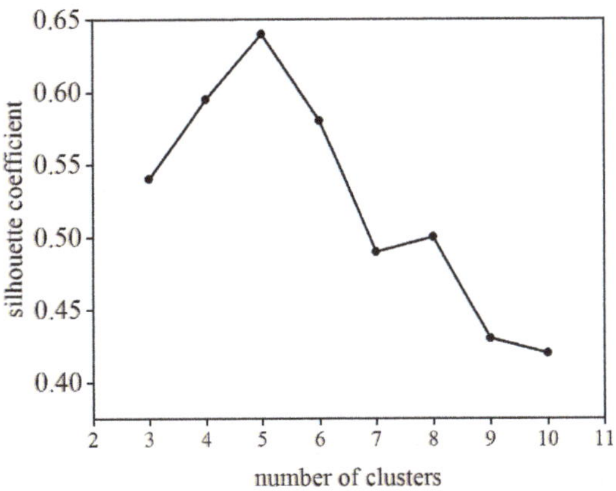

Figure 5. Silhouette coefficients for different numbers of clusters.

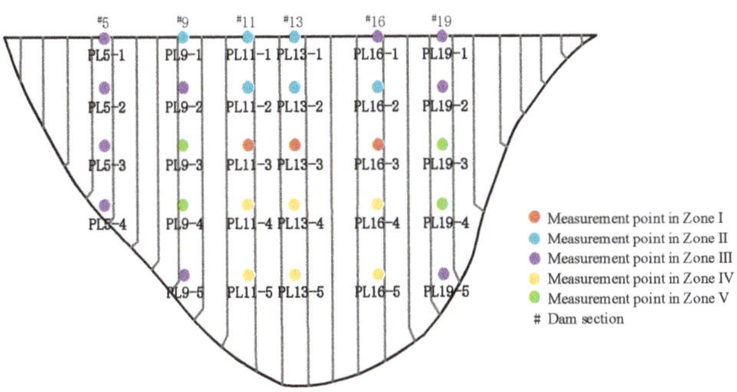

Figure 6. Clustered zones resulting from deformation monitoring points in the arch dam.

4.3. Gross Errors Identification

This section selects displacement monitoring data from typical monitoring points PL13-1 and PL13-2 in Zone II, PL13-3 and PL16-3 in Zone I, as well as PL11-5 and PL16-5 in Zone IV of the dam for the purpose of outlier (gross error) identification verification. In each displacement dataset, 15 artificial gross error data points are introduced. The specific locations and values of the introduced outliers for PL13-1 and PL13-2 are presented in Table 1 (the remaining four monitoring points are shown in Tables A1 and A2 in Appendix A). The displacement time series for each zone after introducing outliers are illustrated in Figure 7.

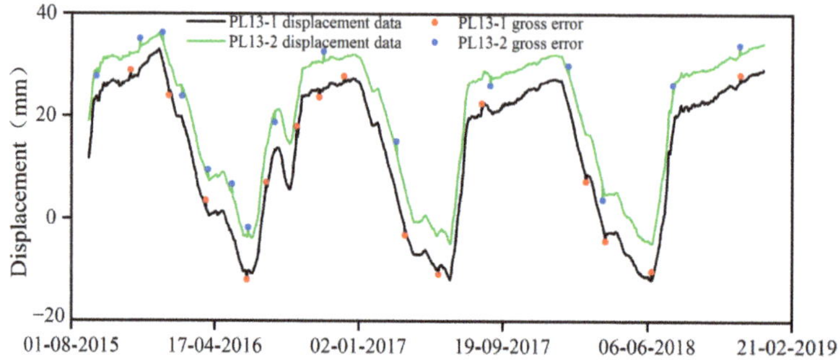

(**a**)

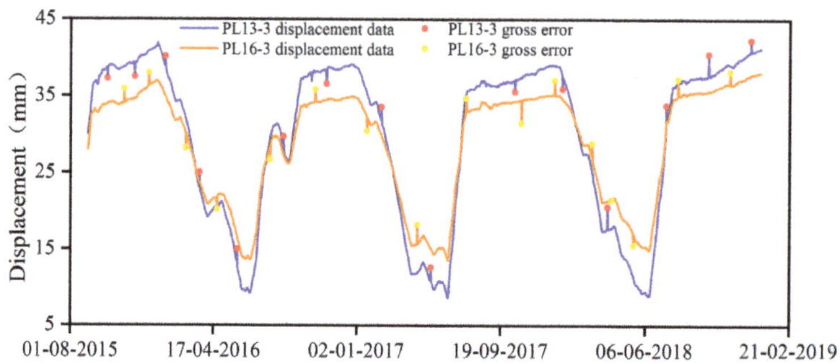

(**b**)

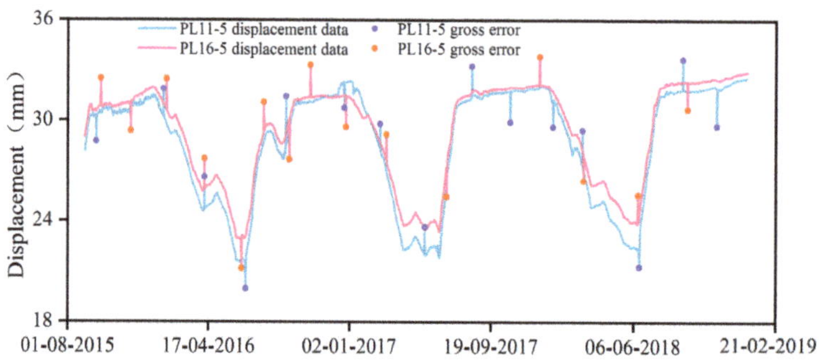

(**c**)

Figure 7. Displacement time histories at each measurement point with the inclusion of gross errors: (**a**) displacement time histories for PL13-1 and PL13-2, (**b**) displacement time histories for PL13-3 and PL16-3, (**c**) displacement time histories for PL11-5 and PL16-5.

Table 1. Locations and magnitudes of gross errors introduced in monitoring points PL13-1 and PL13-2.

PL13-1				PL13-2			
Data	Raw Data/mm	Gross Error Size/mm	the Data after Adding the Gross Error/mm	Data	Raw Data/mm	Gross Error Size/mm	the Data after Adding the Gross Error/mm
14 November 2015	27.62	1.38	29	14 September 2015	29.34	−1.5	27.84
23 January 2016	25.15	−1.15	24	1 December 2015	33.51	1.7	35.21
31 March 2016	2.1	1.4	3.5	11 January 2016	34.59	1.8	36.39
14 June 2016	−10.5	−1.4	−11.9	16 February 2016	25.76	−1.9	23.86
19 July 2016	5.78	1.22	7	4 April 2016	7.82	1.76	9.58
12 September 2016	15.67	2.33	18	18 May 2016	4.7	2	6.7
22 October 2016	25.24	−1.54	23.7	16 June 2016	−3.39	1.66	−1.73
5 December 2016	26.7	1.2	27.9	3 August 2016	20.91	−2.1	18.81
26 March 2017	−1.74	−1.46	−3.2	29 October 2016	30.83	1.88	32.71
26 May 2017	−9.4	−1.40	−10.8	10 March 2017	12.94	2.25	15.19
11 August 2017	20.91	1.59	22.5	26 August 2017	27.98	−1.89	26.09
15 February 2018	8.74	−1.44	7.3	14 January 2018	28.36	1.65	30.01
22 March 2018	−2.74	−1.56	−4.3	17 March 2018	6.72	−3	3.72
12 June 2018	−11.86	1.66	−10.2	20 July 2018	24.32	1.86	26.18
19 November 2018	26.89	1.31	28.2	18 November 2018	32.25	1.77	34.02

Utilizing the OPTICS algorithm, clustering analysis is performed on the displacement data from monitoring points PL13-1 and PL13-2. Given that gross errors constitute only a small portion of the data, the preliminary gross errors are selected based on the top 4% [41] of reachability-distance. The reachable distance plots for the two monitoring points are shown in Figure 8. The LOF values for each data point in the preliminary gross errors are then computed. Notably, some data points exhibited LOF values significantly greater than 1, designating them as gross errors. The scatter plot of LOF values for the monitoring points is illustrated in Figure 9.

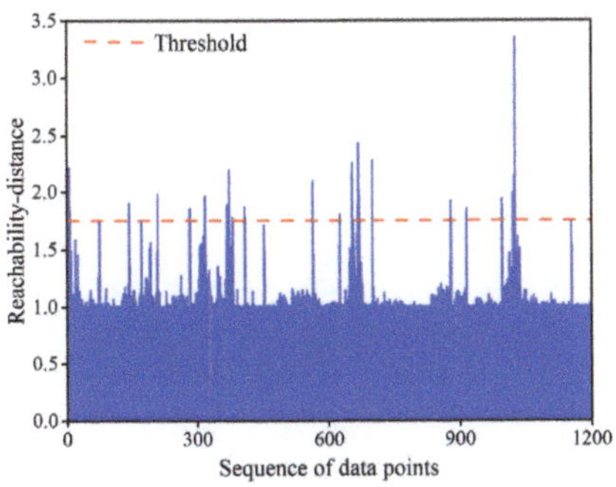

(a)

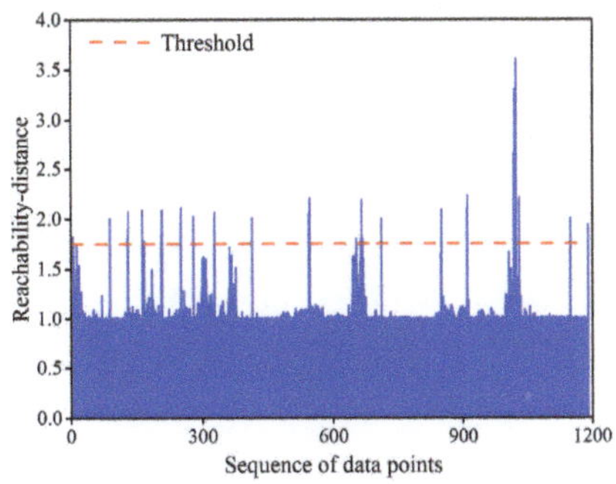

(b)

Figure 8. (**a**) Reachability distance for measurement point PL13-1; (**b**) Reachability distance for measurement point PL13-2.

Considering the quantity of data and the distribution of LOF values, the first 20 data points are identified as outliers. For monitoring point PL13-1, aside from the 15 intentionally introduced anomalies, data points on 7 September 2015, 11 September 2016, 12 July 2017, 11 July 2018, and 12 July 2018 are also flagged as outliers. Similarly, intentional anomalies introduced to PL13-2 are successfully identified, and additional outliers are detected on 7 September 2015, 27 June 2017, 11 July 2017, 12 July 2017, and 11 July 2018. To ascertain the final genuine outliers, a comparison of the occurrence time of abnormal data between PL13-1 and PL13-2 is conducted. Considering the slight lag in the response of monitoring points at different positions on the dam when subjected to environmental factors, the displacement data on 7 September 2015, 12 July 2017, 11 July 2018, and 12 July 2018 for PL13-1, as well as the data on 7 September 2015, 11 July 2017, 12 July 2017, and 11 July 2018 for PL13-2, are determined not to be outliers. Instead, they represent measured data reflecting actual dam displacement changes due to environmental variations and require no further processing.

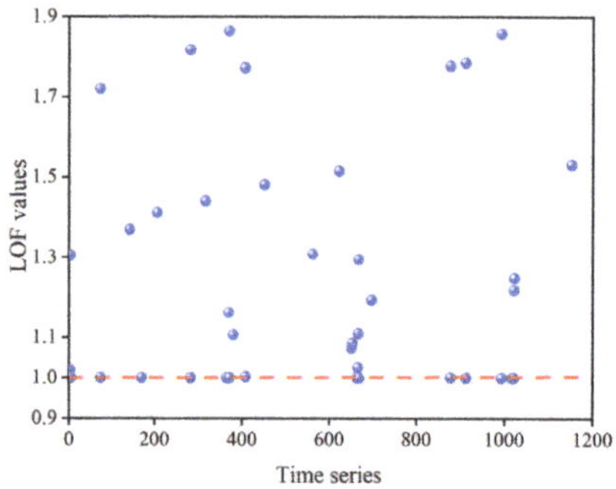

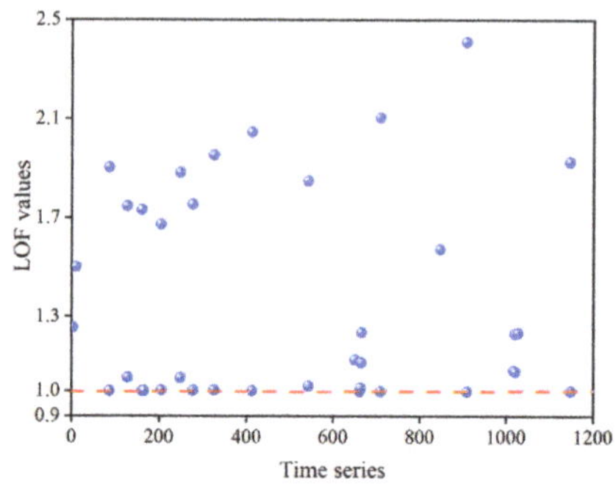

Figure 9. (**a**) LOF values for preliminary gross errors at measurement point PL13-1; (**b**) LOF values for preliminary gross errors at measurement point PL13-2.

Similarly, gross error identification is conducted for monitoring points PL13-3 and PL16-3 in Zone I, as well as PL11-5 and PL16-5 in Zone IV (see Figures A1–A4 in Appendix A). For all monitoring points, the 15 intentionally introduced outlier data points are accurately identified. Furthermore, at PL13-3, data points on 27 June 2017, 28 June 2017, and 11 July 2017 are identified as gross errors. Similarly, at PL16-3, data points on 6 September 2015, 28 June 2017, and 11 July 2017 are flagged as gross errors. However, through comparative analysis, it is determined that the three data points for PL13-3, as well as the data points on 28 June 2017, and 11 July 2017, for PL16-3, are not gross errors. For Zone IV, outlier identification is performed on monitoring points PL11-5 and PL16-5. After the comparison of gross error positions, three points marked as gross error in each of the two measuring points are subsequently identified as normal points.

The above analysis reveals that the FCM-OPTICS-LOF gross error identification method can accurately detect intentionally introduced anomalous data. By employing a comparative approach with similar monitoring points, the anomalous data can be effectively categorized into gross errors and points reflecting the dam's operational state, thereby minimizing misclassifications. This validates the effectiveness and accuracy of the proposed gross error identification method in this study.

4.4. Comparison Analysis with Other Identification Methods

To prove the superiority of the FCM-OPTICS-LOF gross error identification method proposed in this paper, we compare and analyze the gross error identification effects of FCM-LOF, FCM-DBSCAN, and the method proposed in this paper. The *correct rate* is used as the evaluation index:

$$correct\ rate = \frac{T}{N} \tag{12}$$

where N represents the number of gross errors identified by the algorithm; T is the number of gross errors intentionally added.

The detection results of the FCM-LOF and FCM-DBSCAN algorithms for monitoring points PL13-1 and PL13-2 in Zone II are illustrated in Figures 10 and 11, respectively. Using the FCM-LOF algorithm, LOF values are computed for each data point. Based on the distribution, data points with LOF values exceeding 1.2 for PL13-1 and PL13-2 are identified as gross errors. For PL13-1, 25 gross errors are identified using the FCM-LOF algorithm, of which 14 are intentionally added as anomalous data points. Through comparative analysis between the two monitoring points, four points are not gross errors, resulting in an identification accuracy of 66.67%. For PL13-2, in addition to the 15 intentionally added anomalous data points, seven additional data points are identified as outliers. Upon comparison, there are 4 points that reflect the dam's operational state, yielding an identification accuracy of 83.33%.

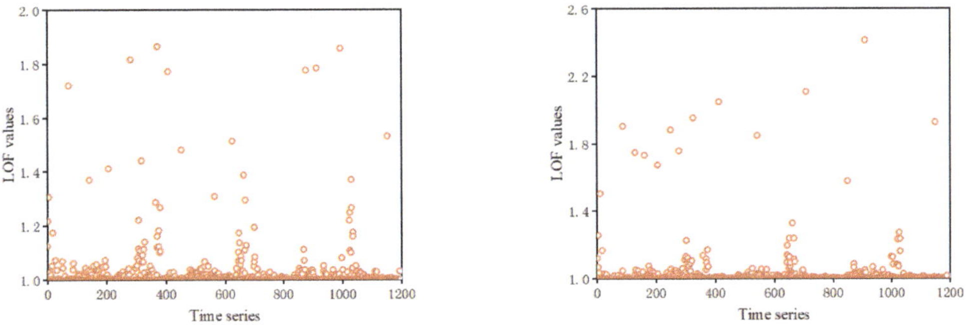

(a) (b)

Figure 10. (**a**) LOF values for measurement point PL13-1; (**b**) LOF values for measurement point PL13-2.

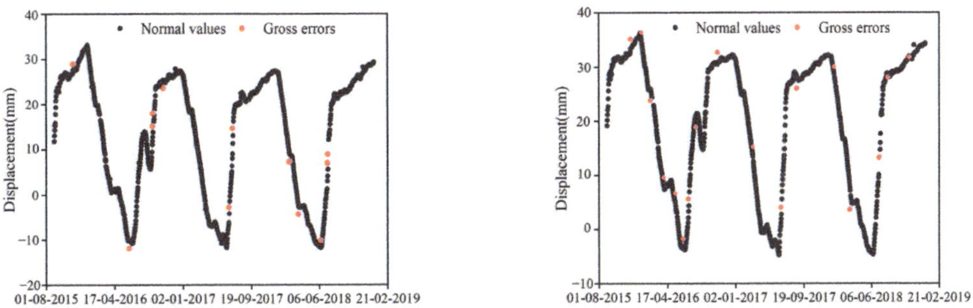

(a) (b)

Figure 11. (**a**) Gross error identification results for measurement point PL13-1 using FCM-DBSCAN algorithm; (**b**) Gross error identification results for measurement point PL13-2 using FCM-DBSCAN algorithm.

Using the FCM-DBSCAN algorithm, 19 data points for PL13-1 are classified as gross errors, with seven being intentionally added as anomalous values. Upon comparison with PL13-2, seven of these data points are not gross errors, resulting in an identification accuracy of 58.33%. For PL13-2, 24 data points are classified as gross errors, with 14 being intentionally added as anomalous values. Seven of these data points are identified as data reflecting the dam's operational state, resulting in an identification accuracy of 82.35%.

The same method is applied to the remaining four monitoring points for gross error detection. The specific identification results are detailed in Table 2 and Comparative Figure 12.

Table 2 lists the correct rate of gross error detection for each monitoring point based on the three methods. The average values of the correct rate are 77.70%, 70.21%, and 95.83% based on FCM-LOF, FCM-DBSCAN, and the proposed method, respectively. Specifically, the correct rate of all six monitoring points is between 66.67% (PL13-1) and 83.33% (PL13-2 and PL11-5) based on the FCM-LOF method, between 53.33% (PL11-5) and 82.35% (PL13-2) based on the FCM-DBSCAN method, and between 93.75% (PL13-1, PL13-2, PL16-3, and PL11-5) and 100% (PL13-3 and PL16-5) based on the proposed method. It is evident that the FCM-LOF and FCM-DBSCAN algorithms exhibit a notable occurrence of omission or misjudgment when identifying gross errors. In comparison, this study combines OPTICS with the LOF algorithm and employs secondary discrimination based on the time of occurrence of outliers at similar monitoring points. This approach effectively categorizes anomalous data into gross errors and points reflecting the dam's operational state, accurately eliminating genuine gross errors. This indicates that the FCM-OPTICS-LOF algorithm possesses the strongest gross error identification capabilities. The proposed method can be applied to the gross error detection of dam monitoring data. In this way, real and reliable monitoring data representing dam health status can be obtained to provide more accurate guidance for dam management and maintenance.

Table 2. The correct rate of different methods in identifying gross errors.

Monitoring Points	Gross Error Identification Method		
	FCM-OPTICS-LOF	FCM-LOF	FCM-DBSCAN
PL13-1	93.75%	66.67%	58.33%
PL13-2	93.75%	83.33%	82.35%
PL13-3	100%	78.94%	75%
PL16-3	93.75%	75%	78.94%
PL11-5	93.75%	83.33%	53.33%
PL16-5	100%	78.94%	73.33%

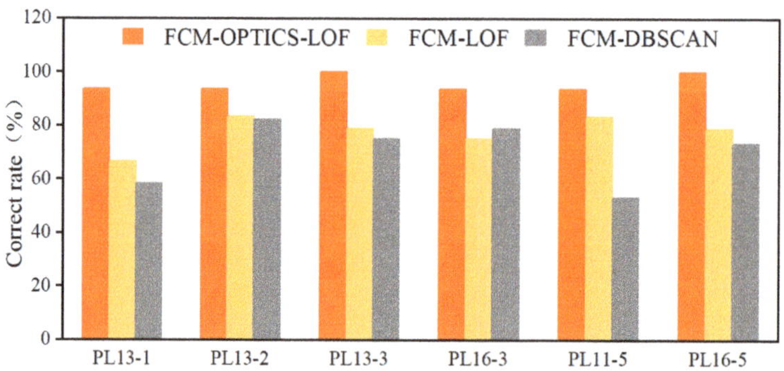

Figure 12. The correct rate of gross error identification using different methods.

5. Conclusions

To address the issue of gross error identification in the safety monitoring data of dams and further enhance accuracy while reducing misjudgments, this paper considers leveraging the deformation similarity between measurement points to distinguish gross errors from genuine data influenced by environmental factors. We propose a gross error identification method based on FCM deformation zoning, combining the OPTICS algorithm with the LOF algorithm. We validate the proposed method using measured displacement data from a concrete arch dam as an example and draw the following conclusions:

(1) The FCM-OPTICS-LOF gross error identification method proposed in this study effectively identifies all gross errors within the dataset. And it will not misjudge abrupt data fluctuations induced by changes in environmental variables as gross errors. This presents a reliable and effective new approach for gross error identification in dam deformation monitoring data.

(2) The proposed FCM-OPTICS-LOF method for gross error identification demonstrates higher accuracy compared to both the standalone LOF method and the widely em-

ployed DBSCAN method, belonging to the same density clustering algorithms. This affirms the superiority of the proposed method. Furthermore, the method's ability to identify gross errors by comparing different measurement points allows simultaneous recognition of two or more points, significantly enhancing the efficiency of gross error detection.

(3) Although the proposed gross error identification method in this paper achieves accurate identification, the threshold selection of the LOF value needs to be manually judged and determined for the specific situation. How to adaptively select the appropriate threshold needs to be further studied and explored.

Author Contributions: Conceptualization, L.C., C.G. and S.Z.; methodology, L.C., C.G. and Y.W.; software, L.C., C.G. and S.Z.; validation, L.C., C.G. and Y.W.; formal analysis, L.C. and C.G.; investigation, C.G.; resources, S.Z.; data curation, L.C. and S.Z.; writing—original draft preparation, L.C. and S.Z.; writing—review and editing, C.G. and Y.W.; supervision, C.G.; project administration, C.G. and S.Z.; funding acquisition, C.G. All authors have read and agreed to the published version of the manuscript.

Funding: This research was funded by the National Natural Science Foundation of China (Grant Nos. U2243223 and 52209159), the Water Conservancy Science and Technology Project of Jiangsu (Grant No. 2022024), the Fundamental Research Funds for the Central Universities (Grant No. B230201011), the Jiangsu Young Science and Technological Talents Support Project (Grant No. TJ-2022-076), and the China Postdoctoral Science Foundation (2023M730934).

Data Availability Statement: Data are contained within the article.

Conflicts of Interest: The authors declare no conflicts of interests.

Appendix A

Table A1 shows the positions and magnitude of gross errors added to PL13-3 and PL16-3. Table A2 shows the positions and magnitude of gross errors added to PL11-5 and PL16-5. Figures A1 and A2 display the reachable distance plots and LOF scatter plots for PL13-3 and PL16-3 in Zone I. Figures A3 and A4 show the reachable distance plots and LOF scatter plots for PL11-5 and PL16-5 in Zone IV.

Table A1. Locations and magnitudes of gross errors introduced in monitoring points PL13-3 and PL16-3.

PL13-3				PL16-3			
Data	**Raw Data/mm**	**Gross Error Size/mm**	**the Data after Adding the Gross Error/mm**	**Data**	**Raw Data/mm**	**Gross Error Size/mm**	**the Data after Adding the Gross Error/mm**
6 October 2015	38.87	−1.56	37.31	5 November 2015	34.09	1.86	35.95
24 November 2015	39.32	−1.78	37.54	20 December 2015	36.35	1.73	38.08
19 January 2016	38.72	1.45	40.17	25 February 2016	30.16	−1.95	28.21
22 March 2016	22.72	2.3	25.02	23 April 2016	21.75	−1.6	20.15
30 May 2016	13.11	1.89	15	29 July 2016	28.71	−2	26.71
22 August 2016	28	1.68	29.68	18 October 2016	34.27	1.59	35.86
7 November 2016	38.52	−1.86	36.66	19 January 2017	32.23	−1.83	30.4
14 February 2017	31.66	1.96	33.62	21 April 2017	15.64	2.56	18.2
16 May 2017	10.84	1.82	12.66	18 July 2017	33.07	1.68	34.75
14 October 2017	37.22	−1.65	35.57	26 October 2017	34.5	−3	31.5
8 January 2018	37.73	−1.76	35.97	25 December 2017	35.14	1.96	37.1
30 March 2018	17.61	2.89	20.5	2 March 2018	27.02	1.82	28.84
13 July 2018	31.9	1.9	33.8	15 May 2018	17.09	−1.49	15.6
27 September 2018	37.72	2.8	40.52	3 August 2018	35.27	2	37.27
13 December 2018	40.77	1.53	42.3	6 November 2018	36.55	1.67	38.22

Table A2. Locations and magnitudes of gross errors introduced in monitoring points PL11-5 and PL16-5.

PL11-5				PL16-5			
Data	**Raw Data/mm**	**Gross Error Size/mm**	**Data after Adding the Gross Error/mm**	**Data**	**Raw Data/mm**	**Gross Error Size/mm**	**Data after Adding the Gross Error/mm**
22 September 2015	30.3	−1.53	28.77	30 September 2015	30.81	1.71	32.52
23 January 2016	30.2	1.69	31.89	24 November 2015	31.18	−1.79	29.39
9 April 2016	24.67	1.98	26.65	29 January 2016	30.65	1.84	32.49
25 June 2016	21.72	−1.76	19.96	9 April 2016	25.86	1.88	27.74
7 September 2016	28.65	2.8	31.45	17 June 2016	23.11	−1.92	21.19
22 December 2016	32.41	−1.64	30.77	28 July 2016	29.32	1.77	31.09
26 February 2017	28.24	1.58	29.82	13 September 2016	30.11	−2.4	27.71
20 May 2017	21.99	1.65	23.64	21 October 2016	31.41	1.92	33.33
14 August 2017	31.54	1.73	33.27	25 December 2016	31.47	−1.85	29.62
23 October 2017	31.74	−1.82	29.92	10 March 2017	27.61	1.58	29.19
10 January 2018	31.29	−1.64	29.65	29 June 2017	27.53	−2.08	25.45
5 March 2018	27.23	2.2	29.43	16 December 2017	32.14	1.7	33.84
16 June 2018	23.03	−1.74	21.29	7 March 2018	27.95	−1.55	26.4
3 September 2018	31.84	1.84	33.68	13 June 2018	23.84	1.72	25.56
5 November 2018	31.99	−2.3	29.69	12 September 2018	32.36	−1.68	30.68

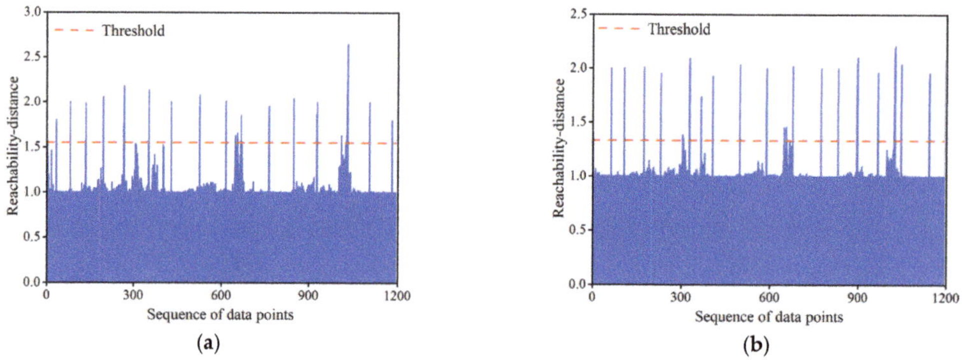

Figure A1. (**a**) Reachability distance for measurement point PL13-3; (**b**) Reachability distance for measurement point PL16-3.

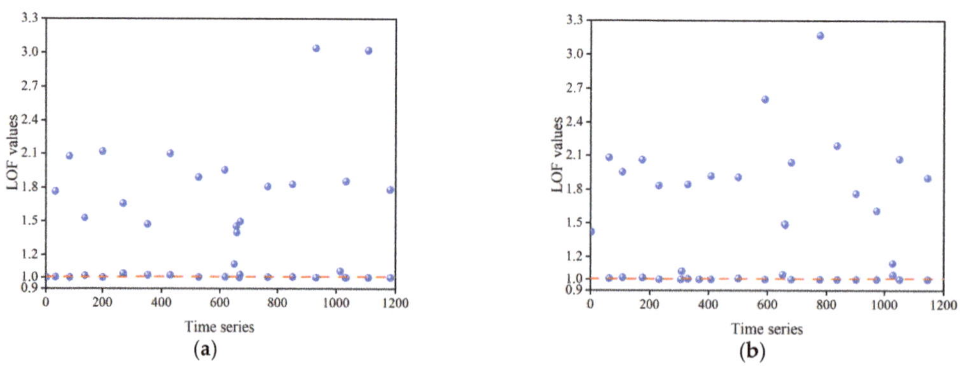

Figure A2. (**a**) LOF values for preliminary gross errors at measurement point PL13-3; (**b**) LOF values for preliminary gross errors at measurement point PL16-3.

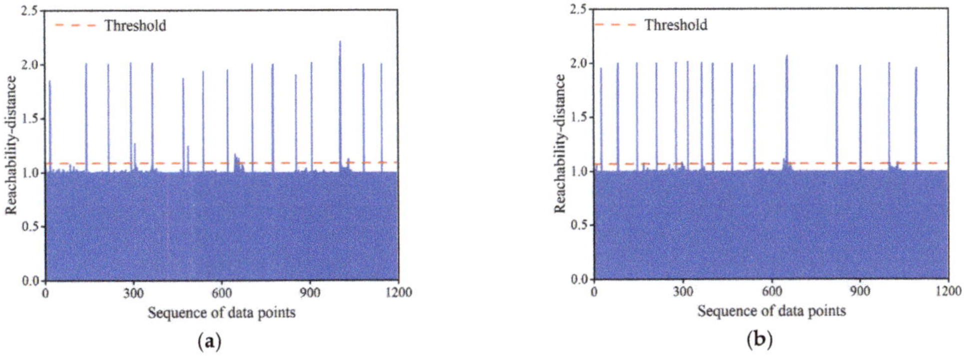

Figure A3. (**a**) Reachability distance for measurement point PL11-5; (**b**) Reachability distance for measurement point PL16-5.

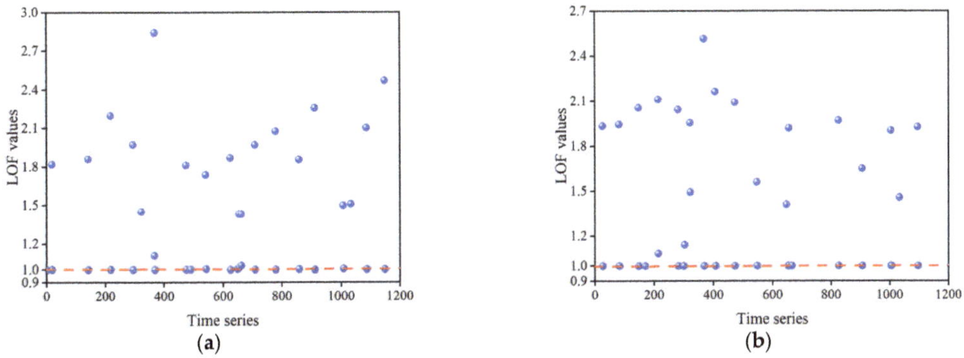

Figure A4. (**a**) LOF values for preliminary gross errors at measurement point PL11-5; (**b**) LOF values for preliminary gross errors at measurement point PL16-5.

References

1. Milillo, P.; Perissin, D.; Salzer, J.T.; Lundgren, P.; Lacava, G.; Milillo, G.; Serio, C. Monitoring dam structural health from space: Insights from novel InSAR techniques and multi-parametric modeling applied to the Pertusillo dam Basilicata, Italy. *Int. J. Appl. Earth Obs. Geoinf.* **2016**, *52*, 221–229. [CrossRef]
2. Gu, C.; Su, H.; Wang, S. Advances in calculation models and monitoring methods for long-term deformation behavior of concrete dams. *J. Hydroelectr. Eng.* **2016**, *35*, 1–14.
3. Yuan, D.; Gu, C.; Gu, H. Displacement behavior analysis and prediction model of concrete gravity dams in cold region. *J. Hydraul. Eng.* **2022**, *53*, 733–746.
4. Wu, Z. *Safety Monitoring Theory and Its Applications in Hydraulic Structures*; Higher Education Press: Beijing, China, 2003.
5. Li, Z.; Hou, H. Dam safety monitoring indices based on motion stability theory. *Eng. J. Wuhan Univ. Eng. Ed.* **2010**, *43*, 581.
6. Zhang, C.; Zhou, X.; Gao, C.; Wang, C. On improving the precision of localization with gross error removal. In Proceedings of the 28th International Conference on Distributed Computing Systems Workshops, Beijing, China, 17–20 June 2008; pp. 144–149.
7. Hu, Y.T.; Shao, C.F.; Gu, C.S.; Meng, Z.Z. Concrete Dam Displacement Prediction Based on an ISODATA-GMM Clustering and Random Coefficient Model. *Water* **2019**, *11*, 714. [CrossRef]
8. Wang, X.; Wang, Z.-Y.; Wen, K.-L.; Chen, H.-C. The development of Matlab toolbox for gross error in measurement. In Proceedings of the 2011 International Conference on System Science and Engineering, Macau, China, 8–10 June 2011; pp. 461–466.
9. Xiong, Z.; Cui, C.; Pei, D. Research on the Processing of the Gross Error of Shock Wave Overpressure Value. *J. Ordnance Equip. Eng.* **2021**, *42*, 94–97.
10. Ge, L.Y.; Wang, Z.Y. Novel uncertainty-evaluation method of virtual instrument small sample size. *J. Test. Eval.* **2008**, *36*, 273–279.
11. Zhao, Z.P.; Chen, K.J.; Zhang, H.; Li, Y.; Wu, Z. The method of gross error identification of dam monitoring data based on robust estimation. *J. Water Resour. Power* **2018**, *36*, 68–71.
12. Li, X.; Li, Y.L.; Zhang, P.; Yang, Z. Research on an improved Pauta criterion based on M-estimation for gross error identification of monitoring data and its application. *China Rural Water Hydropower* **2019**, *8*, 133–136.

13. Qu, X.D.; Yang, J.; Chang, M. A Deep Learning Model for Concrete Dam Deformation Prediction Based on RS-LSTM. *J. Sens* **2019**, *2019*, 4581672. [CrossRef]
14. Al-Samahi, S.S.A.; Ho, K.C.; Islam, N.E. Improving Elliptic/Hyperbolic Localization Under Multipath Environment Using Neura Network for Outlier Detection. In Proceedings of the IEEE Conference on Computer Communications (IEEE INFOCOM), Paris France, 29 April–2 May 2019; pp. 933–938.
15. Bourquin, J.; Schmidli, H.; van Hoogevest, P.; Leuenberger, H. Pitfalls of artificial neural networks (ANN) modelling technique for data sets containing outlier measurements using a study on mixture properties of a direct compressed dosage form. *Eur. J Pharm. Sci.* **1998**, *7*, 17–28. [CrossRef] [PubMed]
16. Chakravarty, S.; Demirhan, H.; Baser, F. Fuzzy regression functions with a noise cluster and the impact of outliers on mainstream machine learning methods in the regression setting. *Appl. Soft Comput.* **2020**, *96*, 106535. [CrossRef]
17. Miao, Y.; Su, H.Y.; Xu, O.G.; Chu, J. Support Vector Regression Approach for Simultaneous Data Reconciliation and Gross Error or Outlier Detection. *Ind. Eng. Chem. Res.* **2009**, *48*, 10903–10911. [CrossRef]
18. Ramachandran, V.; Kishorebabu, V. A Tri- State Filter for the Removal of Salt and Pepper Noise in Mammogram Images. *J. Med Syst.* **2019**, *43*, 40. [CrossRef] [PubMed]
19. Mohandoss, D.P.; Shi, Y.; Suo, K. Outlier Prediction Using Random Forest Classifier. In Proceedings of the IEEE 11th Annua Computing and Communication Workshop and Conference (CCWC), Las Vegas, NV, USA, 27–30 January 2021; pp. 27–33.
20. Song, J.T.; Chen, Y.C.; Yang, J. A Novel Outlier Detection Method of Long-Term Dam Monitoring Data Based on SSA-NAR. *Wirel Commun. Mob. Comput.* **2022**, *2022*, 6569367. [CrossRef]
21. Qi, Z.Y.; Sun, F.T.; Mao, Y.P.; Zhou, J.B.; Zhang, C.H.; Li, Q.Y. Research on Gross Error Detecting Method of Monitored Dam Deformation Data Based on Fully Convolutional Networks. *Water Resour. Power* **2023**, *41*, 87–90.
22. Wang, L.; Zheng, D. Anomaly Identification of Dam Safety Monitoring Data Based on Convolutional Neural Network. *J. Yangtze River Sci. Res. Inst.* **2021**, *38*, 72–77.
23. Hu, J.; Ma, F.H.; Wu, S.H. Anomaly identification of foundation uplift pressures of gravity dams based on DTW and LOF. *Struct Control Health Monit.* **2018**, *25*, e2153. [CrossRef]
24. Song, J.T.; Zhang, S.F.; Tong, F.; Yang, J.; Zeng, Z.Q.; Yuan, S. Outlier Detection Based on Multivariable Panel Data and K-Means Clustering for Dam Deformation Monitoring Data. *Adv. Civ. Eng.* **2021**, *2021*, 3739551. [CrossRef]
25. Shao, C.F.; Zheng, S.; Gu, C.S.; Hu, Y.T.; Qin, X.N. A novel outlier detection method for monitoring data in dam engineering *Expert Syst. Appl.* **2022**, *193*, 116476. [CrossRef]
26. Bao, Y.Q.; Tang, Z.Y.; Li, H.; Zhang, Y.F. Computer vision and deep learning-based data anomaly detection method for structural health monitoring. *Struct. Health Monit. Int. J.* **2019**, *18*, 401–421. [CrossRef]
27. Harrou, F.; Zeroual, A.; Hittawe, M.M.; Sun, Y. Chapter 6—Recurrent and convolutional neural networks for traffic management In *Road Traffic Modeling and Management*; Harrou, F., Zeroual, A., Hittawe, M.M., Sun, Y., Eds.; Elsevier: Amsterdam, The Netherlands, 2022; pp. 197–246.
28. Harrou, F.; Zeroual, A.; Hittawe, M.M.; Sun, Y. *Road Traffic Modeling and Management: Using Statistical Monitoring and Deep Learning* Elsevier: Amsterdam, The Netherlands, 2022.
29. Hittawe, M.M.; Langodan, S.; Beya, O.; Hoteit, I.; Knio, O. Efficient SST prediction in the Red Sea using hybrid deep learning based approach. In Proceedings of the 20th IEEE International Conference on Industrial Informatics (INDIN), Perth, Australia 25–28 July 2022; pp. 107–114.
30. Zhang, J.; Xie, J.M.; Kou, P.G. Abnormal Diagnosis of Dam Safety Monitoring Data Based on Ensemble Learning. In Proceedings of the 3rd International Workshop on Renewable Energy and Development (IWRED), Guangzhou, China, 8–10 March 2019.
31. Gu, C.S.; Wang, Y.B.; Gu, H.; Hu, Y.T.; Yang, M.; Cao, W.H.; Fang, Z. A Combined Safety Monitoring Model for High Concrete Dams. *Appl. Sci.* **2022**, *12*, 12103. [CrossRef]
32. Li, N.; Li, P.; Shi, X.L.; Yan, K.; Ren, W.P. Outlier Identify Based on BP Neural Network in Dam Safety Monitoring. In Proceedings of the 2nd International Asia Conference on Informatics in Control, Automation and Robotics (CAR), Wuhan, China, 6–7 March 2010; pp. 210–214.
33. Liu, J.; Lian, J.J. Outliers Detection of Dam Displacement Monitoring Data Based on Wavelet Transform. In Proceedings of the International Conference on Green Building, Materials and Civil Engineering (GBMCE 2011), Shangri La, China, 22–23 August 2011; pp. 4590–4595.
34. Li, H.J.; Li, J.J.; Kang, F. Risk analysis of dam based on artificial bee colony algorithm with fuzzy c-means clustering. *Can. J. Civ Eng.* **2011**, *38*, 483–492. [CrossRef]
35. Yang, C.; Bao, T. Dam Deformation Prediction Model Based on FCM-XGBoost. *J. Yangtze River Sci. Res. Inst.* **2021**, *38*, 66–71.
36. Liu, W.; Chen, B.; Ge, P.; Zhang, X. Deformation prediction model of a high arch dam based on clustering and MO-LSSVR. *Adv Sci. Technol. Water Resour.* **2023**, *43*, 102–108.
37. Li, Z.K.; Li, W.; Ge, W. Weight analysis of influencing factors of dam break risk consequences. *Nat. Hazards Earth Syst. Sci.* **2018**, *18*, 3355–3362. [CrossRef]
38. Kumar, K.M.; Reddy, A.R.M. A fast DBSCAN clustering algorithm by accelerating neighbor searching using Groups method *Pattern Recognit.* **2016**, *58*, 39–48. [CrossRef]

39. Kalita, H.K.; Bhattacharyya, D.K.; Kar, A. A new algorithm for Ordering of Points To Identify Clustering Structure Based On Perimeter of Triangle: OPTICS (BOPT). In Proceedings of the 15th International Conference on Advanced Computing and Communications, Indian Inst Technol Guwahati, Guwahati, India, 18–21 December 2007; pp. 523–528.
40. Deng, Z.; Hu, Y.Y.; Zhu, M.; Huang, X.H.; Du, B. A scalable and fast OPTICS for clustering trajectory big data. *Clust. Comput. J. Netw. Softw. Tools Appl.* **2015**, *18*, 549–562. [CrossRef]
41. Xiao, X.; Xue, S. An outlier detection algorithm based on improved OPTICS clustering and LOPW. *Comput. Eng. Sci.* **2019**, *41*, 885–892.

water MDPI

Article

The Forecast of Streamflow through Göksu Stream Using Machine Learning and Statistical Methods

Mirac Nur Ciner [1], Mustafa Güler [2,*], Ersin Namlı [3], Mesut Samastı [4], Mesut Ulu [5], İsmail Bilal Peker [6] and Sezar Gülbaz [6]

[1] Engineering Faculty, Department of Environmental Engineering, Istanbul University-Cerrahpaşa, Avcılar, 34320 Istanbul, Türkiye; mirac.ciner@iuc.edu.tr
[2] Engineering Faculty, Department of Engineering Sciences, Istanbul University-Cerrahpaşa, Avcılar, 34320 Istanbul, Türkiye
[3] Engineering Faculty, Department of Industrial Engineering, Istanbul University-Cerrahpaşa, Avcılar, 34320 Istanbul, Türkiye; enamli@iuc.edu.tr
[4] Turkish Institute of Management Science, TUBITAK Tusside Campus Gebze, 41400 Kocaeli, Türkiye; mesutsamasti@gmail.com
[5] Occupational Health and Safety Department, Bandırma Onyedi Eylül University, Bandırma, 10250 Balıkesir, Türkiye; mulu@bandirma.edu.tr
[6] Engineering Faculty, Department of Civil Engineering, Istanbul University-Cerrahpaşa, Avcılar, 34320 Istanbul, Türkiye; pekerbilal@iuc.edu.tr (İ.B.P.); sezarg@iuc.edu.tr (S.G.)
* Correspondence: m.guler@iuc.edu.tr

Abstract: Forecasting streamflow in stream basin systems plays a crucial role in facilitating effective urban planning to mitigate floods. In addition to employing intricate hydrological modeling systems, machine learning and statistical techniques offer an alternative means for streamflow forecasts. Nonetheless, the precision and dependability of these methods are subjects of paramount importance. This study rigorously investigates the effectiveness of three distinct machine learning techniques and two statistical approaches when applied to streamflow data from the Göksu Stream in the Marmara Region of Turkey, spanning from 1984 to 2022. Through a comparative analysis of these methodologies, this examination aims to contribute innovative advancements to the existing methodologies used in the prediction of streamflow data. The methodologies employed include machine learning methods such as Extreme Gradient Boosting (XGBoost), Random Forest (RF), and Support Vector Machine (SVM) and statistical methods such as Simple Exponential Smoothing (SES) and Autoregressive Integrated Moving Average (ARIMA) model. In the study, 444 data points between 1984 and 2020 were used as training data, and the remaining data points for the period 2021–2022 were used for streamflow forecasting in the test validation period. The results were evaluated using various metrics, such as the correlation coefficient (r), mean absolute error (MAE), root mean square error (RMSE), mean absolute percentage error (MAPE), coefficient of determination (R^2), and Nash–Sutcliffe efficiency (NSE). Upon analyzing the results, it was found that the model generated using the XGBoost algorithm outperformed other machine learning and statistical techniques. Consequently, the models implemented in this study demonstrate a high level of accuracy in predicting potential streamflow in the river basin system.

Keywords: streamflow forecast; machine learning; statistical techniques; Göksu Stream

1. Introduction

Accurate and reliable streamflow forecasting plays a vital role in many hydrological and environmental studies such as flood risk assessment, hydroelectric energy, irrigation planning, timely and effective water resource management [1–6]. It is constantly emphasized in the literature that streamflow forecasting requires continuous research and progress due to its impact on the river basin system [7]. Streamflow forecasting is extremely important in making decisions about a project and in retrospectively completing the data of a newly established observation station or one that has not been able to make measurements for a certain period of time [8]. In addition, streamflow forecasting plays a crucial role in influencing drought mitigation strategies and decision-making for water managers and policy makers [9].

The complexity, non-linearity and non-stationarity of the process render streamflow forecasting very challenging. However, it is difficult to predict both in the short and long-term due to the variability in spatial and temporal domains [10,11]. The challenges encountered in accurately forecasting streamflow have made it an interesting research area among hydrologists [12]. Especially in recent years, artificial intelligence (AI) techniques have been increasingly employed alongside hydrological models for streamflow forecast [13]. Moreover, accurate and continuous data forecasting is critical for effective decision-making in flood mitigation activities [14].

Some artificial intelligence and statistical models that can be used for streamflow forecast have appeared in the literature in recent years [15–25]. Artificial neural networks (ANN) are often favored in numerous studies due to their non-empirical nature and their ability to achieve high accuracy rates [19,20,24]. In this context, streamflow forecasting is often favored because it does not necessitate field surveys or physical assessments, unlike other hydrological models. It can also accurately account for non-linear processes such as temporal changes in streamflow [26]. Machine learning algorithms such as Decision Tree (DT) and Support Vector Machine (SVM) have been successfully used in both monthly [27–29] and daily streamflow forecasting [30–32]. Various studies in the literature show that machine learning methods have shown that they can numerically capture nonlinear processes without knowledge of the underlying physical processes [3,33]. Among these machine learning methods, ANNs stand out as a self-learning function approximation tool in modeling non-linear hydrological data [10]. By identifying non-linear relationships between inputs and outputs, ANNs can reproduce strong nonlinear relationships well, especially in cases where these relationships are unknown or cannot be explained in advance [34]. Therefore, ANNs provide advantages over other methods due to greater accuracy, reduced testing time and faster implementation [35].

Streamflow modeling in hydrological studies can be developed by statistical methods. Forecasting studies using the internal dependency between time series values in hydrological models are within the scope of autoregressive modeling. Time series values to be used in these modeling studies can be daily, weekly, monthly, seasonal and annual. Generally, natural events do not have stationary model features. In order for the series to be modeled using autoregressive models, they must have a normal distribution and stationary structure. This comes in three forms: linear stationary stochastic models, autoregressive, moving average and autoregressive moving average model [36]. Moreover, many models containing time series such as wind speed, precipitation, evaporation, and flowrate can be modeled with statistical methods. Simple exponential smoothing techniques (SES), renowned for their simplicity and effectiveness, find extensive application in forecasting and are similarly employed in the forecast of rainfall [37]. This is because SES facilitates the production of accurate and short-term forecasts with robust forecasting capabilities on time series [38]. However, the Autoregressive Integrated Moving Average (ARIMA) method is a widely used method for time series analysis in hydrology and streamflow estimation studies. This model utilizes the structural properties of historical time series to forecast future values. Usually, in hydrological data, variables such as rainfall, runoff or water depth are represented as time series and the ARIMA model can be used to analyze these data.

The present study examines the applicability of three different machine learning methodologies and two statistical approaches in analyzing streamflow data from the Göksu Stream in the Marmara Region of Türkiye. The data cover the period from 1984 to 2022. Machine learning techniques, including Extreme Gradient Boosting (XGBoost), Random Forest (RF), and Support Vector Machine (SVM) are utilized alongside statistical techniques, including Simple Exponential Smoothing (SES) and the Autoregressive Integrated Moving

Average model (ARIMA). Therefore, the results obtained by the methods used in this study were assessed using various metrics. The significance of this study emanates from its inclusion of future streamflow forecasts for a dam projected to be constructed within the study area. The application of five diverse methodologies demonstrates clear advantages for water resources projects within this basin in the upcoming years. This research will act as a pivotal guide for future investigative efforts, offering a comparative analysis among these methodologies and contributing to the advancement of innovative practices in hydrological forecasting.

2. Materials and Methods

2.1. Data and Site Description

Göksu Stream is a stream originating from the Marmara Region of Türkiye and flowing into the Black Sea. Göksu Basin is located within the borders of Istanbul and Kocaeli provinces in Türkiye. The total basin area is approximately 482 km^2. Göksu Stream flows into the Black Sea from the Ağva neighborhood of Istanbul Province. Ağva is an important touristic coastal town located on the Black Sea coast of Istanbul. That is why the population increases threefold in summer [39]. The region is confronted with a critical scarcity of water resources, a situation exacerbated by the concentration of tourist accommodations, beaches, and a marked population surge during the summer months. There are two artificial lakes and one regulator on Göksu Stream. In addition, strategic planning is in progress for the construction of a dam, an initiative aimed at securing a sustainable water supply for Istanbul in the foreseeable future. This proactive measure underscores a comprehensive approach to addressing the water scarcity challenge, highlighting the innovative strategies adopted to ensure the adequacy of water resources in response to both current needs and future demands. In the scope of this research, the study area was defined to encompass the Göksu Basin outlet, which is represented by flow station named D02A004 within the basin's stream network. Figure 1 describes the region determined as the study area. Accordingly, the drainage area at the flow station (D02A004) is about 395 km^2 [40]. Approximate mean elevation is 295 m, varying from 19 m to 647 m. The mainstream length is 43 km with many tributaries feeding the river. The flow data for station D02A004 were obtained from the head office of the General Directorate of State Hydraulic Works (DSİ) in Istanbul, Turkey. Monthly streamflow data covers the period 1984–2022.

Figure 2 shows that monthly average flowrates of Göksu Stream between 1984 and 2022. Considering the highest and lowest averages, although 2014 was quite dry in terms of precipitation, the flow rate reached its highest level in the following year with the effect of precipitation. Although the Göksu Basin varies each year that receives the highest rainfall in the December-February period and the lowest rainfall in the June-August period. The amount of precipitation in the winter months is more intense than in the spring. Natural events and precipitation may not be consistent over the same period each year. As a result, various environmental challenges arise. Due to occasional flooding, especially the surrounding agricultural lands and biological ecosystems can be damaged. For these reasons, it is quite important to prevent flooding, mitigate its impacts and implement flood management strategies.

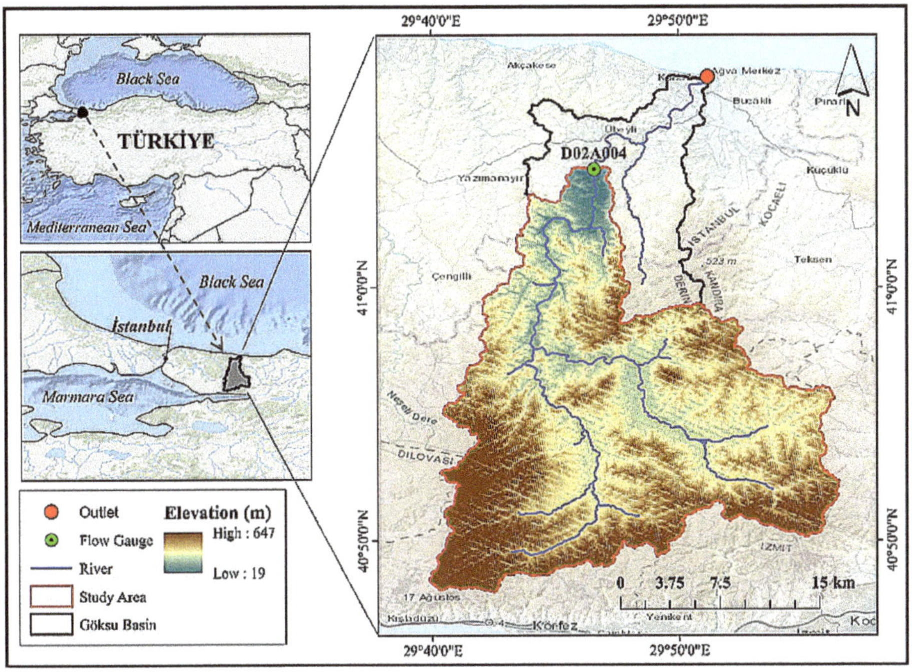

Figure 1. The location, elevation, stream network, and streamflow observation stations within the Göksu Basin.

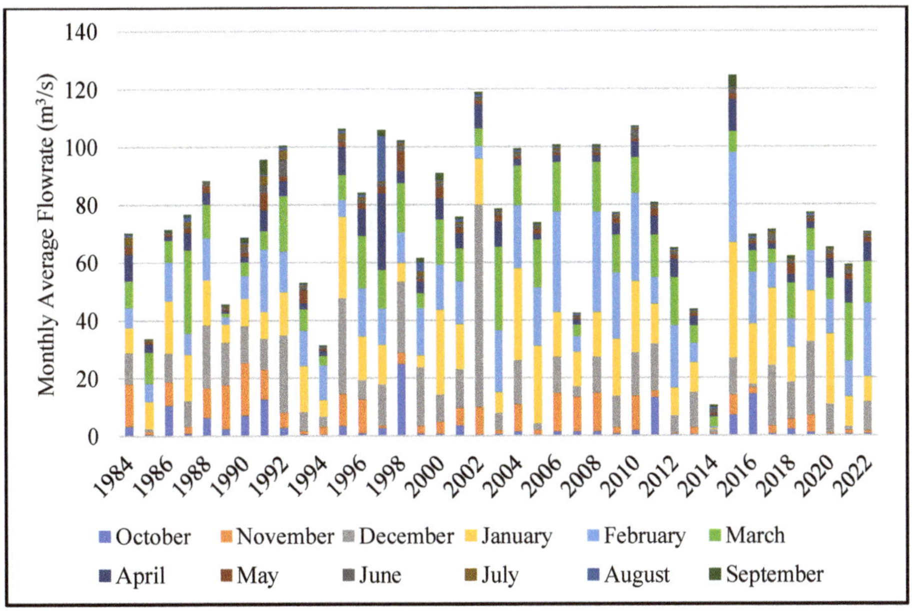

Figure 2. The monthly average flowrate between 1984–2022 obtained from General Directorate of State Hydraulic Works (DSI).

2.2. Machine Learning Methods

The machine learning models used in this study were selected based on their ability to handle the complexity of nonlinear dataset in hydrological systems affected by various environmental factors. Several different machine learning models were tested in the feasibility studies prior to the study. Considering the data structure, XGBoost, RF and SVM algorithms were selected in this study.

2.2.1. Extreme Gradient Boosting (XGBoost)

ANN finds use in various fields, especially classification, modeling and prediction processes. The XGBoost method, one of the improved algorithms of the artificial neural networks set, is an innovative machine learning algorithm whose article was first published by Chen and Guestrin (2016) [41]. The published article was met with interest by data scientists. The XGBoost algorithm is an optimized variant of the Gradient Boosting algorithm. The advantages it provides over previous versions are the most important reason for the widespread use of XGBoost. XGBoost uses the maximum depth value when building the tree. If the created tree shows excessive downward progress, pruning is performed. In this way, overlearning is prevented. While the Gradient Boosting algorithm uses a first-order function to calculate the loss function, XGBoost performs these calculations using second-order functions [41].

XGBoost operates within the boosting framework, an ensemble technique that combines weak learners sequentially to form a strong learner, typically using decision trees [42]. The core of the algorithm is the optimization of a regularized objective function that balances the model's fidelity to the data against its complexity (Equation (1)) [41].

$$L(\theta) = \sum_{i=1}^{n} l(Y_i - \acute{y}_i) + \sum_{k=1}^{K} \Omega\,(f_k) \tag{1}$$

where, $L\ (\theta)$ is the total loss to be minimized, l represents a differentiable convex loss function measuring the discrepancy between the prediction $\acute{y}_i$ and the actual target Y_i, f_k denotes the regularization term, defined as $\Omega, f_k = \gamma\, T + ½\, \lambda\, ||W||^2$, with T representing the tree's number of leaves, W is the scores on leaves, γ is the complexity.

2.2.2. Random Forest (RF)

RF, proposed as a combination of decision trees, is used as an improved version of the bagging method by adding the randomness feature [43]. RF algorithm is an algorithm that helps solve regression problems within the decision tree. It also enables creating a subtree with randomly selected features within the created tree. There is no pruning process for the trees within the algorithm. The random tree algorithm can have a high accuracy rate in evaluations with the large number of random trees created [43,44]. One of the most important advantages of the algorithm is that it solves the overfitting problem in decision trees. The random tree method uses tree type classifier as follows (Equation (2)):

$$\begin{gathered}\{h(x,\ \theta_k)k = 1,\ \ldots\} \\ \Sigma \sum_{j \neq i} (f(Ci\ ,\ T)/|T|)(f(Cj\ ,\ T)/|T|)\end{gathered} \tag{2}$$

The input data is x and the random vector is represented by θ_k [44]. The random tree algorithm uses the GINI index given below to determine the best branch among the available branches [45]. T is the training data set, C_i is the class to which the data belongs, $f\ (C_i, T)/|T|$. It shows the probability that the selected data belongs to the C_i class [46,47]. The random tree algorithm provides better generalization and accurate prediction than other algorithms because it includes random sampling and an improved structure of ensemble algorithms techniques [47].

2.2.3. Support Vector Machine (SVM)

SVM is a machine learning algorithm based on statistical learning theory. This algorithm, which can classify and predict both linear and non-linear data, is frequently used in regression and classification problems. This method creates an optimal separating hyperplane that divides the data into two separate categories using support vectors and class intervals. The main purpose that is optimally separate the data points with a hyperplane. Then, the original training data is transformed using a non-linear mapping in a higher dimension. This transformation ensures that a hyperplane can always be used to separate data into two different classes [48,49]. Mathematically, the SVM decision function for binary classification can be expressed as follows:

$$f(x) = sign + \sum_{i=0}^{N} yi\ \alpha i\ K\ (x,\ xi) + b \tag{3}$$

where, $f(x)$ is the prediction of the example class, x is the example to be predicted, yi is the label of the i-th support vector, αi is the Lagrange multiplier of the i-th support vector, $K(x,xi)$ is the Kernel function (a transformed version of the dot product between x and xi), b is the bias term of the decision function.

The kernel function used here allows SVM to operate in higher dimensional spaces with a technique called kernel trick.

2.3. Statistical Methods

The statistical methods used in this study are methods that aim to estimate the flowrate in a certain time period by using seasonality and time series for the analysis of flow data. These methods have been successfully applied in similar studies in the literature. Although there are many different statistical methods in the literature, it is possible to analyze seasonal and sudden changes in stream with time series. SES and ARIMA methods are preferred in this study thanks to the suitability of the available data structure.

2.3.1. Simple Exponential Smoothing (SES)

SES is a statistical method used to predict future values of a time series. This method aims to predict future values by making heavy use of historical data. SES method includes three basic components that form time series smoothing, smoothed forecasts and future forecasts. These are level, trend and seasonality. SES predicts future values using a combination of these three components. Essentially, the impact of historical data diminishes relative to the past. However, it allows important features such as level, trend and seasonality to be taken into account over time [50]. Here, the parameter values of SES are used as follows in the literature (Equation (4)) [51]:

$$\acute{y}_t = \alpha \cdot \gamma_{t-1} + (1-\alpha) \cdot \acute{y}_{t-1} \tag{4}$$

where, $\acute{y}_t$ is the predicted value at time t, γ_{t-1} is the actual value of the previous time, $\acute{y}_{t-1}$ is the predicted value of the previous time, α is the parameter used between values 0–1.

2.3.2. Autoregressive Integrated Moving Average Model (ARIMA)

ARIMA (p, d, q) Box Jenkins Model proposed by Box and Jenkins is one of the common methods used to create a univariate time series forecasting model [52]. An ARIMA process is a mathematical model used for prediction. Box-Jenkins modeling involves defining an appropriate ARIMA process, fitting it to the data, and then using the appropriate model for prediction. One of the attractive features of the Box Jenkins approach to forecasting is that ARIMA processes include a very rich class of models and generally provide adequate explanation of the data [53]. A non-seasonal ARIMA model is denoted by ARIMA (p, d, q), which is a combination of Autoregressive (AR) and Moving Average (MA) and the order of integration or divergence [54]. In general, the ARIMA model is as follows (Equation (5)):

$$\begin{gathered}\Delta d\ Zt = c + (\varnothing 1\ \Delta d\ Zt - 1 + \ldots + \varnothing p\ \Delta d\ Zt - p) - (\theta 1\ et - 1 + \ldots + \theta q\ et - q) + et \\ \Delta Zt = Zt - Zt - 1 \quad \Delta 2\ Zt - 1 = \Delta Zt - \Delta Zt - 1\end{gathered} \tag{5}$$

where, c is the constant term, θq is the q-th moving average parameter, and $et - k$ is the error term at time tk, $\varnothing p$ is the p-autoregressive parameter, and et is the error term at time t, Δ denotes the difference as shown below. $Zt - 1$ and $Zt - p$ are the values of the past series with delays of 1 and p, respectively.

2.4. Model Performance Metrics

In this study, some performance criteria were used to show the goodness of fit of the models. These are the correlation coefficient (r), mean absolute error (MAE), root mean square error (RMSE), mean absolute percentage error (MAPE), coefficient of determination (R^2), and Nash–Sutcliffe efficiency (NSE).

The correlation coefficient (r) quantifies the degree of closeness of points in a scatter plot to a linear regression line formed by those points (Equation (6)). It varies between −1 and +1. A value of −1 signifies a perfectly linear negative correlation (sloping downward), while a value of +1 signifies a perfectly linear positive correlation (sloping upward). The correlation coefficient, r, is directly related to the coefficient of determination R^2 in an obvious way (Equation (7)). The R^2 value is a measure of how well the regression line approximates the observed data, taking a value between 0 and 1. When $R^2 = 1$, it suggests a perfect fit. And, when $R^2 = 0$, it indicates no discernible relationship between the two variables.

$$r = \frac{n\left(\sum xy\right) - \left(\sum x\right)\left(\sum y\right)}{\sqrt{\left[n\sum x^2 - \left(\sum x\right)^2\right]\left[n\sum y^2 - \left(\sum y\right)^2\right]}} \tag{6}$$

where, n is the number of observations, $\sum xy$ is the Sum of the products of x and y values, $\sum x$ and $\sum y$ are the sum of x and y values, $\sum x^2$ and $\sum y^2$ are the sum of the squares of x and y values.

$$R^2 = 1 - \frac{\Sigma(y_i - \acute{y}_i)^2}{\Sigma(y_i - \overline{y})^2} \tag{7}$$

where, y_i is the actual y values, $\acute{y}_i$ is the predicted y values (output of the regression equation), $\overline{y}$ is the mean of y values.

The MAE is the sum of the absolute value of the differences between the actual values and the predicted values in the data set divided by the number of samples (Equation (8)). The mean absolute error takes values between 0 and ∞. The lower the value, the better the performance.

$$MAE = \frac{1}{n} = \sum_{i=1}^{n} |X_i - X| \tag{8}$$

The RMSE is the sum of the squares of the differences between the actual and predicted values in the data set, divided by the number of samples (Equation (9)). If there are outliers, the mean square error may be high. However, RMSE is calculated by taking the square root of the value found.

$$RMSE = \sqrt{\frac{1}{n}\sum_{i=1}^{n} x_t^2} \tag{9}$$

The MAPE is a statistical measure of the prediction accuracy of a prediction method (Equation (10)). It is also used as a loss function in machine learning problems. MAPE scales the size of the error as a percentage [55].

$$MAPE = \frac{1}{N}\sum_{t=1}^{n}\left|\frac{A_t - F_t}{A_t}\right| \tag{10}$$

where, n is the number of fitted points, A_t is the actual value, F_t is the forecast value.

The model first proposed by Nash and Sutcliffe approached calibrations from a linear regression perspective. Particularly in hydrology, Nash–Sutcliffe efficiency (NSE) is one of the most important measurements used to perform model comparison and validation [56] (Equation (11)).

$$NSE = 1 - \frac{\sum_i (O_i - S_i)^2}{\sum_i (O_i - \overline{O}_i)^2} \quad (11)$$

where, S_i and O_i denote simulations and observations, $\overline{O}_i$ is the observed mean.

3. Results

Forecasting of flowrates was conducted employing machine learning methods XGBoost, RF and SVM, and statistically ARIMA and SES methods, using streamflow data obtained from the Göksu Stream between 1984 and 2022. The dataset of this study was divided into 70% training and 30% testing. Then, the amount of data was divided equally by cross-validation method. Separate success percentages were calculated to determine the performance of each model used separately. The data used in this study was first trained with artificial neural networks and validated with 1984–2020 data. The dataset covering the years 1984 to 2020 was utilized as the training set, while the data from 2021 to 2022 served as the test set for evaluation purposes.

3.1. Machine Learning Methods Results

Figures 3–5 depict the comparative analysis between actual and predicted data for XGBoost, RF and SVM modeling, respectively. Figure 6 illustrates a comparative graphic presenting all ANN model simulation results alongside the actual data.

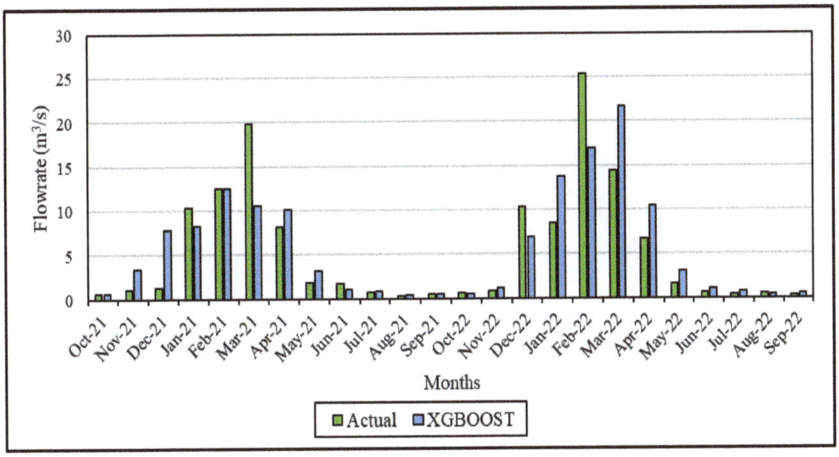

Figure 3. Comparative flow rate graph between actual and predicted data for XGBoost in the modeling period (2021–2022).

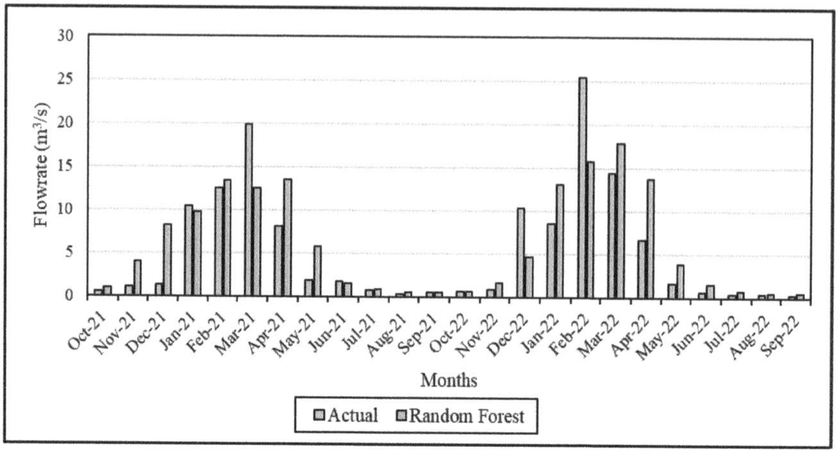

Figure 4. Comparative flow rate analysis between actual and predicted data for RF in the modeling period (2021–2022).

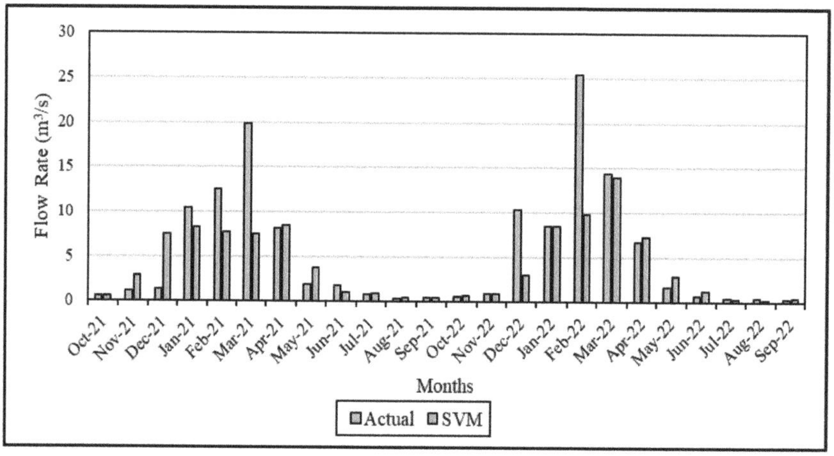

Figure 5. Comparative flow rate graph between actual and predicted data for SVM in the modeling period (2021–2022).

In Figure 3, a notable observation is the highest percentage of error occurring between the actual and predicted values in February and March of 2022. In addition, since the expected precipitation in April is lower than in previous years, there is a difference between the predicted value and the actual value. Similarly, in 2021, the highest difference is again observed in March. The reason for this is that two years in a row, floods occurred in this region as a result of excessive rainfall. The sudden increase in water volume led to anomalies in the observations, a situation evident in the dataset. The difference between the actual and predicted values in November and December in the same year is due to the fact that the region receives rainfall below seasonal norms. The prediction model employed is unable to anticipate such abrupt fluctuations. However, a good agreement was observed between the actual and predicted values in January and February 2021. The margin of error appears to be negligible in this context. While there were substantial differences, particularly in December and March, close estimates were achieved with errors below 5% from June to November. It is noticeable that the calculated error is significantly lower, especially during the summer months or when rainfall amounts are diminished. Therefore, both the data set and the mathematical structure of the XGBoost algorithm directly affect the results.

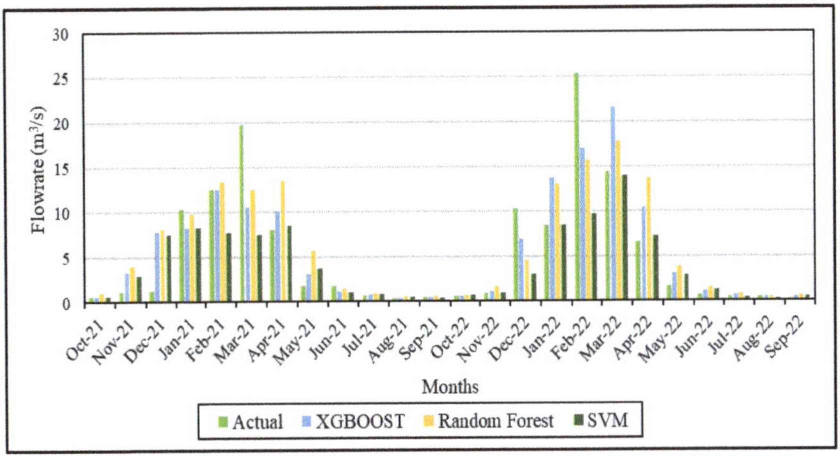

Figure 6. The comparison between actual and predicted data for XGBOOST, RF, SVM in the modeling period (2021–2022).

Figure 4 shows that the RF model shows quite consistent results during the summer months within every two years. However, there is a difference between actual and predicted values due to sudden natural events in November–December and April–May 2021, April 2022. Unusual weather patterns not captured well by the model's training data lead to inaccuracies in predictions. In summer seasons, the margin of error is calculated below 5%. According to the XGBoost model, it is seen that the predictions made in the summer months when precipitation is low and the data are consistent are closer to the reality.

Figure 5, shows that forecasts are far from the actual values in November-December-March 2021 and February 2022, when natural phenomena change rapidly. Like other machine learning algorithms, it is evident that the results obtained in periods with low precipitation are close to reality. Especially in the months with high flow rates, the predicted error rates increase. March-2021 and February-2022 are the months with the highest difference between actual and predicted values. Compared to the XGBoost algorithm, in these two machine learning algorithms, as the amount of streamflow increases, the difference between the actual and predicted values begins to increase. This allows us to observe the performance of each dataset across various models.

In Figure 6, the results of three machine learning models are provided alongside the actual values. For a more precise assessment of the model's performance, it is advisable to utilize metrics such as the R^2 value or other performance indicators based on these test data. The R^2 values calculated for XGBoost, RF, and SVM are 0.72, 0.68, and 0.61, respectively. A R^2 value of 0.72 for XGBoost indicates that the model explains a substantial portion of the variance in the dependent variable. This implies that the model consistently responds to changes in the independent variables associated with the dependent variable, thereby showcasing its effectiveness as a predictive tool in real-world scenarios. R^2 value exceeding 60% signifies that the model effectively explains the independent variables associated with the dependent variable. When the R^2 value is above 60%, it generally implies that the model makes forecasts better than the mean value and that the flow rate forecasts are reliable.

The highest values for the correlation coefficients of the estimated data are 0.845, 0.825, and 0.778 for XGBOOST, RF, and SVM, respectively. The correlation coefficient seriously affects the accuracy of the forecasts. Therefore, the correlation scale is always taken into consideration in experimental studies that require measurement. Hence, the closer the forecasts of the model are to the actual data, the more reliable the model is for these data.

The highest MAPE percentage RF value found for the predicted 2021 and 2022 values is 87.08%. MAPE is used to determine the degree to which a model or forecast is correct or incorrect. It is used as an indicator of how accurately a model can predict future forecasts. It is therefore often used as a measure of accuracy. The results of other evaluation criteria are given in Table 1.

The performance metrics, as shown in Table 1, unequivocally demonstrate how well the XGBoost model outperforms rival forecasting models on a number of important fronts. The robust association between the predicted and actual values is demonstrated by the XGBoost model, which has a correlation coefficient of 0.845. It's remarkable accuracy in forecasting values is demonstrated by its lowest MAE of 2.294 and lowest RMSE of 3.664. With a MAPE of 62.28%, the XGBOOST model likewise obtains the lowest, demonstrating its capacity to produce forecasts with little percentage difference from the actual data. Moreover, XGBoost has the maximum score of 0.711 for the NSE, which measures predictive capability.

3.2. *Statistical Methods Result*

Figures 7 and 8 show the data for the actual and predicted values for 2021–2022 generated by statistical methods. When Figure 7 is analyzed, it is evident that the largest disparity between the actual and predicted values occurs in March 2021 and February 2022. In November-December 2021 and January–February 2022, there is an inconsistency in the predicted values as a result of the anomaly caused by natural phenomena. The fact that the highest percentage of error is observed in the winter months in both years seems to be related to precipitation and thus the increase in the amount of data. Also, Figure 8 shows that the predicted and actual error percentages are quite low, especially in the summer months. Considering the performance metrics in Table 1, the SES model achieves the highest MAPE value. It shows that statistical methods used in time series analysis cannot make successful predictions in the case of sudden changes.

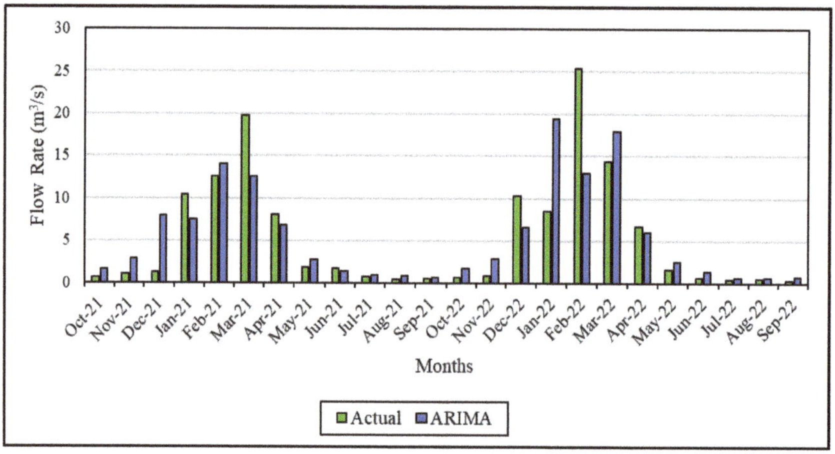

Figure 7. Comparative flow rate graph between actual and predicted data for ARIMA in the modeling period (2021–2022).

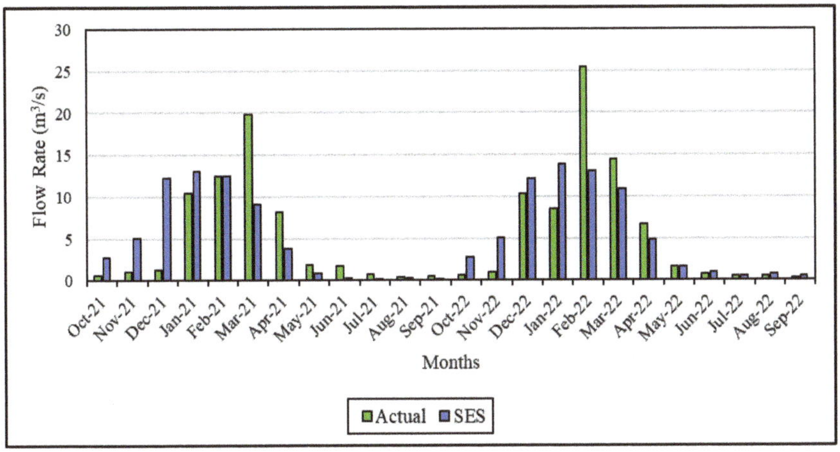

Figure 8. Comparative flow rate graph between actual and predicted data for SES in the modeling period (2021–2022).

When analyzing Figure 7, similar to Figure 8, it can be observed that the error rates between actual and predicted values during the winter months across the two years. In the predictions made in October-December and March-April 2021, the predictions were made with much larger numbers compared to the ARIMA method. The highest error percentages are found in the SES method. The most significant disparity emerges in February 2022, where despite the actual flow rate being 25 m^3/s, the model predicted a considerably lower value. This discrepancy resulted in an error percentage of 44%, marking the highest error rate in this study.

In Figure 9, the results of two statistical models are presented alongside the actual values. The R2 values calculated for ARIMA and SES are 0.63 and 0.55, respectively. The R^2 value of 0.55 suggests that the model insufficiently explains the relationship between independent variables and the dependent variable in the case of SES. However, it is seen that statistical forecasts except for the summer months have high error percentages and lower forecast performances compared to machine learning methods.

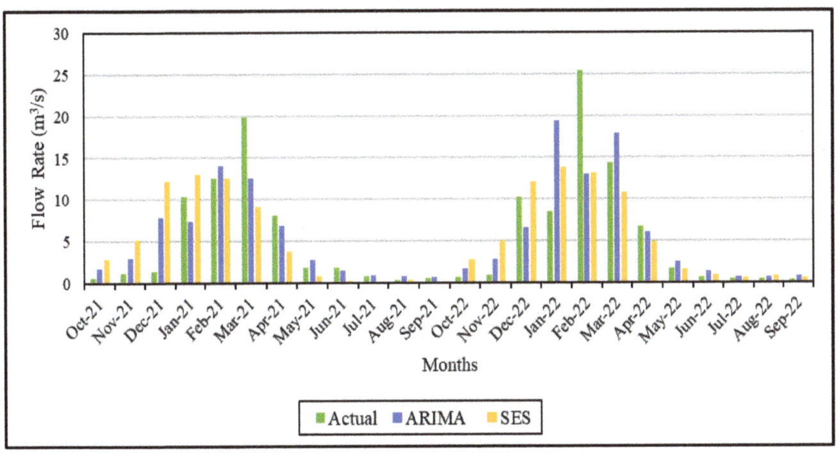

Figure 9. The comparison between actual and predicted data for ARIMA and SES in the modeling period (2021–2022).

Table 1. Performance metrics for the simulations of machine learning and statistical methods.

	Machine Learning Methods			Statistical Methods	
Performance Metrics	XGBOOST	RF	SVM	SES	ARIMA
R^2	0.72	0.68	0.61	0.55	0.63
r	0.845	0.825	0.778	0.74	0.79
MAE	2.294	2.659	2.36	2.94	2.55
RMSE	3.664	3.919	4.683	4.59	4.19
MAPE	62.28%	87.08%	68.40%	133.40%	92.70%
NSE	0.711	0.669	0.528	0.546	0.622
Total Number of Instances	24	24	24	24	24

4. Discussion

The predictive performance of three different machine learning techniques and two statistical methods employed in this study was evaluated using various success metrics. The irregular behavior of some natural phenomena involved in hydrological studies requires some pre-processing to ensure these conditions before modeling. However, in modeling using artificial intelligence models, all natural phenomena can be modeled in a simple way without any pre-processing. In the evaluation made for the correlation coefficient depending on the amount of data in all models, it was observed that all models produced prediction data that converged to the targeted data in high correlation. As the general evaluation criterion, the correlation coefficients of the three different machine learning models were found to be at least 0.78. This shows the confidence rate of the model and the correct relationship between the data. Additionally, XGBoost model has the lowest error rate with MAE and RMSE ratios of 2.294 and 3.664, respectively. Examining the actual values from 2021 to 2022 alongside the forecasts generated by machine learning methods reveals that particularly during the summer months, the forecasts closely align with the actual values, resulting in very low error rates. When we look at the figure of three models, it is seen that all actual and predicted values are consistent except for March 2021 and February 2022. The reason for the inconsistencies for these two months is the flooding and inundation explained separately in the result section. As shown in Table 1, the consistency rate of the forecasts is high. According to the MAPE results, the XGBoost algorithm performed better than RF and SVM. One of the main reasons for this is that the algorithm is suitable for complex hydrological modeling tasks due to its ability to analyze non-linear relationships and high-dimensional data. Furthermore, the NSE, a commonly utilized metric in studies of hydrology and water resources management, is employed to assess the accuracy of model predictions. XGBoost model achieved the highest NSE accuracy rate of 0.711. Considering the results of machine learning models, XGBoost model suitable forecast models for hydrological studies.

Considering the outcomes derived from the two different statistical methods employed in the study, the performance metrics have reached lower success percentages compared to machine learning methods. Although the correlation coefficients of both methods are at acceptable levels of 0.74 and 0.79 for ARIMA and SES, respectively, MAPE values are well above the acceptance limits. Especially the MAPE value of the SES model is around 133% and the predicted values are significantly different from the actual values. Therefore, this is not acceptable as the percentage error between actual and predicted values is much higher than expected. The reason why the margin of error in statistical analysis using time series is considerably higher than in machine learning methods is that sudden changes (floods, storms, etc.) cannot be calculated statistically. As a result, all statistical data is based on historical data and attempts to draw meaningful inferences from it. In artificial intelligence algorithms, on the other hand, it is understandable that the data is divided into a test and training set, first trained and then tested with real data, resulting in realistic values and high accuracy percentages. Then, all the findings of our study are analyzed and brought together and the data set is taken into consideration, the analysis obviously indicates that the XGBoost model is the best choice for generating accurate forecasts.

5. Conclusions

In this study, it was aimed to forecast the monthly average streamflows measured on Göksu Stream with machine learning and statistical methods. The average monthly streamflow data obtained from DSI for the years 1984–2022 were processed and estimated by machine learning (XGBoost, RF and SVM) and statistical (SES and ARIMA) methods. The accuracy of the models was demonstrated by statistical comparison of actual (observed) and predicted values. Moreover, models were compared in terms of monthly streamflow forecast. Finally, various performance metrics such as correlation coefficient (r), mean absolute error (MAE), root mean square error (RMSE), mean absolute percentage error (MAPE), coefficient of determination (R^2) and Nash–Sutcliffe efficiency (NSE) were calculated to assess model accuracy and examine error percentages. Therefore, the following conclusions have been reached based on the outcomes of this study:

i. The study confirmed the effectiveness of selected machine learning and statistical methods in predicting streamflows, demonstrating their utility in hydrological studies.
ii. The findings indicated the potential of these methods to support decisions related to water resource systems.
iii. For a variety of hydrological and environmental applications, including flood flowrate estimation, hydroelectric generation, and water resource management, the study highlights the need of precise streamflow forecast.
iv. The prediction results calculated in the study coincide with the actual streamflow data. After analyzing all the findings from our study and considering the dataset, it is concluded that the XGBoost model is the best choice for making accurate forecasts.
v. The study successfully demonstrates the effectiveness of machine learning and statistical methods in forecasting monthly average streamflows, highlighting their crucial role in hydrological studies and related applications.
vi. It underscores the importance of continuous innovation in modeling techniques and dataset diversification for improving flood predictions, essential for effective basin management and urban planning. Overall, the study contributes to advancing the field of hydrology and provides valuable insights for future research and application.

Author Contributions: Conceptualization, M.N.C., M.G., E.N., İ.B.P. and S.G.; methodology, E.N. and M.S.; software, E.N., M.G., M.U. and M.S.; validation, E.N. and M.G.; investigation, M.N.C., M.G., E.N., İ.B.P. and S.G.; writing—original draft preparation, M.N.C., M.G., E.N., İ.B.P. and S.G.; writing—review and editing, M.G., İ.B.P. and S.G. All authors have read and agreed to the published version of the manuscript.

Funding: This research received no external funding.

Data Availability Statement: Data are contained within the article.

Acknowledgments: The authors would also like to thank the General Directorate of State Hydraulic Works (DSI) and Turkish State Meteorological Service (TSMS) and General Directorate of Water Management (SYGM) for their data support and valuable discussions in undertaking this work.

Conflicts of Interest: The authors declare no conflicts of interest.

References

1. Erdal, H.I.; Karakurt, O. Advancing monthly streamflow prediction accuracy of CART models using ensemble learning paradigms. *J. Hydrol.* **2013**, *477*, 119–128. [CrossRef]
2. Ni, L.; Wang, D.; Singh, V.P.; Wu, J.; Wang, Y.; Tao, Y.; Zhang, J. Streamflow and rainfall forecasting by two long short-term memory-based models. *J. Hydrol.* **2020**, *583*, 124296. [CrossRef]

3. Yaseen, Z.M.; El-Shafie, A.; Jaafar, O.; Afan, H.A.; Sayl, K.N. Artificial intelligence-based models for stream-flow forecasting 2000–2015. *J. Hydrol.* **2015**, *530*, 829–844. [CrossRef]
4. Fathian, F.; Mehdizadeh, S.; Sales, A.K.; Safari, M.J.S. Hybrid models to improve the monthly river flow prediction: Integrating artificial intelligence and non-linear time series models. *J. Hydrol.* **2019**, *575*, 1200–1213. [CrossRef]
5. Tongal, H.; Booij, M.J. Simulation and forecasting of streamflows using machine learning models coupled with base flow separation. *J. Hydrol.* **2018**, *564*, 266–282. [CrossRef]
6. Shafizadeh-Moghadam, H.; Valavi, R.; Shahabi, H.; Chapi, K.; Shirzadi, A. Novel forecasting approaches using combination of machine learning and statistical models for flood susceptibility mapping. *J. Environ. Manag.* **2018**, *217*, 1–11. [CrossRef]
7. Dalkılıç, H.Y.; Yeşilyurt, S.N.; Samui, P. Daily flow modeling with random forest and k-nearest neighbor methods. *Erzincan Univ. J. Sci. Technol.* **2021**, *14*, 914–925.
8. Nazimi, N.; Saplıoğlu, K. Monthly streamflow prediction using ANN, KNN and ANFIS models: Example of Gediz River Basin. *Tek. Bilim. Derg.* **2023**, *13*, 42–49. [CrossRef]
9. Stakhiv, E.; Stewart, B. Needs for climate information in support of decision-making in the water sector. *Procedia Environ. Sci.* **2010**, *1*, 102–119. [CrossRef]
10. Nourani, V.; Komasi, M. A geomorphology-based ANFIS model for multi-station modeling of rainfall–runoff process. *J. Hydrol.* **2013**, *490*, 41–55. [CrossRef]
11. Xiao, Z.; Liang, Z.; Li, B.; Hou, B.; Hu, Y.; Wang, J. New flood early warning and forecasting method based on similarity theory. *J. Hydrol. Eng.* **2019**, *24*, 04019023. [CrossRef]
12. Latifoğlu, L. A novel approach for prediction of daily streamflow discharge data using correlation-based feature selection and random forest method. *Int. Adv. Res. Eng. J.* **2022**, *6*, 1–7. [CrossRef]
13. Kisi, O.; Latifoğlu, L.; Latifoğlu, F. Investigation of empirical mode decomposition in forecasting of hydrological time series. *Water Resour. Manag.* **2014**, *28*, 4045–4057. [CrossRef]
14. Petty, T.R.; Dhingra, P. Streamflow hydrology estimate using machine learning (SHEM). *J. Am. Water Resour. Assoc.* **2018**, *54*, 55–68. [CrossRef]
15. Lin, Y.; Wang, D.; Wang, G.; Qiu, J.; Long, K.; Du, Y.; Xie, H.; Wei, Z.; Shangguan, W.; Dai, Y. A hybrid deep learning algorithm and its application to streamflow prediction. *J. Hydrol.* **2021**, *601*, 126636. [CrossRef]
16. Xiang, Z.; Demir, I. Distributed long-term hourly streamflow predictions using deep learning–A case study for State of Iowa. *Environ. Model. Softw.* **2020**, *131*, 104761. [CrossRef]
17. Delaney, C.J.; Hartman, R.K.; Mendoza, J.; Dettinger, M.; Delle Monache, L.; Jasperse, J.; Martin Ralph, F.; Talbot, C.; Brown, J.; Reynolds, D.; et al. Forecast informed reservoir operations using ensemble streamflow predictions for a multipurpose reservoir in Northern California. *Water Resour. Res.* **2020**, *56*, e2019WR026604. [CrossRef]
18. Malik, A.; Tikhamarine, Y.; Souag-Gamane, D.; Kisi, O.; Pham, Q.B. Support vector regression optimized by meta-heuristic algorithms for daily streamflow prediction. *Stoch. Environ. Res. Risk Assess.* **2020**, *34*, 1755–1773. [CrossRef]
19. Asadi, H.; Shahedi, K.; Jarihani, B.; Sidle, R.C. Rainfall-runoff modelling using hydrological connectivity index and artificial neural network approach. *Water* **2019**, *11*, 212. [CrossRef]
20. Snieder, E.; Shakir, R.; Khan, U.T. A comprehensive comparison of four input variable selection methods for artificial neural network flow forecasting models. *J. Hydrol.* **2020**, *583*, 124299. [CrossRef]
21. Ikram, R.M.A.; Ewees, A.A.; Parmar, K.S.; Yaseen, Z.M.; Shahid, S.; Kisi, O. The viability of extended marine predators algorithm-based artificial neural networks for streamflow prediction. *Appl. Soft Comput.* **2022**, *131*, 109739. [CrossRef]
22. Niu, W.J.; Feng, Z.K.; Chen, Y.B.; Zhang, H.R.; Cheng, C.T. Annual streamflow time series prediction using extreme learning machine based on gravitational search algorithm and variational mode decomposition. *J. Hydrol. Eng.* **2020**, *25*, 04020008. [CrossRef]
23. Hussain, D.; Hussain, T.; Khan, A.A.; Naqvi, S.A.A.; Jamil, A. A deep learning approach for hydrological time-series prediction: A case study of Gilgit river basin. *Earth Sci. Inform.* **2020**, *13*, 915–927. [CrossRef]
24. Adnan, R.M.; Yuan, X.; Kisi, O.; Yuan, Y. Streamflow forecasting using artificial neural network and support vector machine models. *Am. Sci. Res. J. Eng. Technol. Sci.* **2017**, *29*, 286–294.
25. Le, X.H.; Nguyen, D.H.; Jung, S.; Yeon, M.; Lee, G. Comparison of deep learning techniques for river streamflow forecasting. *IEEE Access* **2021**, *9*, 71805–71820. [CrossRef]
26. Saraiva, S.V.; de Oliveira Carvalho, F.; Santos, C.A.G.; Barreto, L.C.; Freire, P.K.D.M.M. Daily streamflow forecasting in Sobradinho Reservoir using machine learning models coupled with wavelet transform and bootstrapping. *Appl. Soft Comput.* **2021**, *102*, 107081. [CrossRef]
27. Kagoda, P.A.; Ndiritu, J.; Ntuli, C.; Mwaka, B. Application of radial basis function neural networks to short-term streamflow forecasting. *Phys. Chem. Earth* **2010**, *35*, 571–581. [CrossRef]
28. Yonaba, H.; Anctil, F.; Fortin, V. Comparing sigmoid transfer functions for neural network multistep ahead streamflow forecasting. *J. Hydrol. Eng.* **2010**, *15*, 275–283. [CrossRef]
29. Zaini, N.; Malek, M.A.; Yusoff, M.; Mardi, N.H.; Norhisham, S. Daily River Flow Forecasting with Hybrid Support Vector Machine–Particle Swarm Optimization. *IOP Conf. Ser. Earth Environ. Sci.* **2018**, *140*, 012035. [CrossRef]
30. Farias, C.A.; Santos, C.A.; Lourenço, A.M.; Carneiro, T.C. Kohonen neural networks for rainfall-runoff modeling: Case study of Piancó River Basin. *J. Urban Environ. Eng.* **2013**, *7*, 176–182. [CrossRef]

31. Danandeh Mehr, A.; Kahya, E.; Şahin, A.; Nazemosadat, M.J. Successive-station monthly streamflow prediction using different artificial neural network algorithms. *Int. J. Environ. Sci. Technol.* **2015**, *12*, 2191–2200. [CrossRef]
32. Adhikary, S.K.; Muttil, N.; Yilmaz, A.G. Improving streamflow forecast using optimal rain gauge network-based input to artificial neural network models. *Hydrol. Res.* **2018**, *49*, 1559–1577. [CrossRef]
33. Prasad, R.; Deo, R.C.; Li, Y.; Maraseni, T. Input selection and performance optimization of ANN-based streamflow forecasts in the drought-prone Murray Darling Basin region using IIS and MODWT algorithm. *Atmos. Res.* **2017**, *197*, 42–63. [CrossRef]
34. Alvisi, S.; Franchini, M. Fuzzy neural networks for water level and discharge forecasting with uncertainty. *Environ. Model. Softw.* **2011**, *26*, 523–537. [CrossRef]
35. Aslan, M.F.; Sabanci, K.; Durdu, A. Different wheat species classifier application of ANN and ELM. *J. Multidiscip. Eng. Sci. Technol.* **2017**, *4*, 8194–8198.
36. Dickey, D.A.; Fuller, W.A. Likelihood ratio statistics for autoregressive time series with a unit root. *Econom. J. Econom. Soc.* **1981**, *49*, 1057–1072. [CrossRef]
37. Kazemi, M. A comparative study of singular spectrum analysis, neural network, ARIMA and exponential smoothing for monthly rainfall forecasting. *J. Math. Model.* **2023**, *11*, 783–803.
38. Murat, M.; Malinowska, I.; Hoffmann, H.; Baranowski, P. Statistical modelling of agrometeorological time series by exponential smoothing. *Int. Agrophys.* **2016**, *30*, 57–65. [CrossRef]
39. Fırat, A. Estimation of Average Flow and Maximum Precipitation by Artificial Neural Networks Case of Istanbul Göksu Stream. Master's Thesis, Sakarya University, Serdivan, Türkiye, 2019.
40. SYGM (General Directorate of Water Management, in Turkish: Su Yönetimi Genel Müdürlüğü). *Marmara Basin Flood Management Plan*; Republic of Türkiye Ministry of Agriculture and Forestry—General Directorate of Water Management: Ankara, Türkiye, 2023; pp. 1–1540.
41. Chen, T.; Guestrin, C. Xgboost: A Scalable Tree Boosting System. In Proceedings of the 22nd ACM SIGKDD International Conference on Knowledge Discovery and Data Mining, San Francisco, CA, USA, 13–17 August 2016.
42. Freund, Y.; Schapire, R.E. A decision-theoretic generalization of on-line learning and an application to boosting. *J. Comput. Syst. Sci.* **1997**, *55*, 119–139. [CrossRef]
43. Breiman, L. Random forests. *Mach. Learn.* **2001**, *45*, 5–32. [CrossRef]
44. Ali, J.; Khan, R.; Ahmad, N.; Maqsood, I. Random forests and decision trees. *Int. J. Comput. Sci. Issues* **2012**, *9*, 272.
45. Pal, M. Random forest classifier for remote sensing classification. *Int. J. Remote Sens.* **2005**, *26*, 217–222. [CrossRef]
46. Gislason, P.O.; Benediktsson, J.A.; Sveinsson, J.R. Random forests for land cover classification. *Pattern Recognit. Lett.* **2006**, *27*, 294–300. [CrossRef]
47. Qi, Y. Random forest for bioinformatics. In *Ensemble Machine Learning: Methods and Applications*; Springer: New York, NY, USA, 2012; pp. 307–323.
48. Atallah, R.; Al-Mousa, A. Heart Disease Detection Using Machine Learning Majority Voting Ensemble Method. In Proceedings of the 2nd International Conference on New Trends in Computing Sciences (ICTCS), Amman, Jordan, 9–11 October 2019.
49. Shevade, S.K.; Keerthi, S.S.; Bhattacharyya, C.; Murthy, K.R.K. Improvements to the SMO algorithm for SVM regression. *IEEE Trans. Neural Netw.* **2000**, *11*, 1188–1193. [CrossRef] [PubMed]
50. Billah, B.; King, M.L.; Snyder, R.D.; Koehler, A.B. Exponential smoothing model selection for forecasting. *Int. J. Forecast.* **2006**, *22*, 239–247. [CrossRef]
51. Corberán-Vallet, A.; Bermúdez, J.D.; Segura, J.V.; Vercher, E. A forecasting support system based on exponential smoothing. In *Handbook on Decision Making: Vol 1: Techniques and Applications*; Springer: Berlin/Heidelberg, Germany, 2010; pp. 181–204.
52. Mensah, E.K. *Box-Jenkins Modelling and Forecasting of Brent Crude Oil Price*; Munich Personal RePEc Archive; University Library of Munich: Munich, Germany, 2015; pp. 1–10.
53. Hyndman, R.J. Box-Jenkins Modelling. In *Regional Symposium on Environment and Natural Resources*; John Wiley & Sons: San Francisco, CA, USA, 2001; pp. 10–11.
54. Box, G.E.; Jenkins, G.M.; Reinsel, G.C.; Ljung, G.M. *Time Series Analysis: Forecasting and Control*; John Wiley & Sons: Hoboken, NJ, USA, 2015.
55. Lewis, C.D. *Industrial and Business Forecasting Methods*; Butterworths Publishing: London, UK, 1982; ISBN 978-0408005593.
56. Nash, J.; Sutcliffe, J.V. River flow forecasting through conceptual models' part I—A discussion of principles. *J. Hydrol.* **1970**, *10*, 282–290. [CrossRef]

water

MDPI

Article

A Deformation Analysis Method for Sluice Structure Based on Panel Data

Zekai Ma [1,2], Benxing Lou [1,2,3,*], Zhenzhong Shen [1,2], Fuheng Ma [3,*], Xiang Luo [1,2,3], Wei Ye [3], Xing Li [3] and Dongze Li [1,2]

[1] The National Key Laboratory of Water Disaster Prevention, Hohai University, No. 1 Xikang Road, Nanjing 210024, China
[2] College of Water Conservancy and Hydropower Engineering, Hohai University, No. 1 Xikang Road, Nanjing 210024, China
[3] Department of Dam Safety and Management, Nanjing Hydraulic Research Institute, No. 223 Guangzhou Road, Nanjing 210098, China
* Correspondence: bxlou@hhu.edu.cn (B.L.); fhma_nhri@163.com (F.M.)

Abstract: Deformation, as the most intuitive index, can reflect the operation status of hydraulic structures comprehensively, and reasonable analysis of deformation behavior has important guiding significance for structural long-term service. Currently, the health evaluation of dam deformation behavior has attracted widespread attention and extensive research from scholars due to its great importance. However, given that the sluice is a low-head hydraulic structure, the consequences of its failure are easily overlooked without sufficient attention. While the influencing factors of the sluice's deformation are almost identical to those of a concrete dam, nonuniform deformation is the key issue in the sluice's case because of the uneven property of the external load and soil foundation, and referencing the traditional deformation statistical model of a concrete dam cannot directly represent the nonuniform deformation behavior of a sluice. In this paper, we assume that the deformation at various positions of the sluice consist of both overall and individual effects, where overall effect values describe the deformation response trend of the sluice structure under external loads, and individual effect values represent the degree to which the deformation of a single point deviates from the overall deformation. Then, the random coefficient model of panel data is introduced into the analysis of sluice deformation to handle the unobservable overall and individual effects. Furthermore, the maximum entropy principle is applied, both to approximate the probability distribution function of individual effect extreme values and to determine the early warning indicators, completing the assessment and analysis of the nonuniform deformation state. Finally, taking a project as an example, we show that the method proposed can effectively identify the overall deformation trend of the sluice and the deviation degree of each measuring point from the overall deformation, which provides a novel approach for sluice deformation behavior research.

Keywords: sluice; nonuniform deformation; random coefficient model; individual effect values; maximum entropy principle

1. Introduction

The sluice is a common hydraulic structure that plays a very important role in controlling flooding, reducing disaster, optimizing water resource allocation, and protecting the ecological environment [1,2]. During service periods, the sluice's safety will be threatened by multiple risks such as external loads and emergencies, and it is necessary to grasp the work state of the project through structural safety evaluation. Deformation, as the most intuitive index, can comprehensively reflect the operation status of the sluice [3], and therefore, reasonable analysis and evaluation of sluice deformation behavior have important guiding significance for its long-term service [4].

Structural deformation is an important research topic in safety evaluation areas of hydraulic engineering at all times. Much research to date has focused on the analysis methods of dam deformation behavior [5,6], in which the statistical model is most commonly used to establish the relationship between deformation and influencing factors, including water level, temperature, and concrete creep [7–9]. In order to improve the accuracy and robustness of the statistical model, scholars introduced some artificial intelligence methods [10,11], including artificial neural networks [12,13], support vector machines [14,15], and deep learning model [16,17], into the analysis of dam deformation. However, because it is a low-head hydraulic structure, the safety of the sluice is easily overlooked without adequate attention [18]. Although the deformation analysis method of a concrete dam can be referenced for the sluice, there exist some significant differences in their deformation characteristics. For a concrete dam, the deformation of a certain measuring point or dam section is the focus of attention because of the high water pressure. For the sluice, the nonuniform deformation between different measuring points should attract special attention, owing to the uneven feature of ground properties and load distribution [19]. Especially before the deformation develops stably, the sluice structure is likely to incur nonuniform deformation, even leading to serious accidents such as waterstop failure, buildings tilting, and concrete slab fracture [20].

In the actual project, the deformations of measuring points in different parts of the sluice are different due to the nonuniformity of external loads and soft soil foundations, but the overall deformation trend is still similar because of the similar service environment. Hence, the deformation of a sluice could be divided into two parts: overall effect values and individual effect values, where overall effect values describe the deformation response trend of the sluice structure under external loads, and individual effect values represent the deviation degree between the deformation of a single point and the overall deformation. The panel data theory was first put forward by Mundlak in 1961, and has been introduced widely into the fields of econometrics, sociology, and market research. Several studies on dam deformation have been published using panel data theory, which paid greatest attention to the spatial correlation as well as the determination of abnormal areas for dam deformation. For instance, Shi et al. [21] proposed a variable intercept panel model based on deformation zoning of ultra-high arch dams, and subsequently, two panel models were established to distinguish the fixed and random modes of special effects of dam deformation. Shao et al. [22] proposed a random coefficient model using panel data theory, which can solve the serious multicollinearity problem among influence factors in the traditional regression method. Zhao et al. [23] established a functional relationship between the measured values and the real-time risk probability based on the estimation of robustness of the panel model. Cui et al. [24] proposed a novel imputation model for missing concrete dam deformation data based on their evaluation of the accuracy of the panel model. In fact, one of the main uses of panel data theory is to handle the unobservable individual or overall effects [25], and we can utilize panel data theory to identify the individual effects of each measurement point, and further to evaluate the nonuniform deformation status of the sluice.

In light of the above analysis, this article introduces the panel data theory into sluice deformation analysis for identifying the overall effect values of sluice structure and the individual effect values of each measurement point. Then, the maximum entropy principle is employed to estimate the probability distribution function of individual effect extreme values, and subsequently an analysis method for the nonuniform deformation state of the sluice is developed. Finally, the deformation data of a sluice obtained through wire alignment transducer offer an example to validate the approach proposed.

2. Influencing Factors of Sluice Displacement

Referring to the statistical model of dam deformation, the influencing factors of sluice deformation δ can be divided into three parts, including water pressure component δ_H, temperature component δ_T, and aging component δ_θ, as shown in Equation (1).

$$\delta = \delta_H + \delta_T + \delta_\theta \tag{1}$$

2.1. Water Pressure Component

The sluice's deformation under upstream and downstream water pressure consists of three parts (see Figure 1): (a) horizontal sliding deformation caused by the water pressure difference; (b) bending deformation of the structure due to movement; and (c) rotation deformation of the foundation because of water gravity difference. The horizontal displacement can be expressed as:

$$\delta_H = \delta_H^a + \delta_H^b + \delta_H^c \tag{2}$$

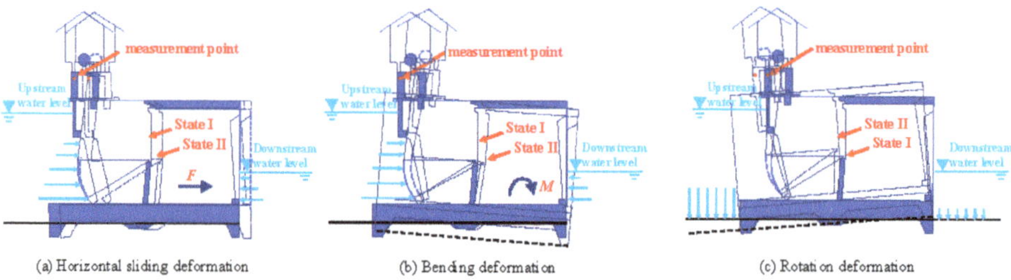

Figure 1. Schematic diagram of the water pressure component of sluice displacement.

The relationship between water pressure and water depth is quadratic, and the deformation caused by water pressure can be described as the 1–4th power of the upstream and downstream water depth, referring to the expression of the water pressure component for concrete dam deformation. Therefore, the water pressure component of sluice deformation can be expressed as:

$$\delta_H = \sum_{i=1}^{4} \left(a_{1i}h^i + a_{2i}H^i\right) \tag{3}$$

where h is downstream water depth, H is upstream water depth, and a_{1i} and a_{2i} are both regression coefficients.

2.2. Temperature Component

The sluice's deformation is also affected by the external environmental temperature. Changes in external environmental temperature will cause temperature differences in concrete structure, resulting in temperature stress and deformation. The relationship between sluice deformation and environmental temperature is shown in Figure 2, where T_i represents the average temperature of the previous i days. It can be seen that the sluice deformation directly follows environmental temperature. Therefore, the environmental temperatures within 7 days before the observation date are selected as the influencing factors, and the temperature component can be presented as:

$$\delta_T = \sum_{i=1}^{m} b_i T_i \tag{4}$$

where b_i is the regression coefficient.

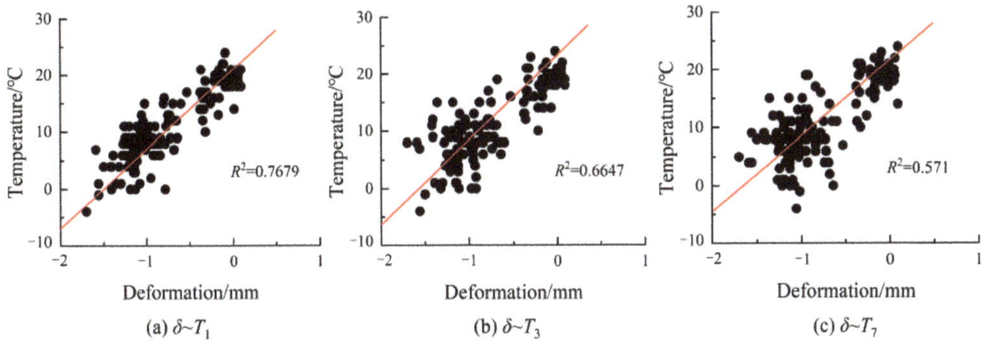

Figure 2. The correlation between deformation and temperature.

2.3. Aging Component

Due to the creep of concrete and soil foundation, the deformation of the sluice structure exhibits a certain characteristic trend, which can be depicted as:

$$\delta_\theta = c_1\theta + c_2 \ln \theta \tag{5}$$

where θ represents the difference of observation date minus start date divided by 100, and c_1 and c_2 are regression coefficients.

Thus, the statistical equations of sluice displacement can be expressed as follows:

$$\delta = \sum_{i=1}^{4} (a_{1i}h^i + a_{2i}H^i) + \sum_{i=1}^{7} b_i T_i + c_1\theta + c_2 \ln \theta + d \tag{6}$$

where d is a constant.

3. An Analysis Method for Sluice Displacement Behavior Based on Panel Data

3.1. Random Coefficient Statistical Model Based on Panel Data

Due to the nonuniformity of external loads and soft soil foundations, the influencing factors for each part of the sluice structure are different, which is the main reason for individual effects or uneven deformation. If we use traditional statistical models to analyze the deformation state of the sluice structure, some potential influencing factors for each part could not be included as independent variables in the model, and we would be unable to take into account the individual differences of measurement points. Therefore, in addition to considering the main factors, we should consider the unmeasurable or unknown influencing factors when modeling, so that the deformation analysis model can absorb the influence of these uncertain factors. Since traditional deformation analysis models cannot detect the individual effects caused by the hidden factors, panel data theory is introduced to identify and reveal the overall features and individual characteristics among multiple measurement points.

Panel data are two-dimensional data that include time series data and cross-sectional data [26]. Time series data reflect the observation data of a measurement point at different times, and cross-sectional data reflect the observation data of different measurement points at the same time. Firstly, we studied the variable coefficient statistical model, in which the regression coefficients of panel data do not change over time but vary cross-sectionally, formulated as follows:

$$y_{it} = \sum_{k=1}^{K} \beta_{ki} x_{kit} + u_{it} \tag{7}$$

where y_{it} represents the panel data form of sluice deformation data, t is the ordinal number of time data, i is the ordinal number of cross-sectional data, k is the ordinal number of explanatory variables, x_{kit} is the explanatory variable, β_{ki} is the regression coefficient, and u_{it} is the random error.

The variable coefficient statistical model consists of both a fixed coefficient model and random coefficient model. In the fixed coefficient model, β_{ki} is considered a fixed constant and varies with individual changes. By stacking up all the deformation data, the model formula is as follows [27]:

$$\begin{bmatrix} y_1 \\ y_2 \\ \vdots \\ y_N \end{bmatrix} = \begin{bmatrix} \mathbf{X}_1 & 0 & \cdots & 0 \\ 0 & \mathbf{X}_2 & \cdots & 0 \\ \vdots & \vdots & \ddots & \vdots \\ 0 & 0 & \cdots & \mathbf{X}_N \end{bmatrix} \begin{bmatrix} \beta_1 \\ \beta_2 \\ \vdots \\ \beta_N \end{bmatrix} + \begin{bmatrix} u_1 \\ u_2 \\ \vdots \\ u_N \end{bmatrix} \tag{8}$$

where $\boldsymbol{y}_i = (y_{i1}, \ldots, y_{iT})$ represents the time series of sluice deformation; $\boldsymbol{\beta}_i = (\beta_0, \beta_1, \ldots, \beta_k)'$ is the regression coefficients to be estimated of the i-th observation point; $\boldsymbol{X}_i = \begin{bmatrix} x_{11} & x_{12} & \cdots & x_{1k} \\ x_{21} & x_{22} & \cdots & x_{2k} \\ \vdots & \vdots & \ddots & \vdots \\ x_{t1} & x_{t2} & \cdots & x_{tk} \end{bmatrix}$ is the explanatory variables matrix of the i-th observation point, and $x_{tk} = (1, H_t, H_t^2, H_t^3, H_t^4, h_t, \cdots, \theta_t, \ln \theta_t)'$; and $\boldsymbol{u}_i$ is the random error of the i-th observation point.

In a fixed coefficient model, the coefficients $\boldsymbol{\beta}_i$ are regarded as fixed constant and independent of each other. However, because the deformation patterns change with measuring points, the parameters of $\boldsymbol{\beta}_i$ can be set as random variables with a common average value $\boldsymbol{\beta}$. Swamy [26] proposed the following hypothesis:

$$\begin{aligned} &E(\gamma_i) = 0 \\ &E(\gamma_i \gamma_j') = \begin{cases} \Delta & (i = j) \\ 0 & (i \neq j) \end{cases} \\ &E(x_{it} \gamma_j') = 0 \\ &E(\gamma_i \gamma_j') = \begin{cases} \sigma_i^2 I_T & (i = j) \\ 0 & (i \neq j) \end{cases} \end{aligned} \tag{9}$$

Then, the random coefficients model is written with a matrix form as follows:

$$\boldsymbol{y} = \boldsymbol{X\beta} + \widetilde{\boldsymbol{X}}\boldsymbol{\gamma} + \boldsymbol{u}. \tag{10}$$

where $\boldsymbol{y} = (y_1', \ldots, y_N')'$; $\boldsymbol{X} = (X_1, \ldots, X_N)'$; $\widetilde{\boldsymbol{X}} = \begin{bmatrix} X_1 & & 0 \\ & \ddots & \\ 0 & & X_N \end{bmatrix}$; $\boldsymbol{\beta} = (\beta_0, \beta_1, \ldots, \beta_k)'$ is the common mean value's coefficient vector; $\gamma = (\gamma_1', \ldots, \gamma_N')'$ is the deviation between individual coefficient and common average coefficient; and $\boldsymbol{u} = (u_1', \ldots, u_N')'$ is the random error vector.

Based on the random coefficient model, we can use $\boldsymbol{X\beta}$ to represent the overall deformation for multiple measurement points, and utilize $\widetilde{\boldsymbol{X}}\gamma + \boldsymbol{u}$ to illustrate the degree to which actual deformation of observation points deviates from overall deformation, that is, nonuniform deformation or individual effect values of observation points. For random effects models, it should be noted that the parameters obtained by least squares are not the valid parameters, and generalized least squares estimation should be selected to estimate the model coefficients.

3.2. Maximum Entropy Principle for the Probability Distribution Function of Individual Effect Extreme Values

To evaluate the health situation of nonuniform deformation for an observation point, the probability distribution function of annual or monthly individual effect extreme values should first be determined. Jaynes [28] proposed that the probability distribution with the maximum entropy should be selected while using local information to infer the global probability distribution, which is also called the maximum entropy principle.

For discrete random variables, the information entropy $H(x)$ can be expressed as:

$$H(x) = -\sum_{i=1}^{n} p_i \ln p_i \tag{11}$$

where p_i is the probability of variable x, and $H(x)$ is the amount of information, representing the uncertainty of random variables.

For continuous random variables, the information entropy $H(x)$ can be expressed as:

$$H(x) = -\int_R f(x) \ln f(x) dx \tag{12}$$

where $f(x)$ is the probability distribution function of variable x, and R is the integration interval.

The maximum entropy principle is used to solve the probability distribution with the maximum amount of information in light of the known sample data. Its objective function is:

$$\max H(x) = -\int_R f(x) \ln f(x) dx \tag{13}$$

The constraint conditions are:

$$\int_R f(x) dx = 1 \tag{14}$$

$$\int_R x^i f(x) dx = \mu_i \tag{15}$$

where μ_i is the i-order origin moment calculated by the known sample data. Usually, selecting a total order of 4 can describe the basic characteristics of a random variable.

To maximize the value of $H(x)$, it is necessary to continuously adjust the form of the probability distribution function $f(x)$, and the following function can be established by Lagrange multiplier method.

$$L = H(x) + (\lambda_0 + 1)\left[\int_R f(x) dx - 1\right] + \sum_{i=1}^{N} \lambda_i \left[\int_R x^i f(x) dx - \mu_i\right] \tag{16}$$

Then the analytical form of the probability distribution function can be solved as follows:

$$f(x) = \exp\left(\lambda_0 + \sum_{i=1}^{N} \lambda_i x^i\right) \tag{17}$$

By substituting Equation (17) into Equation (14), Equation (18) can be found.

$$\lambda_0 = -\ln\left(\int_R \exp\left(\sum_{i=1}^{N} \lambda_i x^i\right) dx\right) \tag{18}$$

Equation (19) can be obtained by substituting Equations (17) and (18) into Equation (15).

$$\int_R x^i f(x)dx = \int_R x^i \exp\left(\lambda_0 + \sum_{j=1}^{N} \lambda_j x^j\right) dx = \frac{\int_R x^i \exp\left(\sum_{j=1}^{N} \lambda_j x^j\right) dx}{\int_R \exp\left(\sum_{j=1}^{N} \lambda_j x^j\right) dx} = \mu_i \tag{19}$$

For the convenience of numerical calculation, Equation (19) can be transformed into Equation (20).

$$1 - \frac{\int_R x^i \exp\left(\sum_{j=1}^{N} \lambda_j x^j\right) dx}{\mu_i \int_R \exp\left(\sum_{j=1}^{N} \lambda_j x^j\right) dx} = r_i \tag{20}$$

where r_i is the residual under the i-order origin moment. When solving the coefficients λ_0, $\lambda_1, \lambda_2, \ldots, \lambda_N$, it is necessary to minimize the residual value, and the objective function can be expressed as:

$$\min r = \sum_{i=1}^{N} r_i^2 = \sum_{i=1}^{N} 1 - \frac{\int_R x^i \exp\left(\sum_{j=1}^{N} \lambda_j x^j\right) dx}{\mu_i \int_R \exp\left(\sum_{j=1}^{N} \lambda_j x^j\right) dx} \tag{21}$$

Therefore, the initial objective function, that is Equation (13), is converted to the form of Equation (21), and the probability distribution function of random variables can be obtained by solving Equation (21).

3.3. *The Analysis Method for Nonuniform Deformation State*

According to Section 3.2, the probability distribution function $f(x)$ of individual effect extreme values for each measurement point can be calculated based on the maximum entropy principle, where the annual or monthly maximum and minimum values are seen as random variables [29]. Excessive deviations from overall displacement, whether towards downstream or upstream, are detrimental to the sluice, and thus we assume that x_α^+ and x_α^- are the early warning indicators of individual effect values. When the individual effect values of the sluice exceed x_α^+ and x_α^-, the sluice will be in a warning state, and the corresponding exceeding or non-exceeding probability can be determined by Equation (22).

$$P_\alpha = \begin{cases} P[x > x_\alpha^+] = \int_{x_\alpha^+}^{+\infty} f(x)dx \\ P[x < x_\alpha^-] = \int_{-\infty}^{x_\alpha^-} f(x)dx \end{cases} \tag{22}$$

where P_α is the probability of sluice failure, which can be determined based on the importance of the sluice, generally ranging from 1% to 5%.

Based on the early warning indicators of x_α^+ and x_α^-, the following criteria can be used to determine whether the nonuniform deformation of the sluice is normal [30].

(1) When the individual effect values meet with $x_\alpha^- \leq x \leq x_\alpha^+$, the nonuniform deformation is in a normal state;
(2) When the individual effect values meet with $x > x_\alpha^+$ or $x < x_\alpha^-$, the nonuniform deformation is in an abnormal or warning state.

In summary, an analysis flow for nonuniform deformation behavior of the sluice is established and shown in Figure 3.

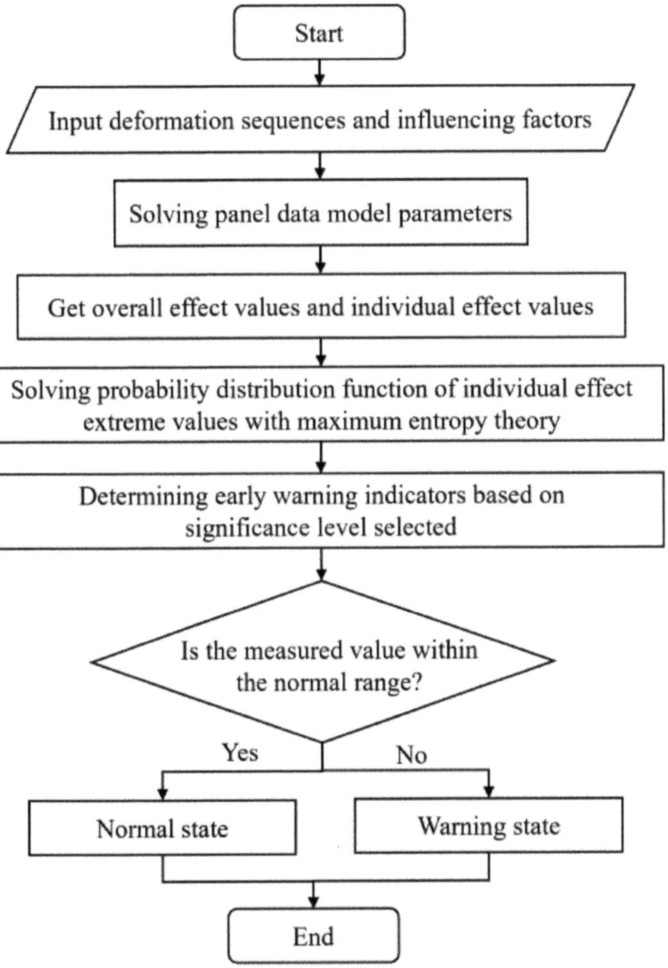

Figure 3. Flowchart of the analysis method for the sluice's deformation state based on panel data.

4. Case Study

4.1. General Situation

A sluice with a reinforced concrete structure is located in Jiangsu Province, China. It has 15 sluice holes with the size of 10 m × 6.2 m, and 5 concrete bottom slabs with the size of 33.2 m × 18 m. Figure 4 shows the layout of observation points for horizontal displacement set on each bottom slab, which are labeled as SP1-1, SP1-2, ..., SP5-1, and SP5-2, respectively. The deformation data of the sluice from 17 October 2022 to 5 April 2023 are presented in Figure 5, and the corresponding upstream water levels, downstream water levels, and environmental temperature are shown in Figure 6.

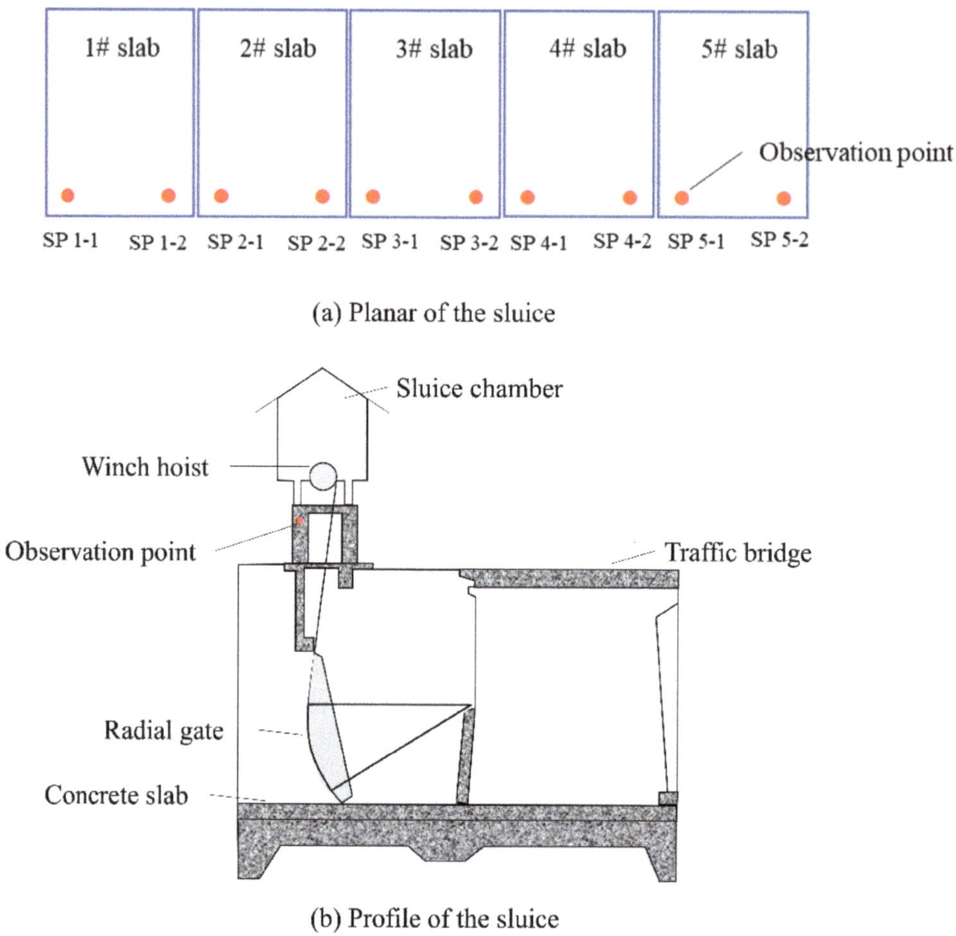

Figure 4. Layout of deformation measurement points for the sluice.

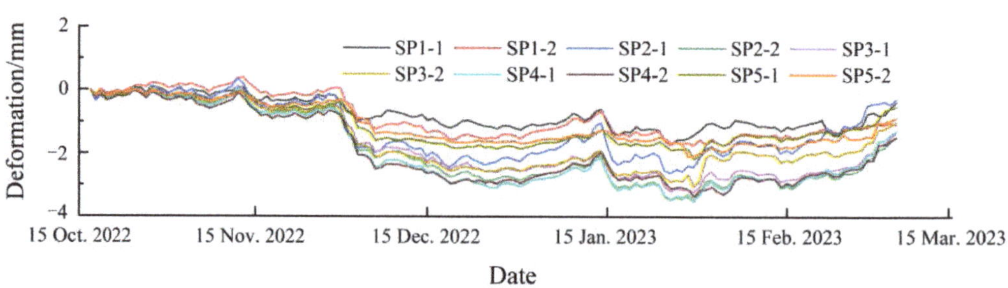

Figure 5. Deformation sequence of each measurement point of the sluice.

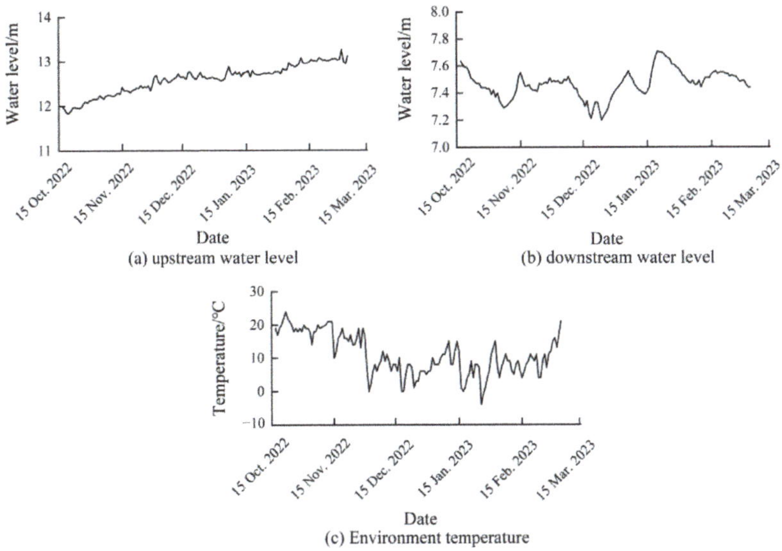

Figure 6. Environmental factor sequences.

4.2. Random Coefficient Statistical Model Based on Panel Data

4.2.1. Fitting Results

In order to verify the fitting effect of the model proposed in Section 3.1, the deformation data of the above-mentioned sluice will be taken as an example. Considering the general form of the random coefficient model, the sluice deformation statistical model can be expressed as:

$$\begin{bmatrix} y_1 \\ y_2 \\ \vdots \\ y_N \end{bmatrix} = \begin{bmatrix} \mathbf{X}_1 \\ \mathbf{X}_2 \\ \vdots \\ \mathbf{X}_N \end{bmatrix} \beta + \begin{bmatrix} \mathbf{X}_1 & 0 & \cdots & 0 \\ 0 & \mathbf{X}_2 & \cdots & 0 \\ \vdots & \vdots & \ddots & \vdots \\ 0 & 0 & \cdots & \mathbf{X}_N \end{bmatrix} \begin{bmatrix} \gamma_1 \\ \gamma_2 \\ \vdots \\ \gamma_N \end{bmatrix} + \begin{bmatrix} u_1 \\ u_2 \\ \vdots \\ u_N \end{bmatrix} \tag{23}$$

Firstly, each explanatory variable is normalized before modeling. Then it is possible to fit all measurement points at once based on Equation (23), and Figure 7 shows the fitted values of the random coefficient model and measured values for each measurement point. In addition, the fitting effects of the random coefficient model and stepwise regression method are compared by correlation coefficients R and standard deviation values S, as shown in Table 1. From the parameters R and S in Table 1, the results calculated by the random coefficient model are almost identical to the results calculated by the traditional stepwise regression model for each single point, with a very high fitting accuracy.

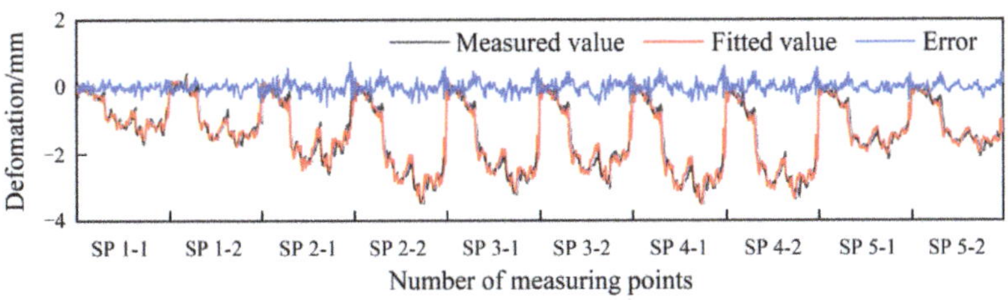

Figure 7. The fitted values of the random coefficient model and measured values of each point.

Table 1. The parameters *R* and *S* of the random coefficient model and stepwise regression method.

Random Coefficients Model						Stepwise Regression Method					
Points	***R***	***S***	**Points**	***R***	***S***	**Points**	***R***	***S***	**Points**	***R***	***S***
SP 1-1	0.974	0.108	SP 3-2	0.980	0.185	SP 1-1	0.974	0.108	SP 3-2	0.980	0.184
SP 1-2	0.980	0.127	SP 4-1	0.984	0.199	SP 1-2	0.981	0.127	SP 4-1	0.984	0.199
SP 2-1	0.970	0.206	SP 4-2	0.986	0.178	SP 2-1	0.970	0.206	SP 4-2	0.986	0.178
SP 2-2	0.986	0.186	SP 5-1	0.980	0.128	SP 2-2	0.986	0.186	SP 5-1	0.980	0.127
SP 3-1	0.985	0.179	SP 5-2	0.979	0.128	SP 3-1	0.985	0.179	SP 5-2	0.979	0.127

4.2.2. Identification of Overall Effect Values and Individual Effect Values

As shown by the high fitting accuracy in Section 4.2.1, it is feasible to use the random coefficient model to extract the overall effect values and individual effect values of sluice deformation. The common mean coefficients and the corresponding test statistics can be estimated by generalized least squares, as shown in Table 2. According to Table 2, the overall test result, i.e., Wald chi (12) = 209.75, and the significance test result, i.e., Prob > chi2 = 0.0, mean that using a random coefficient model to describe the sluice's deformation is reasonable. The coefficients of h^3, h^4, and H^3 are all equal to 0, as hey are completely collinear with other explanatory variables of water pressure components. Therefore, h^3, h^4, and H^3 can be removed from the model.

Table 2. Results of common mean coefficients.

Random-Coefficients Model						
R-sq: Overall = 0.986		**Wald chi2(12) = 209.75**			**Prob > chi2 = 0.00**	
var	**Coefficient**	**Std. Err.**	***z***	***p* > *z***	**[95% Conf. Interval]**	
h	−20.63	8.85	−2.33	0.020	−37.97	−3.29
h^2	1.40	0.60	2.34	0.019	0.23	2.57
h^3	0.00	(omitted)				
h^4	0.00	(omitted)				
H	833.66	108.05	7.72	0.000	621.89	1045.43
H^2	−50.00	6.49	−7.71	0.000	−62.71	−37.29
H^3	0.00	(omitted)				
H^4	0.053	0.007	7.70	0.000	0.04	0.07
T_1	0.035	0.004	8.02	0.000	0.03	0.04
T_2	0.018	0.002	7.41	0.000	0.01	0.02
T_3	0.012	0.002	5.10	0.000	0.01	0.02
T_4	0.021	0.002	9.04	0.000	0.02	0.03
T_5	0.013	0.002	5.43	0.000	0.01	0.02
T_6	0.010	0.002	4.86	0.000	0.01	0.01
T_7	0.015	0.002	6.64	0.000	0.01	0.02
θ	−0.445	0.173	−2.57	0.010	−0.78	−0.11
$\ln\theta$	−0.006	0.033	−0.17	0.863	−0.07	0.06
d	−3.200	0.258	−12.4	0	−3.71	−2.69

The overall effect values of sluice deformation can be obtained by substituting the common mean coefficients in Table 2 into Equation (6), and the individual effect values of each measurement point can be calculated by subtracting the overall effect values from the measured value, which represents the degree of deviation from the overall deformation. Figure 8 displays the values of overall effect and individual effect for the sluice's deformation. It can be seen that for each measuring point, the overall effect values are fixed and do not vary with the position of the measuring point, and the individual effect values are different from each other due to the nonuniformity of the external load, soil foundation, and other factors. In addition, individual effect values represent a pattern with varying measurement point positions. The individual effect values of measurement points on the 1# and 5# slabs, located on both sides of the sluice, are generally positive, implying that individual deformation shows a tendency towards upstream relative to overall deformation, due to the constraint effect at the bank wall. The individual effect values of measurement points on the 2–4# slabs, located in the middle of the channel, are almost negative, indicating that individual deformation shows a tendency towards downstream relative to overall deformation, because the deformations of measurement points are mainly affected by the water pressure. This phenomenon is consistent with the engineering practice, proving that it is rational to distinguish the overall and individual effects of the sluice by random coefficient model.

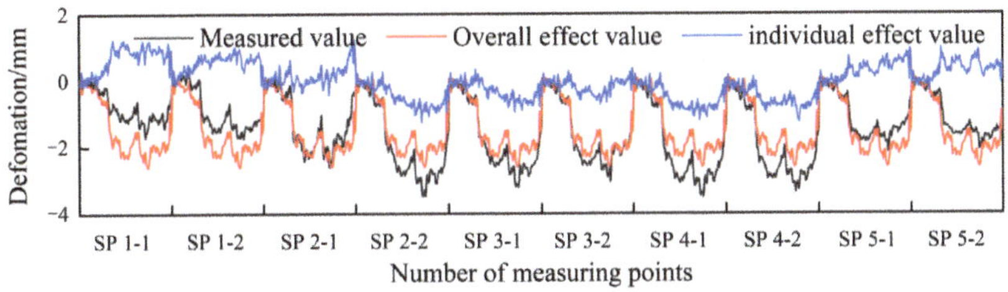

Figure 8. Overall effect values and individual effect values for each measurement point.

4.3. The Probability Distribution Function of Individual Effect Extreme Values

In Figure 8, it can be seen that the individual effect values of each measurement point are different. In order to evaluate the health status of nonuniform deformation, we firstly select the individual effect extreme values for each month, including the maximum and minimum values, and then calculate the probability distribution function of the extreme values for each measurement point separately. The coefficients λ_0, λ_1, λ_2, λ_3, and λ_4 of the probability distribution function for extreme values are shown in Tables 3 and 4, and the corresponding cumulative probability curves are shown in Figure 9. It can be observed that the probability distributions are different and unique for each measurement point.

Table 3. The probability distribution function coefficients of the monthly minimum for each measurement point.

Coefficient	SP1-1	SP1-2	SP2-1	SP2-2	SP3-1	SP3-2	SP4-1	SP4-2	SP5-1	SP5-2
λ_0	−0.59	−0.22	−0.02	−4.43	−2.50	−4.73	−3.75	−10.58	0.65	0.86
λ_1	2.90	2.71	−3.35	−17.96	−8.88	−47.72	−10.45	−29.75	2.34	2.67
λ_2	2.51	0.71	−1.87	−33.56	−20.50	−149.51	−12.97	−22.07	−3.61	−17.26
λ_3	2.47	6.11	10.05	−32.01	−34.38	−199.97	−11.73	−2.24	−11.71	−30.1
λ_4	−19.89	−29.66	−8.64	−11.62	−20.49	−100.20	−5.08	0.99	−109.19	−25.23

Table 4. The probability distribution function coefficients of the monthly maximum for each measurement point.

Coefficient	SP1−1	SP1−2	SP2−1	SP2−2	SP3−1	SP3−2	SP4−1	SP4−2	SP5−1	SP5−2
λ_0	−2.89	−7.96	−0.92	−0.15	−0.11	0.35	0.04	0.11	−1.11	−1.51
λ_1	5.96	12.94	−0.05	−0.05	1.23	1.39	0.20	−1.13	2.44	7.17
λ_2	−7.10	6.02	−0.33	−4.06	−2.31	−2.02	−4.34	−7.48	−3.81	−6.85
λ_3	8.18	−2.09	2.01	−5.05	−5.95	−4.43	−5.07	−10.09	9.92	−0.05
λ_4	−3.85	−9.79	−1.28	−1.92	−3.69	−30.63	−1.79	−5.62	−7.86	0.19

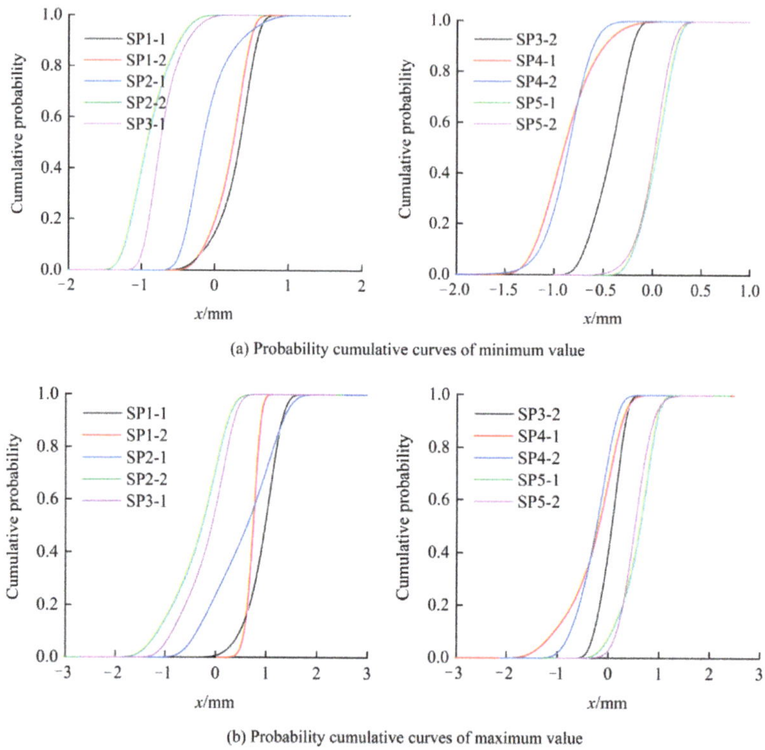

Figure 9. Probability cumulative curves for each measurement point.

4.4. Determination of Early Warning Indicators

Before determining the early warning indicators, we should first select the significance level P_α, that is, the possibility of a small-probability event occurring. In statistics, a small-probability event can be considered an almost impossible event, and if it occurs, this indicates that the deformation of the sluice is in an abnormal or warning state. The significance level is generally taken as 0.01, referring to the literature [30], and then the early warning indicators can be obtained for each measurement point, as shown in Table 5. Based on the overall effect values, we can determine the normal deformation range for each measurement point; for instance, the normal deformation ranges of points SP3-2 and SP5-2 are shown in Figure 10. When the measured values are in the normal range, it indicates that the nonuniform deformation of measurement points is in a normal state, which can assist in the judgment of the uneven deformation state of the sluice structure.

Table 5. Early warning indicator of individual effect values for each measurement point.

Measuring Point	SP1-1	SP1-2	SP2-1	SP2-2	SP3-1	SP3-2	SP4-1	SP4-2	SP5-1	SP5-2
Lower limit/mm	−0.417	−0.363	−0.580	−1.389	−1.084	−0.838	−1.399	−1.451	−0.335	−0.457
Upper limit/mm	1.504	1.020	1.657	0.521	0.599	0.479	0.536	0.368	1.141	1.190

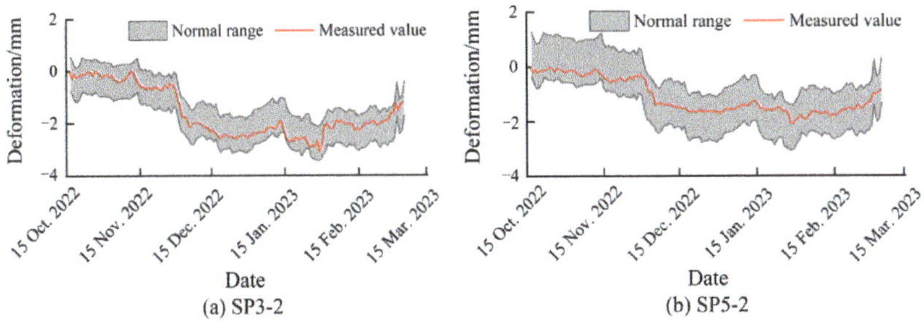

Figure 10. Analysis diagram of deformation behavior for measurement points SP3-2 and SP5-2.

5. Conclusions

Consideration of the sluice structure should focus on the issue of nonuniform deformation due to the uneven property of soft soil foundation. When analyzing the deformation state of a sluice, the traditional methods could not take the nonuniform deformation into account, which is a potential risk in structural safety evaluation. From the perspective of nonuniform deformation, this article proposes a method to determine the deformation behavior of sluice structures. Firstly, we divide the deformation of sluice structures into overall effect and individual effect based on their homogeneity and heterogeneity characteristics, describing the overall deformation trend of the sluice structure and deviation degree between the deformation of single points and the overall deformation separately. Then, in order to explore the influence of hidden factors, panel data theory is introduced to identify the values of overall and individual effects for sluice deformation. Finally, the probability distribution function of individual effect values at each measurement point can be solved based on the maximum entropy principle, and furthermore, the deformation warning indicators for each measurement point can be determined, which provides a novel approach to judging the deformation status of sluice structures.

It should be noted that only the sluice structure is investigated in this study, and future research should consider the potential effects of uneven deformation on other hydraulic structures such as earth dams, embankments, and pump station buildings more carefully.

Author Contributions: Conceptualization, B.L. and Z.M.; Data curation, X.L. (Xiang Luo); Formal analysis, B.L. and Z.M.; Funding acquisition, Z.S. and F.M.; Investigation, D.L.; Methodology, B.L. and Z.M.; Project administration, F.M.; Resources, F.M. and B.L.; Software, W.Y.; Supervision, B.L.; Validation, B.L., Z.M. and Z.S.; Visualization, X.L. (Xing Li); Writing—original draft, B.L. and Z.M.; Writing—review and editing, B.L. and Z.M. All authors have read and agreed to the published version of the manuscript.

Funding: This research was supported by the Modern Multimodal Transportation Laboratory Open Fund Project, grant number MTF2023010, the Yellow River Water Science Research Joint Fund, grant number U2243244, and the National Natural Science Foundation of China, grant number 52179130 and 52209165.

Data Availability Statement: Data is contained within the article.

Conflicts of Interest: The authors declare no conflicts of interest.

References

1. Liang, J.; Li, Z.; Ji, Q.; Lu, W.; Cao, Q.; Ahmed, E. Global Sensitivity Analysis of The Deformation Behavior of Sluice Chamber Structure. *Structures* **2021**, *34*, 4682–4693. [CrossRef]
2. Gu, Z.; Cao, X.; Liu, G.; Lu, W. Optimizing Operation Rules of Sluices in River Networks Based on Knowledge-driven and Data-driven Mechanism. *Water Resour. Manag.* **2014**, *28*, 3455–3469. [CrossRef]
3. Peng, J.; Xie, W.; Wu, Y.; Sun, X.; Zhang, C.; Gu, H.; Zhu, M.; Zheng, S. Prediction for the Sluice Deformation Based on SOA-LSTM-Weighted Markov Model. *Water* **2023**, *15*, 3724. [CrossRef]
4. Zhang, G.; Yu, C.; Guo, G.; Li, L.; Zhao, Y.; Li, H.; Gong, Y. Monitoring Sluice Health in Vibration by Monocular Digital Photography and a Measurement Robot. *KSCE J. Civ. Eng.* **2019**, *23*, 2666–2678. [CrossRef]

5. Xue, B.; Zhang, S.; Fang, H.; Li, M.; Shi, M. Design Method of Polymer Cut-off Wall Density for Earth Dams Based on Mult Objective Optimization. *Structures* **2023**, *53*, 199–204. [CrossRef]
6. Xu, W.; Niu, X.; Zhu, Y. Deformation Behavior and Damage Evaluation of Fly Ash-Slag Based Geopolymer Concrete Under Cycli Tension. *J. Build. Eng.* **2024**, *86*, 108664. [CrossRef]
7. Li, F.; Wang, Z.; Liu, G.; Fu, C.; Wang, J. Hydrostatic Seasonal State Model for Monitoring Data Analysis of Concrete Dams. *Struc Infrastruct. Eng.* **2015**, *11*, 1616–1631. [CrossRef]
8. Gamse, S.; Oberguggenberger, M. Assessment of Long-term Coordinate Time Series Using Hydrostatic-Season-Time Model fo Rock-Fill Embankment Dam. *Struct. Control Health Monit.* **2017**, *24*, e1859. [CrossRef]
9. Song, J.; Zhang, S.; Chen, Y. Long-term Deformation Safety Evaluation Method of Concrete Dams Based on The Time-Varying Stability of Concrete Material. *Mater. Today Commun.* **2023**, *36*, 106468. [CrossRef]
10. Salazar, F.; Morán, R.; Toledo, M.; Oñate, E. Data-Based Models for The Prediction of Dam Behaviour: A Review and Some Methodological Considerations. *Arch. Comput. Methods Eng.* **2017**, *24*, 1–21. [CrossRef]
11. Ma, L.; Ma, F.; Cao, W.; Lou, B.; Luo, X.; Li, Q.; Hao, X. A Multi-Strategy Improved Sooty Tern Optimization Algorithm fo Concrete Dam Parameter Inversion. *Water* **2024**, *16*, 119. [CrossRef]
12. Cao, M.; Qiao, P.; Ren, Q. Improved Hybrid Wavelet Neural Network Methodology for Time-Varying Behavior Prediction o Engineering Structures. *Neural Comput. Appl.* **2009**, *18*, 821–832. [CrossRef]
13. Mata, J. Interpretation of Concrete Dam Behaviour with Artificial Neural Network and Multiple Linear Regression Models. *Eng Struct.* **2011**, *33*, 903–910. [CrossRef]
14. Ranković, V.; Grujović, N.; Divac, D.; Milivojević, N. Development of Support Vector Regression Identification Model fo Prediction of Dam Structural Behaviour. *Struct. Saf.* **2014**, *48*, 33–39. [CrossRef]
15. Wang, S.; Xu, C.; Liu, Y.; Gu, H.; Xu, B.; Hu, K. Spatial Association-Considered Real-time Risk Rate Assessment of High Arch Dams Using Observed Displacement and Combination Prediction Model. *Structures* **2023**, *53*, 1108–1121. [CrossRef]
16. Lu, T.; Gu, C.; Yuan, D.; Zhang, K.; Shao, C. Deep Learning Model for Displacement Monitoring of Super High Arch Dams Basec on Measured Temperature Data. *Measurement* **2023**, *222*, 113579. [CrossRef]
17. Li, Y.; Bao, T.; Chen, H.; Zhang, K.; Shu, X.; Chen, Z.; Hu, Y. A Large-Scale Sensor Missing Data Imputation Framework for Dams Using Deep Learning and Transfer Learning Strategy. *Measurement* **2021**, *178*, 109377. [CrossRef]
18. Yang, X.; Yuan, C.; Hou, M.; Zhou, C.; Ju, Y.; Qi, F. Law and Early Warning of Vertical Sluice Cluster Displacements in Soft Coasta Soil. *KSCE J. Civ. Eng.* **2023**, *27*, 698–711. [CrossRef]
19. Yang, X.; Wang, D.; Xu, Y.; Hou, M.; Wang, Z. Performance Assessment of InSAR-Based Vertical Displacement Monitoring o Sluices in Coastal Soft Soil Area. *KSCE J. Civ. Eng.* **2022**, *26*, 371–380. [CrossRef]
20. Lou, B.; Ma, F.; Luo, X. Abnormality Diagnosis for Deformation Behavior of Pump Station Buildings Based on Nonlinear Dynamics Model. *J. Hydraul. Eng.* **2023**, *54*, 486–496.
21. Shi, Z.; Gu, C.; Qin, D. Variable-Intercept Panel Model for Deformation Zoning of A Super-High Arch Dam. *SpringerPlus* **2016**, *5* 898. [CrossRef] [PubMed]
22. Shao, C.; Gu, C.; Yang, M.; Xu, Y.; Su, H. A Novel Model of Dam Displacement Based on Panel Data. *Struct. Control Health Monit* **2018**, *25*, e2037. [CrossRef]
23. Zhao, E.; Wu, C. Risk Probabilistic Assessment of Ultrahigh Arch Dams Through Regression Panel Modeling on Deformation Behavior. *Struct. Control Health Monit.* **2021**, *28*, e2716. [CrossRef]
24. Cui, X.; Gu, H.; Gu, C.; Cao, W.; Wang, J. A Novel Imputation Model for Missing Concrete Dam Monitoring Data. *Mathematics* **2023**, *11*, 2178. [CrossRef]
25. Hu, J.; Ma, F. Zoned Deformation Prediction Model for Super High Arch Dams Using Hierarchical Clustering and Panel Data *Eng. Comput.* **2020**, *37*, 2999–3021. [CrossRef]
26. Swamy, P.A. Efficient Inference in a Random Coefficient Regression Model. *Econometrica* **1970**, *38*, 311–323. [CrossRef]
27. Zellner, A. An Efficient Method of Estimating Seemingly Unrelated Regressions and Tests for Aggregation Bias. *J. Am. Stat. Assoc* **1962**, *57*, 348–368. [CrossRef]
28. Jaynes, E.T. On the Rationale of Maximum Entropy Methods. *Proc. IEEE* **1982**, *70*, 939–952. [CrossRef]
29. Zhang, J.; Gu, C. Maximum Entropy Method for Operational Loads Feedback Using Concrete Dam Displacement. *Entropy* **2015** *17*, 2958–2972. [CrossRef]
30. Wu, Z.; Shen, C.; Ruan, H. *Safety Monitoring Theory and Its Application of Hydraulic Structures*, 3rd ed.; Higher Education Press Beijing, China, 2003.

water

MDPI

Article

Integrated Time-Dependent Analysis of a Hydraulic Structure on Soft Foundations during Construction

Chao Xu [1,2], Liang Ye [1,2], Suli Pan [3] and Wen Luo [4,*]

1 School of Civil Engineering and Architecture, Zhejiang University of Science and Technology, Hangzhou 310023, China; xuchao@zust.edu.cn (C.X.); yeliang88@126.com (L.Y.)
2 Zhejiang-Singapore Joint Laboratory for Urban Renewal and Future City, Hangzhou 310023, China
3 School of Water Conservancy and Environment Engineering, Zhejiang University of Water Resources and Electric Power, Hangzhou 310018, China; pansl@zjweu.edu.cn
4 Department of Hydraulic Engineering, Zhejiang Tongji Vocational College of Science and Technology, Hangzhou 311231, China
* Correspondence: z11320080903@zjtongji.edu.cn

Abstract: An integrated model that considers multiphysics is necessary to accurately analyze the time-dependent response of hydraulic structures on soft foundations. This study develops an integrated superstructure–foundation–backfills model and investigates the time-dependent displacement and stress of a lock head project on a soft foundation during the construction period. Finite element analyses are conducted, incorporating a transient thermal creep model for concrete and an elasto-plastic consolidation model for the soil. The modified Cam-clay model is employed to describe the elasto-plastic behavior of the soil. Subsequently, global sensitivity analyses are conducted to determine the relative importance of the model parameters on the system's response, using Garson's and partial derivative algorithms based on the backpropagation (BP) neural network. The results indicate that the integrated system exhibits pronounced time-dependent displacement and stress, with dangerous values appearing during specific periods. These values are easily neglected, highlighting the importance of integrated time-dependent analysis. Construction activities, particularly the backfilling process, could cause a sudden change in stress and significantly impact the stress redistribution of the superstructure. Additionally, the mechanical properties of concrete have a significant impact on the stress on the superstructure, while the mechanical properties of the soil control the settlement of the integrated system.

Keywords: hydraulic structure; soft foundation; integrated numerical analysis; time-dependent behavior; global sensitivity analysis; BP neural network

1. Introduction

In coastal and plain areas, the predominant soil type is typically soft soil; in particular, coastal areas are characterized by abundant soils with high water content and low permeability [1]. Hydraulic projects in these regions are often built along coastlines, such as tidal barriers, or within river channels, such as ship locks. Therefore, they can collectively be referred to as hydraulic structures on soft foundations. These projects are usually considered as an integrated system consisting of three main subsystems: the concrete superstructure, soft soil foundation, and backfilled soil. The interactions between these three subsystems are highly complex. Notably, the load from the superstructure influences the pore water pressure and consolidation settlement of the soil [2]. Simultaneously, the deformation of the soil influences the redistribution of stress in the concrete superstructure [3]. Such a sophisticated system should have integrated analysis to improve its design. Each subsystem has various time-dependent physical fields, such as the transient temperature field and creep deformation in the concrete structure, as well as the pore water pressure and consolidation deformation in the soft soil. These fields contribute to the complexity of the entire project, representing a sophisticated system with a time-dependent process arising from multifield coupling. Consequently, analyses with simplified load and boundary conditions under a limited number of severe load cases cannot comprehensively assess the deformation and stress response of the entire system [4]. This approach proves inadequate to accurately assess the safety of hydraulic structures on soft foundations, particularly during the construction phase, when the complexity of physical field coupling is most significant.

Researchers and engineers have acknowledged the interactions between the structure and soil, as well as their influence on project analyses. Initially, it was challenging to consider the effects of interactions, due to limitations in analytical theories and solution methods. Subsequently, researchers proposed the elastic foundation beam method [5], which considers the coordinated deformation at the interface between the beam structure and the elastic foundation, as well as the stiffness of the structure. However, this method did not account for the influence of the stiffness of the superstructure above the beam on the deformation of the superstructure–beam–foundation system, or the time-dependent effects of the system. Later, soil models that accounted for consolidation settlement or viscoelastic deformation were incorporated into the foundation beam method to analyze the system's time-dependent variations, thereby obtaining historical deformation and stress information. For instance, Elhuni et al. [6] employed a viscoelastic foundation model to analyze the dynamic deformation of Euler–Bernoulli beams resting on multilayered soil under oscillating moving loads, and they obtained a time-dependent response to the problem. Simultaneously, Elhuni et al. [7] also utilized Biot's consolidation theory to study the consolidation settlement of flexible foundations on porous elastic layered soil. Lanes et al. [8] combined the boundary element method, Mindlin's fundamental solution, and Terzaghi's consolidation theory to analyze the response of viscoelastic soil–structure interaction for foundations and frame structures on soft soil. With advancements in computational technology and numerical analysis software, complex numerical models can now be seamlessly integrated into analysis frameworks to consider the impact of the superstructure's stiffness on the superstructure–foundation–soil system. For example, Ai et al. [9] connected the stiffness matrices of the superstructure and slab using substructure technology. They employed a boundary element–finite element coupling method to investigate the time-dependent response of the interaction between the slab and viscoelastic saturated soil. Gorini et al. [10] developed a comprehensive system model for the soil–abutment–superstructure in bridge projects to analyze the impact of the local dynamic response of the bridge abutment on the entire system. Simultaneously, more advanced constitutive models, such as elastoplastic soil models and fractional models, have been introduced into numerical analyses to better understand complex system responses. For example, Ai et al. [11] used a fractional viscoelastic cross-anisotropic saturated soil model to analyze the time-dependent behavior of the interaction between the soil and slab. Shamsi et al. [12] conducted a nonlinear analysis of the interaction between factors such as topography, structure, and soil that affect buildings near slopes. They utilized an elastoplastic hysteretic model with anisotropic hardening.

With contributions from previous research advancements, as indicated by the literature mentioned above, the mechanism and degree of influence of the interactions between the soil and structure have been progressively revealed, significantly enhancing the effectiveness and accuracy of analysis. However, for hydraulic structures characterized by significant concrete consumption, which are typically massive concrete structures often surrounded by backfilled soil to enhance stability, the analysis during the construction stage needs to consider factors such as transient thermal stresses and concrete creep, as well as the impact of backfills on the system. Duncan et al. [13] conducted research using the finite element method to study the deformation and stress response of the structure–backfill–foundation system of the Port Allen lock throughout the entire construction process in its early stages. However, this study did not take into account the consolidation process of the subsoil and the thermal creep stress of the concrete. More related studies, such as the one by Rui et al. [14], have primarily focused on the interactions between backfills and retaining walls, without effectively integrating them into the soil–structure system for comprehensive analysis. The literature review reveals a lack of comprehensive research on various interactions in hydraulic structures. In bridge engineering, research on the integral bridge system [15] has focused on soil–structure interaction, thermal loads, concrete creep and shrinkage, and the influence of backfilled soil [16]. Therefore, it is essential to extract valuable insights from such research and conduct integrated studies on hydraulic structures constructed on soft foundations.

To explore the time-dependent behavior of hydraulic structures on soft foundations under multiphysics and complex interactions, this study uses a lock head project as a case study and establishes an integrated superstructure–foundation–backfills system. The lock head's floor has a large contact area with the foundation, and there is extensive contact between the sidewalls and backfilled soil. Simultaneously, this system exemplifies a typical multifield coupled system, involving factors such as the temperature field, the creep thermal stress field of concrete, and the seepage field of the soil. The deformation and stress development of the entire system can be influenced by these fields. This study constructs an integrated finite element model that includes the lock head, foundation, and backfills. Initially, the transient temperature field of this system is calculated during the construction phase. Subsequently, an integrated numerical analysis is conducted based on the thermal results, considering concrete creep and a coupled elasto-plastic consolidation that incorporates Biot's consolidation theory and the modified Cam-clay model. Furthermore, to investigate the influence of calculation parameters on the deformation and stress of this system, a global multi-parametric sensitivity analysis of the mechanical parameters of concrete and soil is conducted. The sensitivity analysis can assess the relative importance of model parameters on results [17]. Currently, various methods such as grey theory [18], particle swarm optimization algorithm [19], and BP neural network [20] have been used for parametric sensitivity analysis. In this study, sensitivity analyses are conducted using the partial derivative and Garson's algorithm based on the BP neural network. These two methods mainly employ the intrinsic topological structure and parameters of the neural network, leading to better interpretability compared to other methods. The structure of this paper is organized as follows: Section 2 briefly introduces the main mathematical and physical equations used in the integrated numerical model to illustrate how multifield are considered in this study. Section 3 provides a detailed description of the lock head project, including basic geometric information, calculation parameter values, and the construction process considered in the time-dependent analysis. Section 4 demonstrates and discusses the results, focusing on the deformation and stress variation of the structure over time, as well as the findings of the parameter sensitivity analysis. Section 5 presents the primary conclusions of the current study.

2. Methods

2.1. Physical Equations for Finite Element Analyses

The transient temperature distribution of the concrete–foundation–backfills system is calculated initially as the basis for solving the stress states of this system. The governing equation for transient thermal calculation is as follows [21]:

$$c\rho\frac{\partial T}{\partial t} = k\left(\frac{\partial^2 T}{\partial x^2} + \frac{\partial^2 T}{\partial y^2} + \frac{\partial^2 T}{\partial z^2}\right) + \frac{\partial Q(t)}{\partial t} \quad (1)$$

where T is the temperature; t is the time; x, y, and z are spatial coordinates; k is the thermal conductivity coefficient; c is the specific heat capacity; ρ is the mass density; and $Q(t)$ represents the accumulated hydration heat of concrete at time t, and it equals zero for soil.

In hydraulic projects, the commonly considered boundary condition is the convection phenomenon, which can be expressed as follows:

$$k\frac{\partial T}{\partial n} = -\beta(T - T_a) \quad (2)$$

where n denotes the normal direction of the surface; β represents the surface heat transfer coefficient; and T_a refers to the environmental temperature.

For transient thermal calculations, the initial condition specified below in Equation (3) should also be considered.

$$T_0(x,y,z) = T(x,y,z,t=0) \tag{3}$$

After calculating the temperature, the thermal strain can be determined using the following formula:

$$\varepsilon_T = \{\alpha\Delta T\ \alpha\Delta T\ \alpha\Delta T\ 0\ 0\ 0\} \tag{4}$$

where ε_T represents the thermal strain vector; ΔT denotes the temperature difference; and α is the thermal expansion coefficient.

Stress induced by thermal strain is calculated using the creep model described by Equations (5) and (6), as follows:

$$E(\tau) = E_0\left[1 - \exp\left(a\tau^b\right)\right] \tag{5}$$

$$C(t,\tau) = \sum_{s=1} \psi_s(\tau)\{1 - \exp[-r_s(t-\tau)]\} \tag{6}$$

where $E(\tau)$ and $C(t,\tau)$ represent the ultimate elastic modulus and unit creep, respectively; τ denotes the loading time, while t is the calculation time; E_0 stands for the ultimate elastic modulus of the concrete; and other parameters are the material constants that have been tested.

In the mechanical simulation phase during construction, consolidation calculations are carried out for soil using Biot's three-dimensional coupled theory [22], which is based on the effective stress principle and Darcy's law. The governing equations are as follows:

$$\begin{cases} G\nabla^2 u - \left(\frac{3K+G}{3}\right)\frac{\partial\varepsilon_v}{\partial x} - \frac{\partial p}{\partial x} + f_x = 0 \\ G\nabla^2 v - \left(\frac{3K+G}{3}\right)\frac{\partial\varepsilon_v}{\partial y} - \frac{\partial p}{\partial y} + f_y = 0 \\ G\nabla^2 w - \left(\frac{3K+G}{3}\right)\frac{\partial\varepsilon_v}{\partial z} - \frac{\partial p}{\partial z} + f_z = 0 \end{cases}$$
$$\frac{\partial\varepsilon_v}{\partial t} + \frac{k_s}{\gamma_w}\frac{\partial^2 p}{\partial x^2} + \frac{k_s}{\gamma_w}\frac{\partial^2 p}{\partial y^2} + \frac{k_s}{\gamma_w}\frac{\partial^2 p}{\partial z^2} = 0 \tag{7}$$

where u, v, and w are displacements; ε_v represents the volumetric strain; K denotes the volumetric modulus; G is the shear modulus; p signifies the pore pressure; f_x, f_y, and f_z are volumetric forces; k_s is the soil permeability coefficient; and γ_w represents the unit weight of water.

The boundary and initial conditions of the pore pressure are described in Equations (8) and (9), as follows:

$$\begin{cases} p = 0 \text{ on drainage surface} \\ \frac{\partial p}{\partial n} = 0 \text{ on non-drainage surface} \end{cases} \tag{8}$$

$$p_0(x,y,z) = p(x,y,z,t=0) \tag{9}$$

In this paper, the modified Cam-clay (MCC) model [23,24] is adopted to accurately demonstrate the behavior of soil in terms of stress–strain relationships. The MCC model is an elasto-plastic model with the yield function of Equation (10). Theoretical details can be found in the relevant literature [25–27].

$$\left(p' - \frac{p'_c}{2}\right) + \left(\frac{q}{M}\right)^2 = \left(\frac{p'_c}{2}\right)^2 \tag{10}$$

where p' is the mean effective stress; q is the deviator stress; M represents the slope of the critical state line in the $p' - q$ plane as shown in Figure 1; and p'_c is the preconsolidation pressure.

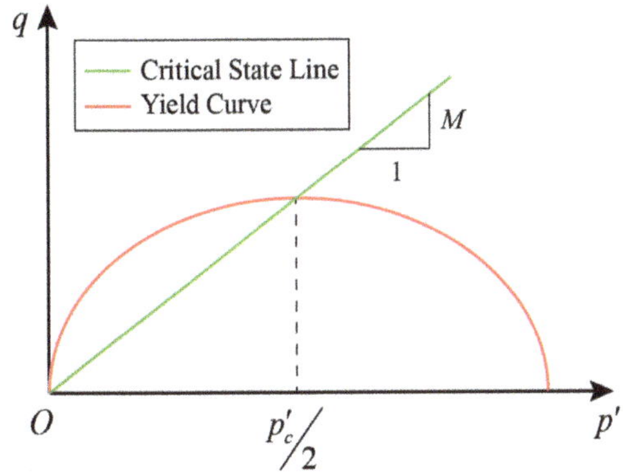

Figure 1. Yield surface of the modified Cam-clay model in the p'–q plane.

2.2. Backpropagation (BP) Neural Network with One Hidden Layer

A BP neural network, which is short for error backpropagation neural network, is a commonly used type of neural network in engineering [28]. It has been proven that any nonlinear relationships can be approximated using a two-layer perceptron network [29]. Hence, the one-hidden-layer backpropagation (BP) neural network, as demonstrated in Figure 2, is utilized in this paper.

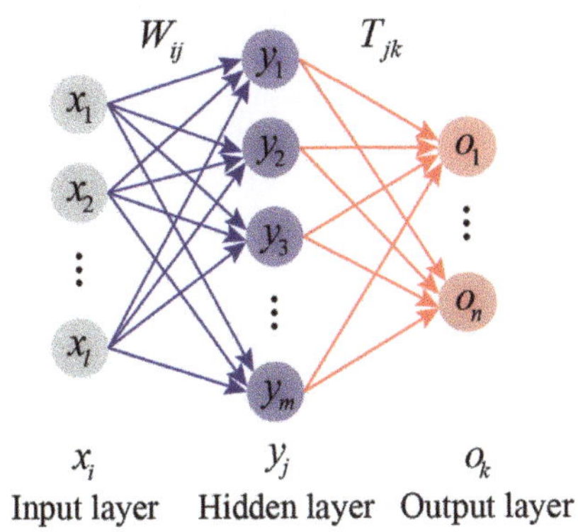

Figure 2. The diagram of the BP neural network model with one hidden layer.

The output of a node in the BP neural network can be calculated using Equation (11) according to the inputs, weights, and thresholds of the layer to which the node belongs, as follows:

$$out_j = f\left(\sum_i w_{ij} in_i + \theta_j\right) \tag{11}$$

where out_j is the output value of the jth output node in this layer; in_i is the input value of the ith input node in this layer; w_{ij} is the weight that connects the ith input node and the jth output node; θ_j is the threshold of the jth output node; and f represents the transfer function of this layer.

There are various types of transfer functions in the BP neural network, including the threshold function, piecewise linear function, sigmoid function, and hyperbolic tangent function. The sigmoid function is adopted as the transfer function in this study due to its advantageous derivative format, as shown in Equations (12) and (13) below:

$$f(x) = \frac{1}{1 + e^{-x}} \tag{12}$$

$$f'(x) = f(x)[1 - f(x)] \tag{13}$$

2.3. *Global Sensitivity Analysis Methods Based on Backpropagation (BP) Neural Network*

Nowadays, there are various types of sensitivity analysis methods, such as derivative-based, distribution-based, variogram-based, regression-based, and response surface-assisted approaches [30]. In this paper, two methods based on the BP neural network are employed for sensitivity analysis.

2.3.1. Garson's Algorithm

Garson's algorithm is a widely used method for sensitivity analysis by many researchers [31]. This method estimates the relative importance of input variables to output variables [32]. Goh [33] improved this method to enhance its applicability to complex systems, as follows:

$$Q_{ik} = \frac{\sum_{j=1}^{m} \left|W_{ij}T_{jk}\right| / \sum_{r=1}^{l} \left|W_{rj}\right|}{\sum_{i=1}^{l} \sum_{j=1}^{m} \left(\left|W_{ij}T_{jk}\right| / \sum_{r=1}^{l} \left|W_{rj}\right|\right)} \tag{14}$$

where Q_{ik} is the sensitivity coefficient that represents the relative importance of the ith input variable to the kth output variable.

Sensitivity coefficients calculated using Garson's algorithm are less than 1.0, and the sum of all sensitivity coefficients in a system equals 1.0. Therefore, the results obtained using this method can be considered as the influence weights of input variables on output variables. However, some comparative studies show that the results from Garson's algorithm may not be very reliable [34]. Therefore, the partial derivative algorithm is also utilized in the current study.

2.3.2. Partial Derivative Algorithm

The convenient calculation of the derivative of the sigmoid function, with respect to the input variable, enables the easy derivation of the analytical partial derivative expression of output variables, with respect to input variables in the BP neural network.

According to the chain rule, the partial derivative of the output variable o_k, with respect to the input variable x_i, is as follows:

$$\frac{\partial o_k}{\partial x_i} = \sum_{j=1}^{m} \frac{\partial o_k}{\partial y_j} \frac{\partial y_j}{\partial x_i} \tag{15}$$

The right-hand side of Equation (15) can be expanded to Equation (16), according to Equation (13).

$$\begin{aligned} \frac{\partial o_k}{\partial y_j} &= T_{jk} o_k (1 - o_k) \\ \frac{\partial y_j}{\partial x_i} &= W_{ij} y_j (1 - y_j) \end{aligned} \tag{16}$$

Then, the formula to calculate the sensitivity coefficients can be obtained as in Equation (17).

$$\frac{\partial o_k}{\partial x_i} = o_k(1 - o_k) \sum_{j=1}^{m} T_{jk} W_{ij} y_j (1 - y_j) \tag{17}$$

2.4. Comprehensive Research Flow

Figure 3 comprehensively illustrates the research flow, which primarily consists of the following steps:

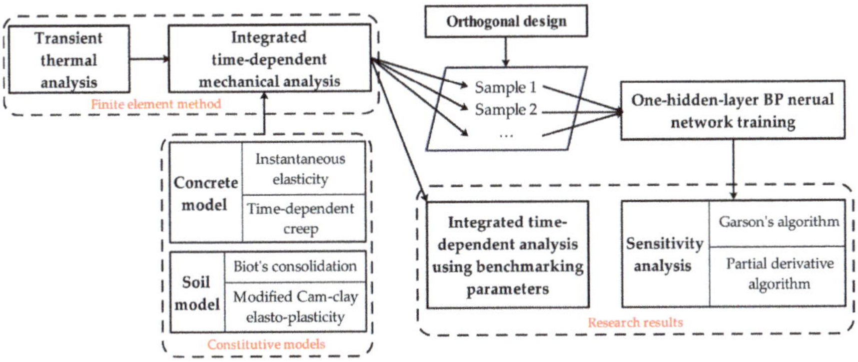

Figure 3. The comprehensive flowchart of the entire research process.

(1) The "finite element method" block is the core of this integrated time-dependent analysis, utilizing the governing equations and constitutive models described in Section 2.1. Initially, transient thermal analysis is conducted to calculate the temperature field of the integrated model, which serves as the foundational data for subsequent mechanical analysis. Subsequently, an integrated time-dependent mechanical analysis is conducted. Concrete utilizes instantaneous elastic and transient thermal creep models, while the soil incorporates Biot's consolidation theory and an elasto-plastic model based on the modified Cam-clay model, as depicted in the "constitutive models" block of Figure 3.

(2) By employing the integrated time-dependent mechanical analysis method, a time-dependent analysis of the integrated model is conducted using benchmarking parameters. The results are then analyzed to study the time-dependent deformation and stress evolution of the hydraulic structure on soft foundations, as indicated in the left part of the "research results" block in Figure 3.

(3) The orthogonal design method is employed to create multiple samples for sensitivity analysis, where the parametric levels in the orthogonal design are multiples of the benchmark parameters. Each sample is computed using the integrated time-dependent mechanical analysis method, and then complete sample data are obtained for training the BP neural network with one hidden layer. Subsequently, the parameters of the BP neural network are obtained.

(4) Based on the trained network parameters, a sensitivity analysis of the mechanical parameters in the integrated time-dependent mechanical analysis is conducted using Garson's and partial derivative algorithms to assess the influence of each parameter on the results (the displacement and stress), as illustrated in the right part of the "research results" block in Figure 3.

3. Project Background

3.1. Project Overview

A ship lock head constructed on a soft foundation is used as a validation case in this study. The total length of the lock head structure along the river is 28.5 m, and the total width across the river is 53.8 m. The total height of the lock head is 11.9 m, and the height of the floor is 2.6 m.

The foundation pit for this project was excavated in two steps with varying slope coefficients, and a platform was installed between the two steps. The size parameters and slope coefficients are shown in Figure 4.

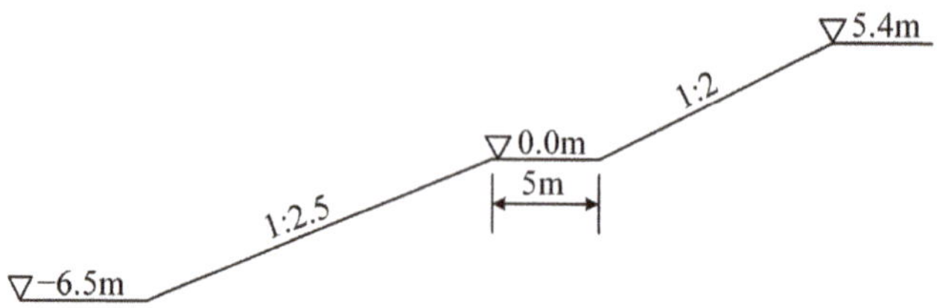

Figure 4. The diagram of the foundation pit excavation.

Structural concrete was cast in three main parts, from bottom to top: (1) the floor; (2) from the top of the floor to the bottom of the hollow box section; and (3) from the bottom of the hollow box section to the top of the head. The backfilling process was also divided into three layers, and the elevations of the top surfaces of the three layers from bottom to top were consistent with the corresponding concrete parts. Concrete was cast in batches and layer by layer. The complete construction process, including concrete casting and soil backfilling, is detailed in Table 1. Overall, the total simulation duration is 377 days. The finite element mesh for the numerical analysis of the concrete–foundation–backfilling system is illustrated in Figure 5.

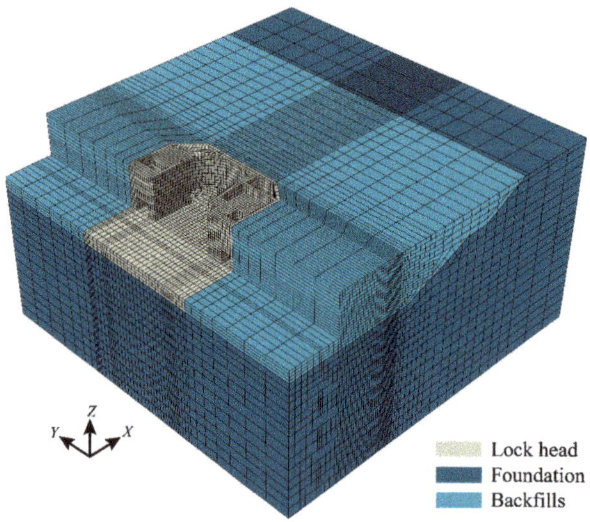

Figure 5. The finite element mesh of the system for numerical calculations.

Table 1. Construction processes of the ship lock head project.

No.	Step Name	Duration (Days)
1st	Casting central floors	8
2nd	Casting side floors	8
3rd	Dismantling formworks of the floor	34
4th	Backfilling the first soil layer	110
5th	Casting water corridor section	10
6th	Dismantling formworks of water corridor section	25
7th	Backfilling the second soil layer	40
8th	Casting hollow box section	10
9th	Dismantling formworks of hollow box section	55
10th	Backfilling the third soil layer	77

3.2. Material Properties

3.2.1. Thermal Parameters

For thermal calculations during construction, the thermal properties of concrete and soil are provided in Table 2. The temperature distribution of the subsoil was calculated for several decades prior to the start of floor casting, taking into account the excavation of the foundation pit. The calculated results were then used as the initial subsoil temperature to simulate the project. The initial temperature of the backfilled soil was 15 °C. The initial temperatures of the three concrete parts mentioned in Section 3.1 were 15 °C, 30 °C, and 20 °C, respectively.

Table 2. Thermal parameters of concrete and soil.

Material	k kJ/(m·d·°C)	c kJ/(kg·°C)	β kJ/(m^2·d·°C)
Concrete	200.145	0.984	413 with formworks 1360 without formworks
Subsoil and backfilled soil	100.63	1.005	500

Figure 6 illustrates the fluctuation of environmental temperature throughout the calculation period. The adiabatic temperature rise model for concrete lock heads is defined as follows:

$$\theta(\tau) = 51.034\left[1 - \exp\left(-0.701\tau^{1.067}\right)\right] \tag{18}$$

where $\theta(\tau)$ represents the adiabatic temperature of concrete at age τ.

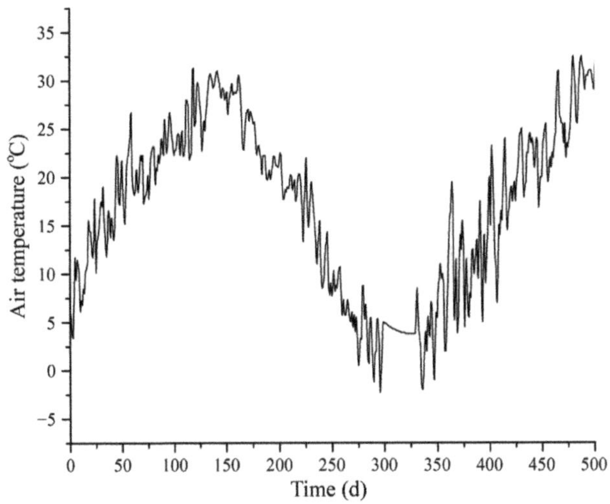

Figure 6. The temperature variation during the calculation period.

3.2.2. Mechanical Parameters

The subsoil beneath the lock head is complex, showing apparent stratifications. According to the prospecting reports, the subsoil can be divided into nine layers for analysis. In this paper, the MCC model is utilized to describe the behaviors of subsoil and backfilled soil. Therefore, the initially anticipated soil properties were transformed into parameters of the MCC model using regression formulas developed by Chen et al. [35], and then slightly adjusted based on other investigations [36–39]. The converted MCC model parameters are listed in Table 3, along with other parameters for calculation, such as initial status parameters, permeability coefficients, and mechanical parameters in Table 4. The calculation parameters of the first layer are used for the backfilled soil. The Poisson's ratios of the subsoil and backfilled soil are both 0.3.

Table 3. MCC model parameters of each subsoil layer.

No.	λ	κ	M
1	0.0836	0.00706	1.351
2	0.0820	0.00697	1.052
3	0.0341	0.00454	1.063
4	0.1661	0.01126	0.993
5	0.1381	0.00983	0.852
6	0.0671	0.00622	1.102
7	0.0919	0.00748	1.067
8	0.1117	0.00848	0.864
9	0.0853	0.00714	0.933

Table 4. Other calculation parameters of each subsoil layer.

No.	K_0	p'_{c0}	OCR	c_u kPa	k_s m/d	ρ kg/m^3
1	1.312	217.667	7.0	65.925	0.5	1880
2	0.962	191.361	4.0	47.436	0.003	1900
3	0.594	158.689	2.0	40.815	0.086	1870
4	0.664	183.466	1.5	45.918	0.776	1900
5	0.587	247.882	1.3	53.009	0.003	2010
6	0.452	243.234	1.1	67.786	0.003	1880
7	0.459	287.498	1.0	77.975	0.776	1900
8	0.483	409.240	1.0	89.043	0.003	1980
9	0.450	450.587	1.0	105.768	0.003	1890

In Tables 3 and 4, ρ represents the mass density; λ denotes the slope of the critical state line in the $e - \ln p'$ plane, where e is the void ratio of the soil; κ stands for the slope of the unloading–reloading line in the $e - \ln p'$ plane; K_0 signifies the static lateral earth pressure coefficient; p'_{c0} indicates the initial yield stress of the soil, OCR refers to the over-consolidation ratio of the soil; and c_u is the undrained shear strength of the soil.

The expansion coefficients of concrete and soil are 0.9×10^{-5}/°C. The mass density of concrete is 2400 kg/m^3, and its Poisson's ratio is 0.167. The elastic modulus and unit creep of the locking head concrete are expressed by Equations (19) and (20), with E_0 set at 34,250 MPa.

$$E(\tau) = E_0\left[1 - \exp\left(-0.4\tau^{0.34}\right)\right] \tag{19}$$

$$\begin{aligned} C(t,\tau) = {} & \tfrac{0.23}{E_0}\left(1 + 9.2\tau^{-0.45}\right)\{1 - \exp[-0.3(t-\tau)]\} + \\ & \tfrac{0.52}{E_0}\left(1 + 1.7\tau^{-0.45}\right)\{1 - \exp[-0.005(t-\tau)]\} \end{aligned} \tag{20}$$

According to Equations (19) and (20), the elastic modulus E and the unit creep C of concrete can be controlled by the parameter E_0. As a result, parameters E and C are considered to vary together. Numerous laboratory investigations have proven that the parameters λ and κ in the MCC model have an approximate relationship ($\kappa \approx 0.1\lambda$). Therefore, the parameters λ and κ can also be considered as a group of parameters that can vary together. Additionally, the permeability coefficient k_s of soil is a parameter that varies independently.

3.3. Sample Design for Sensitivity Analysis

Three different parameter groups (E and C, λ and κ, and k_s) are set at three levels (0.9, 1.0, and 1.1), where the level of 1.0 means that the values of parameters are the same as the aforementioned values in this section, and combined using the orthogonal design method.

Four samples with different level parameter groups can be obtained, as listed in Table 5, for the global sensitivity analysis of this project.

Table 5. Four orthogonally designed samples with different parameter levels.

No.	*E* and *C*	λ and κ	k_s
Sample 1	0.9	0.9	0.9
Sample 2	0.9	1.1	1.1
Sample 3	1.1	0.9	1.1
Sample 4	1.1	1.1	0.9

4. Results and Discussion

Several characteristic points on the floor and main structure of the lock head are defined, as shown in Figure 7, to investigate stresses and deformations of concrete and soil. D1~D6 points are used for analyzing the settlement of the top surface of the subsoil. The variations in the maximum principal stress on the top surface of the floor, the side wall of the water corridor section and the side wall of the hollow box section are described on S1~S6, LD-1 and LD-2 and KX-1 and KX-2 points, respectively.

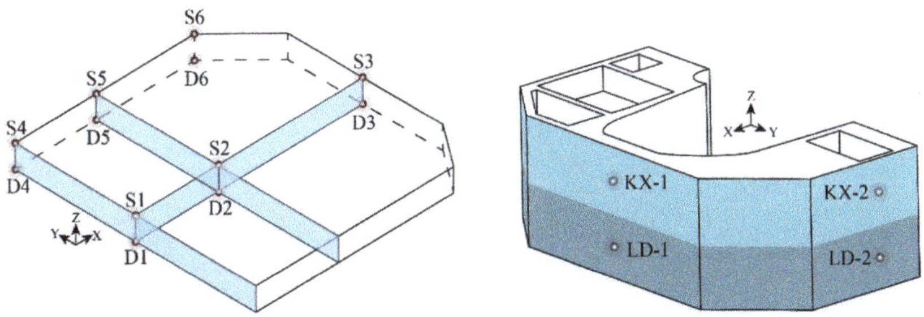

Figure 7. Characteristic points on the floor and main structure of the lock head.

As this study focuses on the variation in displacement and stresses in soil and structure, temperature results are not presented here.

4.1. Variation in Displacement and Stress over Time

The settlements of D1~D6 points are illustrated in Figure 8. It is obvious that the settlement of the foundation surface depends on time. The foundation surface is impermeable during construction; therefore, external loads will generate a significant amount of excess pore pressure that cannot dissipate quickly, leading to increased resistance against external loads. As a result, the instantaneous settlement is small. Even after the completion of backfilling a soil layer, the excessive pore pressure resulting from the weight of the backfilled soil will cause a rebound displacement on the foundation surface, as illustrated in Figure 8 on the 4th, 7th, and 10th lines. The settlement varies with the dissipation of excess pore pressure, and the settlements of central points (D1 and D2) are larger than those of the side points (D3~D6).

Figures 9 and 10 illustrate the differences of settlements between the center line and the boundary line of the floor's bottom surface across the river, as well as the variation curves in the maximum principal stress at points S1~S6 on the upper surface of the floor over time, respectively. In Figure 9, the center line is defined by points D1 and D2, while the boundary line is defined by points D4 and D5. Due to the consolidation of the foundation, settlement differences vary continuously throughout the entire calculation period. The variation range is mainly from 8 mm to 16 mm. This phenomenon leads to the redistribution of stress over time in the superstructure, as shown in Figure 10. Stress variations at both the center and boundary points, aligned along the same line in the river direction, exhibit a similar pattern, as depicted in each subfigure of Figure 10. Construction activities, such as pouring concrete structures and backfilling soil, have a significant impact on the displacement and the stress due to the generated external load, as illustrated on both sides of every step line in these two figures. Meanwhile, the stress at boundary points (points S3, S4, S5, and S6) remains relatively stable due to less influence of the moment induced by the boundary load.

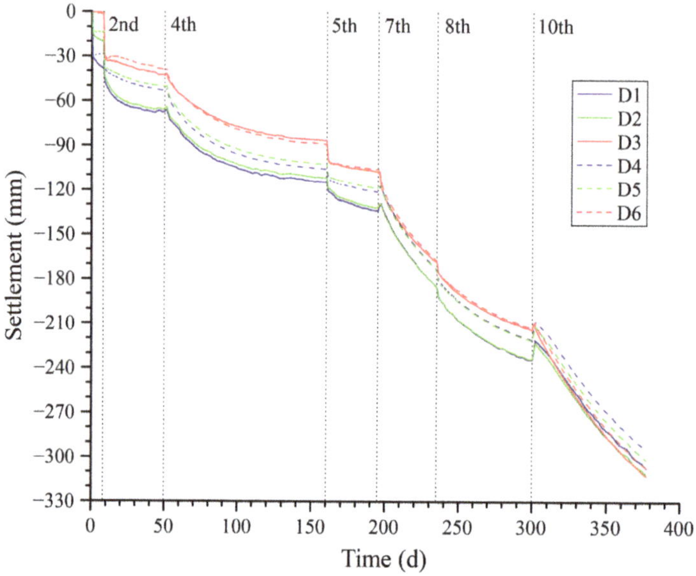

Figure 8. Settlement variations at characteristic points on the center and boundary of the foundation surface (2nd–10th are step numbers listed in Table 1).

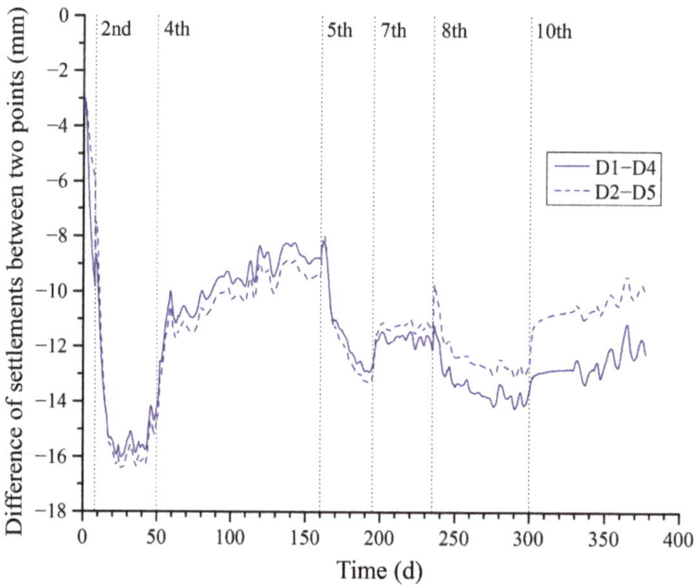

Figure 9. Differences of settlements between center points and boundary points of the floor's bottom surface.

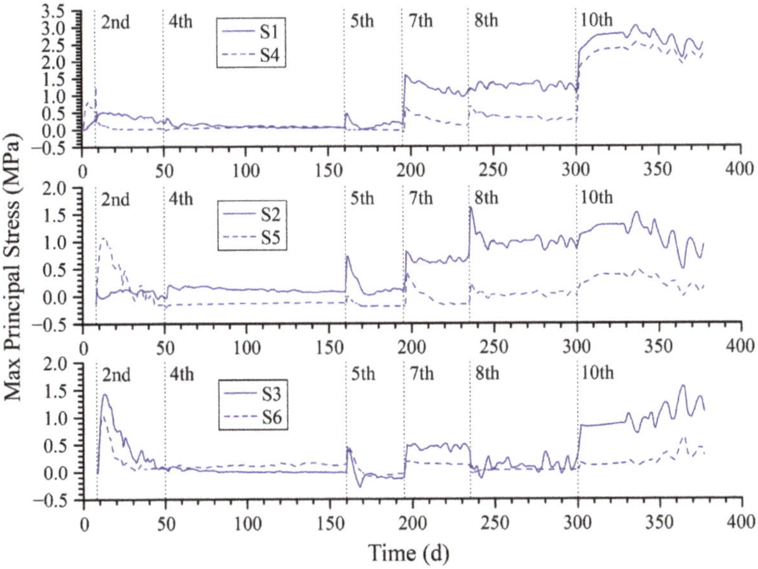

Figure 10. Variations in maximum principal stress at characteristic points on the top surface of the floor.

Furthermore, the curves following the 2nd, 5th, and 8th step lines in Figure 10 indicate a rapid stress reduction after the completion of the concrete superstructure construction. In contrast, after the completion of the backfilling process, the stress either experiences a slight decrease (as observed in the curve following the 4th and 7th step lines) or continues to increase (as observed in the curve following the 10th step line). Based on the analysis of stress variations, it is evident that the backfilling process has a more significant impact on the maximum principal stress of the floor's top surface compared to the construction of the superstructure. Especially after the final backfilling process, the maximum principal stress at point S1, which is the center point of the floor's upper surface, may exceed 3.0 MPa, resulting in cracks. This influence is attributed to the effects of the backfills on the pore pressure and settlement of the subsoil, which consequently result in stress changes within the superstructure.

Variations in maximum principal stress at characteristic points on the sidewalls of the water corridor and hollow box sections are depicted in Figure 11. Points LD-1 and KX-1 correspond to locations on the sidewall along the river, while LD-2 and KX-2 represent points on the upstream sidewall across the river. Generally, the stress in the hollow box section is lower than that in the water corridor. Additionally, the stress difference between LD-1 and KX-1 points is greater than that between LD-2 and KX-2 points. Furthermore, it is observed that, in the water corridor and hollow box sections, the stress variations are significantly more affected by the concrete casting process than the backfilling process. Particularly, it is worth noting that there is a significant stress variation at point LD-1 during the construction of the hollow box (as indicated near the 8th step line), reaching nearly 4 MPa. This sudden change in stress requires attention during phases of design and construction.

The finite element analysis of the lock head project on a soft foundation notably demonstrates the time-dependent results of foundation settlement and the consequent stress variations in the superstructure. During the entire construction process, the applied external load generated from the concrete and backfills induces sharp variations in displacements and soil pore pressure, leading to significant settlements at the foundation surface and stress changes in the superstructure. The magnitude and impact of these changes should be paid attention to and investigated in practical engineering. It is crucial to evaluate these variations to determine necessary engineering measures for mitigating potential adverse effects in the actual construction project.

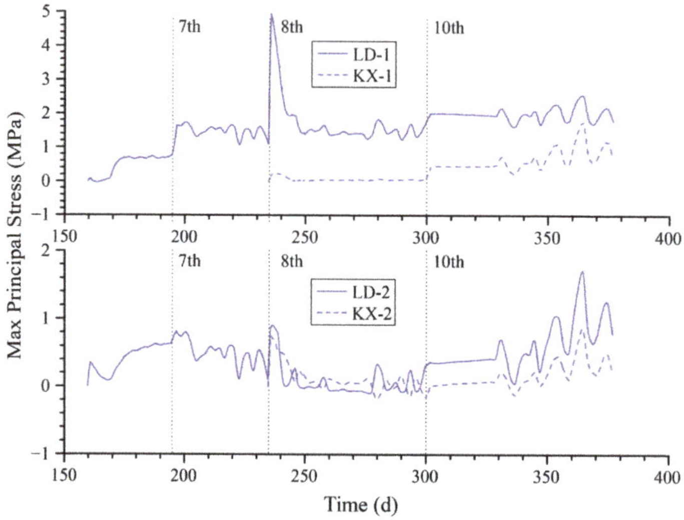

Figure 11. Variations in maximum principal stress at characteristic points in water corridor and hollow box sections.

4.2. Global Sensitivity Analysis of Mechanical Parameters

The settlement and maximum principal stress results obtained for the four designed samples in Section 3.3 at points D1, D4, S1, and S4 are shown in Figures 12 and 13. It is evident that there are clear differences in the calculated results for the four samples. Therefore, this section conducts an in-depth analysis of the impact of multiple parameters to determine the importance of mechanical parameters on the results.

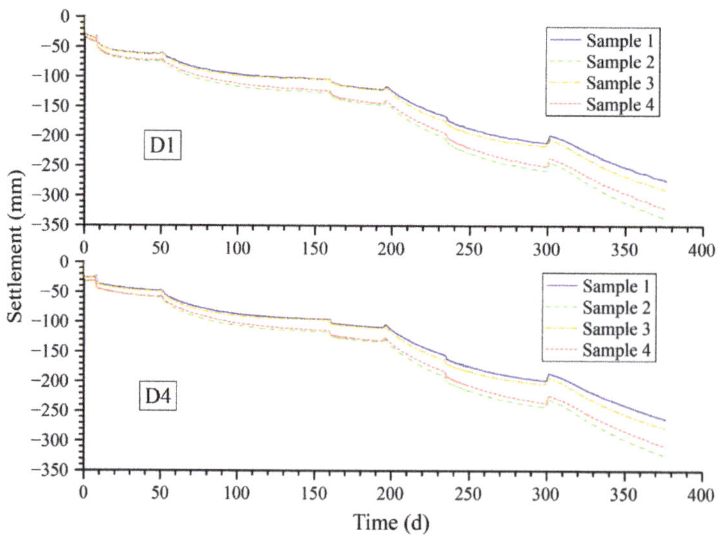

Figure 12. Varying foundation settlement of all samples.

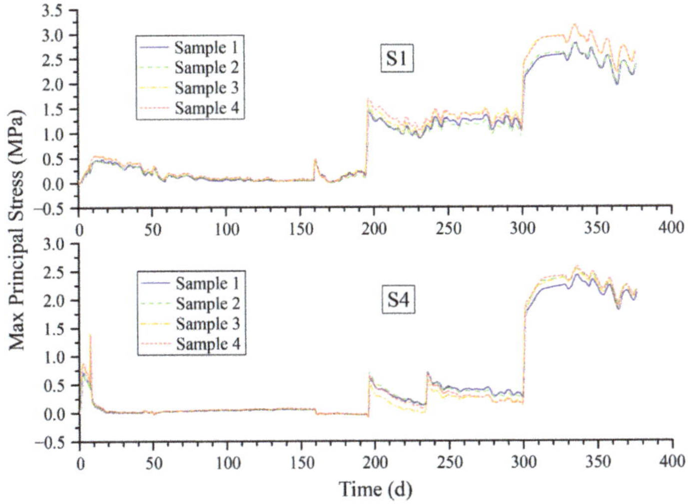

Figure 13. Varying max principal stresses of all samples.

The maximum settlement values at the central point (D1) and boundary point (D4) on the foundation–soil interface, as well as the maximum principal stress values at the top center point (S1) and boundary point (S4) of the floor, have been extracted from the calculation results and are listed in Table 6. Sensitivity analyses of parameter groups were conducted using Garson's method and the partial derivative method based on the BP neural network. The results of the sensitivity analysis are listed in Tables 7 and 8.

Table 6. Maximum results at characteristic points.

No.	D1 (mm)	D4 (mm)	S1 (MPa)	S4 (MPa)
Sample 1	274.456	262.878	2.80315	2.42602
Sample 2	338.13	325.097	2.82619	2.54617
Sample 3	289.682	278.144	3.19282	2.55406
Sample 4	321.78	309.159	3.17707	2.58789

Table 7. Results of the sensitivity analysis based on Garson's algorithm.

Parameter	S-D1	S-D4	σ_{max}-S1	σ_{max}-S4
E	0.2400	0.2471	0.2770	0.2121
C	0.1727	0.1913	0.2413	0.2594
λ	0.2111	0.1984	0.1537	0.1669
κ	0.2003	0.1950	0.1707	0.1956
k_s	0.1759	0.1628	0.1574	0.1661

Note: The full names of S-D1, S-D4, σmax-S1, and σmax-S4 are settlement at D1 point, settlement at D4 point, maximum principal stress at S1 point, and maximum principal stress at S4 point, respectively. The meanings of these abbreviations in Table 8 align with this table.

Table 8. Results of the sensitivity analysis based on the partial derivative algorithm.

Parameter	S-D1	S-D4	σ_{max}-S1	σ_{max}-S4
E	−0.4030	−0.4312	1.0586	0.7369
C	−0.0432	−0.0495	−0.9146	−0.9591
λ	1.2797	1.2568	−0.1500	0.7506
κ	1.1243	1.1175	0.0664	0.8368
k_s	0.9164	0.9036	0.0133	0.4991

The results from Tables 7 and 8 indicate that the rankings of relative importance, based on sensitivity analyses from the two algorithms, are generally consistent. However, there are some differences in the maximum principal stress at point S1, although the determination of the most significant parameter remains consistent. The analysis results from both algorithms suggest that the soil model parameters (λ and κ) have the greatest impact on settlement calculations, with parameter λ having the most significant influence. On the other hand, the concrete model parameters (E and C) have the most significant impact on stress calculations of the superstructure. However, the relative importance rankings of parameters E and C on results at central and boundary points are slightly different. The parameter that has the greatest impact on results at central points is E, while at boundary points, the parameter with the greatest impact is C.

Moreover, the direction of the parameter's influence on the results can be determined based on the sign of the results obtained from the partial derivative algorithm. According to the results, parameter C is negatively correlated with all the outcomes. This indicates that, as C increases, all the calculated results decrease. In contrast, the parameters κ and k_s are positively correlated with the results. Furthermore, different parameters have varying effects on the calculation results. The parameter λ is generally positively correlated with all calculation results. On the other hand, the parameter E of concrete consistently shows a negative correlation with settlement and a positive correlation with stress in all scenarios.

5. Conclusions

This study investigates a lock head project to model and analyze the integrated superstructure–foundation–backfills system. The objective is to examine the time-dependent deformation and stress response of hydraulic structures constructed on soft foundations, taking into account multiphysics and complex interactions. The analysis begins with a transient temperature analysis of the system, followed by a mechanical analysis using time-dependent models. The concrete is modeled using a creep model, and the soft soil foundation undergoes elasto-plastic consolidation analysis using the modified Cam-clay model. Additionally, sensitivity analyses of important mechanical parameters are performed using Garson's and partial derivative algorithms, which are based on the BP neural network. The main achievements of this study can be summarized concisely based on the conducted analyses, as follows:

1. The settlement and stress of the superstructure–foundation–backfills system on soft ground exhibit significant time-dependent characteristics. The settlement difference between the center line and the boundary line of the floor bottom surface primarily fluctuates within a range of 8 mm to 16 mm. The maximum principal stress at the center point on the upper surface of the floor would exceed 3.0 MPa after the final backfilling. Therefore, it is essential to conduct thorough checks and implement appropriate measurements to prevent cracks. The stress variation in the superstructure during the entire computation period is influenced by two factors: transient temperature stress and the continuous settlement of the foundation due to the consolidation of soft soil. Consequently, these factors affect the redistribution of stress in the superstructure. Therefore, when conducting an internal force analysis of hydraulic structures on soft foundations, it is essential to consider the time-dependent interactions.
2. The addition of backfilled soil increases the gravitational load on the foundation pit, causing immediate settlement and an increase in pore water pressure within the subsoil. This increase in pore water pressure prolongs the consolidation process. These immediate and time-dependent processes significantly affect the displacement of the foundation surface, resulting in a sudden change that can reach nearly 4 MPa in the water corridor section and subsequent redistribution of stress in the superstructure. Moreover, when analyzing the structure, it is crucial to carefully consider the stress growth resulting from this sudden change in order to ensure a more accurate assessment of structural safety.

3. Garson's method and the partial derivative algorithm, both based on the BP neural network, can provide a global sensitivity analysis for parameters and their respective rankings of relative importance. The outcomes of these two methods exhibit general consistency, with minor differences that do not have a significant impact on the main findings. This highlights the practicality of using both algorithms to conduct global sensitivity analyses in hydraulic structures on soft foundations. Moreover, the partial derivative algorithm provides insights into the specific direction of each parameter's impact on the results.
4. The settlement of the superstructure–foundation–backfills system is primarily influenced by the soil mechanical parameters, with the parameter λ having the greatest impact. Conversely, stress is mainly affected by the mechanical properties of concrete. However, the relative importance of parameters E and C on stress at different locations in the structure varies slightly. Therefore, specific analyses are necessary for different projects. Additionally, it is important to prioritize the accuracy of values for the more significant parameters in structural and geotechnical analyses.
5. In the context of time-dependent analyses for the proposed system, viscoelasticity is a crucial characteristic of soft soil. Nevertheless, further research is needed to determine the suitable viscoelastic model and apply reasonable parameters for accurate soil displacement analysis in hydraulic structures on soft foundations. Furthermore, the introduction of contact considerations between concrete and soil is necessary to improve the precision of the integrated analysis.

Author Contributions: Conceptualization, C.X. and W.L.; methodology, C.X. and L.Y.; software, L.Y.; investigation, C.X., L.Y., S.P. and W.L.; funding acquisition, S.P.; writing—original draft, C.X.; writing—review and editing, L.Y., S.P. and W.L. All authors have read and agreed to the published version of the manuscript.

Funding: This research was funded by the Zhejiang Provincial Natural Science Foundation, grant number LY22E090010.

Data Availability Statement: Data are available from the corresponding author upon reasonable request.

Conflicts of Interest: The authors declare no conflicts of interest.

References

1. Zhao, C.; Zhao, D. Application of Construction Waste in the Reinforcement of Soft Soil Foundation in Coastal Cities. *Environ. Technol. Innov.* **2021**, *21*, 101195. [CrossRef]
2. Vesterberg, B.; Andersson, M. Settlement and Pore Pressure Behaviour and Predictions of Test Embankments on an Organic Clay. *Int. J. Geotech. Eng.* **2022**, *16*, 1049–1067. [CrossRef]
3. Huang, J. Analysis of the Influence of Differential Consolidation Settlement of Foundation on Redistribution of Internal Forces in Superstructure of Buildings. *IOP Conf. Ser. Earth Environ. Sci.* **2021**, *632*, 022007. [CrossRef]
4. Bordón, J.D.R.; Aznárez, J.J.; Maeso, O.; Bhattacharya, S. Simple Approach for Including Foundation–Soil–Foundation Interaction in the Static Stiffnesses of Multi-Element Shallow Foundations. *Géotechnique* **2021**, *71*, 686–699. [CrossRef]
5. Lamprea-Pineda, A.C.; Connolly, D.P.; Hussein, M.F.M. Beams on Elastic Foundations—A Review of Railway Applications and Solutions. *Transp. Geotech.* **2022**, *33*, 100696. [CrossRef]
6. Elhuni, H.; Basu, D. Dynamic Soil Structure Interaction Model for Beams on Viscoelastic Foundations Subjected to Oscillatory and Moving Loads. *Comput. Geotech.* **2019**, *115*, 103157. [CrossRef]
7. Elhuni, H.; Basu, D. Interaction of Beams with Consolidating Nonlinear Poroelastic Layered Soil. *J. Eng. Mech.* **2022**, *148*, 04021167. [CrossRef]
8. Lanes, R.M.; Greco, M.; Almeida, V.d.S. Viscoelastic Soil–Structure Interaction Procedure for Building on Footing Foundations Considering Consolidation Settlements. *Buildings* **2023**, *13*, 813. [CrossRef]
9. Ai, Z.Y.; Chu, Z.H.; Cheng, Y.C. Time-Dependent Interaction between Superstructure, Raft and Layered Cross-Anisotropic Viscoelastic Saturated Soils. *Appl. Math. Model.* **2021**, *89*, 333–347. [CrossRef]
10. Gorini, D.N.; Callisto, L. A Coupled Study of Soil-Abutment-Superstructure Interaction. In *Geotechnical Research for Land Protection and Development*; Calvetti, F., Cotecchia, F., Galli, A., Jommi, C., Eds.; Lecture Notes in Civil Engineering; Springer International Publishing: Cham, Switzerland, 2020; pp. 565–574. [CrossRef]
11. Ai, Z.Y.; Jiang, Y.H.; Zhao, Y.Z.; Mu, J.J. Time-Dependent Performance of Ribbed Plates on Multi-Layered Fractional Viscoelastic Cross-Anisotropic Saturated Soils. *Eng. Anal. Bound. Elem.* **2022**, *137*, 1–15. [CrossRef]

12. Shamsi, M.; Shabani, M.J.; Vakili, A.H. Three-Dimensional Seismic Nonlinear Analysis of Topography-Structure-Soil-Structure Interaction for Buildings near Slopes. *Int. J. Geomech.* **2022**, *22*, 04021295. [CrossRef]
13. Duncan, J.M.; Clough, G.W. Finite Element Analyses of Port Allen Lock. *J. Soil Mech. Found. Div.* **1971**, *97*, 1053–1068. [CrossRef]
14. Rui, R.; Ye, Y.; Han, J.; Zhang, L.; Zhai, Y. Experimental and Theoretical Investigations on Active Earth Pressure Distributions behind Rigid Retaining Walls with Narrow Backfill under a Translational Mode. *Int. J. Geomech.* **2020**, *20*, 04020178. [CrossRef]
15. Sigdel, L.D.; Al-Qarawi, A.; Leo, C.J.; Liyanapathirana, S.; Hu, P. Geotechnical Design Practices and Soil–Structure Interaction Effects of an Integral Bridge System: A Review. *Appl. Sci.* **2021**, *11*, 7131. [CrossRef]
16. Naji, M.; Firoozi, A.A.; Firoozi, A.A. A Review: Study of Integral Abutment Bridge with Consideration of Soil-Structure Interaction. *Lat. Am. J. Solids Struct.* **2020**, *17*, e252. [CrossRef]
17. Saltelli, A.; Aleksankina, K.; Becker, W.; Fennell, P.; Ferretti, F.; Holst, N.; Li, S.; Wu, Q. Why so Many Published Sensitivity Analyses Are False: A Systematic Review of Sensitivity Analysis Practices. *Environ. Model. Softw.* **2019**, *114*, 29–39. [CrossRef]
18. Xu, J.; Zhao, X.; Yu, Y.; Xie, T.; Yang, G.; Xue, J. Parametric Sensitivity Analysis and Modelling of Mechanical Properties of Normal- and High-Strength Recycled Aggregate Concrete Using Grey Theory, Multiple Nonlinear Regression and Artificial Neural Networks. *Constr. Build. Mater.* **2019**, *211*, 479–491. [CrossRef]
19. Dong, X.; Wang, Y.; Wang, Z. Intelligent Meta-Model Construction and Global Stochastic Sensitivity Analysis Based on PSO-CNN. *Structures* **2022**, *43*, 1516–1529. [CrossRef]
20. Xia, Z.; Quek, S.T.; Li, A.; Li, J.; Duan, M.; Zhou, G.; Shi, H. Sensitivity Analysis in Seismic Reliability of an Urban Self-Anchored Suspension Bridge. *Mech. Syst. Signal Process.* **2022**, *164*, 108231. [CrossRef]
21. Tasri, A.; Susilawati, A. Effect of Cooling Water Temperature and Space between Cooling Pipes of Post-Cooling System on Temperature and Thermal Stress in Mass Concrete. *J. Build. Eng.* **2019**, *24*, 100731. [CrossRef]
22. Liu, X.; Chen, Z. A Virtual Element Method for Overcoming Locking Phenomena in Biot's Consolidation Model. *ESAIM Math. Model. Numer. Anal.* **2023**, *57*, 3007–3027. [CrossRef]
23. Roscoe, K.H. Mechanical Behaviour of an Idealised "Wet Clay". *Proc. 2nd European Conf. Soil Mech.* **1963**, 47–54.
24. Roscoe, K.H.; Schofield, A.N.; Wroth, C.P. On The Yielding of Soils. *Géotechnique* **1958**, *8*, 22–53. [CrossRef]
25. Kadlíček, T.; Janda, T.; Šejnoha, M.; Mašín, D.; Najser, J.; Beneš, Š. Automated Calibration of Advanced Soil Constitutive Models. Part II: Hypoplastic Clay and Modified Cam-Clay. *Acta Geotech.* **2022**, *17*, 3439–3462. [CrossRef]
26. Zhou, X.; Lu, D.; Zhang, Y.; Du, X.; Rabczuk, T. An Open-Source Unconstrained Stress Updating Algorithm for the Modified Cam-Clay Model. *Comput. Methods Appl. Mech. Eng.* **2022**, *390*, 114356. [CrossRef]
27. Miranda, P.A.M.N.; Vargas, E.A.; Moraes, A. Evaluation of the Modified Cam Clay Model in Basin and Petroleum System Modeling (BPSM) Loading Conditions. *Mar. Pet. Geol.* **2020**, *112*, 104112. [CrossRef]
28. Cui, K.; Jing, X. Research on Prediction Model of Geotechnical Parameters Based on BP Neural Network. *Neural Comput. Appl.* **2019**, *31*, 8205–8215. [CrossRef]
29. Paluzo-Hidalgo, E.; Gonzalez-Diaz, R.; Gutiérrez-Naranjo, M.A. Two-Hidden-Layer Feed-Forward Networks Are Universal Approximators: A Constructive Approach. *Neural Netw.* **2020**, *131*, 29–36. [CrossRef]
30. Razavi, S.; Jakeman, A.; Saltelli, A.; Prieur, C.; Iooss, B.; Borgonovo, E.; Plischke, E.; Lo Piano, S.; Iwanaga, T.; Becker, W.; et al. The Future of Sensitivity Analysis: An Essential Discipline for Systems Modeling and Policy Support. *Environ. Model. Softw.* **2021**, *137*, 104954. [CrossRef]
31. Duong, N.T.; Tran, K.Q. Estimation of Seepage Velocity and Piping Resistance of Fiber-Reinforced Soil by Using Artificial Neural Network-Based Approach. *Neural Comput. Appl.* **2023**, *35*, 2443–2455. [CrossRef]
32. Zhou, B.; Vogt, R.D.; Lu, X.; Xu, C.; Zhu, L.; Shao, X.; Liu, H.; Xing, M. Relative Importance Analysis of a Refined Multi-Parameter Phosphorus Index Employed in a Strongly Agriculturally Influenced Watershed. *Water Air Soil Pollut.* **2015**, *226*, 25. [CrossRef]
33. Goh, A.T.C. Back-Propagation Neural Networks for Modeling Complex Systems. *Artif. Intell. Eng.* **1995**, *9*, 143–151. [CrossRef]
34. Olden, J.D.; Joy, M.K.; Death, R.G. An Accurate Comparison of Methods for Quantifying Variable Importance in Artificial Neural Networks Using Simulated Data. *Ecol. Model.* **2004**, *178*, 389–397. [CrossRef]
35. Chen, J.; Sun, H.; Shi, Z.; Shen, M. Estimation of Parameters of Modified Cam-clay Model Coupling Biot Theory. *J. Tongji Univ.* **2003**, *31*, 544–548.
36. Nakase, A.; Kamei, T.; Kusakabe, O. Constitutive Parameters Estimated by Plasticity Index. *J. Geotech. Eng.* **1988**, *114*, 844–858. [CrossRef]
37. Zhou, B.; Wang, J.; Yang, X. Research on Parameters of Modified Cam Clay for Wuhan Clay. *J. Wuhan Univ. Technol.* **2007**, *29*, 103–107.
38. Chang, M.-F.; Teh, C.I.; Cao, L. Critical State Strength Parameters of Saturated Clays from the Modified Cam Clay Model. *Can. Geotech. J.* **1999**, *36*, 876–890. [CrossRef]
39. Doherty, J.; Alguire, H.; Muir Wood, D. Evaluating Modified Cam Clay Parameters from Undrained Triaxial Compression Data Using Targeted Optimization. *Can. Geotech. J.* **2012**, *49*, 1285–1292. [CrossRef]

water

MDPI

Article

Seepage–Deformation Coupling Analysis of a Core Wall Rockfill Dam Subject to Rapid Fluctuations in the Reservoir Water Level

Xueqin Zheng [1], Bin Yan [1], Wei Wang [1], Kenan Du [1] and Yixiang Fang [2,*]

[1] Pumped-Storage Technological & Economic Research Institute, State Grid Xinyuan Company Ltd., Beijing 100761, China; xueqinzheng1987@163.com (X.Z.)
[2] College of Water Conservancy and Hydropower Engineering, Hohai University, Nanjing 210098, China
* Correspondence: hhuyxfang@163.com

Abstract: Core wall rockfill dams are susceptible to cracking at the dam's crest, as well as collapse and settlement of the rockfill during storage and operation periods, particularly due to rapid fluctuations in the water level in pumped storage power stations. Most studies on the impact of fluctuations in the reservoir's water level on dam deformation have considered fluctuations of less than 5 m/d, while pumped storage power stations experience much larger fluctuations. Additionally, the seepage and stress fields within the dam's rock and soil interact and influence each other. Few studies have used the coupling theory of seepage and stress to analyze seepage and deformation in core wall rockfill dams. To address these issues, a finite element model using seepage–stress coupling theory was utilized to investigate the variations in the phreatic line, earth pressure, and deformation of a core wall rockfill dam due to rapid fluctuations in the reservoir's water level. Additionally, the results of the finite element simulation were compared with and analyzed alongside safety monitoring data. The results indicated that, upon a sudden decrease in the reservoir's water level, there was a lag in the decline of the phreatic line in Rockfill I, which created a large hydraulic gradient, resulting in a reverse seepage field on the dam's slope surface and generating a drag force directed upstream. Consequently, a significant concentration of stress occurred on one-third of the upstream slope surface of the dam and the seepage curtain, and the increase in horizontal displacement was substantially greater than the increase in settlement from one-third of the rockfill's height to the dam's foundation. The deformation was more sensitive to the lowest water level of the reservoir rather than to the fastest rate of decline. Sudden rises in the reservoir's water level result in decreased horizontal displacements and settlement of the dam. Amid rapid fluctuations of the reservoir's water level, changes in the vertical earth pressure were more pronounced at the bottom of the core wall than in its midsection. Compared with the core wall, variations in the vertical earth pressure in the upstream and downstream filter layers were minor at similar elevations. A peak horizontal displacement of 6.5 mm was noted at one-third the height of Rockfill I, with the greatest increase in settlement of 3.5 mm at the dam's crest. To ensure a project's safety, it is crucial to control the elevation of the lowest point during a sudden drop in the reservoir's level and to carefully monitor for cracks or voids within approximately one-third of the dam's height in Rockfill I and at the dam crest. This study's results provide a scientific basis for assessing core wall rockfill dams' health and securing long-term safety at pumped storage power facilities.

Keywords: pumped storage power station; core wall rockfill dam; phreatic line; vertical earth pressure; deformation

1. Introduction

Core wall rockfill dams (CWRDs) are distinguished by numerous advantages, including the use of materials sourced from local environments, streamlined construction processes, remarkable adaptability to different foundation types, and superior performance in seismic conditions [1–3]. However, during storage and operation, many CWRDs develop cracks [4,5], experience collapse and settlement of the rockfill [6–9], and settlement of the dam's crest [10–13]. In the case of the Guanyinyan Dam in China, distinguished by its CWRD adjoining a gravity dam via a 75 m high joint. Six cracks were identified at the junction as the reservoir's water level ascended to 1117.00 m on 26 November 2014. The most extensive of these cracks measured 25 m in length, with the greatest width and depth observed at 5 cm and roughly 5.5 m, respectively [14–16]. During the initial filling of the Cherry Valley Dam's reservoir in the United States, the settlement observed in the upstream rockfill was four times greater than that of the core wall. This discrepancy led to the formation of longitudinal cracks at the interface between the rockfill and the core wall along the dam's crest [17].

The roles of pumped storage power stations (PSPSs) in peak shaving, valley filling, emergency backup, frequency modulation, and phase adjustment within the electric grid are globally recognized [18,19]. This recognition has spurred the rapid construction of numerous PSPSs in China [20,21]. Conventional power station reservoirs, with their larger capacities, experience slower rates of decline in the water level. By contrast, the reservoir capacities of PSPSs are significantly smaller than those of conventional stations. Moreover, the substantial flow required for electricity generation or pumping within a few hours leads to significant and frequent fluctuations in the reservoir's water level. Concrete dams and concrete–faced rockfill dams generally adapt more easily to rapid changes in the reservoir's water level [22]. However, sudden changes in the reservoir's water level can exacerbate the uncoordinated deformation and arch effect of the CWRDs, leading to the dam's failure. Therefore, it is of great significance to study the influence of rapid fluctuations in reservoir's water level on the seepage and deformation characteristics of CWRDs.

The seepage characteristics of the core wall are an important aspect of monitoring the safety of CWRDs. Many scholars have studied the seepage characteristics and safety of CWRDs during storage and operation. Larese et al. [23] combined traditional finite element methods with more recent particulate analysis techniques to simulate the influence of overflow of the dam's crest on the rockfill and analyze the formation and evolution of cracks. Mohammad Rashidi et al. [24] found that after the initial water storage, the pore water pressure of the downstream core decreased, and the moisture content of some clay layers exceeded their optimal level. This resulted in an increase in pore water pressure in these areas. Xu et al. [25] established a seepage stability equation to evaluate the safety state of seepage in high CWRDs based on the critical hydraulic gradient safety index within the framework of the noninvasive finite element method. Wu et al. [26] proposed a method of assessing the probability of a percolation failure, combined with Bayesian parameter-based probability estimation method, to study the safety of seepage in a CWRD during sudden drops in the reservoir's water levels.

For decades, many researchers have used finite element software to simulate the deformation of CWRDs under fluctuations in the reservoir's water level. Reza et al. [27] improved the constitutive model for simulating the nonlinear characteristics of rockfill and proposed a method to simulate the collapse and settlement of rockfill dams during the initial filling. They concluded that low water content or poor compactness were the main causes of the collapse and settlement of the rockfill, and that the water storage of rockfill dams should be graded to ensure dam safety. Akhtarpour et al. [28] used a constitutive model of strain hardening to simulate the deformation experienced by the Masjed-e-Soleyman dam during its construction and impoundment phases. They attributed the slow dissipation of the internal excess pressure of pore water, plastic shear deformation, and significant settlement of the upstream body of the dam as the primary factors leading to cracks in the dam's crest. Qiu et al. [29] simulated the deformation of the asphalt core wall during the construction and operation of Quxue Dam using a three-dimensional nonlinear finite element method. Zhou et al. [30] performed an intricate mesh segmentation of a finite element model of a high CWRD. They suggested that a refined mesh facilitates a more accurate analysis of cracks in a dam. Additionally, Zhou et al. [31] observed that during the reservoir's storage process, the horizontal displacement and settlement increment of the dam's crest occur at comparable magnitudes. Liu et al. [32] used the hhu-SH-breakage constitutive model of rockfill to simulate the deformation characteristics of an ultra-high CWRD and found that the crushing settlement of the rockfill could not be ignored. Pan et al. [33] utilized the E^w—v^w wetting model, and using the initial stress and strain methods, simulated the construction and impoundment processes of the Guanyinyan Dam. They concluded that the wetting deformation of the upstream shell materials upon the rise in water level created tensile stress zones at the dam's crest, and identified wetting deformation as the direct cause of cracks in the crest. Furthermore, Pan et al. [34] used a BP neural network to invert the constitutive model parameters of the rockfill and inputted these parameters into a finite element calculation model, improving the accuracy of the results of finite element simulation. Ai et al. [35] proposed an innovative method for updating deformation analysis models using multi-source monitoring data filtering and identification of the model's parameters. This method reduced the redundancy of the information and ensured that the finite element analysis model of the rockfill dam better reflected actual deformation.

Due to the effects of the reservoir's water level, a differential head exists between the upstream and downstream sides of a dam. This difference causes water to flow through the dam's material pores, generating dynamic hydraulic pressure, also known as seepage volume force. The long-term effects of seepage volume force alter the dam's internal stress field, inducing the displacement of soil particles and changes in the porosity and permeability, ultimately modifying the dam's internal seepage field [36,37]. Consequently, the seepage and stress fields within a dam's rock–soil mass interact and influence each other, underscoring the significance of studying their coupled effects on the dam's deformation [38,39]. Chen et al. [40] developed a finite element seepage–stress coupled model for CWRD, investigating a dam's deformation and stress under normal and rapidly declining water levels in the reservoir to dead storage levels. Ri et al. [41] utilized seepage–stress coupling and the strength reduction method to analyze the stability of various earth–rock dams at rates of decline in the reservoir's water level of 0.1 m/d, 0.5 m/d, and 1 m/d. They concluded that core wall dams are safer compared with homogeneous and inclined-core earth dams.

A certain PSPS, operating as a daily regulation facility, experiences significant and frequent daily fluctuations in the reservoir's water levels, with both the maximum daily rise and fall exceeding 15 m/d. Most previous studies on the influence of fluctuations in the reservoir's water level on the dam's deformation characteristics have focused on fluctuations of less than 5 m per day. Additionally, the interaction between the seepage field and the stress field has rarely been considered in seepage and deformation analyses of earth–rock dams. The increased range of fluctuation in the reservoir's water levels leads to greater complexity and changes in the internal stress state of the dam's body, significantly affecting the seepage field. Ignoring the interaction between the seepage field and the stress field will inevitably lead to discrepancies between the results of finite element calculation and the actual results. Moreover, the increased range of fluctuation in the reservoir's water levels will exacerbate the collapse and settlement of the rockfill, the generation of cracks on the dam's crest, and the arch effect of the core wall. The 15-day deformation monitoring cycle for the CWRD of a certain PSPS's lower reservoir is insufficient for accurately understanding the dam's deformation response to rapid fluctuations in the reservoir's water level. In response to these issues, finite element software was used in this study to analyze changes in the seepage, stress, and displacement fields of a CWRD due to rapid fluctuations in the water level. The results of finite element calculation were then compared with monitored data on seepage pressure, vertical earth pressure, and displacement. The research results provide a scientific basis for evaluating the health of CWRDs and ensuring the long-term operational safety of PSPSs.

2. Methodology

2.1. Unsaturated Soil Seepage Equation

The height of the dam's crest exceeds the phreatic line and it comprises unsaturated soil. Since the deformation of the dam's crest area is a critical concern, both saturated and unsaturated seepage flows should be used to simulate the overall deformation of the dam. Initially, Darcy's law was applied exclusively to the seepage analysis of saturated soil. However, Richards' 1931 study demonstrated that Darcy's law is also applicable to unsaturated soil. The primary distinction is that the permeability coefficient of saturated soil remains constant, whereas that of unsaturated soil varies with changes in the water content. According to the principle of mass conservation, the seepage equation for unsaturated soil is as follows

$$\frac{\partial}{\partial x}\left(k_x\frac{\partial H}{\partial x}\right)+\frac{\partial}{\partial y}\left(k_y\frac{\partial H}{\partial y}\right)+Q=m_w\gamma_w\frac{\partial H}{\partial t} \tag{1}$$

where H is the total head, k_x and k_y are the hydraulic conductivity in the x and y directions, respectively, Q is the applied boundary flux, m_w is the slope of the storage curve, γ_w is the unit weight of water, and t is the time.

The definite solution conditions of unsteady seepage analysis are as follows.

(1) Initial condition

$$H(x,y,0)=H_0;(x,y)\in\Omega \tag{2}$$

where H_0 is the initial total head, and Ω is the calculation area of the model.

(2) Boundary condition

In the finite element software used in this study, the water head's boundary condition was defined as the first type or Dirichlet boundary condition. This condition was used to simulate the effect of the reservoir's water level. The governing equation is expressed as follows

$$H(x,y,t)=H_t;(x,y)\in S_1 \tag{3}$$

where S_1 is the head boundary conditions, and H_t is a node head that changes over time.

2.2. Constitutive Equation of Unsaturated Soil

Frelund and Rahardjo extended Terzaghi's effective stress principle by incorporating the gas phase, resulting in the constitutive equation for unsaturated soil as follows

$$\{\Delta\sigma\}=[D]\{\Delta\varepsilon\}-[D]\{m_H\}(u_a-u_w)+\{\Delta u_a\} \tag{4}$$

$$\{m_{\mathrm{H}}\}=\left\langle\frac{1}{H'},\frac{1}{H'},\frac{1}{H'},0\right\rangle \tag{5}$$

where $\{\Delta\sigma\}$ is the stress increment matrix, $[D]$ is the soil constitutive matrix, $\{\Delta\varepsilon\}$ is the strain increment matrix, H is the unsaturated soil modulus for soil structure with respect to matrix suction, u_a is pore gas pressure, u_w is pore water pressure, and $\{\Delta u_a\}$ is the pore gas pressure increment.

2.3. Seepage and Stress Coupling Governing Equation

In seepage-stress coupling calculations, the soil equilibrium equation and the seepage control equation are solved simultaneously. The stress field calculation involves solving for the increments in displacement at the nodes, while the seepage field calculation involves solving for the values of the water head at the nodes. That is, in coupled calculations, the simultaneous solution of Equations (6) and (7) yields the increments in displacement $\{\Delta\delta\}$

and increments in the pore water pressure $\{\Delta u_w\}$, which serve as the fundamental variables for the stress field and the seepage field, respectively.

$$[K]\{\Delta\delta\} + [L_d]\{\Delta u_w\} = \{\Delta F\} \tag{6}$$

$$\beta[L_f]\{\Delta\delta\} - (\frac{\Delta t}{\gamma_w}[K_f] + \omega[M_N])\{\Delta u_w\} = \Delta t(\{Q\}|_{t+\Delta t} + \frac{1}{\gamma_w}[K_f]\{u_w\}|_t) \tag{7}$$

$$\begin{cases} [\ K\ ] = \Sigma[\ B\ ]^{\mathrm{T}}[D][B] \\ [\ L_d\ ] = \Sigma[\ B\ ]^{\mathrm{T}}[D]\{m_H\}\langle N\rangle \\ [\ K_f\ ] = \Sigma[\ B\ ]^{\mathrm{T}}[\ K_w\ ][\ B\ ] \\ [\ M_N\ ] = \Sigma\langle N\rangle^{\mathrm{T}}\langle N\rangle \\ \left[L_f\right] = \Sigma N^T\{m\}[B] \end{cases} \tag{8}$$

where $[K]$ is the stiffness matrix, $[B]$ is the strain matrix, $[L_d]$ is the coupling matrix, $\langle N\rangle$ is the row vector of the shape function, $[\Delta F]$ is the increase in the nodal external load, β is the volumetric strain coefficient, $[L_f]$ is the fluid coupling matrix, $[K_f]$ is the element stiffness matrix, $[K_w]$ is the permeability coefficient matrix, ω is the coefficient of the pore water pressure, and $[M_N]$ is the mass matrix.

3. Numerical Model of a CWRD

3.1. Summary of the Project

A PSPS functions as a daily regulation facility. The power station comprises four units with a total installed capacity of 1000 MW, including the upper and lower reservoirs, a water transmission system, an underground plant, a ground switching station, a central control building, and additional facilities. This research selected the CWRD of the lower reservoir as the research object. The dam section is mainly divided into Rockfill I, the transition layer, the filter layer, the clay core wall, Rockfill II, and the dam's foundation. The reservoir has a capacity of 5.728 million m³, with a normal storage level of 78.90 m, a maximum storage level of 80.60 m, a normal operational water subsidence level for power generation of 58.00 m, and a dead water level of 57.00 m. The crest of the dam stands at an elevation of 83.40 m, with a maximum height of 50.4 m, a span of 485.8 m, and a width of 8.0 m. The upstream slope is designed with a ratio of 1:2.0, and the downstream slope has a ratio of 1:2.5, as illustrated in Figure 1.

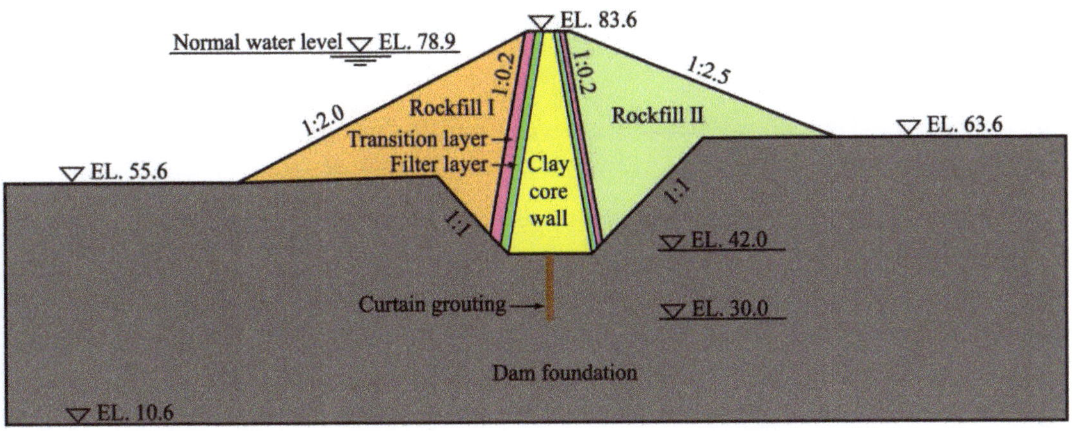

Figure 1. A typical section of the dam.

In order to facilitate monitoring of the safety of the CWRD, a seepage and deformation monitoring control network encompassing the entire project's scope has been established. The monitored data include surface deformation, vertical earth stress, and seepage within both the dam's body and its foundation. This study focused on a typical section for analysis, equipped with three seepage piezometers, four displacement gauges for surface deformation, and four earth pressure gauges, as depicted in Figure 2.

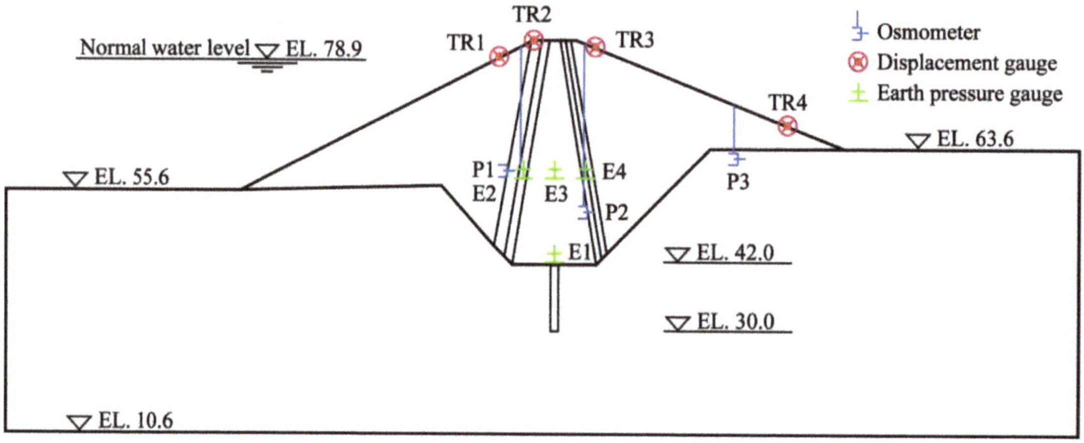

Figure 2. Layout of measurement points on a typical section of the dam.

3.2. Material Parameters and Boundary Conditions

The stress–strain relationships of Rockfill I, the transition layer, the filter layer, the clay core wall, Rockfill II, and the dam's foundation were characterized using the Duncan–Chang E-B model. The materials' physico–mechanical parameters were derived from triaxial compression tests and are detailed in Table 1. The curtain grouting was represented by a linear elastic model, with a bulk density of 24 kN/m³, an elastic modulus of 28 GPa, and a Poisson's ratio of 0.167.

Table 1. Constitutive parameters of the model.

Dam Materials	γ_d, kN•m^{-3}	c, kPa	φ/°	$\Delta\varphi$/°	K	n	R_f	K_b	m	K_{ur}
Rockfill I	22.1	0	39.8	10.5	960	0.49	0.74	490	0.42	2K
Transition layer	22.5	120	40.2	7.7	780	0.47	0.75	475	0.37	2K
Filter layer	18.1	133	43.7	8.5	840	0.42	0.80	450	0.43	2K
Clay core wall	16.2	154	44	7.8	500	0.35	0.77	240	0.35	2K
Rockfill II	21.8	0	43.9	9.6	660	0.49	0.70	258	0.28	2K
Dam's foundation	21.8	55	43.9	9.6	660	0.49	0.70	258	0.28	2K

The time period exhibiting the most significant fluctuation in the reservoir's water level was chosen as the condition for calculation. Consequently, the recorded water levels from 1 May to 10 June 2019 served as the boundary condition for the upstream slope of the dam in the finite element model, with the temporal variation depicted in Figure 3. Analysis of the figure revealed that on 1 May 2019, the reservoir's water level stood at 76.70 m; by 6 May, it had dropped to a minimum of 60.13 m; and by 27 May, it had increased to a maximum of 79.25 m. The sharpest decline in the water level was observed on 2 June, with a rate of 15.53 m/d, reaching a water level of 63.26 m. The most rapid increase occurred on 7 June, at a rate of 17.87 m/d, reaching a water level of 78.93 m.

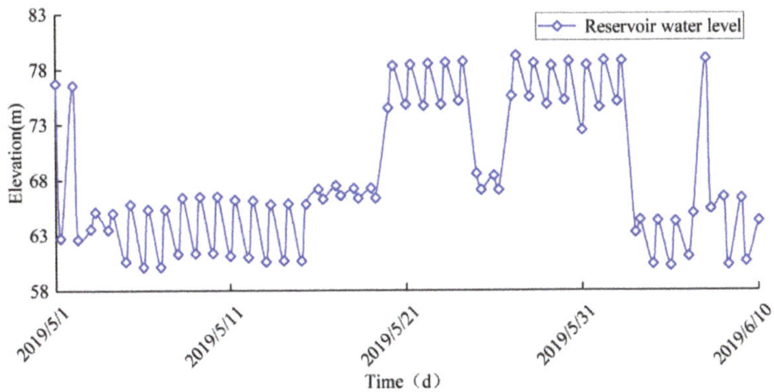

Figure 3. Hydrograph of the reservoir's water level, exhibiting temporal variations.

3.3. Process of Numerical Simulation

Numerical analyses of the seepage, stress, and displacement fields within the CWRD were executed through the SEEP/W and SIGMA/W modules of finite element modeling software. The numerical simulation process entailed, first, inputting the seepage parameters for each zone of the dam into the SEEP/W module, and incorporating the hydraulic boundary of the reservoir's water level for steady-state seepage analysis to determine the initial seepage field. Second, it involved importing the initial seepage field from the previous step into the SIGMA/W module, inputting the stress parameters for each zone of the dam, and adding the stress boundary of the reservoir's water level for in-situ stress analysis. Third, it involved selecting consolidation within the SIGMA/W module as a subcategory of the initial stress field, adding both hydraulic and stress boundary conditions for the reservoir's water level based on operational scenarios, and conducting seepage–stress coupling analysis to acquire the transient seepage and stress fields. Finally, the process involved plotting cloud diagrams of the pore water pressure, displacement, and vertical soil pressure in the results.

4. Results of Numerical Simulation and Analysis of Monitoring Data

4.1. Analysis of the Seepage Field

4.1.1. Analysis of the Results of Simulating Pore Water Pressure

Figure 4 displays the cloud diagrams of pore water pressure at various typical moments. From Figure 4a, it can be observed that at the initial moment with the reservoir's water level at 76.70 m, the pore water pressure exhibited a uniform stratified distribution. The infiltration line in the core wall decreased significantly, aligning with the general principles of seepage. From Figure 4b, it can be noted that the decrease in the infiltration line within the core wall is reduced, and a transient saturated zone formed above the infiltration line in Rockfill I when the reservoir's water level abruptly dropped on 6 May. A hysteresis occurred in the infiltration line's decline within Rockfill I, generating a significant hydraulic gradient in Rockfill I. The reduction in the static water pressure on the dam's slope led to an increase in excess pore water pressure and seepage force, which was directed outward from the slope. This resulted in a reverse seepage field on the surface of the dam's slope, exerting an outward dragging force, thereby amplifying the down-sliding force on the dam's slope and adversely affecting its stability. From Figure 4c, it is evident that a transient saturated zone emerged above the infiltration line in Rockfill II following an abrupt rise in the reservoir's water level. This occurred as rock piles possess inherent porosity and permeability, allowing water to flow through their interstices. An abrupt increase in the reservoir's water level elevated the water pressure, forcing water into the rock piles' interiors and filling the pore spaces, culminating in the transient saturation of the rockfill.

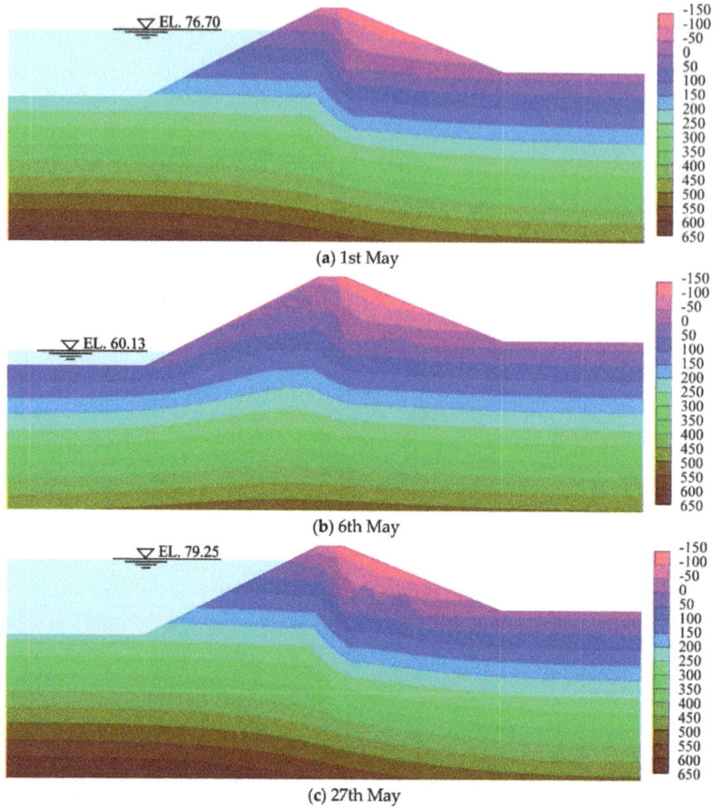

Figure 4. Cloud maps of pore water pressure at different typical moments (unit: kPa).

4.1.2. Analysis of the Monitoring Data of Seepage Pressure

On the upstream side of the dam, one osmometer was installed, while two osmometers were positioned on the downstream side, as detailed in Figure 2, while Figure 5 displays the measured water level of these osmometers. In Figure 5, it is evident that the changes in the water level in the P1 osmometer and the reservoir are positively correlated, with the water level in the P1 osmometer exhibiting only a minor decline. Due to hysteresis in the infiltration line following a drop in the reservoir's water level, on 2 May, the reservoir's water level decreased to 62.61 m, resulting in the P1 osmometer's water level being higher than that of the reservoir. After the reservoir's water level rose to 65.10 m on 3 May, the water level measured by the P1 osmometer was lower than that of the reservoir, subsequently fluctuating in accordance with the changes in the reservoir's water level. The P2 osmometer's water level showed a significant drop compared with the reservoir's water level, with a potential decline of approximately 87%. The correlation between the changes in the P2 osmometer and fluctuations in the reservoir's water level was generally weak. The water level of the P3 osmometer was largely unaffected by the reservoir's water level, primarily due to the anti-seepage effect of the core wall. Given that the P2 osmometer's water level fluctuated with the sudden changes in the reservoir's water level, the P3 osmometer's water level was occasionally higher than that of P2, despite P3 being positioned downstream relative to the reservoir's water level. The head drop was concentrated within the core wall, with a maximum decrease of 14.65 m. The head drop within the upstream and downstream rockfills was only 0.47 to 2.26 m, indicating that the upstream and downstream rockfills have a minimal impact on the head drop.

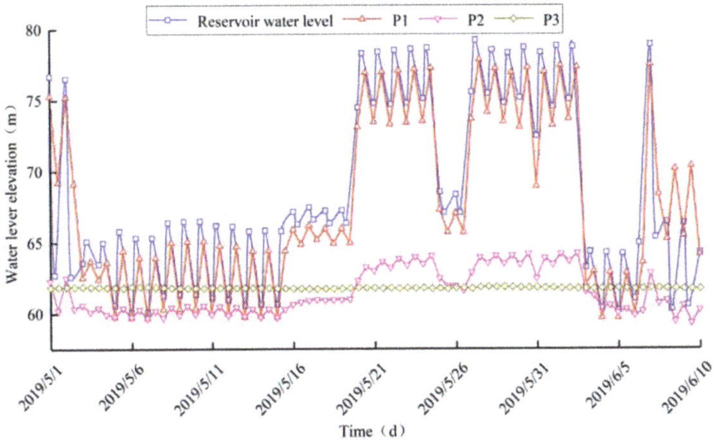

Figure 5. Monitoring data of the seepage pressure.

Table 2 presents a comparison between the osmometers' measured water level and the finite element model's calculation of the phreatic line. It can be observed that the discrepancies between monitored values at various moments and the model's calculations of the phreatic line were less than 0.5 m, indicating the high accuracy of the finite element model.

Table 2. Comparison between the simulated and monitored values of the phreatic line.

Time	Reservoir's Water Level, m	P1		P2		P3	
		Simulated Value, m	Monitored Value, m	Simulated Value, m	Monitored Value, m	Simulated Value, m	Monitored Value, m
5/1	76.70	75.40	75.31	62.60	62.33	61.87	61.85
5/6	60.13	73.60	73.39	62.80	62.31	61.76	61.90
5/27	79.25	77.38	77.87	64.44	63.97	61.94	61.86

4.2. *Analysis of the Stress Field*

4.2.1. Analysis of the Results of Simulating Vertical Soil Pressure

Figure 6 shows the cloud maps of vertical soil pressure at different typical moments. It can be seen from Figure 6a that when the reservoir's water level plummeted, the vertical stress of the curtain grouting experienced a significant concentration of stress, with a maximum value of 1.6 MPa, an increase of about 0.8 MPa compared with the initial moment. This was mainly because the curtain grouting, which originally bore substantial water pressure at a high water level, experienced a rapid reduction in pressure. This led to the redistribution of stress between the curtain and the surrounding body of the dam, resulting in the concentration of stress. The arch effect in the core wall was apparent, a crucial characteristic of the distribution of stress in CWRDs. Because the deformation modulus of the core wall's material was lower than that of the upstream and downstream rockfills, the core wall deformed more, whereas the upstream and downstream rockfills deformed less. This deformation mismatch can lead to a redistribution of stress. The stress of the core wall transferred to the two sides of the rockfill, resulting in the vertical stress of the core wall being lower than the dead weight stress, while the vertical stress of the rockfill on both sides of the core wall was higher than the dead weight stress. The arch effect of the core wall can significantly reduce the compressive stress within the core wall, which is a primary reason for the hydraulic splitting of the core wall. It is evident from Figure 6b that when the reservoir's water level rose abruptly, a significant concentration of stress occurred at approximately one-third the height of the upstream slope of the dam, with a maximum value of 0.9 MPa, an increase of about 0.8 MPa. This concentration of stress continued to increase with repeated sudden rises and drops in the reservoir's water level. The concentration of stress in the curtain grouting disappeared, and the vertical earth pressure decreased from 1.6 MPa to 0.8 MPa.

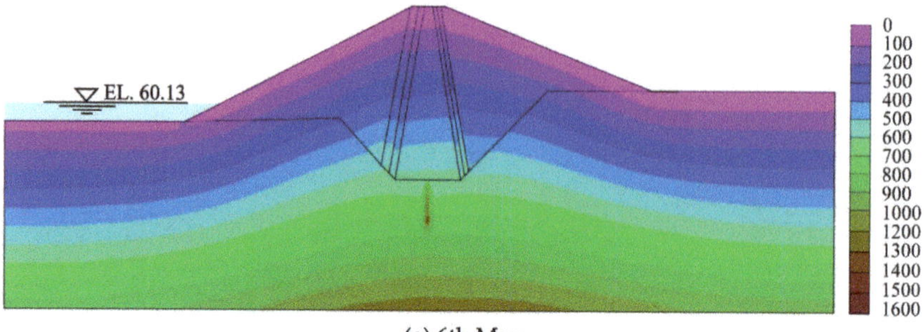

(a) 6th May

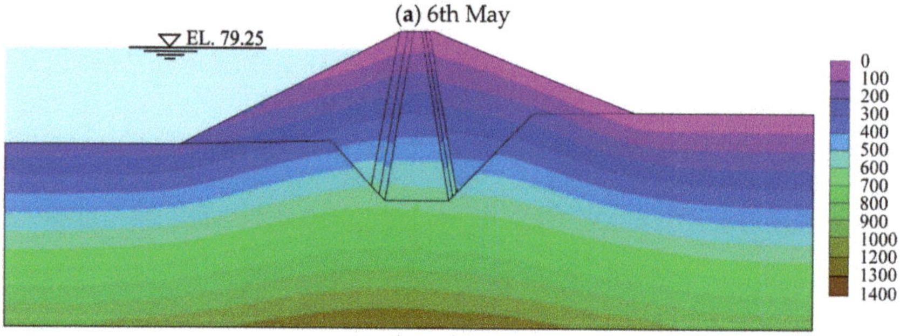

(b) 27th May

Figure 6. Cloud maps of vertical soil pressure at different typical moments (unit: kPa).

4.2.2. Analysis of the Monitoring Data of Vertical Soil Pressure

At the bottom of the core wall (elevation: 42.00 m), the middle of the core wall, the upstream filter layer, and the downstream filter layer (all at an elevation of 61.00 m), earth pressure gauges were installed to monitor the vertical earth pressure of the dam, as depicted in Figure 2. Figure 7 compares the simulated values of vertical earth pressure with the monitored values. The figure indicates that the values of vertical earth pressure simulated by the finite element model followed the same trend as those measured by the earth pressure gauges, with negligible differences. Moreover, the gauges' measured vertical earth pressure at various locations were positively correlated with changes in the reservoir's water level. Comparing the earth pressure gauges (E1, E3) at different elevations showed that the pressure at the core wall's bottom (E1) exceeded that midway up (E3), influenced by rapid fluctuations in the reservoir's water level. The main reason was that the soil at the bottom of the core wall exhibited a higher initial density and smaller porosity due to the influence of gravity and pressure from the overlying soil, and its permeability was lower than that in the middle of the core wall. As a result, when the water level changed, the adjustment speed of pore water pressure in the soil at the bottom of the core wall was slower than in the middle of the core wall, resulting in more significant changes in soil pressure at the bottom of the core wall. At the same elevation, the earth pressure gauges (E2, E4) situated in the upstream and downstream filter layer registered lower measurements compared with the gauge (E3) within the core wall, attributable to the arching effect produced by the lateral compression of the rockfill against the core wall. The fluctuation in earth pressure was most pronounced at gauge E2 and least at gauge E4, indicating that the impact of changes in the reservoir's water level on the dam's vertical earth pressure diminished from upstream to downstream.

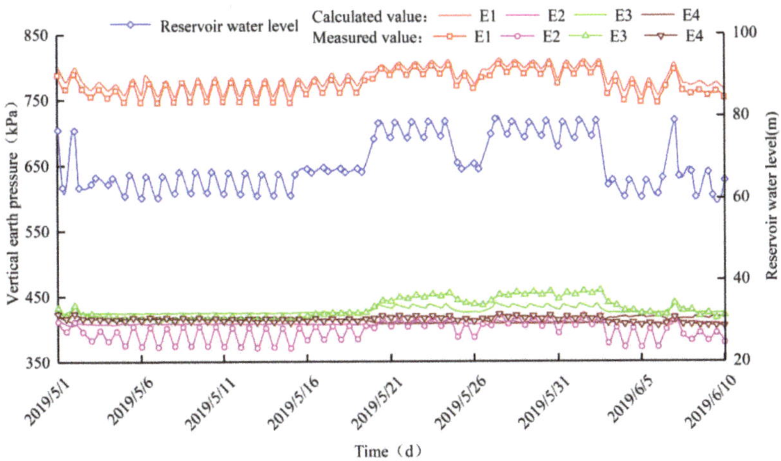

Figure 7. Comparison between the simulated and monitored values of vertical soil pressure.

4.3. *Analysis of the Displacement Field*

4.3.1. Analysis of the Results of Simulating Horizontal Displacement and Settlement

Calculations of the dam's deformation were performed, based on the principle of seepage–stress coupling, identifying feature points on the finite element model that corresponded to the locations of measurement points TR1, TR2, TR3, and TR4. Figure 8 displays the time variation of the increases in horizontal displacement, while Figure 9 presents the time variation of increases in settlement. The figures illustrate that the impact of fluctuations in the reservoir's water level on the dam's horizontal displacement followed a descending order of TR3 ≥ TR1 > TR2 > TR4, while the effect on the dam's settlement was ranked as TR3 > TR2 > TR1 > TR4. The horizontal displacement and settlement of the dam's surface were significantly affected by the reservoir's water level, and there was a lag phenomenon. Depending on the axis of the dam, the dam moves downstream (upstream) as the water level rises (falls). The faster the reservoir's water level fluctuates, the greater the variation in both upstream and downstream sides' horizontal displacement and settlement. Rapid fluctuations in the water level exerted a more pronounced effect on the deformation of the dam's crest. Therefore, the safety inspection should concentrate on identifying and assessing cracks and voids at the dam's crest.

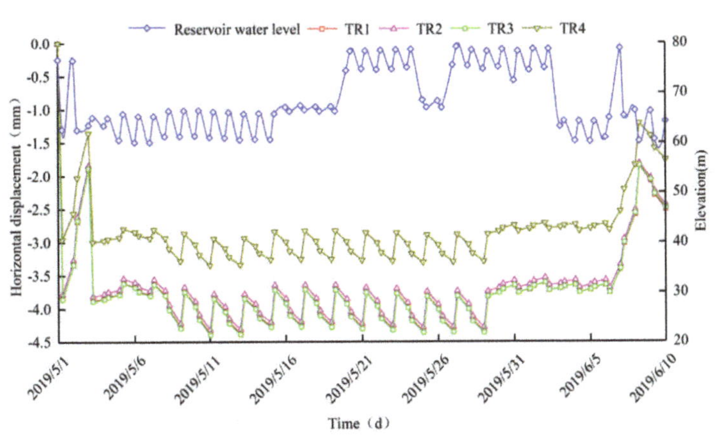

Figure 8. Chart illustrating the time variation of increments in horizontal displacement.

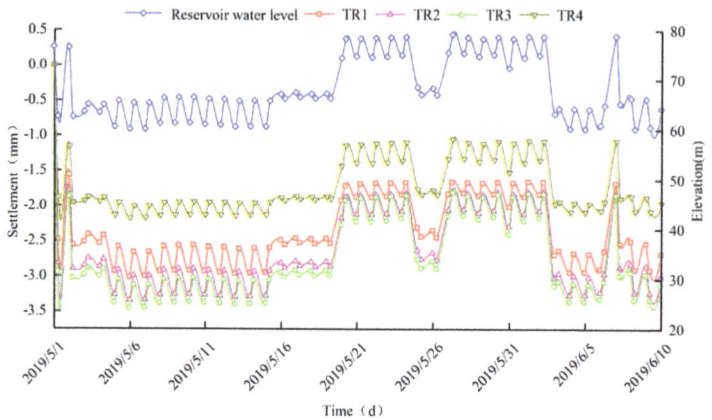

Figure 9. Chart illustrating the time variation of increments in settlement.

Figure 10 presents the cloud diagrams of the dam's horizontal displacement at different typical moments. Figure 10a provides the cloud diagram illustrating the horizontal displacement of the dam when the reservoir water level initially stood at 76.70 m on 1 May. The figure reveals that the horizontal displacement of the dam was symmetrically distributed across the core wall. The maximum horizontal displacement observed on the upstream side of the dam was 25.0 mm, directed upstream, while on the downstream side, the maximum displacement was observed to be 45.0 mm in the downstream direction. The horizontal displacement of Rockfill I was less than that of Rockfill II, due to Rockfill II comprising a greater proportion of soft rock materials. Figure 10b presents the cloud diagram for the increases in the horizontal displacement of the dam when the reservoir's water level was at 60.24 m on 5 June. The figure reveals that the contour of the increase in horizontal displacement formed a closed loop in the body of the upstream rockfill, with the largest horizontal displacement occurring within this loop, and it decreased progressively downstream. The rapid decline in the reservoir's water level caused the entire dam's horizontal displacement to orient upstream, with a maximum displacement of 6.5 mm located approximately one-third up the height of the dam in Rockfill I. The elevation at which maximum displacement occurred was nearly equivalent to the reservoir's water level, located approximately 10 m from the upstream slope of the dam. Figure 10c displays the cloud diagram for the increase in horizontal displacement of the dam when the reservoir's water level reached 79.25 m on 27 May. The figure reveals that the horizontal displacement toward the upstream on the upstream side of the dam diminished following the sudden increase in the reservoir's water level, with the maximum displacement reducing to 3.0 mm.

Figure 11 presents the cloud diagrams of the dam's settlement at different typical moments. From Figure 11a, it can be observed that the influence of fluctuations in the reservoir's water level on the dam's settlement was less than that of horizontal displacement. The contour for the distribution of settlement formed a "U" shape and was approximately symmetrical along the core's wall. A closed loop appeared near the dam's top. The displacement gradient of the upstream slope of the dam was larger than that of the downstream slope of the dam, and the settlement gradually decreased from the dam's crest to the dam's foundation. The maximum settlement occurred at the dam's crest, reaching 160 mm, with the ratio of maximum settlement to the dam's height being less than 0.5%. Figure 11b displays the cloud diagram for the increase in the settlement of the dam on 6 May, when the reservoir's water level was at 60.13 m. The diagram illustrates that the rapid decrease in the reservoir's water level contributed to an ongoing increase in the settlement of the dam's crest, reaching a maximum increase in settlement of 3.5 mm, while the upstream foundation of the dam experienced a lift, peaking at 2.5 mm. From Figure 11c, it is evident that the abrupt rise in the reservoir's water level resulted in a decrease in the maximum incremental settlement at the dam's crest to 2.0 mm, and the maximum lift at the upstream foundation of the dam also reduced to 1.2 mm.

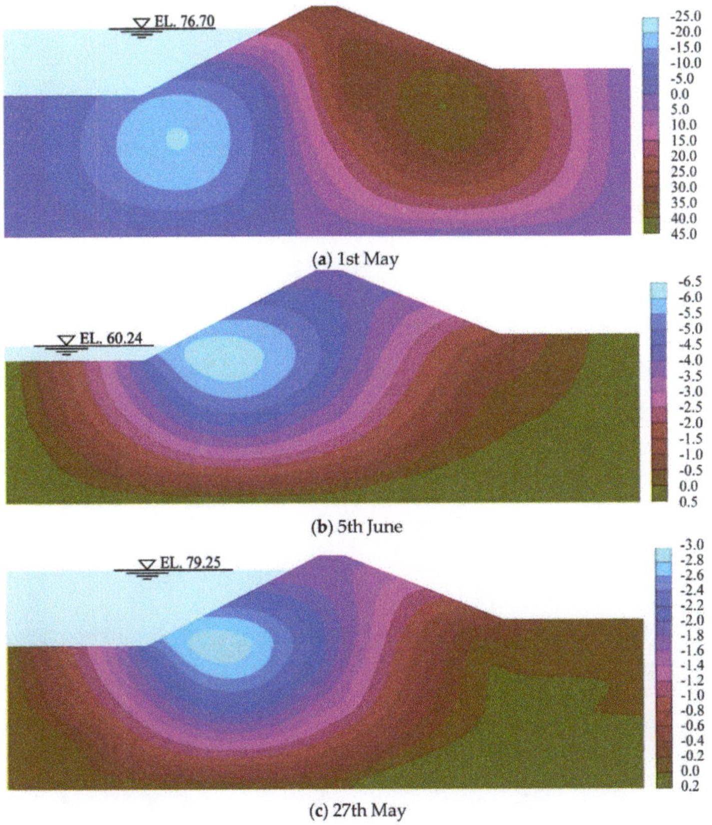

Figure 10. Cloud maps of horizontal displacement at different typical moments (unit: mm).

To analyze the dam's deformation trends during fluctuations in the reservoir's water level, Figure 12 illustrates the vector diagram of the dam's deformation. The red arrows depicts the vector of the increase in deformation when the reservoir's water level dropped sharply to 60.13 m on 6 May. From the diagram, it is evident that the angle between the vector of the increase in deformation and the y-axis increased as the elevation decreased. In the area extending from below one-third of the dam's height in Rockfill I to the dam's foundation, the vector was nearly parallel to the x-axis, indicating that the increase in horizontal deformation significantly exceeds the increase in settlement. If the nonuniform deformation is substantial, it can cause horizontal cracks in the core wall and potentially result in longitudinal cracks separating the core wall from the rockfill. This also significantly affects the stability of the rockfill. Furthermore, the angle between the vector of the increase in deformation and the y-axis in Rockfill II is smaller compared with Rockfill I. The green arrows show the vector of the increase in deformation for the sudden rise in the reservoir's water level to 79.25 m on 27 May. It was observed that the deformation significantly decreased compared with the sudden fall to 60.13 m on 6 May. The angle between the deformation vector and the y-axis reduced, with the angle for Rockfill II being less than 45°. This suggests a diminished tendency towards the upstream movement of the dam's body as the reservoir's water level increases.

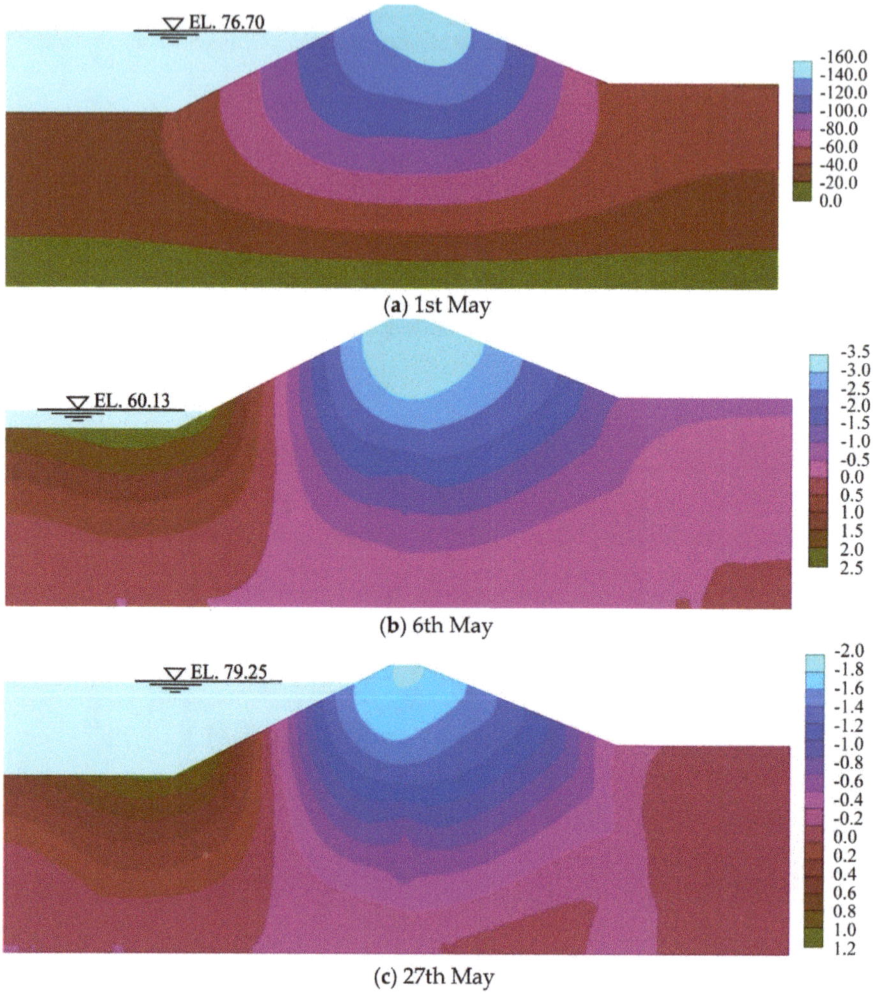

Figure 11. Cloud maps of settlement at different typical moments (unit: mm).

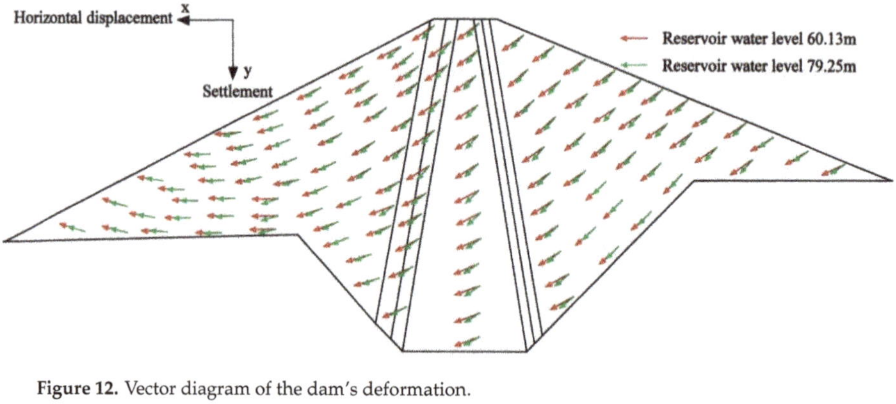

Figure 12. Vector diagram of the dam's deformation.

Table 3 displays the summary of the deformation statistics. According to the table, the maximum values of horizontal displacement and the increase in settlement occurred when the reservoir's water level was at its lowest elevation (5 June and 6 May), rather than at the moment of the fastest decline in the reservoir's water level (2 June). This suggests that the displacement of the CWRD is highly sensitive to the elevation of its lowest point during declines in the reservoir's water level.

Table 3. Summary of the deformation statistics.

Time	Reservoir's Water Level, m	Rate of Decline in the Reservoir's Water Level, (m•d^{-1})	Maximum Increase in Horizontal Displacement, mm		Maximum Increase in Settlement, mm	
			Upstream Point	Downstream Point	Upstream Point	Downstream Point
6/5	60.24	4.06	6.5	0.5	/	/
5/6	60.13	4.38	/	/	2.5	3.5
6/2	63.26	15.53	5.5	0.5	2.0	3.0

4.3.2. Analysis of the Monitoring Data of Displacement

One surface displacement gauge was installed on the upstream side and on the crest of the dam, while two were positioned on the downstream side, as detailed in Figure 2. During the period of working conditions, measurements from the displacement gauges were taken on 1 May, 15 May, and 30 May 2015. Readings from the TR2 and TR3 displacement gauges have been selected for comparison to the simulated values at the corresponding locations within the finite element model, as documented in Table 4. According to the table, the distribution of the simulated values from the finite element model aligned with the monitored values from the displacement gauges at various times. Furthermore, the ratio of the cumulative maximum settlement to the height of the core wall and rockfill was below 0.5%, conforming to the typical deformation pattern observed in dams.

Table 4. Comparison table between the monitored values of the displacement gauges and the simulated values of the finite element model.

Time	TR2 Horizontal Displacement, mm		TR2 Settlement, mm		TR3 Horizontal Displacement, mm		TR3 Settlement, mm	
	Simulated Value, m	Monitored Value, m	Simulated Value, m	Monitored Value, m	Simulated Value, m	Monitored Value, m	Simulated Value, m	Monitored Value, m
5/1	24.88	23.65	154.78	150.73	25.04	24.13	156.35	151.67
5/15	20.60	18.95	151.84	148.32	20.68	19.74	154.47	149.23
5/30	22.87	21.63	152.38	149.30	23.01	22.60	154.83	149.96

5. Conclusions and Prospects

5.1. Conclusions

This study utilized finite element software to investigate the impact of rapid fluctuations in the water level on the seepage, stress, and displacement fields of a PSPS's CWRD. Moreover, comparative analyses were conducted between the results of finite simulation and the integration of monitored data from the dam's osmometers, earth pressure gauges, and displacement gauges. From the analysis, the following conclusions can be inferred.

(1) When the reservoir's water level dropped sharply, the decline in the phreatic line in the rockfill lagged, creating a large hydraulic gradient and a reverse seepage field on the surface of the dam's slope. This resulted in a dragging force directed upstream, with a noticeable deformation trend of the lower core wall and rockfill towards the upstream. One-third of the surface of the upstream slope of the dam and the curtain grouting experienced significant concentrations of stress. Consequently, bending of the core wall could easily produce horizontal cracks, and there was a risk of longitudinal cracks separating from the rockfill, greatly affecting the stability of the rockfill on the upstream slope of the dam.

(2) Upon a sudden decrease in the reservoir's water level, the dam's deformation showed increased sensitivity to the lowest elevation point, compared with the peak rate of decline in the water level. The peak increase in horizontal displacement of 6.5 mm occurred one-third up Rockfill I, while the maximum increase in settlement at the dam's crest was 3.5 mm. Hence, close scrutiny is warranted for cracks or voids at the dam's crest and near one-third of the dam's height within Rockfill I.
(3) A sudden increase in the reservoir's water level led to a reduction in both the upstream and downstream horizontal displacement, as well as decreased in settlement. For both the core wall and the rockfill, the ratio of cumulative maximum settlement to the dam's height was less than 0.5%. Furthermore, the absence of tensile stress zones or cracks at the dam's crest aligned with the established deformation principles of CWRDs.
(4) Under the influence of rapid fluctuations in the reservoir's water level, the variation in vertical earth pressure at the bottom of the core wall was more pronounced than at its midsection. This was because the soil at the bottom of the core wall had a higher initial density, smaller porosity, and lower permeability due to greater gravity and pressure from the overlying soil. Consequently, when the reservoir's water level changed, the pore water pressure in the bottom soil adjusted more slowly than in the middle, resulting in more significant changes in soil pressure at the bottom.
(5) The safety monitoring data for dam seepage pressure, earth pressure, and displacement aligned closely with the finite element model's simulated values for the infiltration line, earth pressure, and deformation. This concordance verified the finite element model's accuracy.

5.2. Prospects

(1) The deformation characteristics of a PSPS's CWRD under earthquake conditions when the reservoir's water level fluctuates during the service period were not discussed in this study. This aspect should be further investigated in future studies.
(2) Due to the lack of detailed 3D geological data, a two-dimensional finite element model was used in this study. This model could not fully capture the distribution of stress and deformation in the dam's three-dimensional space, and should be improved in future research.
(3) This study used the Duncan–Chang E-B model for the stress–strain relationship of rockfill, which could not accurately describe particle breakage and other characteristics of rockfill. Future research should propose a constitutive model that accounts for all the characteristics of rockfill to enhance the accuracy of finite element calculations.

Author Contributions: Conceptualization, X.Z. and Y.F.; methodology, X.Z., B.Y., W.W., K.D. and Y.F.; software, Y.F.; formal analysis X.Z. and Y.F.; investigation, X.Z. and Y.F.; data curation, B.Y., W.W. and K.D.; writing—original draft preparation, X.Z. and Y.F.; writing—review and editing, B.Y., W.W. and K.D. All authors have read and agreed to the published version of the manuscript.

Funding: This research was funded by the National Natural Science Foundation of China (U2243223), the Fundamental Research Funds for the Central Universities of China (B230201011), the Water Conservancy Science and Technology Project of Jiangsu (2022024), and the State Grid Xinyuan (Holding) Company Ltd. Science and Technology Project (SGXYKJ-2023-059).

Data Availability Statement: Data are contained within the article.

Acknowledgments: The authors would like to extend their gratitude to all those who supported this work.

Conflicts of Interest: Authors X.Z., B.Y., W.W. and K.D. were employed by the Pumped-Storage Technological & Economic Research Institute, State Grid Xinyuan Company Ltd. The remaining author declared that the research was conducted in the absence of any commercial or financial relationships that could be construed as a potential conflict of interest.

References

1. Ma, H.; Chi, F. Major Technologies for Safe Construction of High Earth-Rockfill Dams. *Engineering* **2016**, *2*, 498–509. [CrossRef]
2. Wang, X. Discussion on Some Problems Observed in High Earth-Rockfill Dams. *Chin. J. Geotech. Eng.* **2018**, *40*, 203–222. [CrossRef]
3. Rahmani, H.; Panah, A. Effect of Particle Size and Saturation Conditions on the Breakage Factor of Weak Rockfill Materials Under One-Dimensional Compression Testing. *Geomech. Eng.* **2020**, *4*, 315–326. [CrossRef]
4. Wang, Y.; Li, J.; Wu, Z.; Chen, J.; Yin, C.; Bian, K. Dynamic Risk Evaluation and Early Warning of Crest Cracking for High Earth-Rockfill Dams through Bayesian Parameter Updating. *Appl. Sci.* **2020**, *10*, 7627. [CrossRef]
5. Zhang, H.; Jing, Y.; Chen, J.; Gao, Z.; Xu, Y. Characteristics and Causes of Crest Cracking on A High Core-Wall Rockfill Dam: A Case Study. *Eng. Geol.* **2022**, *297*, 106488. [CrossRef]
6. Naylor, D.; Maranha das Neves, E.; Mattar, J.; Veiga Pinto, A. Prediction of Construction Performance of Beliche Dam. *Geotechnique* **1986**, *36*, 359–376. [CrossRef]
7. Naylor, D.; Maranha, J.; Maranha das Neves, E.; Veiga Pinto, A. A Back-Analysis of Beliche Dam. *Geotechnique* **1997**, *47*, 221–233. [CrossRef]
8. Maranha das Neves, E.; Veiga Pinto, A. Modeling Collapse on Rockfill Dams. *Comput. Geotech.* **1988**, *6*, 131–153. [CrossRef]
9. Alonso, E.; Olivella, S.; Pinyol, N. A Review of Beliche Dam. *Geotechnique* **2005**, *55*, 267–285. [CrossRef]
10. Cetin, H.; Laman, M.; Ertunc, A. Settlement and Slaking Problems in the World's Fourth Largest Rock-Fill Dam, the Ataturk Dam in Turkey. *Eng. Geol.* **2000**, *56*, 225–242. [CrossRef]
11. Kartal, M.; Çavusli, M.; Genis, M. 3D Nonlinear Analysis of Atatürk Clay Core Rockfill Dam Considering Settlement Monitoring. *Int. J. Geomech.* **2019**, *19*, 04019034. [CrossRef]
12. Leonards, G.; Davidson, L. Reconsideration of Failure Initiating Mechanisms for Teton Dam. *Int. J. Rock. Mech. Min. Sci. Geomech. Abstr.* **1986**, *23*, A27. [CrossRef]
13. Zhang, X.; Wang, C.; Wong, H.; Jiang, T.; Dong, J. Modeling Dam Deformation in the Early Stage of Internal Seepage Erosion—Application to the Teton Dam, Idaho, Before the 1976 Incident. *J. Hydrol.* **2022**, *605*, 127378. [CrossRef]
14. Jia, Y.; Xu, B.; Chi, S.; Xiang, B.; Xiao, D.; Zhou, Y. Joint Back Analysis of the Creep Deformation and Wetting Deformation Parameters of Soil Used in the Guanyinyan Composite Dam. *Comput. Geotech.* **2018**, *96*, 167–177. [CrossRef]
15. Zhou, X.; Chi, S.; Jia, Y.; Shao, X. A New Wetting Deformation Simulation Method Based on Changes in Mechanical Properties. *Comput. Geotech.* **2020**, *117*, 103261. [CrossRef]
16. Zhou, X.; Chi, S.; Wang, M.; Jia, Y. Study on Wetting Deformation Characteristics of Coarse Granular Materials and Its Simulation in Core-Wall Rockfill Dams. *Int. J. Numer. Anal. Methods Geomech.* **2020**, *44*, 851–873. [CrossRef]
17. Squier, L. Load Transfer in Earth and Rockfill Dams. *J. Soil Mech. Found.* **1970**, *96*, 213–233. [CrossRef]
18. Ding, N.; Duan, J.; Xue, S.; Zeng, M.; Shen, J. Overall Review of Peaking Power in China: Status Quo, Barriers and Solutions. *Renew. Sustain. Energy Rev.* **2015**, *42*, 503–516. [CrossRef]
19. Kong, Y.; Kong, Z.; Liu, Z.; Wei, C.; Zhang, J.; An, G. Pumped Storage Power Stations in China: The Past, the Present, and the Future. *Renew. Sustain. Energy Rev.* **2017**, *71*, 720–731. [CrossRef]
20. Jia, J. A Technical Review of Hydro-Project Development in China. *Engineering* **2016**, *2*, 302–312. [CrossRef]
21. Xu, Z.; Cao, C.; Li, K.; Cai, J.; Xiong, W.; Zhao, J.; Qin, R. Simulation of Drainage Hole Arrays and Seepage Control Analysis of the Qingyuan Pumped Storage Power Station in China: A Case Study. *Bull. Eng. Geol. Environ.* **2019**, *78*, 6335–6346. [CrossRef]
22. Feng, W.; Chi, S.; Jia, Y. Random Finite Element Analysis of A Clay-Core-Wall Rockfill Dam Considering Three-Dimensional Conditional Random Fields of Soil Parameters. *Comput. Geotech.* **2023**, *159*, 105437. [CrossRef]
23. Larese, A.; Rossi, R.; Oñate, E.; Toledo, M.; Morán, R.; Campos, H. Numerical and Experimental Study of Overtopping and Failure of Rockfill Dams. *Int. J. Geomech.* **2015**, *15*, 04014060. [CrossRef]
24. Mohammad Rashidi, S.; Mohsen, H. Evaluation of Behaviors of Earth and Rockfill Dams During Construction and Initial Impounding Using Instrumentation Data and Numerical Modeling. *J. Rock. Mech. Geotech. Eng.* **2017**, *9*, 709–725. [CrossRef]
25. Xu, M.; Pang, R.; Zhou, Y.; Xu, B. Seepage Safety Evaluation of High Earth-Rockfill Dams Considering Spatial Variability of Hydraulic Parameters Via Subset Simulation. *J. Hydrol.* **2023**, *626*, 130261. [CrossRef]
26. Wu, Z.; Chen, C.; Zheng, Y.; Chen, J.; Bian, K.; Li, J. Analysis of Seepage Failure Probability for High Core Rockfill Dams During Rapid Drawdown of Reservoir Water level. *J. Hydrol.* **2024**, *633*, 131046. [CrossRef]
27. Reza, M.; Ali, A.; Behrouz, G. Simulation of Collapse Settlement of First Filling in a High Rockfill Dam. *Eng. Geol.* **2015**, *187*, 32–44. [CrossRef]
28. Akhtarpour, A.; Salari, M. The Deformation Mechanism of A High Rockfill Dam During the Construction and First Impounding. *Sci. Iran.* **2020**, *27*, 566–587. [CrossRef]
29. Qiu, T.; Wang, W.; Höeg, K.; Feng, S.; Zhao, R. 3D Analysis of the 174-m High Quxue Asphalt-Core Rockfill Dam in A Narrow Canyon. *Soils Found.* **2021**, *61*, 1645–1659. [CrossRef]
30. Zhou, X.; Wu, B.; Li, L. Study on Influence of Numerical Simulation Accuracy on High Core Wall Rockfill Dam Deformation and Crack Analysis. *Buildings* **2022**, *12*, 1494. [CrossRef]
31. Zhou, X.; He, J.; Chi, S.; Wang, J. Study on Collapse Settlement and Cracks of Core Wall Rockfill Dams Under Wetting Deformation. *Int. J. Numer. Anal. Methods Geomech.* **2023**, *47*, 106–128. [CrossRef]
32. Liu, S.; He, W.; Sun, Y.; Shen, C.; Wang, L. Analysis of the Behavior of A High Earth-Core Rockfill Dam Considering Particle Breakage. *Comput. Geotech.* **2023**, *157*, 105320. [CrossRef]

33. Pan, L.; Fan, Z.; Wang, D.; Zhou, X. The Wetting Deformation Model of Rockfill and Its Two Methods for Simulating Rock-fil Dam Collapse Settlement. *Sci. World J.* **2023**, *2023*, 5593636. [CrossRef] [PubMed]
34. Pan, L.; Wu, B.; Wang, D.; Zhou, X.; Wang, L.; Zhang, Y. Study on Impoundment Deformation Characteristics and Crack of Hig Core Rockfill Dam Based on Inversion Parameters. *Water* **2024**, *16*, 188. [CrossRef]
35. Ai, Z.; Ma, G.; Zhang, G.; Liu, R.; Deng, S.; Chang, X.; Zhou, W. Multi-Source Monitoring Data Filtering Assisted Deformatior Analysis Model Updating of Ultra-High Rockfill Dam. *Comput. Geotech.* **2024**, *171*, 106323. [CrossRef]
36. Wu, Y.; Zhang, B.; Yu, Y.; Zhang, Z. Consolidation Analysis of Nuozhadu High Earth-Rockfill Dam Based on the Coupling o Seepage and Stress-Deformation Physical State. *Int. J. Geomech.* **2016**, *16*, 04015085. [CrossRef]
37. Dong, Y.; Xing, H.; Liu, J. With FLAC3D Simulating Analysis the Rock Stability by Stretching Anchor Cables Reinforce on th Reservoir Dam Abutment. *Appl. Mech. Mater.* **2011**, *90–93*, 2670–2676. [CrossRef]
38. Li, M.; Xue, B.; Fang, H.; Zhang, S.; Wang, F. Parameter Sensitivity Analysis of Polyurethane Cutoff Walls for Earth Dams unde Multifield Coupling. *Int. J. Geomech.* **2023**, *23*, 04023206. [CrossRef]
39. Liu, S.; Wang, L.; Wang, Z.; Bauer, E. Numerical Stress-Deformation Analysis of Cut-Off Wall in Clay-Core Rockfill Dam on Thic Overburden. *Water Sci. Eng.* **2016**, *9*, 219–226. [CrossRef]
40. Chen, Y.; Lin, G.; Lu, Y. Deformation and Stability Analysis of a Core Rockfill Dam with Leakage. *Shock. Vib.* **2022**, *2022*, 694444 [CrossRef]
41. Ri, Y.; Han, U.; Jang, U.; Jong, D.; Kim, C. Study on Stability Reduction Characteristics of Earth and Rockfill Dams under Rapic Drawdown Using Fully Coupled Seepage-Stress Analysis. *Adv. Civil. Eng.* **2022**, *2022*, 7954991. [CrossRef]

water

MDPI

Article

Stability Study of a Double-Row Steel Sheet Pile Cofferdam Structure on Soft Ground

Yan Jiang [1,2], Fei Guo [1], Wenlong Wang [1], Guanghua Yang [2], Jinchao Yue [1] and Yibin Huang [1,*]

[1] Yellow River Laboratory, Zhengzhou University, Zhengzhou 450001, China
[2] Geotechnical Engineering Research Institute, Guangdong Research Institute of Water Resources and Hydropower, Guangzhou 510610, China
* Correspondence: huangyb@zzu.edu.cn

Abstract: The stability of a double-row steel sheet pile cofferdam structure under soft ground conditions was investigated in this study, using the temporary cofferdam of the Shenzhen–Zhongshan cross-river channel as the engineering background. The stability of the cofferdam design solution was calculated with a model that incorporates factors such as the coordination of independent pile top displacement, as well as the m-value for backfilled sand and the thrown rock body. The internal force and displacement results of the cofferdam under different working conditions are obtained. And the entire construction process was analyzed using the finite element method. The results indicate that the overall stability and overturning stability of the cofferdam satisfy relevant safety requirements, with minimum safety factors of 1.744 and 1.400, respectively. The maximum displacement of the inner and outer steel sheet piles is 34 mm, the maximum bending moment is 249.30 kN·m, and the maximum shear force is 266.66 kN. The displacements of sheet piles were within an acceptable range, and the internal forces remained below the load capacity of the selected sheet pile type for the design. Based on these findings, the cofferdam structure can be considered safe and satisfying the specified requirements. This work may have instructive value for cofferdam design and construction.

Keywords: cofferdam; double-row sheet pile; soft-ground foundation; structural stability; finite element method

1. Introduction

Steel sheet piles have a rich history and have undergone significant development in Europe and Japan since their introduction in the early 20th century. They have become widely utilized in various construction applications, including cofferdams, foundation support in docks, wharves, bridges, immersed tube tunnels, and other projects. Steel sheet piles offer numerous advantages such as high quality, simple construction, durability, and the ability to reduce spatial requirements for construction tasks [1,2]. They are extensively utilized in water conservancy, transportation, municipal, harbor, and navigation projects. In soft soil areas, the double-row steel sheet pile cofferdam structure is particularly favored due to its excellent structural stability, strong resistance to deformation and seepage, and adaptability to engineering challenges [3]. Therefore, it is widely adopted in soft-ground foundation projects.

Scholars have conducted valuable research on the practical applications of double-row steel sheet piles, yielding significant findings. Hou et al. [4] conducted on-site monitoring of a double-row sheet pile cofferdam for a large wharf project in Shanghai, studying its deformation characteristics under construction conditions. Zhu et al. [5] conducted a reliability analysis of double-row steel sheet piles using the Bayesian approach, effectively integrating statistical and field survey data. Mitobe et al. [6] conducted flume model tests on embankment reinforcement using single and double rows of sheet piles to prevent tsunami overflow, observing better performance with double rows of sheet piles. Shen et al. [7]

investigated the effects of different lengths of double-row pile support through model tests to find that structural stability was less affected by the appropriate shortening of the pile's length on the soil-accommodating side. Khan et al. [8] established a centrifugal test model for a double-row sheet pile cofferdam and found that multi-layer tie rods significantly enhanced its stability.

There have been many other valuable contributions to the literature. Zhou et al. [9] found that the displacement of the pile on the back soil side was greater than that of the pile on the facing soil side through small-scale model tests. Sawaguchi et al. [10] simplified the double-row pile structure to a stable frame structure and derived pile displacement curves under different loading conditions. Buhan et al. [11] used the limit equilibrium design theory to calculate the load limit on a cofferdam, considering the weir core fill and sheet piles on both sides as an elastic continuous medium and shell, while also analyzing the deformation based on different boundary conditions. Banerjee et al. [12] analyzed the effect of steady-state seepage on a double-row sheet pile cofferdam structure by adding a seepage safety factor to the theoretical calculations. Lei et al. [13] determined that the damage patterns of slopes reinforced by single and double rows of stabilizing piles differ based on the distribution of potential sliding surfaces, bending moments, and thrusts. Li et al. [14] proposed a simplified analytical model to derive the required resistance of a double-row pile at different rotational angle positions, combining the kinematic approach of limit analysis with the strength discounting technique. Zhang et al. [15] confirmed the feasibility of monitoring the strain of double-row steel pipe piles during foundation excavation through field tests with strain gauges. Zhou et al. [16] reviewed and evaluated the force deformation and earth pressure of double-row pile piles, summarizing current research problems and shortcomings. In a large-scale physical model test based on a deep foundation pit project in Changchun City, Zhou et al. [9] calculated the strain and displacement of double-row piles and soil pressure, finding greater strain and displacement in the front-row piles compared to the rear-row piles. Lefas et al. [17] developed a simplified two-dimensional (2D) model to analyze the steel sheet pile cofferdam structure and investigated its force deformation characteristics during construction according to a calculation program they developed. Byfield et al. [18] conducted experiments and analyses on the calculated deformation of steel plates, providing valuable insights for understanding the forces and pile-soil interaction mechanisms of steel sheet piles.

With advancements in computer technology, numerous scholars have employed numerical analysis methods to investigate the stability of cofferdam structures [19]. Gui et al. [20] analyzed a failed double-row steel sheet pile cofferdam using finite element software and identified low tie weld strength as the cause of damage to the structure. Zhao [21] conducted numerical simulations to analyze the influence of various parameters of double-row steel sheet piles on cofferdam performance, emphasizing the importance of filling materials within the double-row steel pipe pile cofferdam. Xue et al. [22] performed stability analysis on a pile wall frame structure cofferdam through engineering tests and 3D finite element numerical simulation, then proposed a design method based on the limit equilibrium approach. Monika et al. [23] investigated the cofferdam of the main central column foundation of an asymmetric bridge using numerical simulation based on the finite element method with Plaxis geotechnical software. Fujiwara et al. [24] examined the dynamic behavior of a structure with partition walls perpendicular to sheet piles under seismic action through shaking table tests and 2D numerical simulations. Wang et al. [25] used FLAC 3DV6.0 software to numerically simulate the excavation process of steep reinforced riverbank slopes in an effort to predict long-term slope deformation. Xu et al. [26] developed a numerical solution based on the power-law nonlinear damage criterion and the upper bound theorem of limit analysis, deriving the shape of the cofferdam collapse block and landslide thrust under the effects of long-term water waves and residual pore pressure. Hui et al. [27] established a Plaxis 3D finite element model to investigate the relationship between the depth of soil reinforcement between double rows of sheet piles and internal force displacement. Chen et al. [28] studied the hydrodynamic response and seabed response around a dumbbell weir under combined wave and current loads, considering the wave-current interactions. Hu et al. [29] integrated saturated–unsaturated seepage theory and the intensity discounting method into a finite element system to analyze the effects of water level fluctuations on the instability and seepage damage of a landfill weir. Ti et al. [30] evaluated the stochastic wave pressure of a construction cofferdam for a bridge under the action of a tropical cyclone in terms of the pressure spectrum, pressure spectrum characteristics, transfer function, and maximum wave pressure. Wang et al. [31] developed a prediction model for short- and medium-term water levels, enabling dynamic adjustment and adaptation to the actual needs of sheet pile cofferdam construction. Although the above studies have analyzed the structural stability of cofferdams and the safety of the construction process, there is limited research on the structural stability of cofferdams in soft-ground foundations. Additionally, no previous researcher has compared and verified the internal forces and displacements of cofferdam structures using different calculation methods. To this effect, it is difficult to secure accurate predictions of potential adverse conditions during construction, and available techniques for structural reliability analysis are seriously limited.

The safety and stability of double-row steel sheet piles in soft soil areas were analyzed in this study using a Rizheng deep foundation pit and Midas GTS finite element calculation software. The overturning resistance and overall stability of the cofferdam were analyzed, and the minimum internal forces and displacements of the inner and outer steel sheet piles were derived. Additionally, the entire process of cofferdam construction was simulated using the finite element method.

2. Project Overview and Scheme Design

2.1. Project Overview

The Shenzhen–Zhongshan Passage is located approximately 30 km north of Humen Bridge and 38 km south of Hong Kong–Zhuhai–Macau Bridge. To facilitate the construction of a widened bifurcated tunnel, a temporary water retaining structure in the form of a cofferdam will be established in the sea on the Shenzhen side. The cofferdam, classified as grade 4, is designed for the cut and cast structure construction within its boundaries. The cofferdam has a double-row steel sheet pile configuration with a width of 10 m and a total length of approximately 1393.4 m. Once the main tunnel structure is completed and backfilled, the temporary cofferdam above the seabed surface will be dismantled. Figure 1 illustrates the planned layout of the steel sheet pile cofferdam.

Figure 1. Plane position of the cofferdam.

2.2. Scheme Design

2.2.1. Profile Design

The construction section of the weir is designed according to marine and surrounding environmental conditions as well as the construction schedule. The cofferdam is divided into two sides: the pit side (inner side) and the waterfront side (outer side). Steel ties 60 mm in diameter are used to connect the inner and outer rows of steel sheet piles, with a spacing and center elevation of 1.5 m. An "L"-shaped retaining wall is installed on the inside of the outer row of steel sheet piles, measuring 2 m in height and 200 mm in thickness. Support walls that are 150 mm thick are placed at intervals of 1.5 m along the axis of the cofferdam. Concrete beams measuring 400 × 500 mm are used to connect the retaining wall and inner row of sheet piles. To prevent scouring of the steel sheet piles, a sand-ribbed soft drain is installed on the waterward side of the cofferdam. A layer of bagged gravel, 25 cm thick, is placed above the soft drain. On the sea side, a rock throwing layer with minimum thickness of 2.0 m is applied, consisting of rocks weighing between 800 and 1500 kg. Blocks weighing 200–300 kg are used for the remaining sides. A counter pressure soil slope is constructed on the backwater side of the cofferdam, which is not protected by molded concrete. Figure 2 shows a typical cross-section of a steel sheet pile cofferdam.

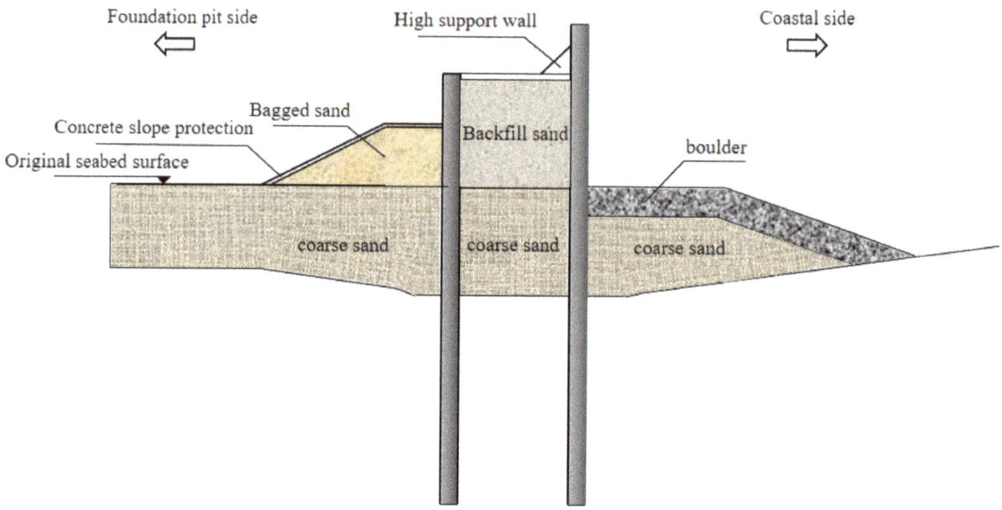

Figure 2. Typical Profile of Steel Sheet Pile.

2.2.2. Steel Sheet Pile Section Design and Wave Force Value

The design under analysis has a steel sheet pile model U 750 × 225 × 14.5. The steel type is Q390BZ and Q345B. The flexural load capacity of the structure is 775 kN·m. The steel sheet pile cross-section and characteristics are shown in Figures 3 and 4 and Tables 1 and 2.

The calculation includes water pressure and wave forces as the primary factors. Water pressure is determined based on each specific working condition, whereas wave crest and wave forces are calculated according to the Harbor Hydrographic Code (JTS 145-2-2013). The design considers extremely unfavorable working conditions, such as encountering waves of the same frequency as the 20-year high tide level. The bed type of the cofferdam project is classified as a low bed, and the wave state is a standing wave. The wave pressure distribution is shown in Figure 5.

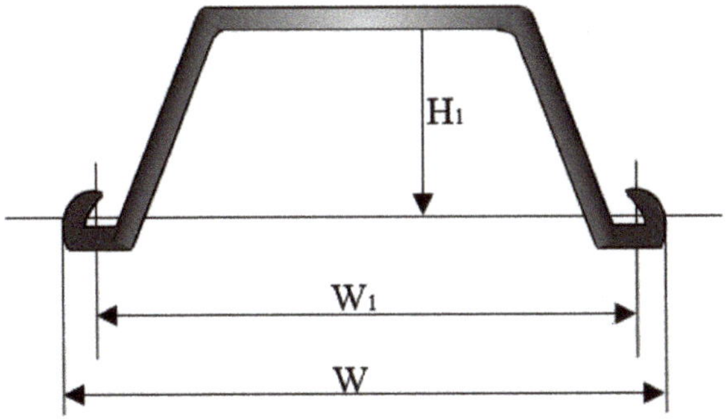

Figure 3. Standard section of steel sheet pile.

Figure 4. Combined section of small locking sheet pile.

Table 1. Steel sheet pile section characteristics.

Model $W \times H \times T_w$	Effective Width W_1 (mm)	Effective Height H_1 (mm)	Board Thickness t (mm)	Cross-Sectional Area per Meter of Sheet (cm^2)	Theoretical Weight per Meter of Sheet(kg/m)
750 × 225 × 14.5	750	225	14.5	188	147.2

Table 2. Steel parameters.

Steel Type	Design Value of Flexural Strength (MPa)	Design Value of Shear Strength (MPa)	Modulus of Elasticity (MPa)	Other (mm)
Q390BZ	350	205	2.06×10^5	Thickness ≤ 16
Q345B	295	170	2.06×10^5	Thickness ≤ 16

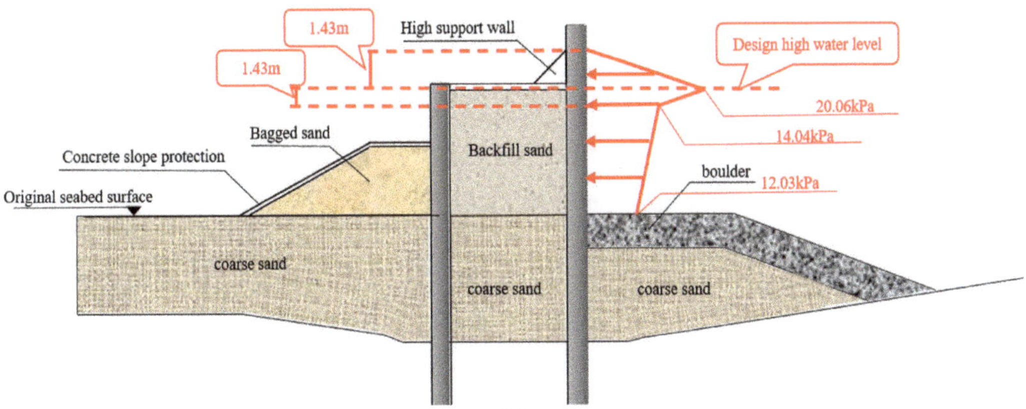

(a)

Figure 5. *Cont.*

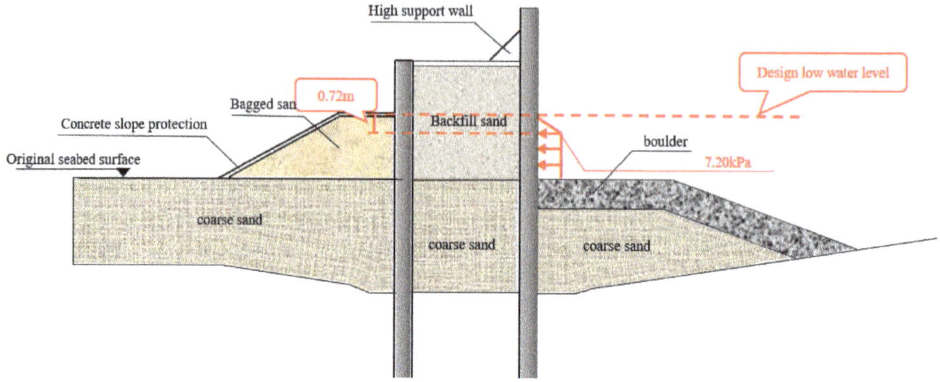

(b)

Figure 5. Cofferdam wave force distribution. (**a**) Ten-year high wave thrust; (**b**) ten-year high wave suction.

3. Structural Stability Analysis

3.1. Overall and Overturning Stability Test

The calculation model was established based on two typical cross sections: K6 + 598 right line and K7 + 030 left line. The overturning and general stability of the cofferdam were calculated using Lizheng Geotechnical Software.

3.1.1. Overall Stability Analysis

The construction period of the temporary cofferdam is expected to span approximately three years. The design considers a flood recurrence period of 5–10 years, and the cofferdam can withstand a level 4 flood. The overall stability calculation requires a safety factor of at least 1.35 for both normal and extraordinary operation conditions. The Swedish circular sliding method is adopted to calculate the overall stability.

$$\min\{K_{s,1}, K_{s,1}, \cdots, K_{s,i} \cdots\} \geq K_s \tag{1}$$

$$K_{s,i} = \frac{\sum\{c_j l_j + [(q_j b_j + \Delta G_j) cos\theta_j - u_j l_j] \, tan\phi_j\}}{\sum (q_j b_j + \Delta G_j) sin\theta_j} \tag{2}$$

where K_s is arc sliding stability safety factor; $K_{s,i}$ is the ratio of the anti-slip moment to the sliding moment of the ith arc sliding body; c_j is the cohesive force (kPa); ϕ_j falls the angle of internal friction (°); b_j is the width of the jth soil strip (m); θ_j is the angle between the normal and the vertical plane at the midpoint of the slip arc of the jth soil strip (°); l_j is slip arc length of the jth soil strip (m); q_j is the standard value of the additional distributed load on the jth soil strip (kPa); u_j is the water pressure on the slip-arc surface of the jth soil strip (kPa); ΔG_j is the self-weight of the jth earth bar (kN).

Two typical cross-sections, the K6 + 598 right line and the K7 + 030 left line, were analyzed for their safety factors under four different working conditions. On the right line of K6 + 598, Condition 1 assumes no precipitation within the cofferdam, and the water level outside the cofferdam sharply drops after its completion. Condition 2 considers the completion of the cofferdam, with the internal water level dropping to 1 m below the surface while the external water level remains at the designed high tide level. Condition 3 assumes no precipitation within the weir, and the external water level rises to the top of the weir's backfill sand. Condition 4 accounts for the internal water level dropping to 1 m below the berm inside the weir, the external water level at the designed high tide level of the weir, and the backfill sand filled to the top of the weir.

On the left line of K7 + 030, Condition 5 assumes no precipitation within the cofferdam, and the water level outside the cofferdam suddenly drops after its completion. Condition 6 considers the completion of the cofferdam, with the internal water level dropping to 1 m below the surface while the external water level remains at the designed high tide level. Condition 7 assumes no precipitation in the weir and the external water level rises to the top of the weir's backfill sand. Condition 8 accounts for the internal water level dropping to 1 m below the berm inside the weir, the external water level at the designed high tide level of the weir, and the backfill sand filled to the top of the weir.

The results of the cofferdam stability calculations for comparison are presented in Figure 6.

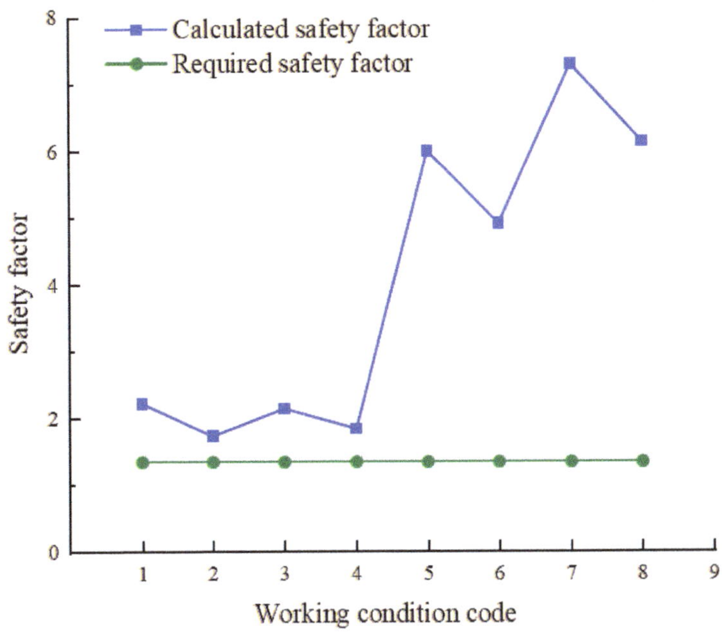

Figure 6. Comparison of overall stability safety factors.

As shown in Figure 6, the overall stability of this cofferdam was evaluated for eight working conditions in two typical design sections. The minimum safety coefficient is 1.744. All the safety coefficients obtained exceed the safety coefficients specified in the regulations, indicating that the design meets the necessary safety requirements.

3.1.2. Overturning Stability Analysis

The overturning stability of the cofferdam is conducted based on a gravity-type cement soil wall, taking into account the discounted internal friction angle of the finite bagged sand and thrown rock. The discount factor is determined by comparing the required width of the passive zone calculated by the 45-degree rupture angle with the actual width of the passive zone. The calculation of overturning stability follows the formula specified in the Technical Regulations for Construction Pit Support (JGJ120-2012), which state that the safety factor should be at least 1.3. The overturning stability of the gravity-type cement soil wall can be calculated as follows:

$$\frac{E_{pk}a_p + (G - u_m B)a_G}{E_{ak}a_a} \geq K_{ov} \quad (3)$$

where K_{ov} is the overturning safety factor; a_a is the vertical distance from the point of combined active earth pressure on the outside of the cement wall to the toe (m), E_{ak} is the standard value of active earth pressure (kN/m); E_{pk} is the passive earth pressure acting on the cement soil wall (kN/m), G is the self-weight of the concrete wall (kN/m); B is the width of the concrete wall's bottom section (m), u_m is the water pressure on the bottom surface of the cement wall (kPa), a_p is the vertical distance from the action point of the combined passive earth pressure on the inside of the concrete wall to the toe (m), a_G is the horizontal distance from the action point of the combined weight of the cement wall and the water pressure at the bottom of the wall to the toe (m).

The safety coefficient of overturning resistance is calculated for two typical cross-sections, the K6 + 598 right line and K7 + 030 left line, under two sets of respective working conditions. As shown in Figure 7, the overturning stability of this cofferdam is calculated for four working conditions of two design typical sections. The minimum safety factor is 1.4, which meets the specified safety requirements.

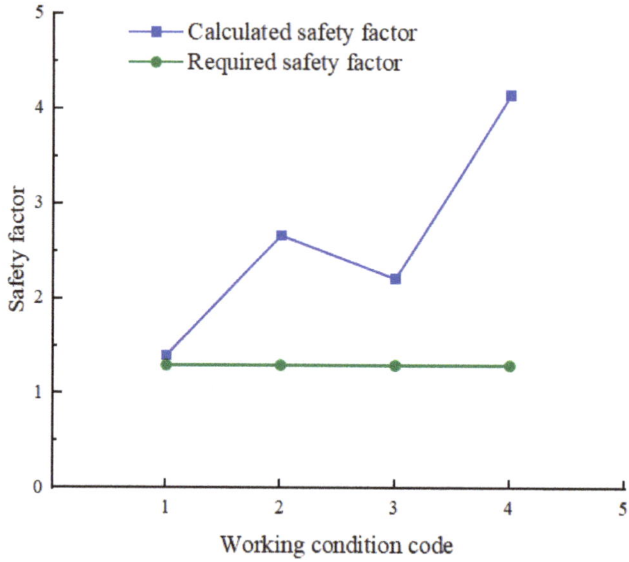

Figure 7. Comparison of safety factors for overturning stability.

3.2. Internal Force and Displacement Analysis

The spacing between the double rows of sheet piles in the cofferdam is 10 m, which is much larger than the width of a single sheet pile. When the sand filling reaches the active limit equilibrium state, the angle between the damaged surface and the sheet pile is considered to be 45° − φ/2. However, at an excavation depth of 7–8 m, the top of the damaged surface is still located between the front and rear rows of the sheet pile, which does not align with the calculation mode for double-row piles. Therefore, the single-row pile mode is considered separately.

The simulation performed in Lizheng deep foundation pit software considers independent pile top displacements for the inner and outer sheet piles, as well as discounts for backfilling sand and the thrown stone body (m-value discounts). To equalize the top displacement of the front and back rows of the pile model, the concentrated pull rod force is progressively adjusted during the calculation. At this stage, the pull rod force is considered more appropriate. This tie force is reapplied to the front pile model to analyze the sheet pile displacements and forces. The material parameters are shown in Table 3 and the calculation model is shown in Figure 8.

Table 3. Internal force and displacement calculation parameters.

Name	Natural Density (g/cm^3)	Compression Modulus (MPa)	Poisson's Ratio	Cohesive Force (kPa)	Internal Friction Angle (°)
Sludge I	1.49	1.53	0.4	3.1	2.0
Sludge II	1.52	1.41	0.4	2.8	1.8
Clay	1.97	6.29	0.3	32.2	8.7
Residual silty clay	1.87	9.6	0.3	15.1	25.1
Completely weathered granite	1.89	18	0.2	18.5	27.4
Sandy strongly weathered granite	1.93	26	0.2	27.5	28.6
Fragmented strongly weathered granite	2.55	/	0.2	3000	30
Moderately weathered granite	2.62	26.2	0.2	15,000	41
Riprap filling	2.1	/	0.2	0	38
Backfilling with medium coarse sand	1.9	/	0.3	0	35
Bagging sand	1.9	/	0.3	12.3	25.96
Mixing pile	1.8	/	0.3	20	20
Concrete (retaining walls, support beams)	2.4	/	0.15	/	/
Steel (steel tie rods, steel sheet pile cofferdams, steel supports)	78.5	/	0.3	/	/

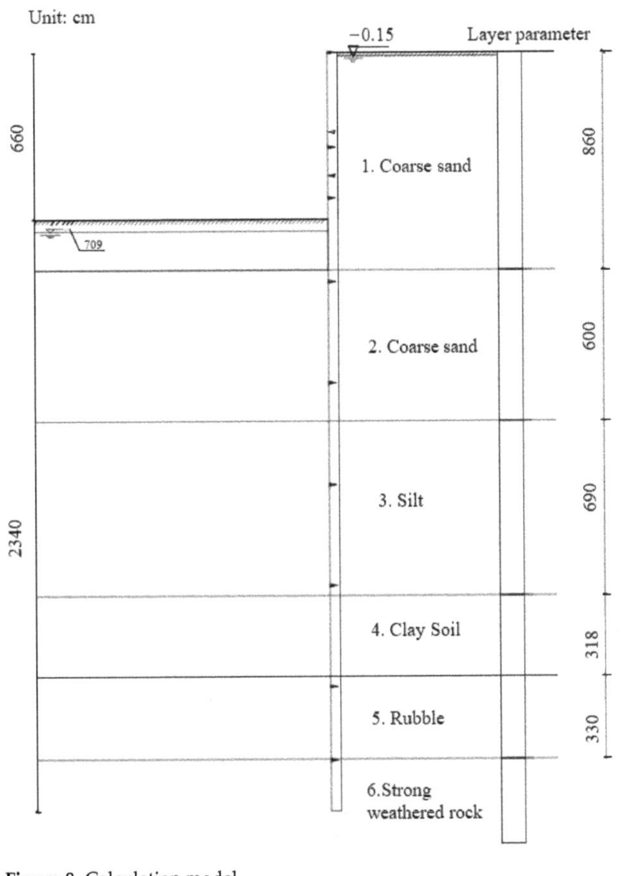

Figure 8. Calculation model.

The m-value method is an approach that utilizes a vertical plane elastic foundation beam method, where a retaining wall is treated as an elastic beam element and the passive earth pressure inside the pit is simulated using soil springs. This method offers several advantages, including a simplified model, fewer calculation parameters, the ability to simulate distributed excavation, and the ability to capture the relationship between passive soil pressure and displacement. It is widely employed for calculating and analyzing the forces acting on retaining structures during foundation pit excavations.

In the calculation model, the m-values finite rock throwing, coarse sand replacement, and bagged sand are discounted. For reinforced soil, the m-values of bagged sand, back-filled sand inside the weir, sand replacement on the outside of the weir, and thrown rock are discounted by 0.20, 0.12, 0.24, and 0.26, respectively. Calculations were performed in Lizheng deep foundation pit software calculation to validate the design by comparing the displacements of semi-infinite and finite bodies. By calculating the stiffness of both bodies, reduction factors and average values for the same soil layer were obtained as depicted in Figure 9.

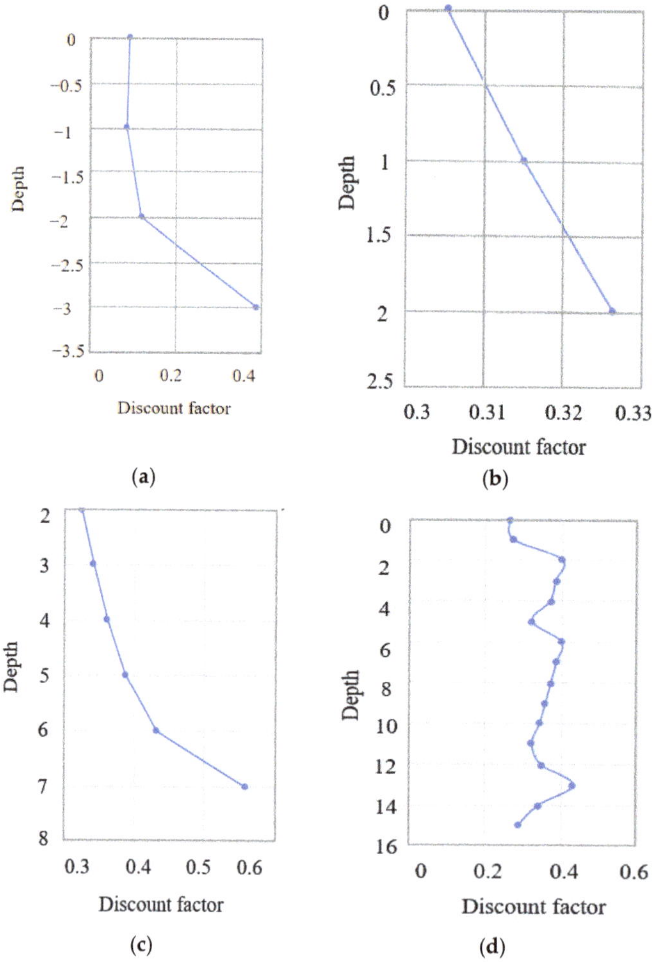

Figure 9. m-value discount result. (**a**) Bagged sand; (**b**) riprap; (**c**) sand backfills; (**d**) sand backfills between piles.

Two working conditions were calculated and analyzed. In Condition 1, the inner water level is the design high tide level and the outer water level drops after completion of the cofferdam. The water level reduction is based on a maximum tidal difference of 2.6 m according to hydrological data, and the corresponding wave suction force is taken into account. In Condition 2, the inner water level is lowered to 1 m below the inner apron of the cofferdam and the outer water level remains at the designed high tide level after the cofferdam is built. The corresponding wave thrust is considered.

As an example, consider the K6 + 598 right-line working condition for Condition 1. The inner and outer piles of the cofferdam are inclined outward. The inner and outer sheet piles are treated as independent models. For the inner side piles of the cofferdam, the pit's inner and outer sides are defined as shown in Figure 10, and a tie rod force pointing out of the cofferdam is applied to the inner piles. The water levels inside and outside the foundation pit are set to high, and the portion above the bagged sand is simplified as a soil layer with c = 0, φ = 0, and γ = 0. The inner bagged sand is considered as a semi-infinite soil layer, and the backfill sand inside the weir is analyzed as reinforced soil in the pit. The excavation depth of the pit is set to 0 m.

As shown in Figure 11, the sheet pile is displaced to the seaward side. The maximum displacement value is −21 mm at the top of the pile, the maximum bending moment value is 150.60 kN·m, and the maximum shear force value is −93.00 kN. The blue line in the figure shows the elastic method and the green line shows the passive earth pressure.

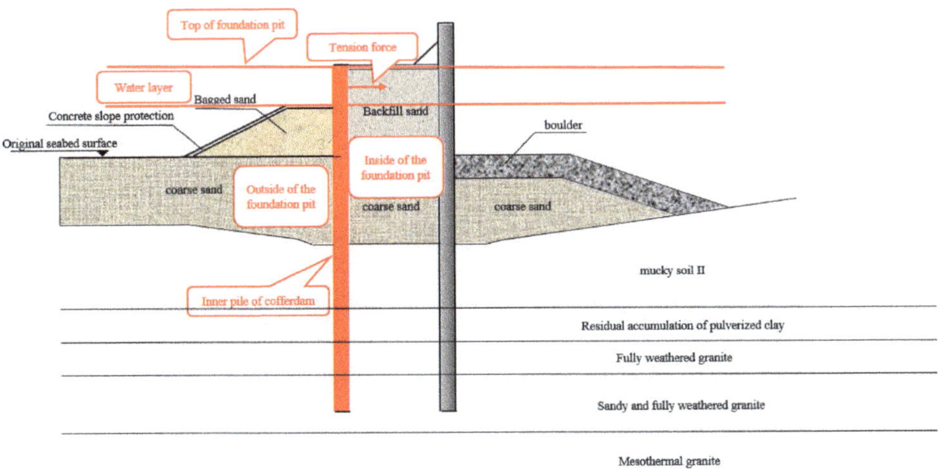

Figure 10. Calculation model of inner pile.

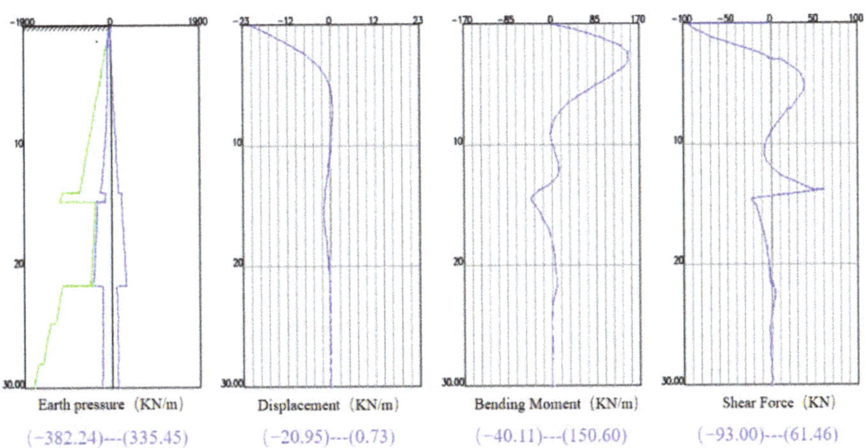

Figure 11. Internal force displacement diagram of the inner pile.

The maximum displacement is 34 mm on the inner and outer sheet piles of the cofferdam, the maximum bending moment is 249.30 kN·m, and the maximum shear force is 266.66 kN. The calculated displacement results are relatively small, and the magnitude of the internal forces satisfies the force requirements for the sheet pile. Therefore, the internal force deformation of the cofferdam meets the original design safety requirements. The internal force deformation results for the two sections of the cofferdam under each working condition are shown in Figure 12.

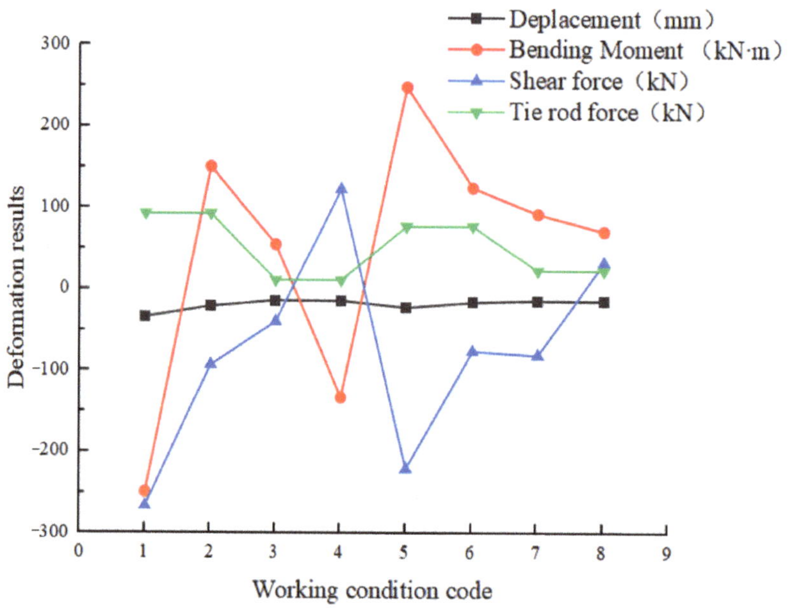

Figure 12. Internal force and deformation results.

The calculation results for the single-row steel sheet pile anchor model indicate that the stress on the sheet pile is significantly lower than the allowable stress of the material. This suggests that the material properties of the sheet pile are not fully utilized in the model calculations. Additionally, the horizontal displacement of the steel sheet pile structure is found to be small. This is because the model calculation does not consider the integrity of the double-row sheet pile structure and the interaction between the pile and soil in terms of force and deformation coordination. Therefore, for the initial design stage of the double-row sheet pile cofferdam, it is more appropriate to use the Lizheng deep foundation software. For later stages, however, it is recommended to optimize the design solution using the finite element calculation method.

3.3. Planar Finite Element Calculation of Construction Process

The finite element method has gained popularity due to its ability to incorporate spatial effects and its adaptability, especially for flexible structures in deep soft-ground foundations. The finite element simulation encompasses the entire construction process and relies on geotechnical investigation and ground profiles provided by the design department. To simplify the analysis here, the Mohr–Coulomb method is applied. The finite element calculation model is shown in Figure 13.

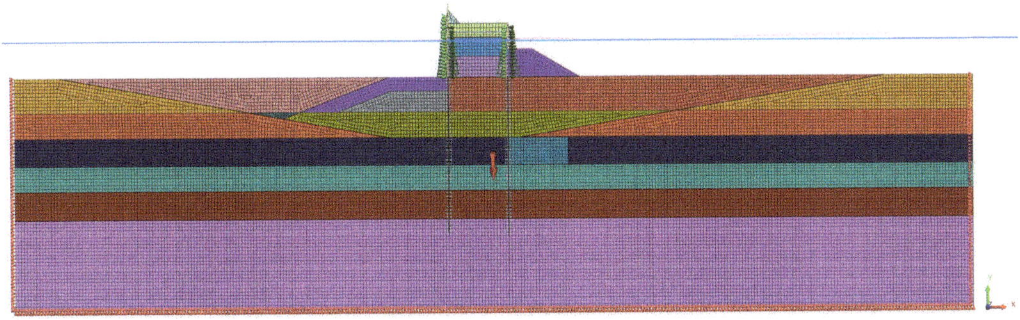

Figure 13. Finite element overall calculation model. Different colors represent different grid divisions.

The cohesion and friction angle of sludge, clay, and weathered granite were measured in the laboratory by the design department for the purposes of this analysis (Table 4).

Table 4. Planar finite element calculation parameters.

Attribute	Material Name	Elastic Modulus (MPa)	Poisson's Ratio	Constitutive	Cohesive Force (kPa)	Internal Friction Angle (°)
Plane Strain (2D)	Sludge I	4.6	0.4	Mohr–Coulomb	3.1	2.0
Plane Strain (2D)	Sludge II	4.2	0.4	Mohr–Coulomb	2.8	1.8
Plane Strain (2D)	Clay	20	0.3	Mohr–Coulomb	32.2	8.7
Plane Strain (2D)	Residual silty clay	20	0.3	Mohr–Coulomb	15.1	25.1
Plane Strain (2D)	Completely weathered granite	50	0.2	Mohr–Coulomb	18.5	27.4
Plane Strain (2D)	Sandy strongly weathered granite	90	0.2	Mohr–Coulomb	27.5	28.6
Plane Strain (2D)	Fragmented strongly weathered granite	200	0.2	Mohr–Coulomb	3000	30
Plane Strain (2D)	Moderately weathered granite	500	0.2	Mohr–Coulomb	15,000	41
Plane Strain (2D)	Riprap filling	100	0.2	Mohr–Coulomb	0	38
Plane Strain (2D)	Backfilling with medium coarse sand	30	0.3	Mohr–Coulomb	0	35
Plane Strain (2D)	Bagging sand	30	0.3	Mohr–Coulomb	12.3	25.96
Plane Strain (2D)	Mixing pile	30	0.3	Mohr–Coulomb	20	20
Plane Strain (2D)	Concrete (retaining walls, support beams)	30,000	0.15	Linear Elasticity	/	/
Beam/Truss (1D)	Steel (steel tie rods, steel sheet pile cofferdams, steel supports)	200,000	0.3	Linear Elasticity	/	/

K6 + 598 right line and K7 + 030 left line cross-sections were selected for calculation. According to the cofferdam design of this project and the site construction conditions, the main construction steps were determined as shown in Table 5.

Table 5. Construction sequence.

Number	Construction Step
1	Calculation of initial crustal stress
2	Excavation of foundation trench
3	Overall replacement of medium-coarse sand to the original seabed surface
4	Driving inner and outer steel sheet piles
5	Outer row steel sheet pile outer sand rib soft row, bagged gravel, and bagged soil for roof protection
6	Install steel tie rods
7	Synchronous layered backfilling of medium to coarse sand in the weir body to an elevation of −2.0~+1.0 m
8	Backfilling inside the weir to an elevation of +1.0 m
9	Continue backfilling the dam body to an elevation of +3.0 m
10	Construction anti-pressure soil slope, inner soft soil reinforcement
10 (a)	The water level inside the cofferdam is constant, while the water level outside the cofferdam drops sharply by 2.6 m + 10-year wave suction
10 (b)	Design a high tide level on the inner side of the cofferdam, with a sudden drop in water level of 2.6 m and a 10-year wave suction force on the outer side of the cofferdam
11	The water level on the inner side of the cofferdam drops, the anti-pressure soil slope is protected by concrete bags, and the top of the cofferdam is supported by a retaining wall
11 (a)	Considering 10-year high water level and 10-year wave thrust
11 (b)	Considering 10-year low water level and 10-year wave suction

Take the left line section of K7 + 030 as an example. The initial horizontal displacement of the steel sheet pile when upon its set-up is not considered. The horizontal displacements of the inner and outer sheet piles at construction step 11 are shown in Figure 14.

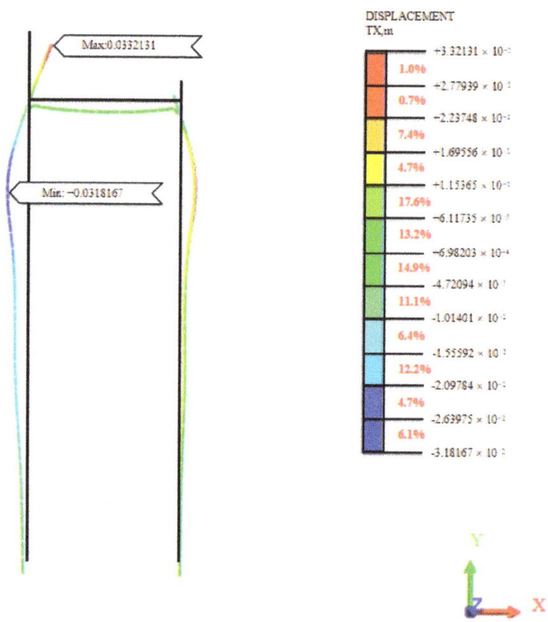

Figure 14. Horizontal displacement of steel sheet pile.

The maximum internal force on the cofferdam occurs during construction step 10 (b). At this point, the maximum bending moment of the steel sheet pile inside and outside the cofferdam is 373.6 kN·m and the maximum shear force is 189.7 kN. The displacement of the cofferdam is largest during construction step 11 (a), with the maximum displacement of the steel sheet pile reaching 82 mm. Out of this displacement, 34 mm is attributed to precipitation.

Based on the calculation results of two typical sections under different construction steps and water level working conditions, it is evident that each construction step has limited influence on the internal force deformation of the steel sheet pile. The internal force acting on the steel sheet pile is well below the bearing capacity of the designed steel sheet pile type.

4. Conclusions

This study investigated the overall stability and overturning stability of a double-row steel sheet pile cofferdam in relation to specific projects. The conclusions of this work can be summarized as follows.

(1) The calculation results show that the minimum safety factor for all working conditions is 1.744, and the minimum safety factor against overturning for all working conditions is 1.40. Both satisfy the code safety requirement of 1.35.
(2) In the analysis of internal forces and displacements of the cofferdam structure, the calculation results of two typical sections revealed that under the water level drop condition, the maximum displacement on the outside of the right-line section at K6 + 598 is 34 mm. The maximum bending moment is 249.30 kN·m, and the maximum shear force is 266.66 kN. The displacements of the steel sheet pile inside and outside the cofferdam under each condition are minimal, and the internal forces are below the design sheet pile type bearing capacity. Therefore, the cofferdam structure is considered to be safe.
(3) The influence of construction procedures on the internal force deformation of the sheet pile is found to be negligible, according to the calculation results for two typical sections at different steps in the construction process and under different water level conditions.

Author Contributions: Methodology and software, Y.J.; writing—original draft preparation and software, F.G.; validation, writing—review and editing, W.W.; conceptualization and investigation, G.Y.; project administration and investigation, J.Y.; validation and data curation, Y.H. All authors have read and agreed to the published version of the manuscript.

Funding: We thank the National Natural Science Foundation of China for the support of this research (52078143) and the Guangdong Provincial Science and Technology Program (2019B020208003).

Data Availability Statement: Details on all data supporting the reported results can be obtained in Tables 1–5 and Figures 5–14 in this original manuscript.

Acknowledgments: We thank Guangdong Research Institute of Water Resources and Hydro-power for providing test materials and fields for the study.

Conflicts of Interest: The authors declare no conflict of interest.

References

1. Yang, Y.; Liu, W.; Hu, A.; Li, F.; Yang, Z. Application of Steel Sheet Pile in Deep Foundation Pit Support of Collapsible Loess Regions. In *IOP Conference Series: Earth and Environmental Science*; IOP Publishing: Philadelphia, PA, USA, 2020; Volume 474, No. 7. [CrossRef]
2. Wu, S.; Hao, W.; Yao, Y.; Li, D. Investigation into durability degradation and fracture of cable bolts through laboratorial tests and hydroge-ochemical modelling in underground conditions. *Tunn. Undergr. Space Technol.* **2023**, *138*, 105198. [CrossRef]
3. Zhu, Y.; Li, X.J.; Shi, Z.M.; Peng, M.; Xuan, L.J.; Gao, J.Y.; Cai, S. Dynamic Behavior of Double Steel Sheet Pile Cofferdam under Different Wave Actions. In *IOP Conference Series: Earth and Environmental Science*; IOP Publishing: Philadelphia, PA, USA, 2021; Volume 861, No. 7. [CrossRef]
4. Hou, Y.-M.; Wang, J.-H.; Gu, Q.-Y. Deformation performance of double steel sheet piles cofferdam. *J. Shanghai Jiaotong Univ.* **2009**, *43*, 1577.
5. Yan, Z.; Qian-Yan, G.; Jie, J.; Ming, P. Reliability analysis for overall stability of large-span double-row steel sheet-piled dock cofferdam based on Bayesian method. *Rock Soil Mech.* **2016**, *37*, 609–615.

6. Mitobe, Y.; Adityawan, M.B.; Roh, M.; Tanaka, H.; Otsushi, K.; Kurosawa, T. Experimental study on embankment reinforcemen by steel sheet pile structure against tsunami overflow. *Coast. Eng. J.* **2016**, *58*, 1–18. [CrossRef]
7. Shen, Y.; Yu, Y.; Ma, F.; Mi, F.; Xiang, Z. Earth pressure evolution of the double-row long-short stabilizing pile system. *Enviror Earth Sci.* **2017**, *76*, 1–11. [CrossRef]
8. Khan, M.R.A.; Takemura, J.; Kusakabe, O. Behavior of double sheet pile wall cofferdam on sand observed in centrifuge tests. *In J. Phys. Model. Geotech.* **2001**, *1*, 1–16. [CrossRef]
9. Zhou, Y.; Yao, A.; Li, H.; Zheng, X. Correction of Earth Pressure and Analysis of Deformation for Double-Row Piles in Foundatio Excavation in Changchun of China. *Adv. Mater. Sci. Eng.* **2016**, *19*, 167–180. [CrossRef]
10. Tschebotarioff, G.P. Lateral behavior of a double sheet pile wall structure. *Soils Found.* **1974**, *14*, 92–93.
11. De Buhan, P.; Corfdir, A. Limit Design of Axisymmetric Shells with Application to Cellular Cofferdams. *J. Eng. Mech.* **1996**, *12* 921–929. [CrossRef]
12. Banerjee, S.; Banerjee, M.S.; Member, A. Design Charts for Double-Walled Cofferdams. *J. Geotech. Eng.* **1993**, *119*, 214–222 [CrossRef]
13. Lei, H.; Liu, X.; Song, Y.; Xu, Y. Stability analysis of slope reinforced by double-row stabilizing piles with different locations. *Na Hazards* **2021**, *106*, 19–42. [CrossRef]
14. Li, C.; Chen, W.; Song, Y.; Gong, W.; Zhao, Q. Optimal Location of Piles in Stabilizing Slopes Based on a Simplified Double-Row Piles Model. *KSCE J. Civ. Eng.* **2020**, *24*, 377–389. [CrossRef]
15. Zhang, M.-Y.; Ma, J.-X.; Yang, S.-J.; Wang, Y.-H.; Bai, X.-Y.; Sun, S.-X. Experimental Study on Bending Moment of Double-Rov Steel Pipe Piles in Foundation Excavation. *Adv. Civ. Eng.* **2020**, *2020*, 1–8. [CrossRef]
16. Zhou, Y.; Luo, L.; Zheng, L. Review of the Deformation Mechanism and Earth Pressure Research on the Double-Row Pile Suppor Structure. *Front. Earth Sci.* **2022**, *10*, 933840. [CrossRef]
17. Lefas, I.; Georgiannou, V. Analysis of a cofferdam support and design implications. *Comput. Struct.* **2001**, *79*, 2461–2469 [CrossRef]
18. Byfield, M.P.; Crawford, R.J. Oblique bending in U-shaped steel sheet piles. *Proc. Inst. Civ. Eng.-Struct. Build.* **2003**, *156*, 255–261 [CrossRef]
19. Wu, S.; Zhang, Z.; Chen, J.; Yao, Y.; Li, D. Characterisation of stress corrosion durability and time-dependent performance of cabl bolts in under-ground mine environments. *Eng. Fail. Anal.* **2023**, *150*, 107292. [CrossRef]
20. Gui, M.-W.; Han, K.-K. An investigation on a failed double-wall cofferdam during construction. *Eng. Fail. Anal.* **2009**, *16*, 421–432 [CrossRef]
21. Zhao, T.; Ding, W.; Wei, L.; Wu, W.; Qiao, Y. The behavior analysis of a cofferdam constructed by double sheet pile wall abov muck. In *Proceedings of the GeoShanghai 2018 International Conference: Tunnelling and Underground Construction, Shanghai, China 27–30 May 2018*; Springer: Singapore, 2018.
22. Xue, R.; Bie, S.; Guo, L.; Zhang, P. Stability Analysis for Cofferdams of Pile Wall Frame Structures. *KSCE J. Civ. Eng.* **2019**, *23* 4010–4021. [CrossRef]
23. Sulovska, M.; Stacho, J.; Kopecky, M. The Stability Analysis of a Cofferdam Using the Numerical Modelling. In *IOP Conferenc Series: Materials Science and Engineering*; IOP Publishing: Philadelphia, PA, USA, 2020; Volume 960, No. 2.
24. Fujiwara, K.; Taenaka, S.; Otsushi, K.; Yashima, A.; Sawada, K.; Ogawa, T.; Takeda, K. Study on levee reinforcement using doubl sheet-piles with partition walls. *Jpn. Geotech. Soc. Spec. Publ.* **2017**, *5*, 11–15. [CrossRef]
25. Wang, K.; Gao, Y.; Jin, Z.; Zhou, X.; Chen, L.; Zhang, C. Research on stability of steep bank slope and reserved thin-walled rock cofferdam during excavation of intake foundation pit. *Eng. Fail. Anal.* **2022**, *141*, 106659. [CrossRef]
26. Xu, F.; Li, S.-C.; Zhang, Q.-Q.; Li, L.-P.; Wang, K.; Liu, H.-L. Analysis and Design Implications on Stability of Cofferdam Subjected to Water Wave Action. *Mar. Georesources Geotechnol.* **2014**, *34*, 181–187. [CrossRef]
27. Hui, S. Research on the Bearing Capacity of Double Row Steel Sheet Pile Reinforcement Depth of the Based on Soil between Piles *Int. J. Eng. Res. Afr.* **2016**, *25*, 127–132. [CrossRef]
28. Chen, L.; Jeng, D.-S. Study on the seabed response around a dumbbell cofferdam under combined wave and current loading *Ocean Eng.* **2022**, *256*, 111456. [CrossRef]
29. Hu, Y.; Huang, J.; Tao, Z.; Zhu, J.; Lv, Q.; Qiao, J.; Cai, W. The Influence of Water-Level Fluctuation on the Instability and Seepage Failure of Dump-Fill Cofferdam. *Adv. Civ. Eng.* **2022**, *2022*, 5845340. [CrossRef]
30. Ti, Z.; Wei, K.; Qin, S.; Mei, D.; Li, Y. Assessment of random wave pressure on the construction cofferdam for sea-crossing bridges under tropical cyclone. *Ocean Eng.* **2018**, *160*, 335–345. [CrossRef]
31. Wang, J.; Jiang, Z.; Li, F.; Chen, W. The prediction of water level based on support vector machine under construction condition o steel sheet pile cofferdam. *Concurr. Comput. Pract. Exp.* **2020**, *33*, e6003. [CrossRef]

water

MDPI

Article

A Novel Artificial Intelligence Prediction Process of Concrete Dam Deformation Based on a Stacking Model Fusion Method

Wenyuan Wu [1,2], Huaizhi Su [1,2,3,*], Yanming Feng [4,5], Shuai Zhang [4,5], Sen Zheng [1,2], Wenhan Cao [1,2] and Hongchen Liu [1,2]

1 The National Key Laboratory of Water Disaster Prevention, Hohai University, Nanjing 210024, China; wenyuanwu@hhu.edu.cn (W.W.); zheng_sen@hhu.edu.cn (S.Z.); caowenhan@hhu.edu.cn (W.C.); liuhongchen1995@163.com (H.L.)
2 College of Water Conservancy and Hydropower Engineering, Hohai University, Nanjing 210024, China
3 Cooperative Innovation Center for Water Safety and Hydro Science, Hohai University, Nanjing 210024, China
4 Yunnan Provincial Key Laboratory of Water Resources and Hydropower Engineering Safety, Kunming 650051, China; fengyanming_kmy@powerchina.cn (Y.F.); zhangshuai_kmy@powerchina.cn (S.Z.)
5 Powerchina Kunming Engineering Corporation Limited, Kunming 650051, China
* Correspondence: su_huaizhi@hhu.edu.cn

Abstract: Deformation effectively represents the structural integrity of concrete dams and acts as a clear indicator of their operational performance. Predicting deformation is critical for monitoring the safety of hydraulic structures. To this end, this paper proposes an artificial intelligence-based process for predicting concrete dam deformation. Initially, using the principles of feature engineering, the preprocessing of deformation safety monitoring data is conducted. Subsequently, employing a stacking model fusion method, a novel prediction process embedded with multiple artificial intelligence algorithms is developed. Moreover, three new performance indicators—a superiority evaluation indicator, an accuracy evaluation indicator, and a generalization evaluation indicator—are introduced to provide a comprehensive assessment of the model's effectiveness. Finally, an engineering example demonstrates that the ensemble artificial intelligence method proposed herein outperforms traditional statistical models and single machine learning models in both fitting and predictive accuracy, thereby providing a scientific and effective foundation for concrete dam deformation prediction and safety monitoring.

Keywords: concrete dam; structure safety monitoring; deformation prediction; artificial intelligence; feature engineering; stacking method; performance indicators

1. Introduction

Concrete dam projects are mostly built in high mountain canyons with significant faults and deep coverage. These structures not only endure the cyclic action of various loads but also face deterioration from a complex and harsh environment. The safety behavior of their service is a nonlinear dynamic process, evolving due to the interactions between materials and structures under the combined influence of various factors [1,2]. Deformation serves as a comprehensive reflection of the structural form of concrete dams and is a crucial indicator for evaluating the transformation of the structural state and long-term service health of the dam. Establishing an accurate and effective deformation prediction model is vital for predicting its structural state evolution, conducting safety monitoring, and ensuring stable and efficient operation.

Three traditional approaches used to describe and assess the performance of concrete dams through the analysis of real-time monitoring data include statistical models, deterministic models, and hybrid models based on causality [3,4]. These models are frequently employed to construct deformation prediction models. In these numerical simulation models, the effects of hydrostatic pressure, ambient temperature, and time are primarily considered. In a statistical model, the effect quantity is divided into components of water pressure, temperature, and aging. A mathematical model is constructed by integrating the measured dam data with statistical methods such as multiple linear regression. The advantages of statistical methods include simplicity of formulation and rapid execution speed. However, when variables exhibit multicollinearity or contain internal outliers, the performance of existing statistical regression models is often unsatisfactory. In conjunction with the real operational behavior of the dam and its foundation, the deterministic model utilizes the finite element method to compute fields impacted by the load, including displacement, stress, and seepage, and then optimizes and fits these with the measured values to refine the physical and mechanical parameters, thus establishing the deterministic model. Deterministic models involve solving nonlinear partial differential equations, which are often complex or impossible to solve analytically, and their accuracy is difficult to ensure [5,6]. The hybrid model utilizes finite element calculations for the water pressure component, while other components continue to employ the statistical model, which is then optimized and fitted with measured values [7].

However, traditional models exhibit several limitations [2]. For instance, due to the strong correlation between water temperature and air temperature among input factors, the traditional statistical model struggles to accurately model this nonlinear effect. Traditional models are usually based on explicit assumptions about data distribution and hypotheses, often employing simpler and more interpretable models. In contrast, artificial intelligence (AI) methods do not impose strict distribution assumptions and are more data-driven. They rely on large datasets and iterative training to continuously optimize model parameters, thereby enhancing prediction accuracy. Additionally, AI methods are typically complex and highly nonlinear, which makes machine learning approaches more effective in handling large-scale, high-dimensional data. In recent years, artificial intelligence methods have been increasingly applied in various engineering fields [8–10]. Techniques such as support vector machines [11–13], random forests [14–16], and neural networks [17–20] have been incorporated into the concrete dam deformation prediction system.

The structure of ordinary neural networks must be modified by a predetermined algorithm during the training process, which does not guarantee that the optimal structure is achieved, often resulting in convergence to local optima. Support vector machines effectively handle practical challenges such as small samples, non-linearity, high dimensionality, and local minima, substantially mitigating the issues of "dimension disaster" and "overlearning". However, due to the numerous and complex factors influencing concrete dam deformation, the accuracy of models based solely on support vectors is often compromised. The random forest algorithm addresses the interactions and nonlinear relationships between variables, enhancing the accuracy of classification and regression tasks without significantly increasing computational demands. Nevertheless, it is susceptible to overfitting in the presence of noisy data. Some scholars [17,21,22] advocate for model integration or strategy combination to improve prediction performance, although the shared training dataset among individual models often limits enhancements in generalization ability and the models do not provide algorithmic synergy, with insufficient theoretical backing to convince users.

Moreover, research on concrete dam deformation prediction models frequently overlooks data preprocessing and comprehensive model performance evaluation. The process of collecting deformation monitoring data is often compromised by factors such as instrument failure and human error, leading to data issues such as missing, repeated, erroneous, and outlier values, which diminish the completeness and reliability of the monitoring data and may even result in misjudgments of the dam's operational state. Therefore, developing effective monitoring data preprocessing methods is essential to enhancing the data's completeness and reliability, thereby facilitating the establishment of prediction models based on accurate and reliable health status information. Additionally, a comprehensive model performance evaluation system is crucial for assessing various prediction models from multiple perspectives [3,18].

To enhance the integrity and reliability of displacement monitoring information, improve the accuracy and generalization ability of prediction models, and comprehensively evaluate model performance, this paper proposes a novel artificial intelligence-based concrete dam deformation prediction process. This process utilizes a stacking model fusion method combined with feature engineering. Initially, missing values in the displacement time series are addressed using the concept of temporal and spatial correlation among dam monitoring points. The wavelet analysis method is then employed to denoise the displacement time series, ensuring complete and accurate monitoring information and facilitating accurate reflection of the dam's deformation safety status and the establishment of a reliable prediction model. Subsequently, using the stacking model fusion method, a two-layer stacking ensemble learning prediction process is constructed with XGBoost, Extremely Randomized Trees (Extra-Trees), and SVR as the first-layer base-learners, and multi-response linear regression (MLR) as the second-layer meta-learner. Finally, the measured deformation data from a concrete dam are used for training and prediction. The basic and novel performance evaluation indicators are applied to assess the predictive capabilities of different methods. The results show that the proposed process markedly enhances both accuracy and generalizability over traditional statistical models and single machine learning models, offering a robust foundation for monitoring deformation safety in concrete dams.

2. Materials and Methods

2.1. Feature Engineering of Deformation Monitoring Data

In the process of acquiring time series of monitoring data, various factors such as environmental conditions, instrumentation, data acquisition systems, and human error can degrade data quality, leading to issues such as missing values and noise, etc. These issues may prevent the data from accurately reflecting the actual operational behavior of the dam [23–25]. Therefore, filling missing values and reducing noise, coupled with performing feature engineering on dam monitoring data, are crucial for accurately constructing safety monitoring models and comprehensively evaluating operational performance.

2.1.1. Spatiotemporal Correlation Weighted Regression Interpolation Method

During the operation of concrete dams, critical monitoring data such as water level and displacement often go missing due to instrument failure and human error. The absence of this data can significantly impact the analysis of deformation behavior and the estimation of parameters for deformation prediction models. The common approaches to address missing data include deletion and filling methods. The deletion method can result in the loss of substantial monitoring information, potentially leading to errors in the evaluation of the dam's safety state. Filling methods typically involve statistical techniques (such as mean and median filling) and machine learning techniques (such as SVR, EM, and KNN). However, statistical filling can reduce the variance of the data set, distort its distribution characteristics, and overlook the time series relationships and physical significance in the dam monitoring data. Machine learning filling may lead to overfitting in the prediction model, compromising the model's robustness. Consequently, this paper proposes a spatiotemporal correlation weighted regression interpolation method.

Concrete dams are a quintessential example of large spatial structures with a certain degree of integrity at different scales. Consequently, the distribution of deformation in dams also exhibits continuity and similarity throughout the structure. Assuming that the material and structure of the dam are homogeneous and consistent, we illustrate the spatial correlation regression interpolation method for handling missing deformation values in concrete dams, using three points as examples. It is assumed that a concrete dam has three spatially adjacent and structurally related monitoring points, A, B, and C, as shown in Figure 1. P_{uh}, P_{uv}, P_{dh}, and P_{dv} represent upstream horizontal water pressure, upstream vertical water pressure, downstream horizontal water pressure, and downstream vertical water pressure, respectively.

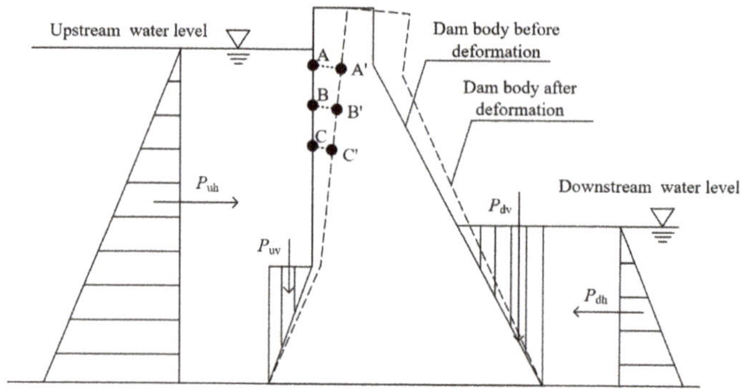

Figure 1. Location of deformation monitoring points of a concrete dam.

Among them, the monitoring data for A and C are complete, while the monitoring data for B are partially missing, as depicted in Figure 2. T_1, T_m, and T_n represent the same time period in each year.

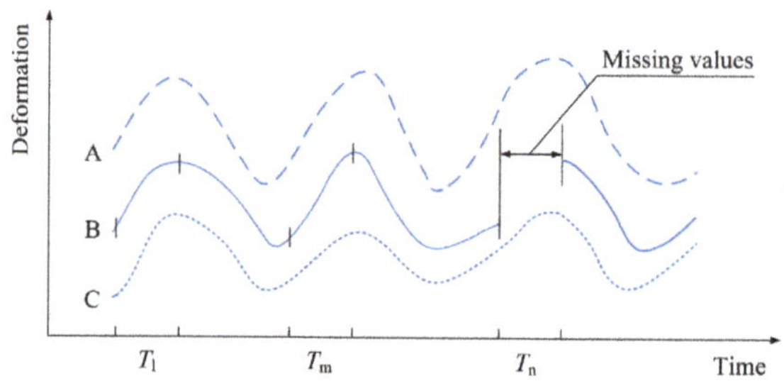

Figure 2. Missing deformation monitoring data of a concrete dam.

Due to the structural integrity of dams, the distribution of deformation values also exhibits continuity in space [26]. If the spatial proximities of measurement points A, B, and C are considered simultaneously, the missing value at measurement point B should be correlated with the deformation values at measurement points A and C during the missing period. Therefore, based on the deformation information from measurement points A and C, the deformation sequence of measurement point B can be expressed as follows:

$$\delta_B = g(\delta_A) + g(\delta_C) + \varepsilon \tag{1}$$

where $g(\delta_A)$ and $g(\delta_C)$ represent the functions of the deformation values at A and C as a function of δ_B, respectively, and ε represents allowable deviation, the same as below. If the relationships between δ_A and δ_B and between δ_C and δ_B are expressed by polynomials, then the following expressions apply:

$$\delta_B = \alpha_A \sum_{i=1}^{K_A} \lambda_{Ai} \delta_A^i + \alpha_C \sum_{i=1}^{K_C} \lambda_{Ci} \delta_C^i + \beta_B + \varepsilon \tag{2}$$

where α_A and α_C are the coefficients of polynomials containing terms δ_A and δ_C, respectively, and K_A and K_C represent the highest degrees of these polynomials. The values of K_A and K_C can be determined by plotting a scatter diagram to analyze the correlation between variable δ_B and variables δ_A, δ_B, and δ_C. The term β_B represents a translation item.

Considering generality, for the deformation sequence δ_{it} of any monitoring point i with missing data, the following expressions of spatial correlation are applicable:

$$\delta_S = \sum_{j=1}^{L} \alpha_{ij} g(\delta_j) + \beta_i + \varepsilon \tag{3}$$

where L represents the number of points adjacent to i, determined by the spatial continuity between the location of the point and the dam section, and the similar trend in deformation value distribution. δ_j represents the deformation sequence of the measurement point j adjacent to point i, $g(\delta_j)$ denotes the correlation function between the deformation sequences δ_{jt} and δ_{it}, and β_i represents the translation term for i.

Based on the sequences δ_{it} and δ_{jt}, the value of α_{ij} is estimated using the least squares method to define the expression for the spatial proximity regression interpolation model.

Through the application of a single spatial proximity regression interpolation method, considerations are given to the spatial continuity of different measurement points and the consistent trend in deformation value distribution. However, the volatility inherent to a single measurement point is often overlooked. The missing sequence of adjacent measurement points in space is often the same. Similarly, considering the periodic distribution of the measurement value at a point over time, the expression for time correlation at a single measurement point is:

$$\delta_T = \sum_{k=1}^{M} \alpha_{tk} g(\delta_k) + \gamma_i + \varepsilon \tag{4}$$

where M represents the number of time periods adjacent to time period t. δ_k represents the deformation sequence of time periods adjacent to t. $g(\delta_k)$ denotes the correlation function between the deformation sequences δ_k and δ_t, and γ_i represents the translation term for i.

In this paper, a spatiotemporal correlation weighted regression interpolation method is proposed.

$$\delta = \frac{\overline{R_S}}{\overline{R_S} + \overline{R_T}} \delta_S + \frac{\overline{R_T}}{\overline{R_S} + \overline{R_T}} \delta_T \tag{5}$$

where $\overline{R_S}$ and $\overline{R_T}$ are the average correlation coefficients of deformation time series between spatially correlated different measurement points in the same time period and target sequence, and between time-correlated different time periods in the same measurement point and target sequence, respectively.

2.1.2. Wavelet Analysis Denoising Method of Monitoring Data

Wavelet transform is a multi-scale analysis method that exhibits strong local recognition capabilities in both the time and frequency domains. In the wavelet domain, the effective signal of dam safety monitoring data is represented by large wavelet coefficients, while noise, characterized by randomness and discontinuity in the time domain, has small coefficients. Based on these characteristics, an appropriate threshold can be determined to nullify noise coefficients while preserving the effective signal, thereby achieving noise reduction. Traditional threshold determination methods include the hard threshold and soft threshold methods. Currently, commonly employed methods for estimating thresholds include sqtwolog, rigrsure, heursure, and minimaxi.

In this paper, the hard threshold and soft threshold methods, along with the four threshold estimation methods mentioned previously, are employed to reduce noise. The root mean square error (*RMSE*) and signal-to-noise ratio (*SNR*) after noise reduction are calculated, the indicators after noise reduction are compared, and the method combination that provides the best noise reduction effect is selected.

2.2. Stacking Ensemble Learning Prediction Process

Stacking is a method known for its simple structure, high performance, and strong classification capability [27]. Essentially, stacking resembles the hierarchical structure characteristic of neural networks, with its effectiveness largely derived from feature extraction [28,29]. Thus, the enhancement of learning ability through stacking primarily results from the cumulative learning capabilities of different classifiers for various features. As depicted in Figure 3, under the ensemble learning approach based on stacking, the construction of the model is divided into two steps. Prediction accuracy is enhanced by cascading the prediction results among learners. In the first step, the original dataset is divided into a training set and a test set in a predetermined proportion, and suitable base learners are chosen to train the training set via cross-validation. Each base learner predicts the outcomes for the validation set and the test set. In the second step, the prediction outcomes from the base learners are used as feature data for training and predicting with the meta-learner. This meta-learner consolidates the features gathered in the earlier stage and the labels from the original training set to construct the model. Ultimately, it delivers the final prediction results of the stacking method.

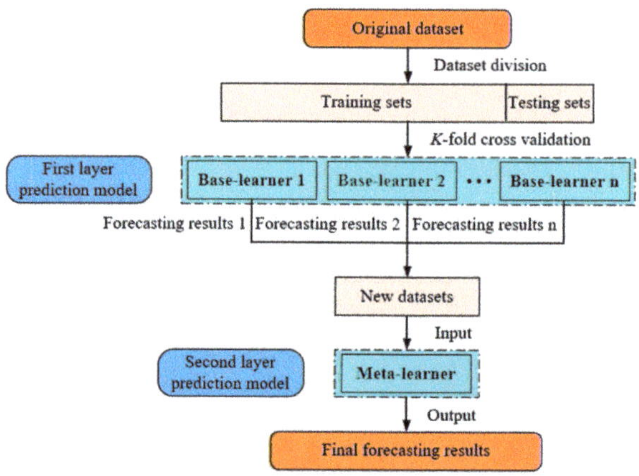

Figure 3. Stacking ensemble learning prediction process.

In the initial step, a machine learning model with excellent predictive performance should be chosen to ensure the diversity of the models. In the subsequent stage, the meta-learner typically selects a straightforward model with robust stability to enhance the overall performance of the model.

The model is based on the concept of K-fold cross-validation, typically setting K to 5. This approach effectively avoids the overfitting phenomenon that can occur due to a limited amount of data. The detailed steps for implementing the stacking algorithm are as follows:

1. The original dataset is split into a training set T and a test set D in a specified proportion.
2. The training set (T) is randomly divided into five equal parts: T_1, T_2, T_3, T_4, and T_5. Each subset T_i (i = 1, 2, 3, 4, 5) serves as the validation set, with the remaining subsets combined as the training set for model training. This setup is used to predict outcomes for both the validation and test sets.
3. After completing the 5-fold cross-validation, the first base learner compiles the predicted values for each validation set and the test set into separate columns. These predictions are then consolidated: the predictions from each validation set are merged into a single column denoted as A_1, and the average of the predictions for the test set is calculated and recorded as B_1, as depicted in Figure 4.
4. Upon completion of the training for all base learners, the input characteristic matrices for the meta-learner model, denoted as $\widetilde{T} = (A_1, A_2, \cdots, A_n)$, and for the final test set, as $\widetilde{D} = (B_1, B_2, \cdots, B_n)$, are established. The label values from the original training set are used as the output matrix for the model. The meta-learner then uses these

matrices to generate the final prediction result after model fusion, as illustrated in Figure 5.

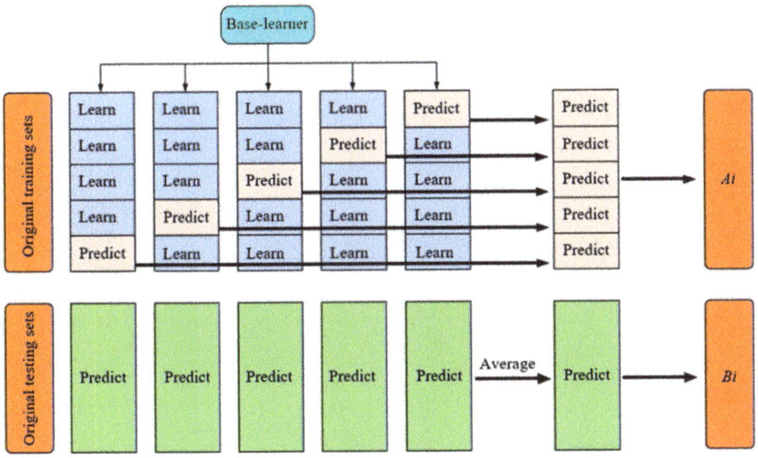

Figure 4. Schematic diagram of the stacking-based learner.

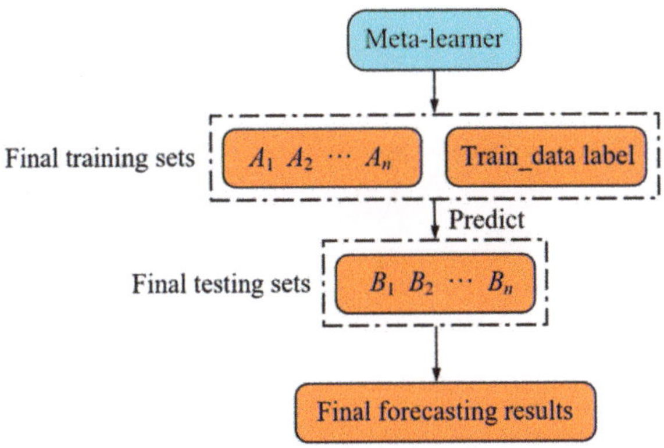

Figure 5. Schematic diagram of the stacking meta-learner.

2.3. *Stacking Ensemble Model-Based Learner*

2.3.1. XGBoost

XGBoost, proposed by Chen and Guestrin [30] in 2016, is an enhancement of the serial ensemble machine learning algorithm known as the gradient-boosting machine (GBM). The core concept is to amalgamate hundreds of tree models, each with low prediction accuracy, into a highly accurate composite model. During the generation of each tree, the gradient descent method is employed to guide the optimization process towards minimizing a specified objective function. Building on the GBM, XGBoost performs a second-order Taylor expansion of the loss function, allowing for customization of the loss function and improved prediction accuracy. Its capability for automatic multithreading on CPUs enhances the algorithm's execution speed. Due to its exceptional performance, XGBoost has been widely adopted in Kaggle competitions and various other machine learning contests, consistently delivering impressive results. Figure 6 illustrates the schematic diagram of XGBoost.

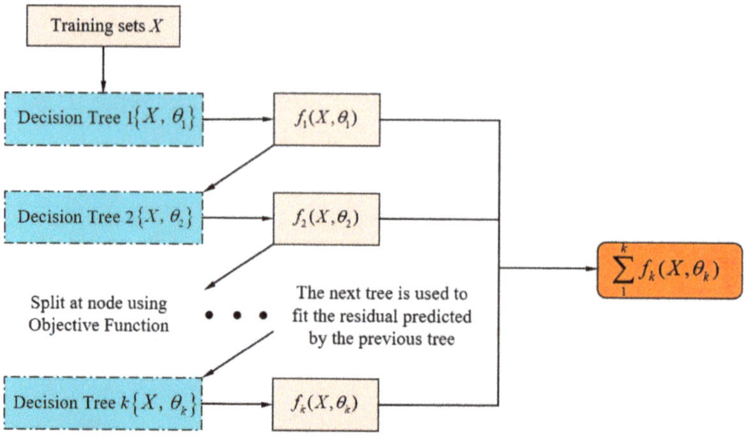

Figure 6. Flow chart of the XGBoost algorithm.

2.3.2. Extra-Trees

Extra-Trees, proposed by Pierre Geurts [31] and other scholars in 2006 after extensive experimental research, is a parallel ensemble machine learning method. The Extra-Trees algorithm is a variant of the Random Forest method. It constructs multiple decision trees using node random splitting technology and utilizes the weighted average of these trees for regression predictions. Unlike Random Forest, Extra-Trees do not search and compare within a subset of features to split nodes; instead, a feature and a threshold are randomly selected as the basis for each node's division. This introduction of dual randomness enhances the generalization capability of Extra-Trees beyond that of Random Forest. Moreover, the method can handle high-dimensional data and complex nonlinear relationships between multiple variables. It also exhibits robustness against outlier data. In recent years, it has been widely used in areas such as agriculture, engineering, medicine, and finance for selecting features and detecting outliers. Figure 7 illustrates the schematic diagram of Extra-Trees.

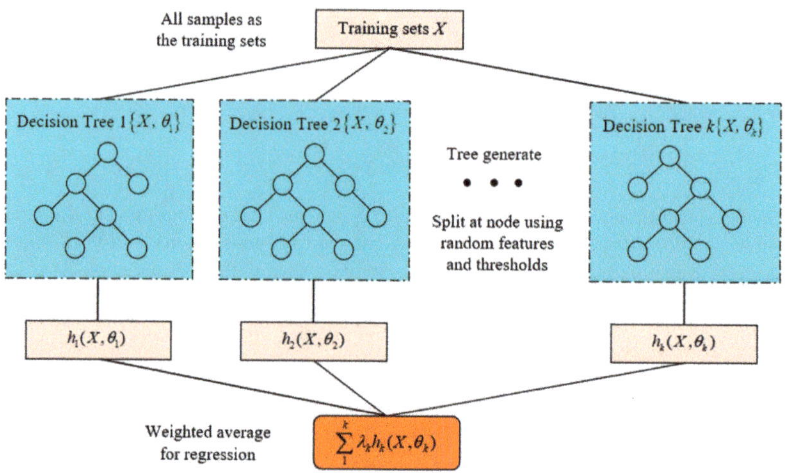

Figure 7. Flow chart of the Extra-Trees algorithm.

2.3.3. SVR

In 1995, Vapnik [32] proposed the regression support vector machine model SVR. Based on the structural risk minimization criterion, SVR exhibits global optimality and robust generalization capabilities. It effectively addresses the issues of overfitting and the "dimension disaster" that are common in methods based on the empirical risk minimization principle, such as the BP neural network. Through the nonlinear transformation enabled by the inner product kernel function, SVR converts the input variables from low-dimensional space to high-dimensional space. This transformation turns the nonlinear problem into a linear one, facilitating the discovery of linear relationships between input and output variables in high-dimensional space, ultimately finding the optimal separation hyperplane. The schematic diagram of SVR is illustrated in Figure 8.

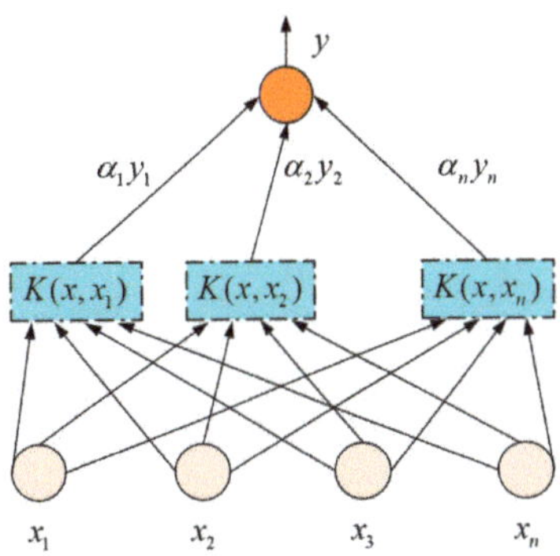

Figure 8. Flow chart of the SVR algorithm.

2.4. Model Performance Evaluation System

Performance evaluation indicators play a crucial guiding role in comparing and selecting different models. However, previous model evaluations often relied solely on single indicators such as the Mean Absolute Error (*MAE*), Mean Squared Error (*MSE*), and Root Mean Square Error (*RMSE*). This approach did not allow for a comprehensive and categorized analysis and discussion of model performance, and often overlooked the generalization abilities of the models, which is detrimental to the comprehensive evaluation of the prediction models [33]. Therefore, designing scientific performance evaluation indicators for dam deformation prediction models and establishing a comprehensive performance evaluation framework are critical challenges that need to be addressed in concrete dam deformation prediction.

2.4.1. Superiority Evaluation Indicator

The superiority of a prediction model is demonstrated through enhanced prediction accuracy compared to previous models. In past research, differences in model performance were qualitatively explained by comparing *MAE* and *MSE* values between models. However, the variations in performance between models were not quantitatively studied. Therefore, from a quantitative comparison perspective, the Superiority Evaluation Indicator *SEI* for concrete dam deformation prediction models is established by selecting two metrics: Relative Mean Absolute Error *RMAE* and Percentage Better Mean Absolute Error *PB(MAE)* [34]. The formula for the indicator is as follows:

$$SEI_1 = RMAE = \frac{MAE}{MAE^*} \tag{6}$$

$$SEI_2 = PB(MAE) = 100\% \cdot \frac{1}{n}\left(\sum_{i=1}^{n} I\{AE_i < AE_i^*\}\right) \tag{7}$$

where MAE and MAE^* represent the average absolute errors of the prediction model and the reference model for comparison, respectively. "I" is an operator that generates a value of 0 or 1 based on the given expression.

$$I(AE) = \left\{ \begin{array}{l} 0,\ if\ AE_i < AE_i^* \\ 1,\ \text{otherwise} \end{array} \right\} \tag{8}$$

2.4.2. Accuracy Evaluation Indicator

The accuracy of a prediction model is determined by the consistency between the predicted values derived from model training and the actual values obtained from measurements. When the monitoring data of a concrete dam include values of 0 or close to 0, MAPE may become excessively large or incalculable. *MSE* is also sensitive to the scale of the data and varies across different data subsets. Consequently, this paper utilizes two improved metrics: the Mean Arctangent Absolute Percentage Error *MAAPE* [34] and the Normalized Root Mean Square Error n*RMSE* to establish the Accuracy Evaluation Indicator *AEI* for concrete dam deformation prediction models. The formula for the indicator is as follows:

$$AEI_1 = nRMSE = \frac{1}{\tilde{\bar{y}}} \cdot RMSE = \frac{1}{\tilde{\bar{y}}} \cdot \sqrt{\frac{1}{n} \sum_{i=1}^{n} (y_i - f_i)^2} \tag{9}$$

$$AEI_2 = MAAPE = \frac{1}{n} \sum_{i=1}^{n} (\arctan(\left| \frac{y_i - f_i}{y_i} \right|)) \tag{10}$$

where $\tilde{y}$ is the normalization factor.

2.4.3. Generalized Evaluation Indicator

The generalization of a prediction model refers to the model's adaptability when applied to a test set after being trained on a training set. Typically, overfitting is mitigated and the model's generalization ability is enhanced by increasing the amount of training data and incorporating a regularization term into the model's loss function. In this paper, the Generalization Evaluation Indicator *GEI* is established using the following formula:

$$GEI = \left| \frac{nRMSE_P}{nRMSE_T} \right| \tag{11}$$

where the subscript P represents the prediction process and the subscript 'T' represents the training process. $nRMSE_P$ and $nRMSE_T$ denote entities within the prediction process and the training process, respectively.

3. Design and Implementation of the Model

3.1. Project Overview

A concrete gravity dam was constructed in 1973 and completed in September 1986. The dam crest elevation is 138.8 m, the lowest bedrock elevation is 86 m, the maximum dam height is 52.8 m, and the main dam length is 1354 m. The normal pool level is 133 m, and the total storage capacity is 320 million cubic meters. The dam is segmented into 82 sections, including 36 non-overflow sections, 29 overflow sections, eight flood discharge bottom outlet sections, one riverbed power station section, two headwork power station sections, two drainage bottom outlet sections, two east headwork power station sections, and two elevator shaft sections.

Currently, deformation monitoring is one of the most important monitoring items of the dam, which includes the monitoring of horizontal and vertical displacement. Among these, three normal vertical lines (measurement points are PL) are positioned at sections 1#, 1 k#, and 30#, seven inverted vertical lines (measurement points are IP) are positioned at sections XT3, 1#, 14#, 19#, 30#, 56#, and 58#, and sets of extension lines are positioned at an elevation of 110 m on the dam body (measurement points are EXL) and 138.8 m on the dam crest (measurement points are EXD). The horizontal displacement monitoring system for the concrete dam comprises normal vertical lines, inverted vertical lines, and extension lines. The horizontal displacement monitoring system is depicted in Figure 9.

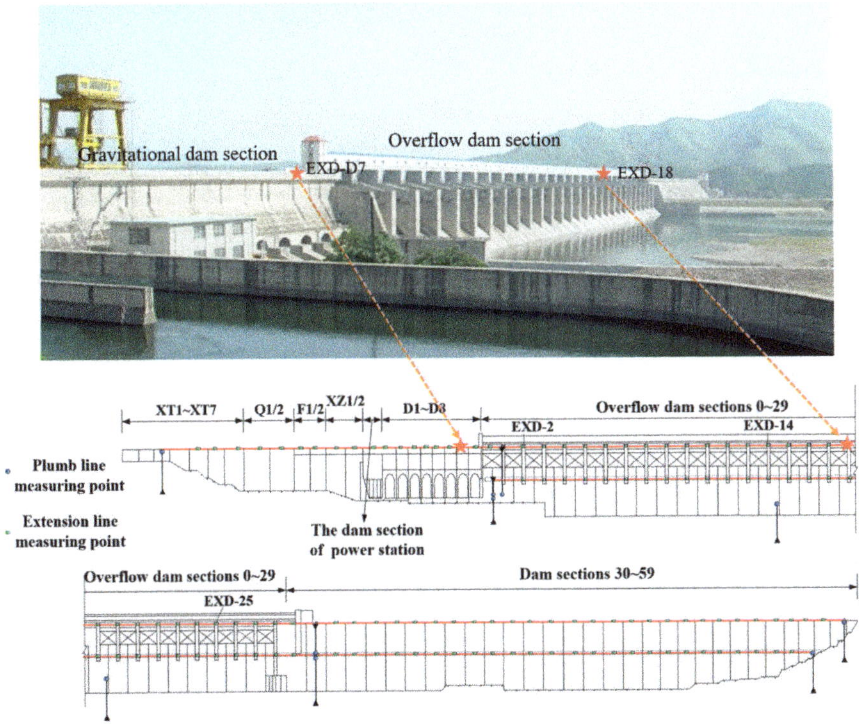

Figure 9. Layout of the horizontal displacement monitoring system of a concrete dam.

3.2. Model Sample and Input Factor Selection

Monitoring data indicate that the horizontal displacement of the concrete dam adheres to the general deformation law of concrete gravity dams. Specifically, the amplitude of variation is large in the riverbed dam sections, smaller in the bank slope dam sections, larger at higher elevations, and smaller at lower elevations. Given the horizontal displacement distribution law of the concrete dam and its hydrogeological conditions, the measurement points EXD-18 at the crest of riverbed overflow dam section 18# and EXD-D7 at the crest of diversion bottom outlet dam section D7 have been selected as the focal points for analysis in this paper. This analysis aims to evaluate the model's superiority, accuracy, and generalization.

The HST analysis model is the most widely used quantitative interpretation model in dam deformation analysis. The model predicts the horizontal displacement of concrete dams using long-term monitoring information and mathematical statistical analysis. Based on the specific engineering characteristics and operational state of the dam discussed in this paper, and in conjunction with the HST model, the displacement is categorized into water pressure component δ_{H}, temperature component δ_{T}, and aging component δ_{θ}.

$$\delta_{\mathrm{H}} = \sum_{i=1}^{3} a_i (H^i - H_0^i) \tag{12}$$

$$\delta_{\mathrm{T}} = \sum_{i=1}^{2}\left[b_{1i}(\sin\frac{2\pi it}{365} - \sin\frac{2\pi it_0}{365}) + b_{2i}(\cos\frac{2\pi it}{365} - \cos\frac{2\pi it_0}{365})\right] \quad (13)$$

$$\delta_{\theta} = c_1(\theta - \theta_0) + c_2(\ln\theta - \ln\theta_0) \quad (14)$$

Combined with the three formulas described previously, the calculation formula of the measured value $\hat{\delta}$ is as follows:

$$\hat{\delta} = a_0 + \sum_{i=1}^{3} a_i(H^i - H_0^i) + \sum_{i=1}^{2}\left[b_{1i}(\sin\frac{2\pi it}{365} - \sin\frac{2\pi it_0}{365}) + b_{2i}(\cos\frac{2\pi it}{365} - \cos\frac{2\pi it_0}{365})\right] + c_1(\theta - \theta_0) + c_2(\ln\theta - \ln\theta_0) \quad (15)$$

where: H is the upstream water depth on the monitoring day; H_0 is the upstream water depth on the starting date; t is the cumulative number of days from the monitoring date to the starting date; t_0 is the cumulative number of days from the first day of the monitoring period to the starting date; θ is the cumulative number of days from the monitoring date to the starting date divided by 100; θ_0 is the cumulative number of days from the first day of the monitoring period to the starting date divided by 100. h_a is the dam foundation elevation of a dam section; a_0 is a constant term; a_i, b_{1i}, b_{2i}, c_1, and c_2 are regression coefficients.

In this paper, 1958 groups of monitoring data in the dam deformation time series, from 1 January 2013 to 12 May 2018, are selected as the training set, and 346 groups of data from 13 May 2018 to 23 April 2019 are selected as the test set.

3.3. Model Implementation

3.3.1. Feature Engineering

Based on the concept of feature engineering, the initial step involves removing gross errors from the concrete dam displacement time series using the widely applied Pauta criterion in practical engineering. This process eliminates apparent outliers and sudden jumps in the data. Given the prevalence of missing values in the time series, which hinder the research and analysis of dam deformation behavior and the estimation of parameters for deformation prediction models, a spatiotemporal correlation weighted regression interpolation model is proposed. This model aims to address missing values by considering fluctuations in space and time of the monitoring data.

Additionally, this paper leverages the unique advantages of multi-scale localization offered by wavelet transform for noise reduction in dam safety monitoring data. Two threshold determination methods—hard threshold and soft threshold—and four threshold estimation methods, namely Sqtwolog, Rigrsure, Heursure, and Minimaxi, are employed to reduce noise in concrete dam deformation monitoring signals. The *RMSE* and *SNR* are calculated and compared across two noise reduction indicators to determine the method combination that achieves the most effective noise reduction.

3.3.2. Stacking Ensemble Learning Prediction Process

To ensure the effectiveness of the stacking model fusion method, it is crucial in the first step to select proper models while ensuring diversity and complementarity between models. In the second step, the meta-learner generally opts for a simple model with good stability to enhance the overall performance of the system.

Bagging family algorithms, such as Random Forest and Extra-Trees, enhance performance by reducing the model's variance. In contrast, boosting family algorithms such as Gradient Boosting Decision Trees (GBDT) and XGBoost enhance performance by reducing the model's bias. In this paper, XGBoost (2.0.2), based on the boosting strategy, Extra-Trees, based on the bagging strategy, and SVR, which is widely used in practical engineering, are selected as the base learners for the stacking ensemble learning prediction process. Grey Wolf Optimization is employed to optimize the hyperparameters that influence the performance of the algorithm. A Multiple Linear Regression (MLR) algorithm, known for its strong stability, is selected as the meta-learner. The displacement influence factor from the HST model is used as the input for the model, with the horizontal displacement time series as the output. This study predicts the displacement time series of the concrete dam.

3.3.3. Model Performance Evaluation

To evaluate the performance of the prediction method combined with feature engineering proposed in this paper, we have introduced three evaluation indicators: superiority, accuracy, and generalization. These are designed to provide a thorough and comprehensive assessment of the model.

Following data training that incorporates feature engineering, we compared basic evaluation indicators (*MAE*, *MSE*, and *RMSE*) and the superiority, accuracy, and generalization indicators between each individual model (XGBoost, Extra-Trees, SVR, and HST) and the stacking ensemble learning prediction process. This comparison helps verify the superiority of the proposed process in terms of fitting and forecasting accuracy.

The workflow chart described in this paper is illustrated in Figure 10.

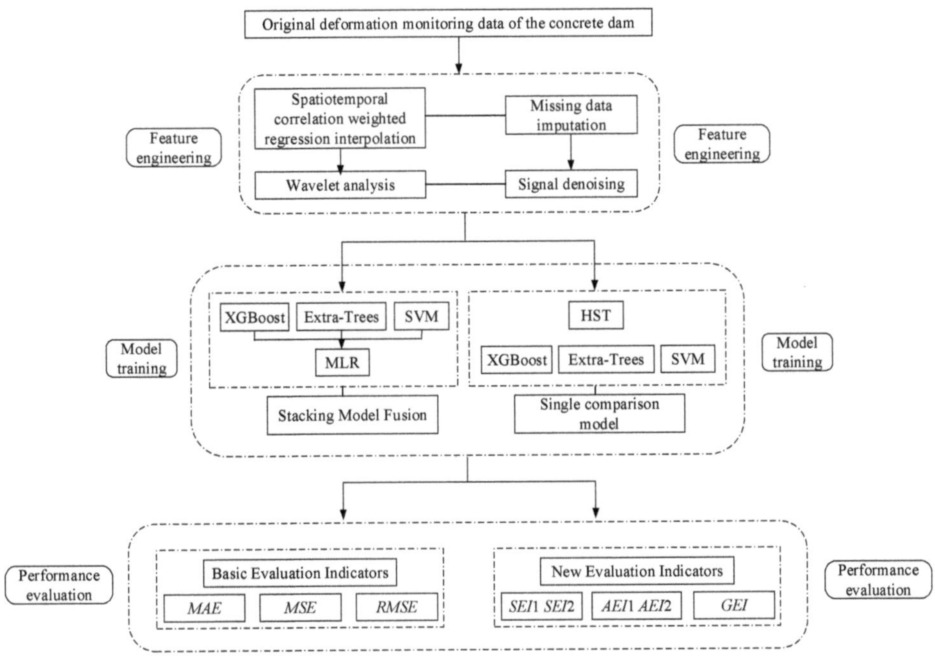

Figure 10. Overall flowchart of the paper.

4. Results and Discussion

4.1. Feature Engineering of Deformation Monitoring Data

4.1.1. Monitoring Data Spatiotemporal Correlation Weighted Regression Interpolation

Taking measurement point EXD-18 as an example, nearby measurement points EXD-2, EXD-3, and EXD-4, which are spatially close to EXD-18 and exhibit roughly the same small-scale fluctuations in measured values, are selected as the spatial reference series for the spatiotemporal correlation regression interpolation method. Additionally, the data from 2013, 2015, and 2017, which have a relatively complete series of measured values, are chosen as the temporal reference points for the 2014 data, which have a higher incidence of missing measurements. The correlation between each reference sequence and the EXD-18 sequence is illustrated in Figures 11 and 12.

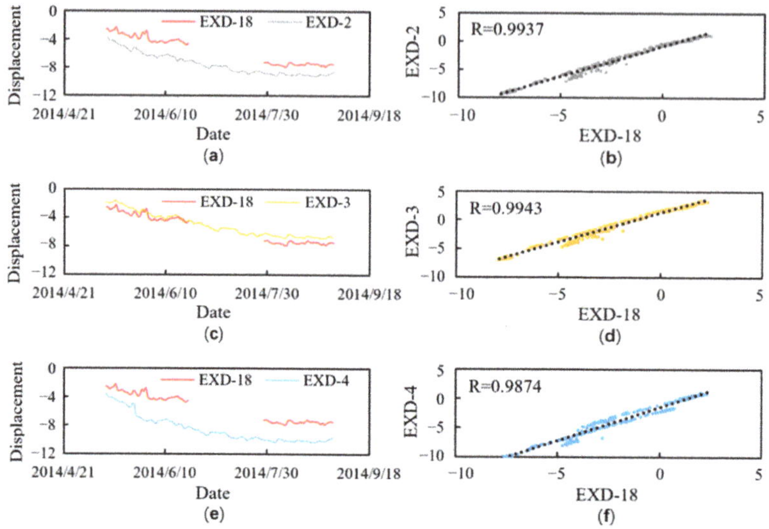

Figure 11. Spatial correlation of EXD-18 measurement point reference sequence: (**a**) Time sequence of EXD-18 and EXD-2; (**b**) Correlation coefficient between time sequences of EXD-18 and EXD-2; (**c**) Time sequence of EXD-18 and EXD-3; (**d**) Correlation coefficient between time sequences of EXD-18 and EXD-3; (**e**) Time sequence of EXD-18 and EXD-4; (**f**) Correlation coefficient between time sequences of EXD-18 and EXD-4.

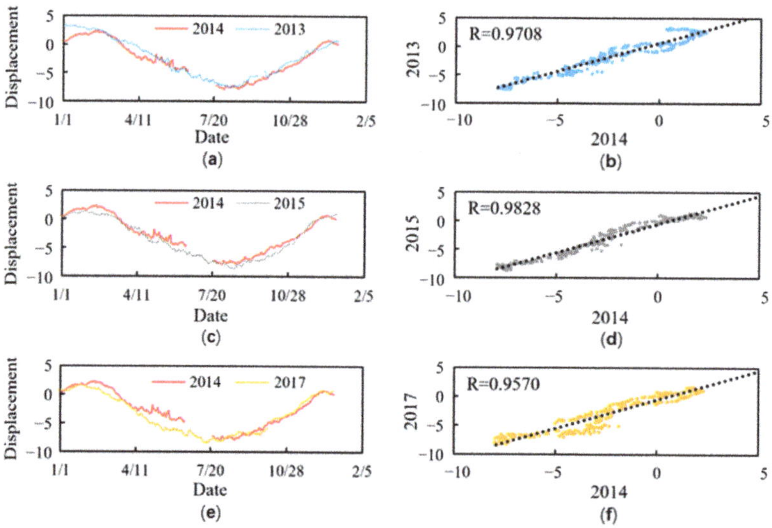

Figure 12. Time correlation of EXD-18 measurement point reference sequence: (**a**) Time sequence of 2014 and 2013; (**b**) Correlation coefficient between time sequences of 2014 and 2013; (**c**) Time sequence of 2014 and 2015; (**d**) Correlation coefficient between time sequences of 2014 and 2015; (**e**) Time sequence of 2014 and 2017; (**f**) Correlation coefficient between time sequences of 2014 and 2017.

During the operation of the dam, each dam section is uniformly affected by environmental factors. Concurrently, the monitoring instruments at each measurement point may fail, resulting in absent measured values. For the time period requiring interpolation, the spatiotemporal correlation regression interpolation method (STCM) is adopted when the referenced measurement points and periods are available, ensuring that both temporal and spatial correlations are considered. If the referenced measurement points are missing, the time correlation regression interpolation method (TCM) is employed; if the referenced periods are missing, the spatial correlation regression interpolation (SCM) method is utilized.

The interpolation expressions and error indicators for each interpolation period of EXD-18 and EXD-D7 measurement points are detailed in Tables 1 and 2. For EXD-18, $x_1 \sim x_6$, respectively, represent the measured value sequences in 2013, 2015, and 2017, and the measured value sequences of EXD-2, EXD-3, and EXD-4 correspond to the year to be interpolated. For EXD-D7, $x_0 \sim x_6$, respectively, represent the measured value sequence of the measurement point in 2013, 2016, 2017, and 2018, and the measured value sequences of EXD-2, EXD-3, and EXD-4 correspond to the year to be interpolated.

Table 1. Interpolation expression and error indicators of EXD-18 measurement points in each period.

Interpolation Periods and Methods	**18 June 2014~23 July 2014**			**26 May 2016~4 June 2016 16 July 2016~26 July 2016**	**22 June 2018~18 July 2018**
	STCM	**TCM**	**SCM**	**TCM**	**TCM**
Equation	$y = 0.084x_1 + 0.003x_2 - 0.166x_3 + 0.552x_4 + 0.351x_5 + 0.091x_6$	$y = 0.328x_1 + 0.642x_2 - 0.030x_3$	$y = 0.238x_4 + 0.434x_5 + 0.247x_6$	$y = -0.021x_1 + 0.409x_2 + 0.579x_3$	$y = 0.367x_1 + 0.178x_2 + 0.516x_3$
MAE	0.222	0.436	0.216	0.399	0.441
MSE	0.080	0.291	0.087	0.246	0.282

Table 2. Interpolation expression and error indicators of EXD-D7 measurement points in each period.

Interpolation Periods and Methods	**27 July 2013~7 August 2013 31 October 2013~19 November 2013**			**26 May 2016~4 June 2016 16 July 2016~26 July 2016**	**31 May 2018~12 June 2018**
	STCM	**TCM**	**SCM**	**TCM**	**SCM**
Equation	$y = -0.447x_1 + 0.173x_2 + 0.528x_3 - 2.396x_4 + 1.221x_5 + 2.047x_6$	$y = 0.249x_1 + 0.652x_2 + 0.074x_3$	$y = -1.556x_4 + 1.915x_5 + 0.722x_6$	$y = 0.140x_0 + 0.991x_2 - 0.219x_3$	$y = 1.386x_4 + 0.822x_5 - 1.382x_6$
MAE	0.196	0.635	0.347	0.485	0.337
MSE	0.072	0.620	0.173	0.333	0.160

The error indicators demonstrate that, generally, the spatiotemporal correlation regression interpolation method outperforms the spatial correlation regression interpolation method, which in turn is superior to the temporal correlation regression interpolation method. From the perspective of concrete dam operations, the operational conditions among different measurement points within the same year are identical, whereas the conditions across different years are broadly similar. This results in better synchronization of displacement curves among different measurement points within the same year compared to displacement curves of the same measurement point across different years. From the perspective of the measured data, it is evident from the missing sequence at the first section of the EXD-18 measurement point that the coefficient before the displacement time series of the same measurement point in different years is significantly smaller than that before the time series of different measurement points in the same year. This indicates that the correlation between displacement time series and target series at the same measurement point across different years is much weaker than that between time series at different measurement points in the same year and the target series.

Therefore, when selecting independent variable time series for regression analysis, those with a more pronounced correlation to the target series should be chosen, considering both temporal correlation and spatial continuity.

4.1.2. Denoising of Monitoring Data

The hard threshold method and soft threshold method are employed to reduce the noise in sequences. Utilizing the four threshold determination methods, the noise reduction thresholds for different high-frequency subsequences are calculated, as detailed in Table 3.

Table 3. Comparison of noise reduction effects of different wavelet denoising methods.

Noise Reduction Method	EXD-18		EXD-D7	
	RMSE	*SNR*	*RMSE*	*SNR*
Sqtwolog-hard	0.1006	32.9741	0.0859	32.3884
Sqtwolog-soft	0.1435	29.8774	0.1186	29.5752
Rigorous-hard	0.0336	42.494	0.0260	42.7755
Rigorous-soft	0.0517	38.7613	0.0408	38.8422
Heursure-hard	0.0590	37.6046	0.0502	37.0474
Heursure-soft	0.0675	36.4320	0.0557	36.1532
Minmaxi-hard	0.0706	36.0454	0.0609	35.3709
Minmaxi-soft	0.1104	32.1618	0.0926	31.7294

It is evident from Table 3 that the Rigorous-hard wavelet denoising method yields the smallest *RMSE*, the highest *SNR*, and the most effective denoising result. Consequently, the Rigorous-hard wavelet denoising method has been chosen to denoise the original observational signal.

4.2. Analysis of Modeling Characteristics of Prediction Models

Using measurement point EXD-18 as an example, an upstream water level–temperature–displacement hydrograph has been drawn, as depicted in Figure 13. Analysis of this Figure reveals that the horizontal displacement at this measurement point in the water flow direction exhibits a clear periodic variation pattern, which correlates strongly with changes in environmental temperature and upstream reservoir water levels. As the reservoir water level rises, the displacement of the dam towards the downstream direction increases; conversely, as the water level decreases, the displacement of the dam towards the downstream direction diminishes. Similarly, as the temperature rises, the displacement of the dam upstream increases; inversely, as the temperature decreases, the downstream displacement increases. Overall, the horizontal displacement in the water flow direction of the dam demonstrates a stable and significant periodic variation pattern.

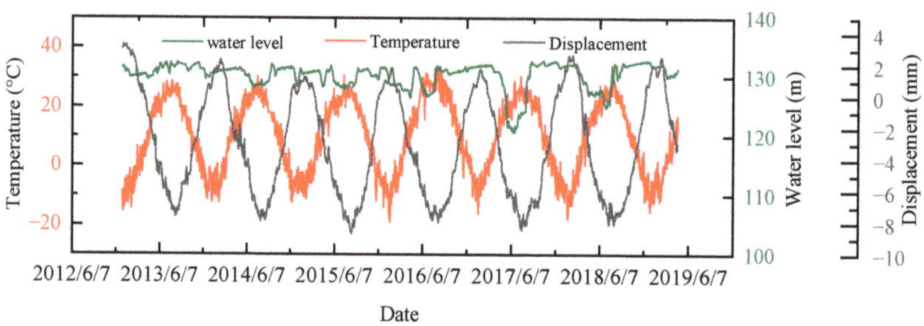

Figure 13. Water level–temperature–displacement hydrograph.

The horizontal displacement measurement points, EXD-18 in the water flow direction of the 18# section of the riverbed overflow dam and EXD-D7 in the water flow direction of the diversion bottom outlet dam section, are designated as the focal points of analysis in this article. A total of 1958 sets of measured data from 1 January 2013 to 12 May 2018 have been selected as the training set, while 346 sets of measured data from 13 May 2018 to 23 April 2019 constitute the test set. The models were trained using the statistical model HST and the machine learning models XGBoost, Extra-Trees, and SVR, as well as the stacking model fusion method proposed in this paper. Employing the Grey Wolf optimization algorithm from biomimetic intelligent algorithms, the hyperparameter space of each sub-model within the stacking method was optimized. The optimal hyperparameter list is presented in Table 4.

Table 4. List of optimal hyperparameters for each based-learner model.

Model	Parameters	Optimal Parameter EXD-18	Optimal Parameter EXD-D7
XGBoost	Gamma	0.05	0.04
	Alpha	5.05	4.07
	Eta	0.20	0.18
	Max_depth	7	8
	Min_child_weight	7	9
	Lambda	0.65	0.53
	Colsample_bytree	0.98	0.85
	Subsample	0.77	0.64
Extra-trees	N_estimators	89	69
	Max_depth	6	8
	Min_samples_split	2	4
	Min_samples_leaf	4	6
SVR	C	1.26	1.03
	Gamma	0.11	0.18

4.2.1. Training and Prediction Process Analysis of Dam Deformation Prediction Models

The training and prediction processes of the time series for EXD-18 and EXD-D7, corresponding to the HST, Extra-Trees, XGBoost, SVR, and the stacking method proposed in this paper, are illustrated in Figures 14 and 15, respectively.

From Figures 14 and 15, it is apparent that the fitting and prediction residuals of the four machine learning models at the two measurement points are relatively small, and their performance exceeds that of the traditional HST model. The efficacy of different models varies across various time series datasets. For instance, during the fitting process for the time series at the two measurement points, the Extra-Trees model and the prediction process proposed in this paper exhibit the best fitting capabilities for the EXD-18 measurement point. At the EXD-D7 measurement point, the Extra-Trees, XGBoost, and the proposed process all demonstrate strong fitting abilities. These differences can be attributed to the distinct distribution and statistical characteristics of the datasets. Moreover, it can be clearly seen from the figures that the prediction residuals of the prediction process proposed in this paper are relatively small, indicating the best prediction performance and a significant improvement over the other four models.

To better demonstrate and compare the prediction performance of various models, the prediction results at EXD-18 of each model and their linear regression analysis graphs were produced. Additionally, to more clearly showcase the level of prediction accuracy, a histogram of the absolute values of the prediction residuals was created.

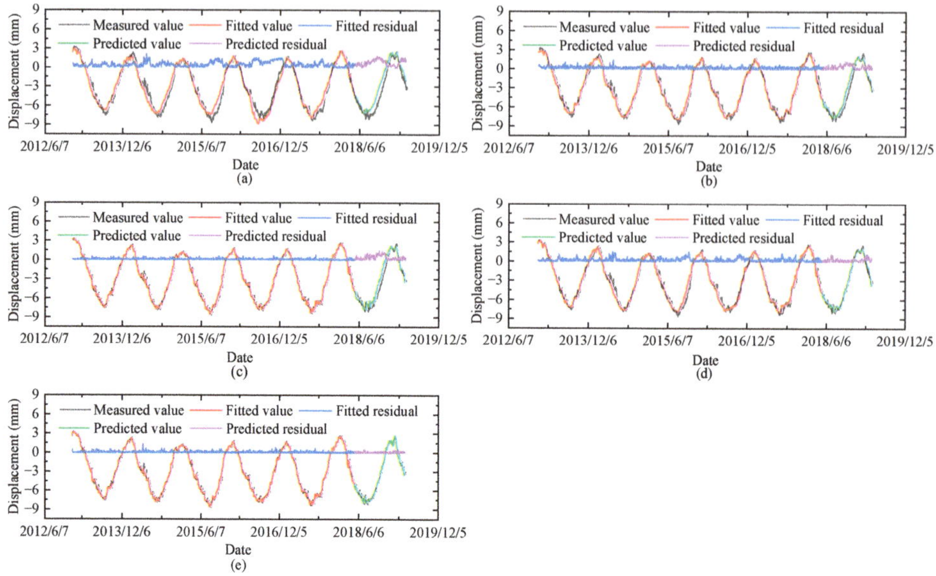

Figure 14. Training and prediction processes of different models for horizontal displacement time series of EXD-18 measurement points: (**a**) HST, (**b**) Extra-Trees, (**c**) XGBoost, (**d**) SVR, and (**e**) Proposed.

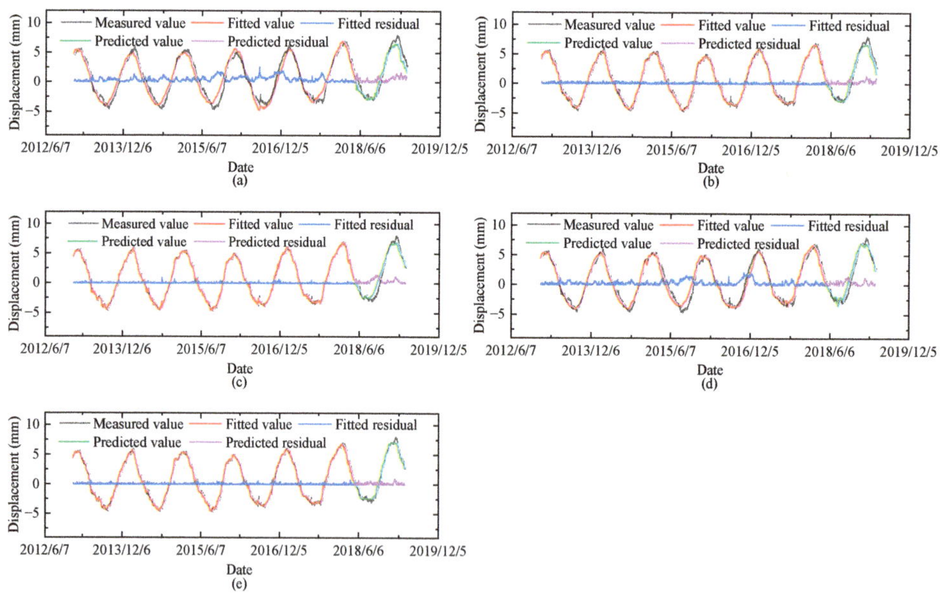

Figure 15. Training and prediction processes of different models for horizontal displacement time series of EXD-D7 measurement points: (**a**) HST, (**b**) Extra-Trees, (**c**) XGBoost, (**d**) SVR, and (**e**) Proposed.

From Figure 16, it is clear that the HST model exhibits the highest prediction residual and the poorest prediction performance. This paper introduces a prediction process with the smallest prediction residual and the most superior prediction performance. The residual distribution of individual machine learning models (XGBoost, Extra-Trees, and

SVR) is relatively uneven, and their predictive abilities are considered average. Regarding deformation trends, HST, XGBoost, Extra-Trees, and SVR all demonstrate significant deviations in their predictions at various stages, which the prediction process proposed in this paper substantially improves. In terms of deformation peaks and local extremes, HST, XGBoost, Extra-Trees, and SVR show large prediction errors and limited characterization ability. This paper presents a prediction process with minimal prediction errors and robust characterization ability. The time series of dam deformation is non-stationary and nonlinear, and its peaks and local extremes contain valuable information about the underlying impact mechanisms, which are crucial for capturing the peaks and local extremes of dam deformation time series. Overall, the prediction process proposed in this paper exhibits outstanding performance compared to other individual models in terms of both overall trend prediction and local detail capture, and it can accurately and effectively predict the changing laws of dam deformation.

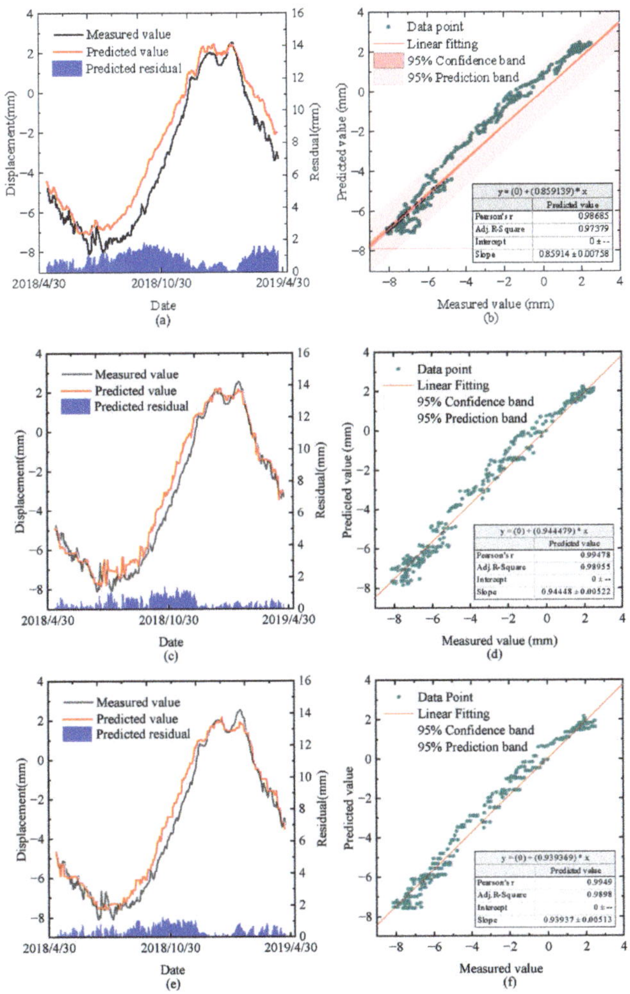

Figure 16. *Cont.*

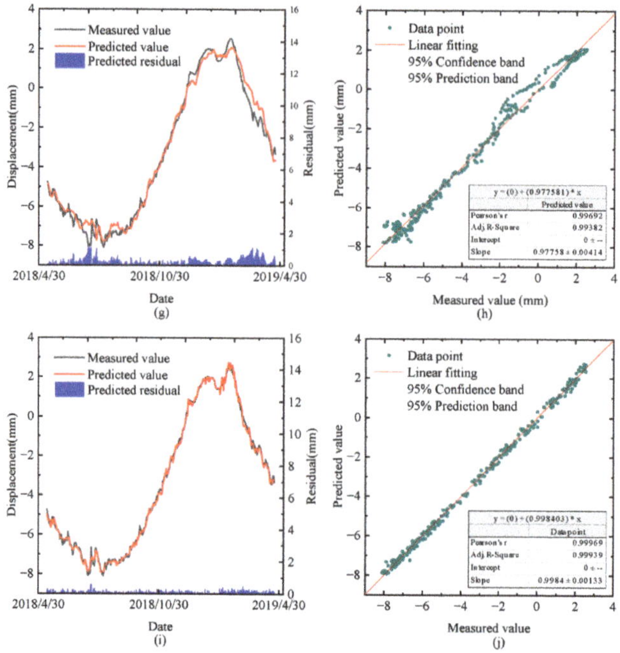

Figure 16. The prediction results at EXD-18 for different models and their linear analysis: (**a**) prediction results of HST; (**b**) linear analysis of HST prediction; (**c**) prediction results of Extra-Trees; (**d**) linear analysis of Extra-Trees prediction; (**e**) prediction results of XGBoost; (**f**) linear analysis of XGBoost prediction; (**g**) prediction results of SVR; (**h**) linear analysis of SVR prediction; (**i**) prediction results of the proposed process; (**j**) linear analysis of HST and the proposed process.

The prediction process for the results at measurement points EXD-2, EXD-14, and EXD-25 were created, as depicted in Figure A1 of Appendix A.

4.2.2. Training and Prediction Residual Analysis of Dam Deformation Prediction Models

Figures 17 and 18 display the Raincloud plots for residuals and the absolute value of residuals, respectively. From Figure 17, it is apparent that the residuals of the prediction process proposed in this paper are more concentrated than those of other models in terms of time series prediction performance at the two measurement points. Almost all residuals are distributed within 1.5 times of the interquartile range, with only a few outliers, and the median residual is close to zero. Figure 18 illustrates more clearly that the absolute value of residuals of the prediction process in this paper is relatively minimal, indicating superior prediction accuracy at each point in the time series prediction process compared to other models. Therefore, the prediction process proposed in this paper demonstrates better prediction accuracy and stability relative to other models.

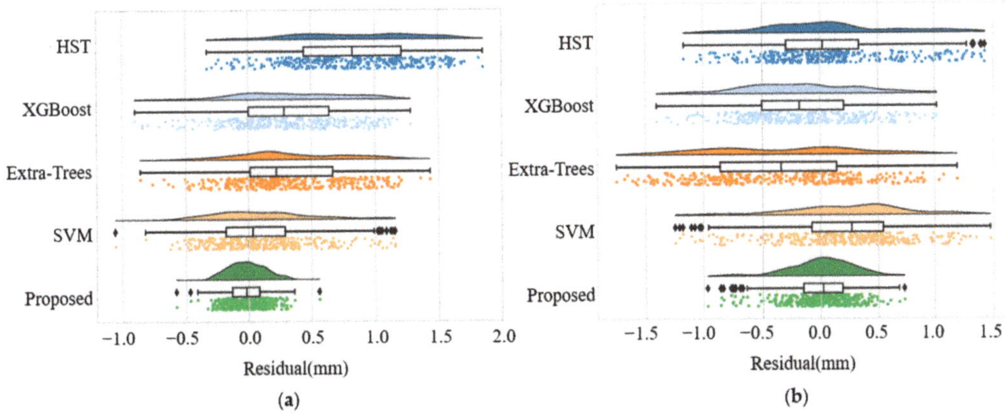

Figure 17. Raincloud plots for residuals of different models: (**a**) EXD-18 prediction residuals; (**b**) EXD-D7 prediction residuals.

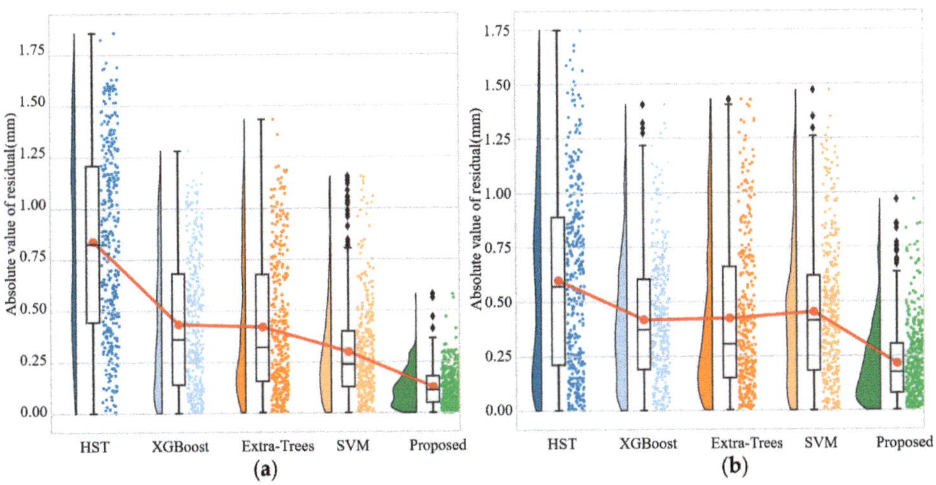

Figure 18. Raincloud plots for the absolute value of residuals of different models: (**a**) Absolute value of residuals of EXD-18 prediction; (**b**) Absolute value of residuals of EXD-D7 prediction. The red lines connect the various averages.

4.2.3. Analysis of Evaluation Indicators of Dam Deformation Prediction Models

Tables 5 and 6 display the basic performance evaluation indicators (*MAE*, *MSE*, and *RMSE*) for different prediction models at the EXD-18 and EXD-D7 measurement points, respectively. These three indicators are utilized to assess the model's performance at the two points. According to the tables, the prediction process proposed in this article has *MAE* values of 0.1007 and 0.0906 for EXD-18 and EXD-D7 during the training process, respectively, while the *MSE* values are 0.0207 and 0.0164, respectively. During the prediction process, the *MAE* values are 0.1249 and 0.2153 and the *MSE* values are 0.0242 and 0.0784, respectively. This demonstrates that the prediction process's fitting and prediction accuracies are superior to those of the HST model and other single machine learning models.

Table 5. List of basic performance evaluation indicators of different prediction models at EXD-18.

EXD-18		HST	XGBoost	Extra-Trees	SVR	Proposed
Training Set	*MAE*	0.4705	0.2213	0.0946	0.2538	0.1007
	MSE	0.3705	0.0813	0.0171	0.1225	0.0207
	RMSE	0.6087	0.2851	0.1308	0.3500	0.1439
Test Set	*MAE*	0.8389	0.4341	0.4209	0.2990	0.1249
	MSE	0.9183	0.2951	0.2889	0.1488	0.0242
	RMSE	0.9583	0.5432	0.5375	0.3857	0.1555

Table 6. List of basic performance evaluation indicators of different prediction models at EXD-D7.

EXD-D7		HST	XG-Boost	Extra-Trees	SVR	Proposed
Training Set	*MAE*	0.5416	0.1250	0.0511	0.3699	0.0906
	MSE	0.4852	0.0260	0.0075	0.2944	0.0164
	RMSE	0.6966	0.1612	0.0866	0.5426	0.1281
Test Set	*MAE*	0.5979	0.4172	0.4233	0.4516	0.2153
	MSE	0.5398	0.2572	0.3067	0.3107	0.0784
	RMSE	0.7347	0.5071	0.5538	0.5574	0.2800

Due to the strong numerical scale dependence of the *MAE* and *MSE* indicators, comparing the performance of the same model at different measurement points based solely on these values is not straightforward. The three model evaluation indicators proposed in this paper provide a more comprehensive and objective evaluation of the model's performance.

Tables 7 and 8 present the proposed evaluation indicators for different prediction models at the EXD-18 and EXD-D7 measurement points, respectively.

Table 7. List of proposed performance evaluation indicators of different prediction models at EXD-18.

EXD-18	HST	XG-Boost	Extra-Trees	SVR	Proposed
SEI_1	1.0000	0.5175	0.5017	0.3564	0.1489
SEI_2		0.1272	0.1243	0.1734	0.0607
AEI_1	0.1000	0.0559	0.0535	0.0391	0.0146
AEI_2	0.3179	0.1862	0.1798	0.1646	0.0808
GEI	1.9564	2.1021	4.7960	1.2484	1.1906

Table 8. List of proposed performance evaluation indicators of different prediction models at EXD-D7.

EXD-D7	HST	XG-Boost	Extra-Trees	SVR	Proposed
SEI_1	1.0000	0.6978	0.7080	0.7553	0.3601
SEI_2		0.3035	0.3613	0.4827	0.2312
AEI_1	0.0779	0.0530	0.0586	0.0531	0.0298
AEI_2	0.2653	0.2042	0.2207	0.2326	0.1128
GEI	1.3052	3.6880	7.7606	1.0624	2.6485

SEI_1 and SEI_2 utilize a set of indicators for comparing the performance against the HST model, which is the most commonly used model for predicting the deformation of concrete dams and serves as the benchmark. These indicators demonstrate the relative advancement of the comparative models against this benchmark. The value range for SEI_1 and SEI_2 is [0, 1], with values closer to 0 indicating a model that is more advanced relative to the benchmark. SEI_1 and SEI_2 have threshold values of 1 and 0.5, respectively. Values below these thresholds suggest that the model's predictive ability is more advanced compared to the benchmark; otherwise, it suggests reduced capability.

AEI_1 and AEI_2 are designed to avoid numerical scale dependence, with a value range also of [0, 1]. Values closer to 0 signify greater accuracy. *GEI* represents the generalization ability of the model, with a *GEI* value closer to 1 when the accuracy and superiority of the model are maximized, indicating stronger generalization ability.

From the tables, it is evident that SEI_1, SEI_2, AEI_1, and AEI_2 for the prediction process in this paper are 0.1489, 0.0607, 0.0146, and 0.0808 for EXD-18 and 0.3601, 0.2312, 0.0298, and 0.1128 for EXD-D7, respectively. These values demonstrate that the prediction process's accuracy is superior to that of other single models and exhibits strong progressiveness. Notably, the SEI_2 values are very small for both measurement points, indicating that the prediction process's accuracy at most time points of series predictions surpasses that of the HST model. The *GEI* values for the prediction process are 1.1906 and 2.6485, respectively, signifying robust generalization. Although the *GEI* of the SVR model is closer to 1, this does not necessarily indicate superior generalization compared to the prediction process proposed in this paper, as strong generalization must also be founded on accuracy.

Figures 19 and 20 display the radar chart of the basic evaluation indicators and the histogram of the proposed evaluation indicators, respectively. From these figures, it is evident that the *MAE*, *MSE*, and *RMSE* of the prediction process discussed in this paper are all lower than those of other models, indicating the highest prediction accuracy. Similarly, the values of AEI_1, AEI_2, SEI_1, and SEI_2 are low, demonstrating the accuracy and progressiveness of the prediction process discussed in this paper.

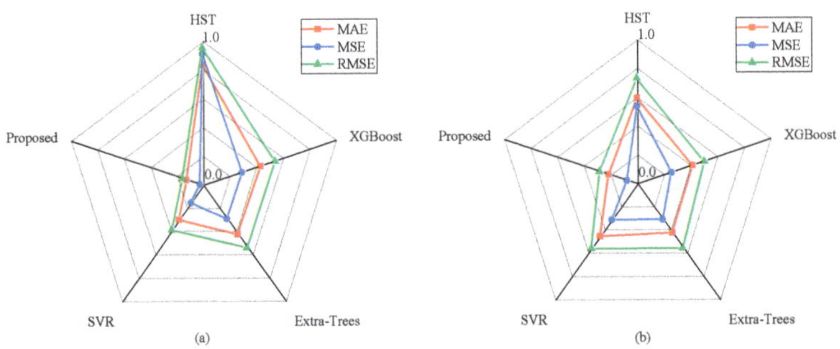

Figure 19. Radar chart of the basic evaluation indicators for different models. (**a**) EXD-18 (**b**) EXD-D7.

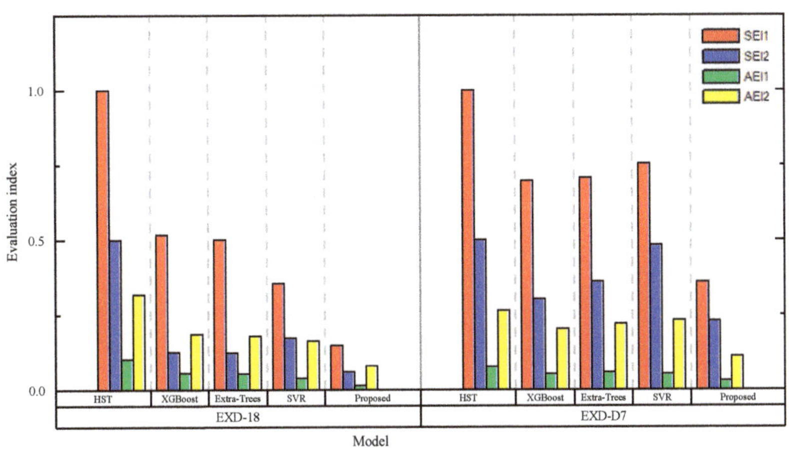

Figure 20. Histogram of the proposed evaluation indicators for different models.

5. Conclusions

This paper proposes a novel artificial intelligence prediction process for concrete dam deformation, based on a stacking model fusion method. Initially, considering the spatiotemporal correlation, effective filling of dam deformation monitoring data was achieved by using the proposed spatiotemporal correlation weighted regression interpolation method. Using widely accepted wavelet analysis theory, the monitoring signal was denoised. The preprocessing of the original observation signals for concrete dam deformation was accomplished. Subsequently, employing the stacking method, three regression prediction artificial intelligence algorithms—XGBoost, Extra-Trees, and SVR—were utilized as the first layer of base learners, and MLR was employed as the second-layer meta-learner. The Grey Wolf optimization algorithm was used to optimize the hyperparameters of the base learners, and a concrete dam displacement prediction process was constructed by integrating multiple machine learning algorithms within the stacking model fusion framework. Additionally, a multi-dimensional evaluation framework for model performance was established, incorporating indicators such as superiority, accuracy, and generalization, thereby creating a holistic assessment system for prediction effectiveness.

An engineering example demonstrates that the proposed preprocessing for the displacement monitoring information of concrete dams is accurate and effective, avoiding issues such as data loss and noise in monitoring data. Compared to traditional statistical models and commonly used single machine learning models, the stacking ensemble learning prediction process significantly enhances fitting accuracy, prediction accuracy, and generalization capabilities. Whether in predicting overall trends or capturing local details, the proposed process exhibits outstanding performance. The model performance evaluation system introduced in this paper can comprehensively and accurately evaluate the performance of prediction models. The prediction process based on the stacking model fusion method provides a scientific and effective basis for deformation prediction and the safety monitoring of concrete dams.

Author Contributions: Conceptualization, W.W. and Y.F.; methodology, W.W. and S.Z. (Shuai Zhang); software, W.W. and S.Z. (Sen Zheng); validation, W.W. and S.Z. (Sen Zheng); formal analysis, W.W. and W.C.; investigation, W.W. and H.L.; resources, H.S.; data curation, H.S.; writing—original draft preparation, W.W.; writing—review and editing, W.W. and H.S.; visualization, W.W.; supervision, H.S.; project administration, W.C. and H.S.; funding acquisition, H.S. All authors have read and agreed to the published version of the manuscript.

Funding: This research was funded by the National Natural Science Foundation of China, grant No. 52239009, 51979093, the National Key Research and Development Program of China, grant No. 2019YFC1510801, the Open Foundation of The National Key Laboratory of Water Disaster Prevention, grant No. 523024852 and the Fundamental Research Funds for the Central Universities, grant No. 2019B69814.

Data Availability Statement: The data presented in this study are available on request from the corresponding author. The data are not publicly available due to laboratory's policy and project collaboration agreements.

Conflicts of Interest: Authors Yanming Feng and Shuai Zhang are employed by the company Powerchina Kunming Engineering Corporation Limited. The remaining authors declare that the research was conducted in the absence of any commercial or financial relationships that could be construed as a potential conflict of interest.

Appendix A. Prediction at Measurement Points EXD-2, EXD-14, and EXD-25

To more effectively illustrate the prediction performance of the proposed process across different measurement point time series, line graphs of the prediction process and linear correlation graphs for the results at measurement points EXD-2, EXD-14, and EXD-25 were created, as depicted in Figure A1.

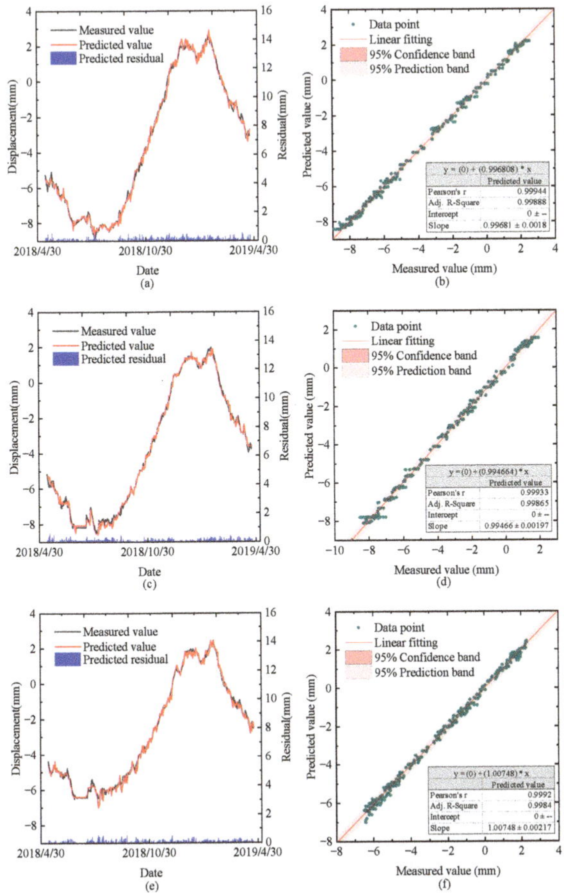

Figure A1. Proposed process prediction results and their linear analysis at different points: (**a**) prediction results of EXD-2; (**b**) linear analysis of EXD-2 prediction; (**c**) prediction results of EXD-14; (**d**) linear analysis of EXD-14 prediction; (**e**) prediction results of EXD-25; (**f**) linear analysis of EXD-25 prediction.

References

1. Su, H.; Wen, Z.; Wang, F.; Hu, J. Dam Structural Behavior Identification and Prediction by Using Variable Dimension Fractal Model and Iterated Function System. *Appl. Soft. Comput.* **2016**, *48*, 612–620. [CrossRef]
2. Bukenya, P.; Moyo, P.; Beushausen, H.; Oosthuizen, C. Health Monitoring of Concrete Dams: A Literature Review. *J. Civil Struct. Health Monit.* **2014**, *4*, 235–244. [CrossRef]
3. Salazar, F.; Moran, R.; Toledo, M.A.; Onate, E. Data-Based Models for the Prediction of Dam Behaviour: A Review and Some Methodological Considerations. *Arch. Comput. Method Eng.* **2017**, *24*, 1–21. [CrossRef]
4. Kang, F.; Liu, J.; Li, J.; Li, S. Concrete Dam Deformation Prediction Model for Health Monitoring Based on Extreme Learning Machine. *Struct. Control Health Monit.* **2017**, *24*, e1997. [CrossRef]
5. Salazar, F.; Toledo, M.A.; Oñate, E.; Morán, R. An Empirical Comparison of Machine Learning Techniques for Dam Behaviour Modelling. *Struct. Saf.* **2015**, *56*, 9–17. [CrossRef]
6. Bernier, C.; Padgett, J.E.; Proulx, J.; Paultre, P. Seismic Fragility of Concrete Gravity Dams with Spatial Variation of Angle of Friction: Case Study. *J. Struct. Eng.* **2016**, *142*, 05015002. [CrossRef]
7. Wang, S.; Xu, Y.; Gu, C.; Bao, T.; Xia, Q.; Hu, K. Hysteretic Effect Considered Monitoring Model for Interpreting Abnormal Deformation Behavior of Arch Dams: A Case Study. *Struct. Control Health Monit.* **2019**, *26*, e2417. [CrossRef]
8. Nguyen, T.T.; Le, V.D.; Huynh, T.Q.; Nguyen, N.H.T. Influence of Settlement on Base Resistance of Long Piles in Soft Soil—Field and Machine Learning Assessments. *Geotechnics* **2024**, *4*, 447–469. [CrossRef]

9. Huynh, T.Q.; Nguyen, T.T.; Nguyen, H. Base Resistance of Super-Large and Long Piles in Soft Soil: Performance of Artificia Neural Network Model and Field Implications. *Acta Geotech.* **2023**, *18*, 2755–2775. [CrossRef]
10. Hariri-Ardebili, M.A.; Mahdavi, G.; Nuss, L.K.; Lall, U. The Role of Artificial Intelligence and Digital Technologies in Dar Engineering: Narrative Review and Outlook. *Eng. Appl. Artif. Intell.* **2023**, *126*, 106813. [CrossRef]
11. Rankovic, V.; Grujovic, N.; Divac, D.; Milivojevic, N. Development of Support Vector Regression Identification Model fo Prediction of Dam Structural Behaviour. *Struct. Saf.* **2014**, *48*, 33–39. [CrossRef]
12. Su, H.; Li, X.; Yang, B.; Wen, Z. Wavelet Support Vector Machine-Based Prediction Model of Dam Deformation. *Mech. Syst. Signa Proc.* **2018**, *110*, 412–427. [CrossRef]
13. Yao, K.; Wen, Z.; Yang, L.; Chen, J.; Hou, H.; Su, H. A Multipoint Prediction Model for Nonlinear Displacement of Concrete Dan *Comput.-Aided Civil Infrastruct. Eng.* **2022**, *37*, 1932–1952. [CrossRef]
14. Zhang, S.; Zheng, D.; Liu, Y. Deformation Prediction System of Concrete Dam Based on IVM-SCSO-RF. *Water* **2022**, *14*, 373 [CrossRef]
15. Gu, C.; Wu, B.; Chen, Y. A High-Robust Displacement Prediction Model for Super-High Arch Dams Integrating Wavele De-Noising and Improved Random Forest. *Water* **2023**, *15*, 1271. [CrossRef]
16. Dai, B.; Gu, C.; Zhao, E.; Qin, X. Statistical Model Optimized Random Forest Regression Model for Concrete Dam Deformatio Monitoring. *Struct. Control Health Monit.* **2018**, *25*, e2170. [CrossRef]
17. Cao, E.; Bao, T.; Yuan, R.; Hu, S. Hierarchical Prediction of Dam Deformation Based on Hybrid Temporal Network an Load-Oriented Residual Correction. *Eng. Struct.* **2024**, *308*, 117949. [CrossRef]
18. Qu, X.; Yang, J.; Chang, M. A Deep Learning Model for Concrete Dam Deformation Prediction Based on RS-LSTM. *J. Sens.* **201** *2019*, 4581672. [CrossRef]
19. Song, S.; Zhou, Q.; Zhang, T.; Hu, Y. Automatic Concrete Dam Deformation Prediction Model Based on TPE-STL-LSTM. *Wate* **2023**, *15*, 2090. [CrossRef]
20. Zhang, C.; Fu, S.; Ou, B.; Liu, Z.; Hu, M. Prediction of Dam Deformation Using SSA-LSTM Model Based on Empirical Mod Decomposition Method and Wavelet Threshold Noise Reduction. *Water* **2022**, *14*, 3380. [CrossRef]
21. Cao, E.; Bao, T.; Gu, C.; Li, H.; Liu, Y.; Hu, S. A Novel Hybrid Decomposition-Ensemble Prediction Model for Dam Deformatior *Appl. Sci.* **2020**, *10*, 5700. [CrossRef]
22. Li, Y.; Bao, T.; Gong, J.; Shu, X.; Zhang, K. The Prediction of Dam Displacement Time Series Using STL, Extra-Trees, and Stacke LSTM Neural Network. *IEEE Access* **2020**, *8*, 94440–94452. [CrossRef]
23. Liu, M.; Wen, Z.; Zhou, R.; Su, H. Bayesian Optimization and Ensemble Learning Algorithm Combined Method for Deformatior Prediction of Concrete Dam. *Structures* **2023**, *54*, 981–993. [CrossRef]
24. Song, J.; Chen, Y.; Yang, J. A Novel Outlier Detection Method of Long-Term Dam Monitoring Data Based on SSA-NAR. *Wire Commun. Mob. Comput.* **2022**, *2022*, 6569367. [CrossRef]
25. Shao, C.; Zheng, S.; Gu, C.; Hu, Y.; Qin, X. A Novel Outlier Detection Method for Monitoring Data in Dam Engineering. *Exper Syst. Appl.* **2022**, *193*, 116476. [CrossRef]
26. Cao, W.; Wen, Z.; Su, H. Spatiotemporal Clustering Analysis and Zonal Prediction Model for Deformation Behavior of Super-Higl Arch Dams. *Expert Syst. Appl.* **2023**, *216*, 119439. [CrossRef]
27. Gao, Y.; Yu, X.; Su, Y.; Yin, Z.; Wang, X.; Li, S. Intelligent Identification Method for Drilling Conditions Based on Stacking Mode Fusion. *Energies* **2023**, *16*, 883. [CrossRef]
28. Zhang, Q.; Wu, J.; Ma, Y.; Li, G.; Ma, J.; Wang, C. Short-Term Load Forecasting Method with Variational Mode Decompositior and Stacking Model Fusion. *Sustain. Energy Grids Netw.* **2022**, *30*, 100622. [CrossRef]
29. Yu, F.; Liu, X. Research on Student Performance Prediction Based on Stacking Fusion Model. *Electronics* **2022**, *11*, 3166. [CrossRef
30. Chen, T.; Guestrin, C. XGBoost: A Scalable Tree Boosting System. In Proceedings of the KDD'16: 22nd ACM SIGKDD Internationa Conference on Knowledge Discovery and Data Mining, San Francisco, CA, USA, 13–17 August 2016; Assoc Computing Machinery New York, NY, USA, 2016; pp. 785–794.
31. Geurts, P.; Ernst, D.; Wehenkel, L. Extremely Randomized Trees. *Mach. Learn.* **2006**, *63*, 3–42. [CrossRef]
32. Cortes, C.; Vapnik, V. Support-Vector Networks. *Mach. Learn.* **1995**, *20*, 273–297. [CrossRef]
33. Mehdiyev, N.; Enke, D.; Fettke, P.; Loos, P. Evaluating Forecasting Methods by Considering Different Accuracy Measures. *Procedia Comput. Sci.* **2016**, *95*, 264–271. [CrossRef]
34. Shcherbakov, M.V.; Brebels, A.; Shcherbakova, N.L.; Tyukov, A.P.; Janovsky, T.A.; Kamaev, V.A. A Survey of Forecast Error Measures. *World Appl. Sci. J.* **2013**, *24*, 171–176.

water

MDPI

rticle

Vison Transformer-Based Automatic Crack Detection on Dam Surface

ian Zhou [1], Guochuan Zhao [1,2,*] and Yonglong Li [3,4]

1 School of Information Engineering, Southwest University of Science and Technology, Mianyang 621010, China; jianzhou@swust.edu.cn
2 Sinograin Chendu Storage Research Institute Co., Ltd., Chendu 610000, China
3 Sichuan Energy Internet Research Institute, Tsinghua University, Chengdu 610213, China; liyonglong@tsinghua-eiri.org
4 Tianfu Yongxing Laboratory, Chengdu 610213, China
* Correspondence: guochuanzhao@163.com

Abstract: Dam is an essential structure in hydraulic engineering, and its surface cracks pose significant threats to its integrity, impermeability, and durability. Automated crack detection methods based on computer vision offer substantial advantages over manual approaches with regard to efficiency, objectivity and precision. However, current methods face challenges such as misidentification, discontinuity, and loss of details when analyzing real-world dam crack images. These images often exhibit characteristics such as low contrast, complex backgrounds, and diverse crack morphologies. To address the above challenges, this paper presents a pure Vision Transformer (ViT)-based dam crack segmentation network (DCST-net). The DCST-net utilizes an improved Swin Transformer (SwinT) block as the fundamental block for enhancing the long-range dependencies within a SegNet-like encoder–decoder structure. Additionally, we employ a weighted attention block to facilitate side fusion between the symmetric pair of encoder and decoder in each stage to sharpen the edge of crack. To demonstrate the superior performance of our proposed method, six semantic segmentation models have been trained and tested on both a self-built dam crack dataset and two publicly available datasets. Comparison results indicate that our proposed model outperforms the mainstream methods in terms of visualization and most evaluation metrics, highlighting its potential for practical application in dam safety inspection and maintenance.

Keywords: Swin Transformer; Vision Transformer; feature fusion; concrete dam; crack detection

1. Introduction

Dams are essential hydraulic structures of hydropower stations, playing a significant role in economic development and societal advancement. However, prolonged exposure to environmental erosion, internal chemical reactions, and various loads can precipitate damage within the concrete dam structure [1]. Among the foremost concerns are cracks, which pose a significant threat to the integrity, durability, strength, and stability of concrete dams [2,3]. For instance, both the Koelnbrein Dam in Australia and the Chencun Dam in China have experienced breaches in their grouting curtains due to the presence and propagation of dam cracks [4]. Hence, conducting routine and methodical inspections to identify and assess cracks in concrete dams is paramount to guaranteeing their safe and stable operation. The current primary approach for detecting dam surface cracks relies on traditional visual inspection [5], which is inherently subjective in both quantitative and qualitative assessments of cracks. Additionally, this method presents drawbacks such as time-consuming procedures, labor-intensive efforts, safety concerns, and limitations in monitoring accuracy and coverage due to the environmental conditions and scale of dams. Moreover, achieving continuous crack monitoring and longitudinal comparisons over time proves unrealistic under manual monitoring practices. Tragically, errors in crack detection contributed to the failure of the Canyon Lake Dam, resulting in the deaths of 236 individuals, as documented [6]. Additionally, the tragic loss of 76 lives occurred in the failure of the Austin Dam due to insufficient attention and maintenance of cracks and defects [7]. Therefore, the formulation and implementation of a highly precise and efficacious crack detection methodology is essential for the autonomous evaluation of dam integrity.

With the rapid advancements in computer hardware and computer vision technology, vision-based automated detection methods have demonstrated notable success in detecting concrete surface cracks in civil infrastructure [8–10]. In these vision-based approaches, a crucial step is to extract features from the images that are sensitive to cracks. Previous studies [11–13] have proposed utilizing image processing techniques to extract crack-sensitive features. For instance, Fan et al. [12] designed CrackLG for underwater dam crack detection. First, a k-means clustering algorithm was employed to distinguish the image blocks including cracks. Then, the final crack areas were extracted using global feature information. However, handcrafted feature-based crack detection approaches suffer from limitations in terms of accuracy and generalization.

Deep learning (DL), with its ability to automatically extract relevant features from data, has become a preferred approach for crack detection [14], yielding promising results in various applications [15]. DL-based concrete surface crack detection methods can be classified into classification-oriented, object-oriented, and segmentation-oriented methods. For example, Zhang et al. [16] proposed a dam crack detection method based on an improved ResNet algorithm using knowledge distillation. Initially, the improved residual neural networks were trained on mini-ImageNet for multi-classification. Subsequently, a parameter/model transfer method was employed to achieve crack detection. However, the image classification-based algorithms just ascertain whether cracks are present or absent within images, without providing specific information about the structural characteristics of the cracks [17], such as their width, length, or orientation.

One potential solution is the object recognition-based method that enables the direct acquisition of both the positional coordinates and categorical labels of objects through the utilization of bounding boxes. The most popular architecture used for performing object recognition in crack detection area is faster R-CNN [18], such as for concrete crack detection [19] and road crack detection [20]. For dam crack detection, Xu et al. [21] proposed AF-RCNN (Attention-based Faster-RCNN), achieving an mAP (mean Average Precision) of 81.07% on an expanded dam crack dataset, surpassing the performance of the original Faster-RCNN. YOLO families are also used as the main architectures in object recognition tasks. For example, YOLOX [22] and YOLOv5 [23] are used for dam crack detection. However, these object recognition-based methods, with the aid of bounding boxes, still suffer from the drawback of coarse crack localization and struggle to encompass a full longitudinal crack within a single bounding box [5].

Another alternative approach performing crack classification is the semantic segmentation-based method. By distinguishing between the background and the cracks at the pixel level, the semantic segmentation-based method inherently confers a distinctive merit in terms of achieving superior accuracy in demarcating the spatial boundaries of cracks within images. Considering the crack images with high resolution, Zhang et al. [24] introduced a dam crack detection method. Initially, a CNN was trained for crack classification, followed by the utilization of an FCN to achieve crack segmentation. Similarly, Pang et al. [25] firstly employed a target detection method to identify cracks using bounding boxes. Subsequently, crack segmentation was performed using image-processing techniques. Within the domain of crack detection research, the majority of crack segmentation tasks are predominantly executed through the utilization of encoder–decoder architectures [26–28]. Various well-known architectures and their improvements such as UNet [29,30] and DeepLabv3+ [31,32] were also proposed to conduct dam crack detections. These deep learning-driven semantic segmentation methodologies have been found to offer enhanced detection results under challenging and noisy environmental scenarios, and precision measurement in cracks [27,31]. However, due to the inherent bias in the convolutional structure, it cannot fully understand global semantic information, which to some extent limits the accuracy and robustness of crack detection.

Comparative to CNNs, the transformer architecture has emerged as a formidable paradigm, exhibiting notable prowess in the domain of Natural Language Processing (NLP) by leveraging its capacity to capture global and long-range information. Google introduced the Vision Transformer (ViT) in 2020 [33], employing transformer structures for image classification and establishing itself as the leading network at that juncture [34]. The ViT-based methods introduce novel model design concepts for CV tasks. In the research field of crack detection, it has been demonstrated that ViT-based methods can be employed for autonomous and efficient dam crack segmentation [35]. For example, CrackFormer [36] is designed for fine-grained carack detection. Ref. [37] introduced a ViT-based framework for crack segmentation on concrete surfaces, showcasing the robust performance of ViT across various types of noise signals. ViT utilizes a self-attention mechanism to extract and integrate contextual information. Nevertheless, ViT's tokens are fixed in number and dimensions, limiting its ability to train and predict at multiple scales. Additionally, employing the ViT algorithm for crack identification entails substantial computational expenses. To address these constraints, Liu et al. [38] introduced the Swin Transformer, which utilizes non-overlapping shift windows for self-attention computation, enabling distinct windows to interact with one another. Swin Transformer has demonstrated significant success in pavement crack detection tasks [39–41]. Some studies utilized the Swin Transformer as an encoder for extracting deep feature representations [42,43]. However, unlike pavement cracks, dam surface cracks in the real world exhibit distinct characteristics such as significant image noise, high background complexity, and considerable scale diversity. When applying these advanced methods directly for dam crack detection purposes, there is a tendency for suboptimal performance outcomes [44].

This paper introduces a new semantic segmentation network for dam crack detection, named DCST-net. To the best of our knowledge, DCST-net is a pioneering instance of a pure ViT-based SegNet-like structure for dam surface crack segmentation. Specifically, its symmetric encoder and decoder are both constructed based on the improved Swin Transformer blocks (SwinT blocks). In addition, it employs a weighted attention block on the encoder and the corresponding decoder features for activating crack features and suppressing the non-crack ones. The principal contributions of this paper can be summarized as follows:

1. It is the first attempt to perform dam surface crack detection with a pure ViT-based encoder–decoder network (DCST-net); our approach yields superior crack segmentation performance on the dam crack dataset collected from a real dam surface as well as two open benchmark crack datasets, outperforming state-of-the-art models;
2. To establish long-range pixel interaction, we propose an improved SwinT-block as the fundamental unit of the DCST-net; this block efficiently extracts contextual information across feature channels through the utilization of depth-wise separable convolution kernels (DWConv); moreover, it integrates spatial domain contextual information through a proficient position-encoding scheme, thereby capturing a wide receptive field;
3. To alleviate the loss of semantic details, we introduce a weighted attention module; it utilizes features from the encoder to produce an attentive mask, which serves as attention coefficients; these coefficients are then multiplicatively applied element-wise to the corresponding features in the decoder, thus suppressing non-crack features while enhancing crack features.
4. To facilitate the training of deep networks, we propose a multi-level label supervision training method, which directly supervises different depth feature layers with crack labels; in addition, we design a hybrid loss function to overcome the problem of class imbalance in crack images.

The rest of the paper is organized as follows: Section 2 presents the proposed DCST-net and elaborates its each block; Section 3 introduces the implementation details, datasets and evaluation metrics; Section 4 demonstrates the extensive experiments results and companions; and Section 5 summarizes the findings and the superiority of our model.

2. Methodology

Generally, crack segmentation networks use feature extractors to reduce feature resolution and extract high-dimensional semantic features to identify cracks. Subsequently, the low-resolution and high-dimensional features are restored to the original image resolution through interpolation or deconvolution, thus producing semantic segmentation results. Convolutional structure-based crack segmentation methods extract features by stacking convolutional layers [16,17,26,32]. In shallower layers, convolution kernels have a smaller receptive field, observing only local features of cracks, which is not favorable for crack orientation detection. Larger receptive fields are only available in deeper layers. In contrast, ViT-based crack segmentation networks [35,44] utilize a self-attention mechanism [21,27,36] to construct a feature extractor and acquire the global semantic features of cracks. This approach addresses the issue of inadequate global information in low-dimensional features, thereby enhancing the prediction of crack direction. In addition, the encoding–decoding structure [26,29–31,36] is a commonly employed design in semantic segmentation networks, where encoding corresponds to feature extraction, and decoding involves resolution recovery. Moreover, skip connections [16,26,32] are implemented between the encoding and decoding layers of the identical resolution to mitigate the loss of fine-grained details and boost crack segmentation. Motivated by these effective mechanisms, we propose DCST-net to enhance the segmentation performance for dam cracks in real-world scenarios.

2.1. *Architecture of DCST-Net*

The DCST-net embraces a SegNet-like encoder–decoder design, as depicted in Figure 1, comprising an encoder, decoder, and a weighted attention module. To effectively capture global and long-range semantic interdependencies, the encoder–decoder module is built upon an improved SwinT-block (presented in Section 2.2), serving as its fundamental unit. In order to refine the segmentation results in detail-rich regions, a weighted attention module (presented in Section 2.4) is adopted between the symmetric encoder and decoder, serving as a filter suppressing other interfering features.

Firstly, the input image, dimension as $[H, W, 3]$, undergoes a patch-embedding module, dividing it into non-overlapping patches (size of 4×4). This results in each patch having a feature dimension of $4 \times 4 \times 4 = 48$. Meanwhile, a local detail module (the upper right of Figure 1) is designed to down-sample the original image by a factor of 2 to obtain local detail features with a dimension of $[H/2, W/2, 48]$. Subsequently, these patch tokens are inputted into a five-stage Swin Transformer-based encoder network, to extract multi-scale features. Within this process, the patch-merging layer [38,45] is tasked with down-sampling and expanding the feature dimension, while the improved SwinT-block focuses on feature representation learning. In contrast, the corresponding Swin Transformer-based decoder network works to enable the restoration of the resolution of the encoder output feature. Specifically, a patch-expanding layer [38,45] is employed for the purpose of up-sampling. This is achieved stage by stage, until the feature map resolution is regained as one-half of the authentic image size. Finally, at each stage, the weighted attention module is designed to concatenate the features from the corresponding encoder and decoder, generating an attentive mask to refine the predicted crack. Subsequently, the fused feature at the current stage undergoes up-sampling by a factor of 2 before being passed to the next stage. This process iterates five times, yielding the final predicted results with identical dimensions as the original input images. Each block of the DCST-net is elaborated in the following.

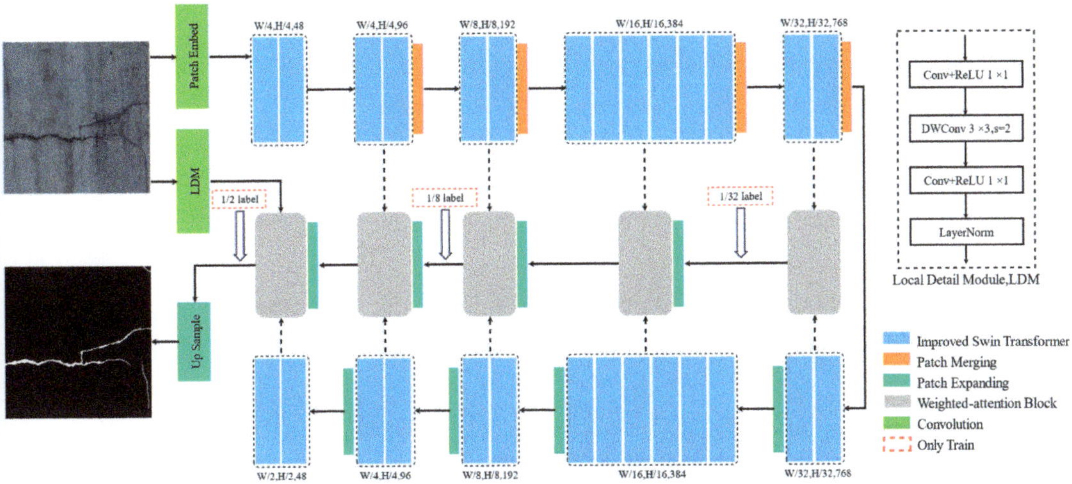

Figure 1. Overview of DCST-net architecture.

2.2. *The improved Swin Transformer Block*

In Figure 2, the replacement of the MLP in the original SwinT-block [38] with a 1×1 DW-Conv results in the two sequential improved SwinT-blocks. Each block comprises a multi-head self-attention module, layer normalization (LN), residual connection and 1×1 depth-wise separable convolutions (DW-Conv). The W-MSA module and the SW-MSA module, adopted in the two consecutive improved SwinT-blocks, utilize improved self-attention computation, as illustrated in Figure 3. These modifications lead to an enlarged receptive field and the acquisition of cross-channel semantic features.

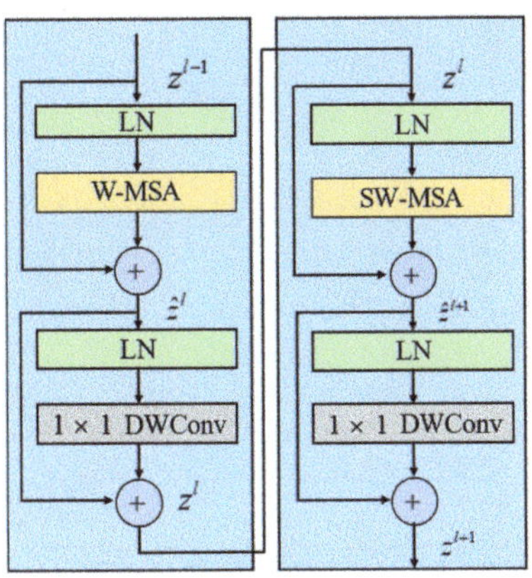

Figure 2. Two successive improved Swin Transformer blocks.

Let $X \in R^{d_{in} \times W \times H}$ be the input tensor, where W, H and d_{in} are the width, height and dimension of the input tensor, respectively. Similar to the work [45], the computation of self-attention proceeds as follows:

$$F^c = SoftMax\left(\frac{qk^T}{\sqrt{d}} + b\right)v \tag{1}$$

where q, k and v, are generated by 1×1 DW-Conv and denote the query, key and value matrices, respectively. The dimension of the query or key is set as d; b presents a relative position vector and is a learnable weight parameter.

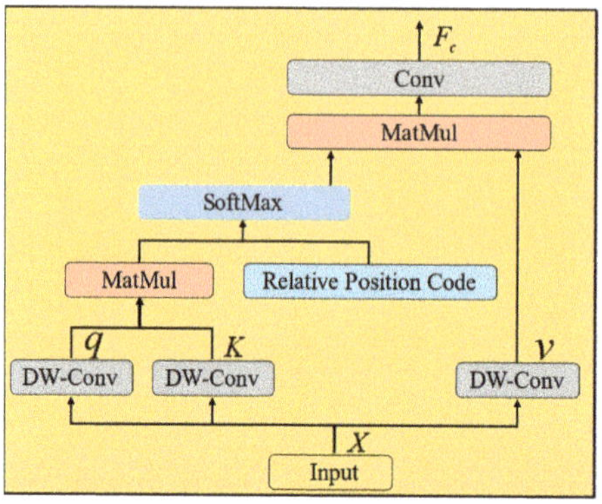

Figure 3. The improved self-attention block.

The two successive improved SwinT-blocks utilize *W-MSA* and *SW-MSA* in pairs. *W-MSA* efficiently reduces the computational cost during self-attention computation, while *SW-MSA* conducts cross-region self-attention computation to obtain a global perspective, effectively enhancing the segmentation performance of dam surface cracks. The computation of continuous SwinT-blocks is given as follows:

$$\hat{z}^l = W - MSA\left(LN\left(z^{l-1}\right)\right) + z^{l-1}, \tag{2}$$

$$z^l = DWConv\left(LN\left(\hat{z}^l\right)\right) + \hat{z}^l, \tag{3}$$

$$\hat{z}^{l+1} = SW - MSA\left(LN\left(z^l\right)\right) + z^l, \tag{4}$$

$$z^{l+1} = DWConv\left(LN\left(\hat{z}^{l+1}\right)\right) + \hat{z}^{l+1}. \tag{5}$$

where $\hat{z}^l$ and z^l are the outputs of the *(S)W-MSA* module and the DW-Conv module of the lth block, respectively.

2.3. Swin Transformer-Based Encoder

The encoder of DCST-net comprises improved SwinT-blocks and patch merging layers, organized into 5 stages based on the {2,2,2,6,2} layout. As depicted in Figure 1, except for the initial stage, which lacks patch-merging layers, the subsequent stages consist of either 2 or 6 improved SwinT-blocks succeeded by a patch-merging layer. The patch tokens undergo representation learning within the two successive improved blocks, preserving the consistency of feature dimension and resolution. Simultaneously, the patch-merging layer [38,45] down-samples the token number by a factor of 2 and increases the feature dimension to double its original dimension. Therefore, the feature sizes outputted by the 5 stages are, respectively, $1/4$, $1/4$, $1/8$, $1/16$ and $1/32$ of the original image size.

2.4. Swin Transformer-Based Decoder

Referring to the encoder, the decoder of the DCST-net has a symmetrical layout of {2,6,2,2,2}. On the contrary, the patch-merging layer located behind the improved SwinT-blocks is replaced by the patch-expanding layer. It serves as the inverse process of patch merging, doubling the resolution of the input feature tensor while halving the number of channels. Finally, the feature sizes outputted in decoder are, respectively $1/32$, $1/16$, $1/8$, $1/4$ and $1/2$ of the original image size.

2.5. Weighted Attention Block

The SwinT-blocks have been demonstrated to be effective for modeling global dependencies, making them valuable for segmenting long cracks [21]. However, relying solely on SwinT-blocks may not be sufficient for addressing cases involving fine-grained cracks with strong background noise. To address this problem, drawing inspiration from the attention gate in U-Net [46] and the scaling attention in SegNet [36], it is evident that suppressing irrelevant regions while highlighting salient features crucial for the segmentation task offers a straightforward yet effective solution. To achieve this, similar to [36], we adopt a weighted attention model between the symmetric encoder and decoder at the same stage. The features from the encoder are utilized to generate a self-attention mask as attention coefficients ranging from 0 to 1, which are then element-wise multiplied with the related features in the symmetric decoder. This process, depicted in Figure 4, functions as a filter that activates crack-related features while suppressing the non-crack ones.

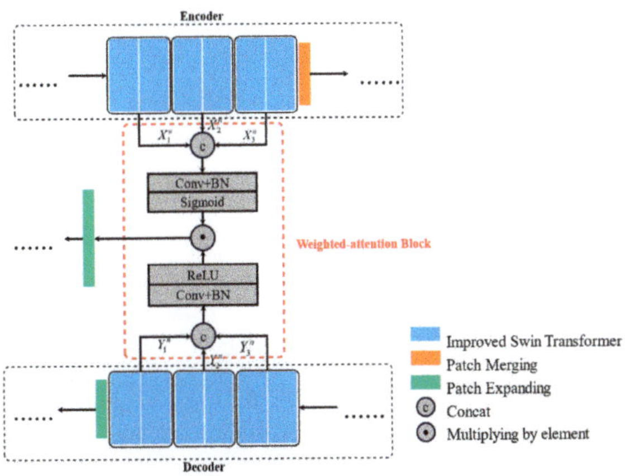

Figure 4. The weighted attention block.

Let us take the feature fusion with the two continuous Swin Transformers in stage n as an example. Features in the symmetric encoder and decoder are denoted as $\{X_1^n, X_2^n, X_3^n\}$ and $\{Y_1^n, Y_2^n, Y_3^n\}$, respectively. As presented in Figure 4, the mask L_{mask}^n is then given by (Equation (6)) as follows:

$$L_{mask}^n = \delta(BN(\otimes 3 \times 3(Concat(X_1^n, X_2^n, X_3^n)))) \quad (6)$$

where $Concat()$ is the concatenation operation, $\otimes 3 \times 3$ represents the convolution operation with a 3×3 convolution kernel, BN is the batch normalization operation, δ denotes the sigmoid activation function. Subsequently, the fusion result, S_{side}^k denoted as side output, is obtained by the weighted attention block as follows:

$$S_{side}^k = L_{mask}^k \odot BN(\otimes 3 \times 3(Concat(Y_1^n, Y_2^n, Y_3^n))), \quad (7)$$

where $\odot$ denotes an element-wise multiplication operation. The side output is then up-sampled by a factor of 2 using the patch-expending operation. As shown in Figure 1, this procedure is repeated 5 times, ultimately obtaining a semantic segmentation prediction image that matches the dimensions of the original image.

2.6. Loss Function

In general, crack segmentation tasks suffer from a severe imbalance between positive and negative samples [47]. For example, through the analysis of our self-built dataset (Section 3.1), it can be found that the average proportion of crack pixels is only 1.34%, while background pixels reach 98.66%. A substantial quantity of background samples facilitate the model's fast learning of background prediction, resulting in a more rapid reduction in background prediction loss compared to that of crack samples. Additionally, despite the small loss for each background pixel, the overall background loss is substantially higher than the crack loss due to the high proportion of background samples. Consequently, the network prioritizes learning background information, leading to challenges such as low accuracy and suboptimal performance in crack segmentation. To tackle the issue of class imbalance between crack and background, one intuitively simple method is to randomly exclude some background pixels from the training process. However, its effectiveness in improving the crack segmentation is limited. Essentially, the poor performance in crack segmentation is caused by the overwhelming effect of crack loss being dominated by background loss. Therefore, we seek the help from hybrid loss function.

In this work, Weighted Cross-Entropy loss and Dice loss are employed to form the hybrid loss function. It is calculated as follows:

$$loss = loss_{wce} + \lambda * loss_{dice} \tag{8}$$

where $loss_{wce}$ and $loss_{dice}$ present Weighted Cross-Entropy loss and Dice loss, and λ is the weight ratio between these two loss functions. As the network progresses through multiple epochs and gains preliminary understanding of both background and crack foreground information, the Weighted Cross-Entropy loss tends to be a certain multiple of the Dice loss. Accordingly, λ is set to an appropriate multiple to optimize the performance of the hybrid loss function.

By introducing weight coefficients for positive and negative samples separately, the Weighted Cross-Entropy loss function becomes more sensitive to the higher-weighted parts, thereby enhancing the learning focus on the target category. Weighted Cross-Entropy loss is defined as follows:

$$loss_{wce} = \frac{1}{n}\sum_{i}^{n} w_f true_f \log\left(pred_f\right) + w_b true_b \log(pred_b), \tag{9}$$

where w_f is the weight for crack loss; w_b is the weight for background loss; $ture_f$ and $true_b$ refer to the crack label and the background label, respectively; $pred_f$ and $pred_b$ denote the predicted probabilities of cracks and background, respectively; and n presents the total number of pixels.

The Dice loss quantifies the dissimilarity between crack prediction values and the true values by directly utilizing Intersection over Union (IoU), aiming for model optimization. Its formulation is represented by Equation (10) as follows:

$$loss_{dice} = \frac{2 * true \cap pred}{true \cup pred + true \cap pred} \tag{10}$$

where *true* is the ground truth value and *pred* is the predicted value.

2.7. Multiple-Level Label Supervision

To enhance the utilization of features at different levels, the implementation in this paper adopts three additional low-resolution labels during the training phase to supervise the parameter updates of deep networks. These low-resolution labels are obtained by down-sampling the original image labels by 1/2, 1/8, and 1/32, respectively (as shown in Figure 1). The rationale behind incorporating these low-resolution labels is based on their unique information content; the 1/2 scale contains richer details of crack information, the 1/32 scale encompasses broader global crack backbone information, and the 1/8 scale lies between these two, exhibiting larger discrepancies. This comprehensive approach enhances model performance and generalization ability, striking a balance between accuracy and speed.

Training the model at different resolutions facilitates a better understanding and adaptation to images of various scales and resolutions, thereby improving the model's generalization ability. Additionally, introducing extra low-resolution labels can increase the training data volume, reducing the risk of overfitting. Implementing multi-level label supervision during training increases the training burden and diminishes training speed, while the multi-level label structure is removed during testing, thus exerting no influence on inference speed.

3. Experiment Preparation

3.1. Dataset Description

In order to evaluate the effectiveness of our proposed crack segmentation network in real-world engineering scenarios, we compiled a dataset named DamSCrack, specifically focusing on dam surface cracks. Furthermore, two openly accessible datasets, namely DeepCrack [48] and Crack500 [49], were employed to validate the generalization capability of the proposed network.

DamSCrack: The dam surface crack images were initially captured using a drone at a hydropower station located in Sichuan, China. As shown in Figure 5, the hydroelectric power station employs a concrete gravity dam design, incorporating eight surface spillways in the overflow section and five bottom outlets in the non-overflow section. Spanning a length of 995.4 m, the dam rises to a height of 465 m. Then, these real-world images underwent cropping and selection. We further meticulously screened, diagnosed, and manually annotated them under the guidance of domain experts, utilizing the ImgAnnotation software as our annotation tool. The annotation process is detailed as follows: firstly, we employed the 1×1 pixel-sized annotation pen from ImgAnnotation to outline the crack's edge and refine its shape; secondly, based on the crack width, we utilized an annotation pen ranging from 3×3 to 10×10 pixels to fill the crack, achieving pixel-level labeling; finally, DamSCrack was generated, comprising 1000 crack images with a resolution of 448×448, and each crack image had a binary label image, with red denoting the crack and black representing the background.

In our experiments, the DamSCrack dataset was randomly divided into training, validation, and test set with the ration of 8:1:1, respectively. Then, five image augmentation strategies, including random brightness variation, random rotation, erasing, blending, and shear blending, were applied to the training set images. These strategies aimed to provide the network with more challenging samples, thereby improving the model's robustness. Regarding every crack image within the training dataset, the five strategies independently occurred with a probability of 0.5. This way, the training dataset expanded from 800 images to 2800 images.

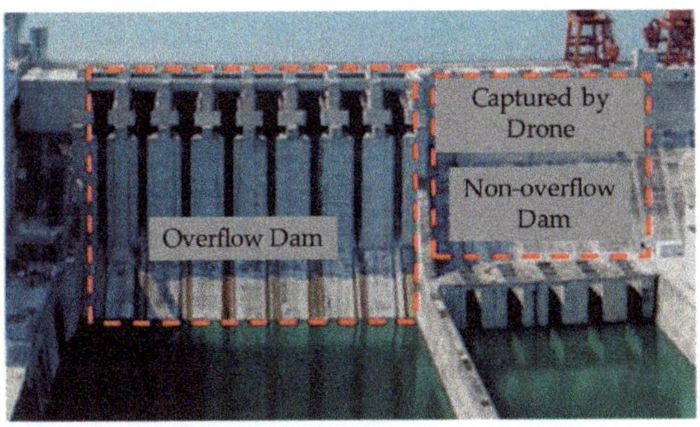

Figure 5. The on-site situation of the dam.

Crack500: This dataset consists of 500 RGB images depicting pavement cracks. Each image with a resolution of 2000×1500 is accompanied by a pixel-level binary label image, featuring a black background and a white representation of the crack. Due to the limited number and high resolution of this dataset, we cropped each image into 12 non-overlapping sub-images with a resolution of 448 × 448. Following the approach in [49], we retained the sub-images containing a minimum of 1000 pixels of cracks. Subsequently, a total of 3368 crack images were selected and subsequently partitioned into distinct subsets for training, validation, and testing purposes, with the ratio of 8:1:1, in our experiments.

DeepCrack: There are 537 RGB crack images with a resolution of 544 × 384 in this dataset, covering concrete surface cracks with various scenes and multiple scales. Each crack image corresponds to a manually annotated pixel-level mask label, where the background is white, and the cracks are black. In our experiments, the original images in the DeepCrack dataset underwent a scaling operation, where the long side was scaled to 448, and the short side was padded with black pixels to achieve an RGB image with a resolution of 448 × 448. Similarly, training images, verification images, and test images were set with the ratio of 8:1:1.

3.2. Experiment Settings

All experiments in this paper operate in an identical hardware and software environment. The details of experiment settings are listed in Table 1. The DCST-net was optimized by using the stochastic gradient descent (SGD) method. The initial learning rate was 1×10^{-5} and the batch size was 8. In addition, the parameters of the comparative methods (Section 4.3) were set to be the same as those in the original paper.

Table 1. Software and hardware configuration.

Hardware/Software	Parameters/Version
CPU	Intel(R) Xeon(R) CPU E5-2620 v4 @ 2.10 GHz
GPU	2 × NVIDIA GTX TITAN Xp, 24 GB memory
RAM	DDR4, 32 GB
Operating System	Ubuntu18.04
Python	3.6
CUDA	10.2
Cudnn	8.0.2
TensorRT	7.0

3.3. Evaluation Metrics

Three commonly used evaluation metrics for binary classification, namely precision (*Pre*), recall (*Rec*), and F1 score (*F1s*) were employed in this work. Meanwhile, the commonly used intersection over union (*Iou*) metric for measuring the segmentation effect was also adopted in this work. Similar to other works, calculation formulas of these four evolution metrics are as follows:

$$Pre = \frac{TP}{TP + FP}, \tag{11}$$

$$Rec = \frac{TP}{TP + FN}, \tag{12}$$

$$F1s = \frac{2 \times Pre \times Rec}{Pre + Rec}, \tag{13}$$

$$Iou = \frac{TP}{TP + FN + FP}. \tag{14}$$

where TP (True Positive) and FP (False Positive) mean that a crack is correctly predicted as a crack and a non-crack is wrongly detected as a crack at pixel level, respectively; FN (False Negative) denotes that a crack is wrongly detected as a non-crack at pixel level.

4. Experimental Results and Discussion

4.1. Analysis of Training Results

Figure 6 illustrates the loss curve of training and validation during the training process of the DCST-net on DamSCrack dataset. It can be observed that the model's loss rapidly decreased from an initial value of 51 to below 10 in just over 10 epochs, with a decrease of over 80%. This indicates that the selection of training parameters and the design of the loss function in this paper are appropriate, leading to a rapid convergence of the network. As the training epoch hits the 600th round, the network has essentially reached the convergence limit, maintaining stability in loss with an average intersection over union of 67.21%. Throughout the entire training process, the validation loss curve maintains a consistent and proximate alignment with the training loss curve, indicating that the model avoids overfitting and possesses strong generalization capabilities.

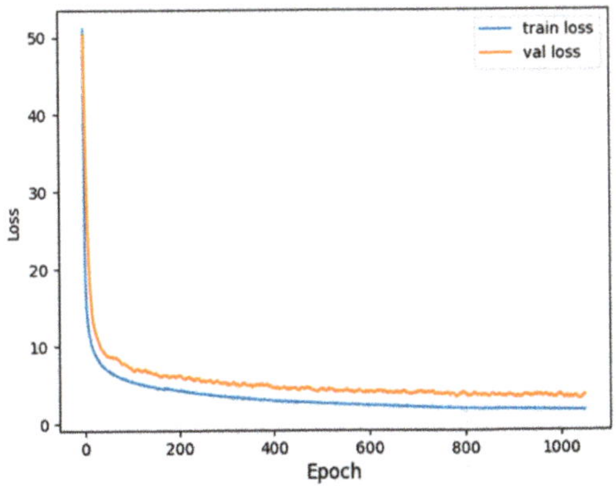

Figure 6. Loss decay curve during training.

Figure 7 presents the curves of precision, recall, and average intersection over the union on validation dataset during the training process. These curves clearly show that in the early stages of network training, precision and average intersection over union rapidly increase. Meanwhile, recall, starting from a relatively high value, quickly optimizes downwards. This indicates that the network is rapidly learning crack segmentation knowledge, improving its ability to segment cracks and outlining the general framework of cracks. By the 600th training round, although the network loss remains stable, the three evaluation metrics continue to improve. This suggests that the class-balanced loss function designed in this paper continues to play a role, allowing the network to continuously learn and optimize for crack details, enhancing the segmentation of crack details. By the 900th training round, the three evaluation metrics essentially stop changing, indicating that the network has completed training without overfitting and exhibits high performance in crack segmentation.

Figure 8 shows the prediction results of the DCST-net on the typical dam surface crack images in the test dataset of DamSCrack. These results clearly indicate that the DCST-net can accurately address the practical problem of dam surface crack detection, even when faced with challenging cracks, such as those with complex branching structures, slender characteristics, intensity in-homogeneity, and strong background interference.

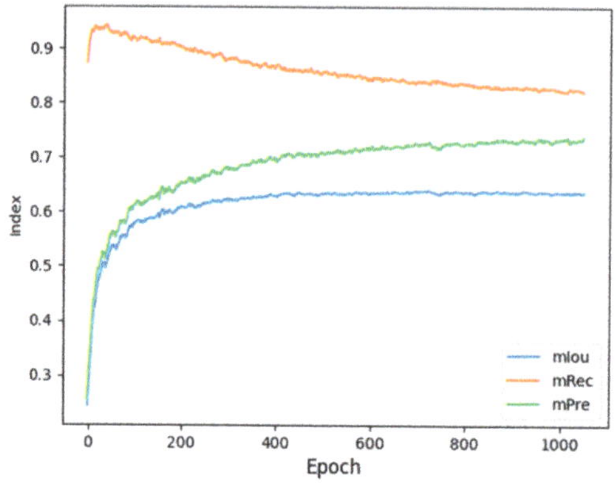

Figure 7. Evaluation index transformation curve during training.

Figure 8. Segmentation results of the DCST-net for typical cracks in the DamSCrack dataset.

4.2. Ablation Study

To further validate the advantage derived from each improved module in DCST-net, an ablation analysis is conducted on the DamSCrack dataset. Specifically, the proposed improvement modules in this paper are added one by one to the Swin Transformer network, resulting in a total of 6 ablation methods (method a, b,. . .,f), as presented in Table 2. The resulting evaluation metrics are presented in Table 3, and the related predictions are visually shown in Figure 9.

Method a: It utilizes the Swin Transformer [38] serving as the backbone for feature extraction, employs up-sampling to restore feature resolution, and thus, outputs the semantic segmentation results. Benefiting from the powerful global representation capability of the self-attention mechanism, method a can segment the crack backbone. However, its segmentation of the local parts of cracks is not satisfactory due to the loss of low-level features, as illustrated in the third column of Figure 9.

Method b: It replaces all SwinT-blocks in method a with improved SwinT-blocks. As indicated in the third row of Table 3, the improved SwinT-blocks considerably boosts the evaluation metrics *Pre*, *Rec*, *F1s* and *mIoU* by 4.50%, 0.80%, 2.92% and 3.24%, respectively, achieving better segmentation performance.

Method c: It introduces only the weighted attention block to method a. The weighted attention module generates a self-attention mask to suppress non-semantic features, leading to the optimized segmentation of the local parts of cracks (the fifth column in Figure 9), with an increase in *Pre*, *Rec*, *F1s* and *mIoU* by 4.78%, 1.69%, 3.30%, and 3.66%, respectively.

Method d: Both the improved SwinT-blocks and the weighted attention module are introduced to method a. These two modules jointly significantly boost the crack segmentation performance, with a corresponding to an increase in *mIoU* by 8.42% and the maximum value of 86.40% for *Rec*.

Method e: Methods a, b, c, and d utilize the cross-entropy loss function provided by PyTorch for loss computation without any weight coefficients. By substituting the loss function in method d with the hybrid loss function, method e demonstrates superior performance over method d (the seventh column in Figure 9), indicating that the problem of class imbalance in crack images is effectively alleviated.

Method f: Finally, the multi-level label supervision training method is implemented to train method e. Through the combined effect of the improved SwinT-blocks, weighted attention module, hybrid loss function and multi-level label supervision training method, method f achieves the optimum segmentation performance, with the best evaluation metrics values.

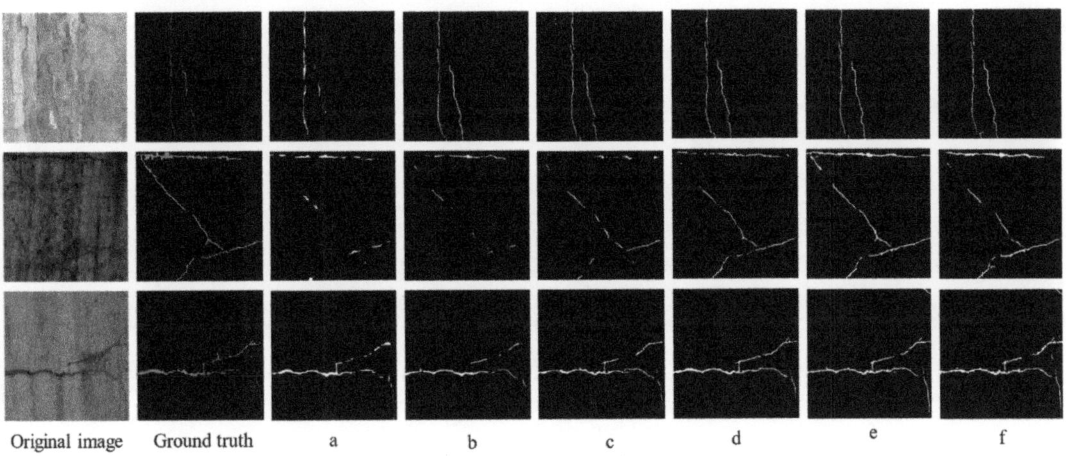

Figure 9. Segmentation results of ablation study of DamSCrack dataset.

Table 2. Methods in ablation study.

Method	Improved SwinT-Blocks	Weighted Attention Block	Hybrid Loss Function	Multiple-Level Label Supervision
a	×	×	×	×
b	√	×	×	×
c	×	√	×	×
d	√	√	×	×
e	√	√	√	×
f(DCST-net)	√	√	√	√

Table 3. Evaluation metrics of methods in ablation study on DamSCrack dataset.

Method	Pre (%)	Rec (%)	F1s (%)	mIou (%)
a	68.16	73.46	69.10	54.42
b	72.66	74.26	72.02	57.66
c	72.94	75.15	72.40	58.08
d	69.80	86.40	76.43	62.84
e	77.81	81.73	79.48	66.65
f	78.41	82.02	79.96	67.21

In conclusion, the experimental results of the ablation study indicate that the crack segmentation network presented in this work, by progressively encoding crack features using the improved SwinT-blocks, enhances the global perception of cracks. In addition, the utilization of the weighted attention module to cross-integrate features from various levels of the encoder and decoder improves the segmentation of local crack parts. Moreover, the combined effects of the crack segmentation loss function and multi-level label supervision effectively alleviate the issue of class imbalance and optimize the network's weight parameters.

4.3. *Comparative Study*

The performance of the DCST-net is empirically validated on the DamSCrack dataset and compared with some classical segmentation models, such as the SegNet [50], FCN-8s [51], DeepLab v3+ [52], U-Net [53], and LR-ASPP [54]. Table 4 presents the quantitative comparison results of performance indices, and Figure 10 demonstrates some typical prediction outcomes generated by the DCST-net and other comparative methods.

Table 4. Evaluation metrics of comparison methods in DamSCrack dataset.

Method	Pre(%)	Rec(%)	F1s (%)	mIou (%)
SegNet [50]	61.05	58.87	57.53	42.26
FCN-8s [51]	54.90	71.17	61.98	44.91
DeepLab v3+ [52]	69.75	79.69	73.49	58.86
U-Net [53]	72.32	76.47	73.02	58.80
LR-ASPP [54]	54.15	66.76	58.18	42.11
DCST-net (ours)	78.41	82.02	79.96	67.21

It can be seen from Table 4 that SegNet [50] and LR-ASPP [54] exhibit the poorest segmentation outcomes for dam surface cracks, achieving an $mIoU$ score of merely 42.26% and 42.11%, respectively. Obvious deficiencies are observed in their segmentation results (column 3 and 7 of Figure 10), such as incorrectly identifying many non-crack areas as crack areas and mis-detecting some of the crack regions. FCN-8s [51] delivers suboptimal segmentation results, with an $mIoU$ of 44.91 and a high recall of 71.17%. While FCN-8s [54] can predict the rough outline of cracks, its performance on fine details remains unsatisfactory. DeepLab v3+ [52] and U-Net [53] offer improved segmentation results for performance metrics, particularly achieving an *mIoU* score of 58.86% and 58.80%, respectively. From columns 5 and 6 of Figure 10, one can discern that these two approaches can effectively predict the crack backbone and provide relatively adequate, detailed information. These observations align well with the existing research literature [55–59], which highlights the superior accuracy of U-Net and DeepLabv3+ compared to other SOTA models (such as FCN and SegNet) in semantic segmentation tasks related to road pavement and concrete cracks. However, CNNs rely on convolutional and pooling layers to process input images, granting them translational invariance by uniformly filtering each patch. While this characteristic is essential for CNNs' effectiveness compared to fully connected networks in vision tasks, prioritizing local pixel connectivity compromises a global context [60,61]. Consequently, CNNs are susceptible to image distortions such as translation and scaling, making them less robust [62]. Hence, U-Net and DeepLabv3+ still face challenges in achieving continuous predictions and struggle in complex regions when dealing with the DamSCrack. On the contrary, ViTs are not limited by local pattern operations and can instead concentrate on information from various distances around the input target area [61]. Therefore, the proposed DCST-net equipped with a ViT-based encoder and decoder demonstrates superior segmentation results, presenting continuous and detailed outputs without the misidentification of crack segmentation. It outperforms the other five mainstream algorithms, as illustrated in the visual comparison figures (last column of Figure 10). In the quantitative comparisons presented in Table 4, the proposed DCST-net achieves superior scores across all performance metrics compared to those related to the classical algorithms. Specifically, its precision is 6.09% higher than that of U-Net [53], and its recall, F1 score, and *mIoU* are 2.33%, 6.47%, and 8.35% higher than those related to DeepLab v3+ [52], respectively. To sum up, in addressing the practical challenge of dam crack detection, the current mainstream crack segmentation algorithms appear unsuitable and cannot be directly applied.

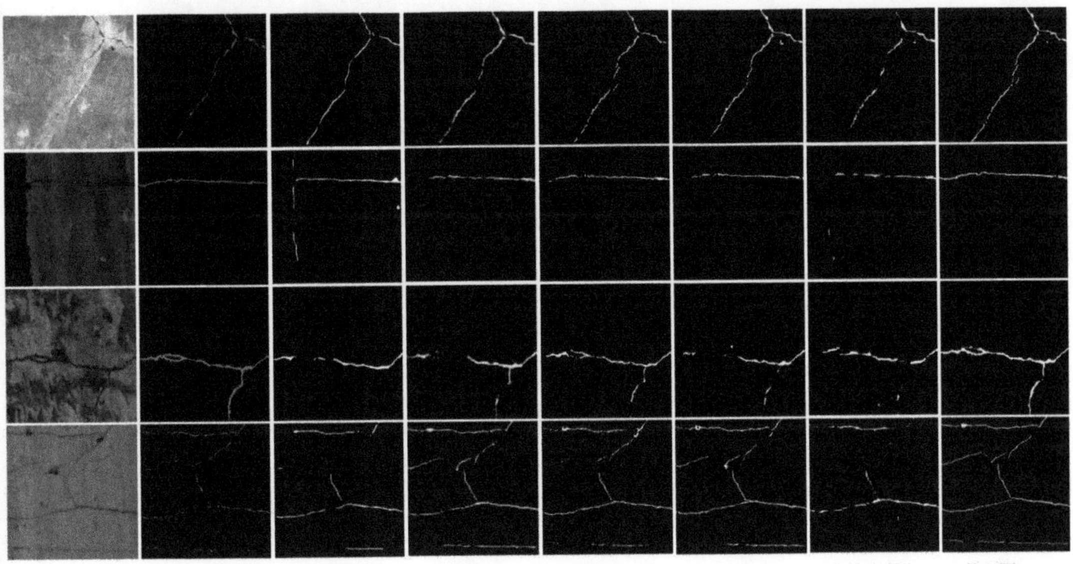

Figure 10. Segmentation results of comparison methods of DamSCrack dataset.

4.4. Generalization Study

To validate the generalizability and further demonstrate the effectiveness of the proposed DCST-net, comparison experiments were performed on openly accessible datasets, DeepCrack [48] and Crack500 [49], against five mainstream segmentation networks.

Comparative results on the DeepCrack dataset: Figure 11 illustrates the prediction results of some typical sample cracks of DeepCrack [48]. Evaluation metrics including inference time are presented in Table 5. It is observable that LR-ASPP [54] has the lowest crack detection accuracy, with an $mIoU$ of 65.19%, but the fastest forward inference speed of 18.27 ms. U-Net [53] achieves more consistent performance across evaluation metrics, compared with those obtained by FCN8s [51] and SegNet [50]. Obviously, DeepLab v3+ [52] demonstrates the strongest crack segmentation performance on DeepCrack [48], achieving Pre of 84.87%, Rec of 82.89%, $F1s$ of 82.46%, and $mIoU$ of 71.41%, respectively, but with the longest inference time of 31.29 ms. The notable performance of DeepLabv3+ can be attributed to its extensive utilization of separable convolutions and spatial pyramid-pooling modules which enable the model to effectively capture multi-scale contextual information while maintaining a wide field of view [63]. However, the proposed DCST-net achieves a balance between performance and speed on the DeepCrack dataset. While its evaluation metrics are slightly lower than those of DeepLab v3+ [52], the forward inference time is approximately 30% faster than that of DeepLab v3+. This suggests that DeepLab v3+ prioritizes higher accuracy at the expense of reduced inference speed. Consequently, the proposed DCST-net achieves better overall performance compared to DeepLab v3+ in the dataset of DeepCrack.

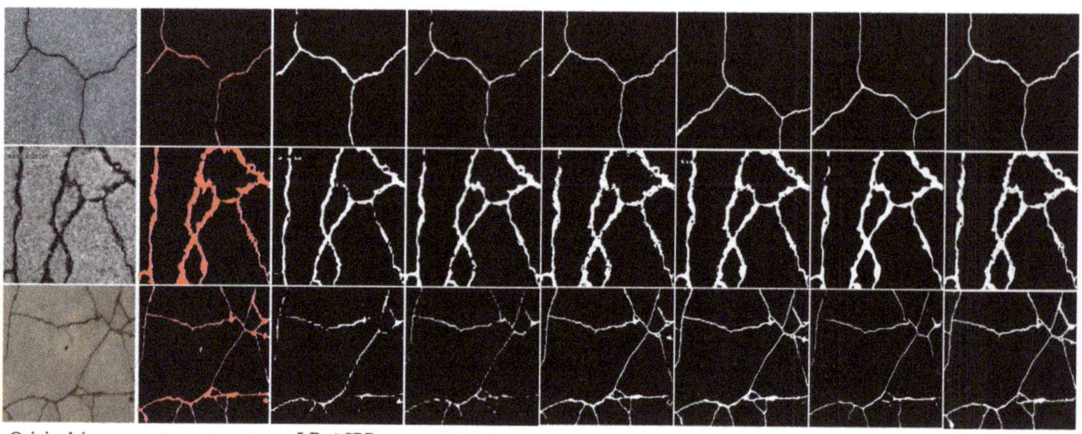

Figure 11. Segmentation results of comparison methods in DeepCrack dataset.

Table 5. Evaluation metrics of comparison methods in DeepCrack dataset.

Method	Pre (%)	Rec (%)	F1s (%)	mIou (%)	Time (ms)
LR-ASPP [54]	81.61	77.34	77.79	65.19	18.27
FCN8s [51]	87.10	74.15	80.10	66.81	20.15
U-Net [53]	83.50	82.58	81.30	68.31	19.52
DeepLab v3+ [52]	84.87	82.89	82.46	71.41	31.29
SegNet [50]	82.71	73.57	75.93	62.98	18.17
DCST-net (ours)	84.64	81.31	80.36	69.46	19.77

Comparative results on the Crack500 dataset: Figure 12 presents the crack segmentation results derived by DCST-net and the comparison methods. It can be observed, in Figure 12, that LR-ASPP [54] achieves the poorest performance, while the detection effect of the DCST-net is better compared with that of comparison methods from visualization aspects. However, all the algorithms incorrectly predict non-crack areas as cracks (row 2 of Figure 12). This is attributed to the presence of image background noise that is challenging even for the human eye to distinguish. As evidenced in Table 6, the DCST-net has achieved the best scores over most of the evaluation indices.

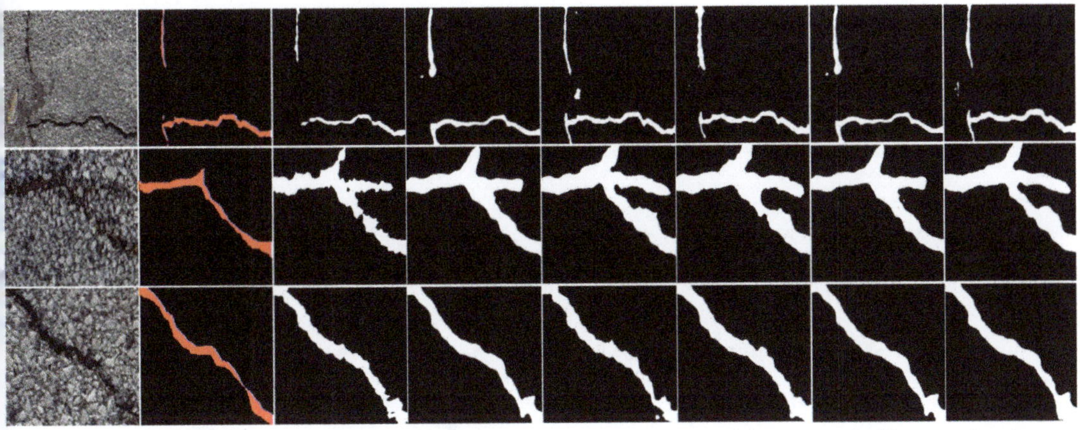

Figure 12. Segmentation results of comparison methods in Crack500 dataset.

Table 6. Evaluation metrics of comparison methods in Crack500 dataset.

Method	Pre (%)	Rec (%)	F1s (%)	mIou (%)	Time (ms)
LR-ASPP [54]	65.68	73.04	65.85	51.20	16.03
FCN8s [51]	65.44	79.21	69.24	54.83	19.97
U-Net [53]	60.98	80.05	67.03	52.10	18.88
DeepLab v3+ [52]	63.64	81.15	69.13	54.55	26.11
SegNet [50]	64.64	75.86	67.41	52.78	18.14
DCST-net (ours)	70.78	75.35	73.00	57.48	22.42

The typical sample cracks in DamSCrack, DeepCrack, and Crack500 are presented in the first column of Figures 10–12, respectively. Obviously, DamSCrack, collected from dam scenes, presents more challenges compared to the other two datasets. It includes weak cracks, diverse crack patterns, low contrast, and significant noise. In contrast, DeepCrack possesses relatively high contrast, making crack regions visually distinguishable even in complex scenes. The primary challenge in Crack500 is that some cracks closely resemble the background. From the evaluation metrics, it is evident that LR-ASPP [54], FCN8s [51], U-Net [53], and SegNet [50] perform adequately when the background noise is weak (Table 5), but exhibit relatively lower performance in the presence of strong background noise (Tables 4 and 6). In contrast, DCST-net and DeepLabv3+ demonstrate distinctive capabilities in extracting global contextual layers, resulting in higher precision in crack segmentation. However, due to dilated convolutions leading to the loss of very fine-grained information, DeepLabv3+ falls short of achieving the high accuracy attained by DCST-net equipped with a ViT-based encoder–decoder architecture in precisely discerning the extremely sharp boundaries of cracks and restoring the connectivity among crack pixels. It is worth mentioning that the proposed algorithm outperforms the other five advanced algorithms on these three datasets, even though it is specifically designed for dam surface crack segmentation.

5. Conclusions

In this paper, a pure ViT-based dam surface crack segmentation network, named DCST-net, is developed for automatic crack detection at pixel level to enhance dam assessment and maintenance in practical engineering applications. Specifically, the improved SwinT-block serving as the basic component of the DCST-net is proposed for characterizing features and learning cross-channel and long-range semantic information. Additionally, to enhance segmentation performance and sharpen crack detection, a weighted attention block is incorporated to combine features between the symmetric encoder and decoder. Moreover, to address the challenge of class imbalance in crack segmentation, we propose a hybrid loss function combining Weighted Cross-Entropy loss and Dice loss. Furthermore, in the training process, we adopt a multi-level label supervision training method to enhance the utilization of features at different levels.

Based on the methodology and experimental findings presented by this work, several main conclusions can be outlined. (1) The comparison results for the self-built dam crack dataset demonstrate that our DCST-net outperforms mainstream methods such as SegNet [50], FCN-8s [51], DeepLab V3+ [52], U-Net [53], and LR-ASPP [54] for pixel-level crack segmentation. In particular, its precision surpasses that of U-Net by 6.09% [53], while its Rec, $F1s$ and $mIoU$ exceed those of DeepLab v3+ by 2.33%, 6.47%, and 8.35%, respectively [52]. This superiority is particularly evident when dealing with various categories of dam cracks, such as elongated, slender, and complex geometry cracks under heavy noise. These findings highlight the potential of the adoption of ViT-based approaches for dam crack detection at pixel level. (2) The comparison results of ablation study reveal that the improved strategies adopted in this work, including the improved SwinT-block, the weighted attention block, the hybrid loss function and the multiple-level label supervision training method, contribute to the performance enhancement of the model to varying degrees. (3) The experimental comparison results both on pavement cracks (Crack500 [49]) and concrete surface cracks (DeepCrack [48]) further demonstrate that the DCST-net possesses good generalization, which is a necessity for real-world applications [37].

The precise segmentation of crack width and the effective elimination of crack-like patterns pose significant challenges in crack detection automation [56,57,64]. Despite the proposed DCST-net demonstrating the best performance in both visualization and quantitative indicators across three crack datasets, it occasionally misidentifies defects or backgrounds as cracks under conditions characterized by extremely weak or significantly disturbed backgrounds. In future work, it is essential to consider the impact of other crack-like defects on crack detection to further enhance accuracy. One possible approach would be to create a more diverse and extensive dataset by collecting images from various dams worldwide to meet the requirements of ViTs [65]. In our study, due to the lack of a rigorous calibration procedure and the shaking of the drone during the collection of crack images, the quantification of segmented cracks cannot be achieved with a unified model. However, precise geometric features of dam cracks are crucial for monitoring and assessing dam safety. Additionally, adopting cost-effective, rapid, and highly precise online systems for crack segmentation in real-world scenarios is imperative. Therefore, developing an encoder–decoder architecture integrated with a Vision Transformer (ViT) capable of enabling real-time applications on edge devices is another research direction.

Author Contributions: Conceptualization, J.Z.; methodology, G.Z.; software, G.Z.; validation, J.Z. and G.Z.; formal analysis, J.Z.; investigation, J.Z.; resources, J.Z. and Y.L.; data curation, G.Z. and Y.L.; writing—original draft preparation, J.Z. and G.Z.; writing—review and editing, J.Z.; visualization, G.Z.; supervision, J.Z.; project administration, J.Z. and Y.L.; funding acquisition, J.Z. All authors have read and agreed to the published version of the manuscript.

Funding: This research was funded by Sichuan Science and Technology Program, grant number 2022YFG0242, and the Open Fund of Robot Technology Used for Special Environment Key Laboratory of Sichuan Province, grant number 21kftk01.

Data Availability Statement: Publicly available datasets were analyzed in this study. This data can be found here: [https://github.com/yhlleo/DeepCrack (accessed on 9 May 2023); https://github.com/fyangneil/pavement-crack-detection (accessed on 11 June 2023)].

Conflicts of Interest: Author Guochuan Zhao was employed by the company Sinograin Chendu Storage Research Institute Co., Ltd., Chendu, China. The remaining authors declare that the research was conducted in the absence of any commercial or financial relationships that could be construed as a potential conflict of interest.

References

1. Kang, F.; Li, J.; Zhao, S.; Wang, Y. Structural health monitoring of concrete dams using long-term air temperature for thermal effect simulation. *Eng. Struct.* **2019**, *180*, 642–653. [CrossRef]
2. Zhang, G.; Liu, Y.; Zheng, C.; Feng, F. Simulation of influence of multi-defects on long-term working performance of high arch dam. *Sci. China Technol. Sci.* **2011**, *54*, 1–8. [CrossRef]
3. Ye, X.W.; Jin, T.; Li, Z.X.; Ma, S.Y.; Ding, Y.; Ou, Y.H. Structural crack detection from benchmark data sets using pruned fully convolutional networks. *J. Struct. Eng.* **2021**, *147*, 04721008. [CrossRef]
4. Li, Y.; Bao, T.; Shu, X.; Gao, Z.; Gong, J.; Zhang, K. Data-driven crack behavior anomaly identification method for concrete dams in long-term service using offline and online change point detection. *J. Civ. Struct. Health* **2021**, *11*, 1449–1460. [CrossRef]
5. Hamishebahar, Y.; Guan, H.; So, S.; Jo, J. A comprehensive review of deep learning-based crack detection approaches. *Appl. Sci.* **2022**, *12*, 1374. [CrossRef]
6. Graham, W. *A Procedure for Estimating Loss of Life Caused by Dam Failure*; Bureau of Reclamation, Dam Safety Office: Denver, CO, USA, 1999; p. 10.
7. Rich, T.P. Lessons in social responsibility from the Austin dam failure. *Int. J. Eng. Educ.* **2006**, *22*, 1287–1296.
8. Chen, B.; Zhang, H.; Wang, G.; Huo, J.; Li, Y.; Li, L. Automatic concrete infrastructure crack semantic segmentation using deep learning. *Autom. Constr.* **2023**, *152*, 104950. [CrossRef]
9. Shi, P.; Shao, S.; Fan, X.; Zhou, Z.; Xin, Y. MCL-CrackNet: A Concrete Crack Segmentation Network Using Multi-level Contrastive Learning. *IEEE T. Instrum. Meas.* **2023**, *72*, 5030415. [CrossRef]
10. Bhowmick, S.; Nagarajaiah, S.; Veeraraghavan, A. Vision and deep learning-based algorithms to detect and quantify cracks on concrete surfaces from UAV videos. *Sensors* **2020**, *20*, 6299. [CrossRef]
11. Shi, P.; Fan, X.; Ni, J.; Wang, G. A detection and classification approach for underwater dam cracks. *Struct. Health Monit.* **2016**, *15*, 541–554. [CrossRef]
12. Fan, X.N.; Wu, J.J.; Shi, P.F.; Zhang, X.W.; Xie, Y.J. A Novel Automatic Dam Crack Detection Algorithm Based on Local-Global Clustering. *Multimed. Tools Appl.* **2018**, *77*, 26581–26599. [CrossRef]
13. Mohan, A.; Poobal, S. Crack detection using image processing: A critical review and analysis. *Alex. Eng. J.* **2018**, *57*, 787–798. [CrossRef]
14. Cao, W.; Liu, Q.; He, Z. Review of Pavement Defect Detection Methods. *IEEE Access* **2020**, *8*, 14531–14544. [CrossRef]
15. Li, B.; Wang, K.; Zhang, A.; Yang, E.; Wang, G. Automatic classification of pavement crack using deep convolutional neural network. *Int. J. Pavement. Eng.* **2020**, *21*, 457–463. [CrossRef]
16. Zhang, J.; Bao, T. An improved resnet-based algorithm for crack detection of concrete dams using dynamic knowledge distillation. *Water* **2023**, *15*, 2839. [CrossRef]
17. Li, Y.T.; Bao, T.F.; Xu, B.; Shu, X.S.; Zhou, Y.H.; Du, Y.; Wang, R.J.; Zhang, K. A deep residual neural network framework with transfer learning for concrete dams patch-level crack classification and weakly-supervised localization. *Measurement* **2022**, *188*, 110641. [CrossRef]
18. Ren, S.; He, K.; Girshick, R.; Sun, J. Faster R-CNN: Towards real-time object detection with region proposal networks. *IEEE Trans. Pattern Anal. Mach. Intell.* **2017**, *39*, 1137–1149. [CrossRef]
19. Deng, J.; Lu, Y.; Lee, V.C.S. Concrete crack detection with handwriting script interferences using faster region-based convolutional neural network. *Comp. Aided Civ. Infrastruct. Eng.* **2020**, *35*, 373–388. [CrossRef]
20. Ciaparrone, G.; Serra, A.; Covito, V.; Finelli, P.; Scarpato, C.A.; Tagliaferri, R. A deep learning approach for road damage classification. In *Proceedings of Advanced Multimedia and Ubiquitous Engineering*; Springer: Singapore, 2018; pp. 655–661.
21. Xu, G.; Han, X.; Zhang, Y.; Wu, C. Dam crack image detection model on feature enhancement and attention mechanism. *Water* **2022**, *15*, 64. [CrossRef]
22. Ben, H.; Fei, K.; Yu, T. A real-time detection method for concrete dam cracks based on an object detection algorithm. *J. Tsinghua Univ.* **2023**, *63*, 1078–1086.
23. Li, Y.; Bao, T. A real-time multi-defect automatic identification framework for concrete dams via improved YOLOv5 and knowledge distillation. *J. Civ. Struct. Health Monit.* **2023**, *13*, 1333–1349. [CrossRef]
24. Zhang, J.; Zhang, J. An improved nondestructive semantic segmentation method for concrete dam surface crack images with high resolution. *Math. Probl. Eng.* **2020**, *2020*, 5054740. [CrossRef]
25. Pang, J.; Zhang, H.; Feng, C.; Li, L. Research on crack segmentation method of hydro-junction project based on target detection network. *KSCE J. Civ. Eng.* **2020**, *24*, 2731–2741. [CrossRef]
26. Feng, C.; Zhang, H.; Wang, H.; Wang, S.; Li, Y. Automatic pixel-level crack detection on dam surface using deep convolutional network. *Sensors* **2020**, *20*, 2069. [CrossRef] [PubMed]
27. Chen, B.; Zhang, H.; Li, Y.; Wang, S.; Zhou, H.; Lin, H. Quantify pixel-level detection of dam surface crack using deep learning. *Meas. Sci. Technol.* **2022**, *33*, 065402. [CrossRef]

28. Kang, D.; Cha, Y. Efficient attention-based deep encoder and decoder for automatic crack segmentation. *Struct. Health Moni* **2022**, *21*, 2190–2205. [CrossRef]
29. Lv, Z.; Tian, J.; Zhu, Y.; Li, Y. Automatic crack detection of dam concrete structures based on deep learning. *Comput. Concr.* **202** *32*, 615.
30. Li, J.; Lu, X.; Zhang, P.; Li, Q. Intelligent Detection Method for Concrete Dam Surface Cracks Based on Two-Stage Transfe Learning. *Water* **2023**, *15*, 2082. [CrossRef]
31. Wu, Z.; Tang, Y.; Hong, B.; Liang, B.; Liu, Y. Enhanced precision in dam crack width measurement: Leveraging advance lightweight network identification for pixel-level accuracy. *Int. J. Intell. Syst.* **2023**, *2023*, 9940881. [CrossRef]
32. Zhu, Y.; Tang, H. Automatic damage detection and diagnosis for hydraulic structures using drones and artificial intelligenc techniques. *Remote Sens.* **2023**, *15*, 615. [CrossRef]
33. Dosovitskiy, A.; Beyer, L.; Kolesnikov, A.; Weissenborn, D.; Zhai, X.; Unterthiner, T.; Houlsby, N. An image is worth 16 × 1 words: Transformers for image recognition at scale. *arXiv* **2020**, arXiv:2010.11929.
34. Paul, S.; Chen, P.Y. Vision Transformers Are Robust Learners. In Proceedings of the AAAI Conference on Artificial Intelligence Virtually, 22 February–1 March 2022; Volume 36, pp. 2071–2081.
35. Zhao, S.; Kang, F.; Li, J. Intelligent segmentation method for blurred cracks and 3D mapping of width nephograms in concret dams using UAV photogrammetry. *Autom. Constr.* **2024**, *157*, 105145. [CrossRef]
36. Liu, H.; Miao, X.; Mertz, C.; Xu, C.; Kong, H. Crackformer: Transformer network for fine-grained crack detection. In Proceeding of the IEEE/CVF International Conference on Computer Vision (ICCV), Montreal, QC, Canada, 10–17 October 2021; pp. 3783–379
37. Shamsabadi, E.A.; Xu, C.; Rao, A.S.; Nguyen, T.; Ngo, T.; Dias-da-Costa, D. Vision transformer-based autonomous crack detectio on asphalt and concrete surfaces. *Autom. Constr.* **2022**, 104316. [CrossRef]
38. Liu, Z.; Lin, Y.; Cao, Y.; Hu, H.; Wei, Y.; Zhang, Z.; Lin, S.; Guo, B. Swin transformer: Hierarchical vision transformer using shifte windows. In Proceedings of the IEEE/CVF International Conference on Computer Vision (ICCV), Montreal, QC, Canada, 10–1 October 2021; pp. 10012–10022.
39. Huang, H.; Hao, X.; Pei, L.; Ding, J.; Hu, Y.; Li, W. Automated detection of through-cracks in pavement using three-instantaneou attributes fusion and Swin Transformer network. *Autom. Constr.* **2024**, *158*, 105179. [CrossRef]
40. Sun, Z.; Zhai, J.; Pei, L.; Li, W.; Zhao, K. Automatic Pavement Crack Detection Transformer Based on Convolutional and Sequentia Feature Fusion. *Sensors* **2023**, *23*, 3772. [CrossRef] [PubMed]
41. Luo, H.; Li, J.; Cai, L.; Wu, M. STrans-YOLOX: Fusing swin transformer and YOLOX for automatic pavement crack detectio *Appl. Sci.* **2023**, *13*, 1999. [CrossRef]
42. Guo, F.; Qian, Y.; Liu, J.; Yu, H. Pavement crack detection based on transformer network. *Autom. Constr.* **2023**, *145*, 10464 [CrossRef]
43. Guo, F.; Liu, J.; Lv, C.; Yu, H. A novel transformer-based network with attention mechanism for automatic pavement crac detection. *Constr. Build. Mater.* **2023**, *391*, 131852. [CrossRef]
44. Zhang, E.; Shao, L.; Wang, Y. Unifying transformer and convolution for dam crack detection. *Autom. Constr.* **2023**, *147*, 104712 [CrossRef]
45. Cao, H.; Wang, Y.; Chen, J.; Jiang, D.; Zhang, X.; Tian, Q.; Wang, M. Swin-unet: Unet-like pure transformer for medical imag segmentation. In Proceedings of the European Conference on Computer Vision (ECCV), Tel Aviv, Israel, 23–27 October 2022 pp. 205–218.
46. Ozan, O.; Jo, S.; Loic, L.F.; Matthew, L.; Mattias, H.; Kazunari, M.; Kensaku, M.; Steven, M.; Nils, Y.H.; Bernhard, K.; et a Attention u-net: Learning where to look for the pancreas. *arXiv* **2018**, arXiv:1804.03999.
47. Zhao, F.; Chao, Y.; Li, L. A Crack Segmentation Model Combining Morphological Network and Multiple Loss Mechanism. *Sensors* **2023**, *23*, 1127. [CrossRef]
48. Liu, Y.; Yao, J.; Lu, X.; Renping, X.; Li, L. DeepCrack: A deep hierarchical feature learning architecture for crack segmentation *Neurocomputing* **2019**, *338*, 139–153. [CrossRef]
49. Yang, F.; Zhang, L.; Yu, S.; Prokhorov, D.; Mei, X.; Ling, H. Feature pyramid and hierarchical boosting network for pavemen crack detection. *IEEE Trans. Intell. Transp. Syst.* **2019**, *21*, 1525–1535. [CrossRef]
50. Badrinarayanan, V.; Kendall, A.; Cipolla, R. Segnet: A deep convolutional encoder-decoder architecture for image segmentatio *IEEE Trans. Pattern Anal. Mach. Intell.* **2017**, *39*, 2481–2495. [CrossRef]
51. Long, J.; Shelhamer, E.; Darrell, T. Fully convolutional networks for semantic segmentation. In Proceedings of the IEEE Conference on Computer Vision and Pattern Recognition (CVPR), Boston, MA, USA, 7–12 June 2015; pp. 3431–3440.
52. Chen, L.C.; Papandreo, G.; Schroff, F.; Adam, H. Rethinking Atrous Convolution for Semantic Image Segmentation. *arXiv* **2017** arXiv:1706.05587.
53. Ronneberger, O.; Fischer, P.; Brox, T. U-net: Convolutional networks for biomedical image segmentation. In Proceedings of the Medical Image Computing and Computer-Assisted Intervention (MICCAI), Munich, Germany, 5–9 October 2015; pp. 234–241.
54. Howard, A.; Sandler, M.; Chu, G.; Chen, L.C.; Chen, B.; Tan, M.; Adam, H. Searching for mobilenetv3. In Proceedings of the IEEE/CVF International Conference on Computer Vision (ICCV), Seoul, Republic of Korea, 27 October–2 November 2019 pp. 1314–1324.
55. Dung, C.V. Autonomous concrete crack detection using deep fully convolutional neural network. *Autom. Constr.* **2019**, *99*, 52–58

6. Dais, D.; Bal, I.E.; Smyrou, E.; Sarhosis, V. Automatic crack classification and segmentation on masonry surfaces using convolutional neural networks and transfer learning. *Autom. Constr.* **2021**, *125*, 103606.
7. Hsieh, Y.-A.; Tsai, Y.J. Machine learning for crack detection: Review and model performance comparison. *J. Comput. Civ. Eng.* **2020**, *34*, 04020038. [CrossRef]
8. Alipour, M.; Harris, D.K.; Miller, G.R. Robust pixel-level crack detection using deep fully convolutional neural networks. *J. Comput. Civ. Eng.* **2019**, *33*, 04019040. [CrossRef]
9. Liu, Z.; Cao, Y.; Wang, Y.; Wang, W. Computer vision-based concrete crack detection using U-net fully convolutional networks. *Autom. Constr.* **2019**, *104*, 129–139. [CrossRef]
0. Geirhos, R.; Rubisch, P.; Michaelis, C.; Bethge, M.; Wichmann, F.A.; Brendel, W. ImageNet-trained CNNs are biased towards texture; increasing shape bias improves accuracy and robustness. *arXiv* **2018**, arXiv:1811.12231.
1. Tuli, S.; Dasgupta, I.; Grant, E.; Griffiths, T.L. Are Convolutional Neural Networks or Transformers more like human vision? *arXiv* **2021**, arXiv:2105.07197.
2. Azulay, A.; Weiss, Y. Why do deep convolutional networks generalize so poorly to small image transformations? *arXiv* **2018**, arXiv:1805.12177.
3. Chen, L.C.; Zhu, Y.; Papandreou, G.; Schroff, F.; Adam, H. Encoder-decoder with atrous separable convolution for semantic image segmentation. In Proceedings of the European Conference on Computer Vision—ECCV, Munich, Germany, 8–14 September 2018; pp. 833–851.
4. Mei, Q.; Gül, M.; Azim, M.R. Densely connected deep neural network considering connectivity of pixels for automatic crack detection. *Autom. Constr.* **2020**, *110*, 103018. [CrossRef]
5. Touvron, H.; Cord, M.; Douze, M.; Massa, F.; Sablayrolles, A.; Jegou, H. Training data-efficient image transformers & distillation through attention. In Proceedings of the 38th International Conference on Machine Learning, Virtual, 18–24 July 2021; Volume 139, pp. 10347–10357.

MDPI AG
Grosspeteranlage 5
4052 Basel
Switzerland
Tel.: +41 61 683 77 34

Water Editorial Office
E-mail: water@mdpi.com
www.mdpi.com/journal/water

www.ingramcontent.com/pod-product-compliance
Lightning Source LLC
Chambersburg PA
CBHW061808210926
51850CB00006B/640
9783725854455